AF598064

Birkhäuser

Applied and Numerical Harmonic Analysis

For further volumes:
http://www.springer.com/series/4968

Travis D. Andrews • Radu Balan
John J. Benedetto • Wojciech Czaja
Kasso A. Okoudjou
Editors

Excursions in Harmonic Analysis, Volume 1

The February Fourier Talks at the Norbert Wiener Center

Editors
Travis D. Andrews
Norbert Wiener Center
Department of Mathematics
University of Maryland
College Park, MD, USA

Radu Balan
Norbert Wiener Center
Department of Mathematics
University of Maryland
College Park, MD, USA

John J. Benedetto
Norbert Wiener Center
Department of Mathematics
University of Maryland
College Park, MD, USA

Wojciech Czaja
Norbert Wiener Center
Department of Mathematics
University of Maryland
College Park, MD, USA

Kasso A. Okoudjou
Norbert Wiener Center
Department of Mathematics
University of Maryland
College Park, MD, USA

ISBN 978-0-8176-8375-7 ISBN 978-0-8176-8376-4 (eBook)
DOI 10.1007/978-0-8176-8376-4
Springer New York Heidelberg Dordrecht London

Library of Congress Control Number: 2012951313

Mathematics Subject Classification (2010): 26-XX, 35-XX, 40-XX, 41-XX, 42-XX, 43-XX, 44-XX, 46-XX, 47-XX, 58-XX, 60-XX, 62-XX, 65-XX, 68-XX, 78-XX, 92-XX, 93-XX, 94-XX

Printed on acid-free paper

Springer is part of Springer Science+Business Media (www.birkhauser-science.com)

Dedicated to
Tom Grasso,
Friend and Editor Extraordinaire

ANHA Series Preface

The *Applied and Numerical Harmonic Analysis (ANHA)* book series aims to provide the engineering, mathematical, and scientific communities with significant developments in harmonic analysis, ranging from abstract harmonic analysis to basic applications. The title of the series reflects the importance of applications and numerical implementation, but richness and relevance of applications and implementation depend fundamentally on the structure and depth of theoretical underpinnings. Thus, from our point of view, the interleaving of theory and applications and their creative symbiotic evolution is axiomatic.

Harmonic analysis is a wellspring of ideas and applicability that has flourished, developed, and deepened over time within many disciplines and by means of creative cross-fertilization with diverse areas. The intricate and fundamental relationship between harmonic analysis and fields such as signal processing, partial differential equations (PDEs), and image processing is reflected in our state-of-theart *ANHA* series.

Our vision of modern harmonic analysis includes mathematical areas such as wavelet theory, Banach algebras, classical Fourier analysis, time-frequency analysis, and fractal geometry, as well as the diverse topics that impinge on them.

For example, wavelet theory can be considered an appropriate tool to deal with some basic problems in digital signal processing, speech and image processing, geophysics, pattern recognition, biomedical engineering, and turbulence. These areas implement the latest technology from sampling methods on surfaces to fast algorithms and computer vision methods. The underlying mathematics of wavelet theory depends not only on classical Fourier analysis, but also on ideas from abstract harmonic analysis, including von Neumann algebras and the affine group. This leads to a study of the Heisenberg group and its relationship to Gabor systems, and of the metaplectic group for a meaningful interaction of signal decomposition methods. The unifying influence of wavelet theory in the aforementioned topics illustrates the justification for providing a means for centralizing and disseminating information from the broader, but still focused, area of harmonic analysis. This will be a key role of ANHA. We intend to publish the scope and interaction that such a host of issues demands.

Along with our commitment to publish mathematically significant works at the frontiers of harmonic analysis, we have a comparably strong commitment to publish major advances in the following applicable topics in which harmonic analysis plays a substantial role:

Biomedical signal processing
Compressive sensing
Communications applications
Data mining/machine learning
Digital signal processing
Fast algorithms
Gabor theory and applications
Image processing
Numerical partial differential equations
Prediction theory
Radar applications
Sampling theory
Spectral estimation
Speech processing
Time-frequency and time-scale analysis
Wavelet theory

The above point of view for the ANHA book series is inspired by the history of Fourier analysis itself, whose tentacles reach into so many fields.

In the last two centuries, Fourier analysis has had a major impact on the development of mathematics, on the understanding of many engineering and scientific phenomena, and on the solution of some of the most important problems in mathematics and the sciences. Historically, Fourier series were developed in the analysis of some of the classical PDEs of mathematical physics; these series were used to solve such equations. In order to understand Fourier series and the kinds of solutions they could represent, some of the most basic notions of analysis were defined, e.g., the concept of "function". Since the coefficients of Fourier series are integrals, it is no surprise that Riemann integrals were conceived to deal with uniqueness properties of trigonometric series. Cantor's set theory was also developed because of such uniqueness questions.

A basic problem in Fourier analysis is to show how complicated phenomena, such as sound waves, can be described in terms of elementary harmonics. There are two aspects of this problem: first, to find, or even define properly, the harmonics or spectrum of a given phenomenon, e.g., the spectroscopy problem in optics; second, to determine which phenomena can be constructed from given classes of harmonics, as done, e.g., by the mechanical synthesizers in tidal analysis.

Fourier analysis is also the natural setting for many other problems in engineering, mathematics, and the sciences. For example, Wiener's Tauberian theorem in Fourier analysis not only characterizes the behavior of the prime numbers, but also provides the proper notion of spectrum for phenomena such as white light; this latter process leads to the Fourier analysis associated with correlation functions in filtering and prediction problems, and these problems, in turn, deal naturally with Hardy spaces in the theory of complex variables.

Nowadays, some of the theory of PDEs has given way to the study of Fourier integral operators. Problems in antenna theory are studied in terms of unimodular trigonometric polynomials. Applications of Fourier analysis abound in signal processing, whether with the fast Fourier transform (FFT), or filter design, or the adaptive modeling inherent in time-frequency-scale methods such as wavelet theory. The coherent states of mathematical physics are translated and modulated Fourier transforms, and these are used, in conjunction with the uncertainty principle, for dealing with signal reconstruction in communications theory. We are back to the raison d'être of the ANHA series!

University of Maryland
College Park

John J. Benedetto
Series Editor

Preface

The chapters in these two volumes have at least one (co)author who spoke at the February Fourier Talks during the period 2006–2011.

The February Fourier Talks

The February Fourier Talks (*FFT*) were initiated in 2002 as a small meeting on harmonic analysis and applications, held at the University of Maryland, College Park. Since 2006, the *FFT* has been organized by the Norbert Wiener Center in the Department of Mathematics, and it has become a major annual conference. The *FFT* brings together applied and pure harmonic analysts along with scientists and engineers from industry and government for an intense and enriching two-day meeting. The goals of the *FFT* are the following:

- To offer a forum for applied and pure harmonic analysts to present their latest cutting-edge research to scientists working not only in the academic community but also in industry and government agencies,
- To give harmonic analysts the opportunity to hear from government and industry scientists about the latest problems in need of mathematical formulation and solution,
- To provide government and industry scientists with exposure to the latest research in harmonic analysis,
- To introduce young mathematicians and scientists to applied and pure harmonic analysis,
- To build bridges between pure harmonic analysis and applications thereof.

These goals stem from our belief that many of the problems arising in engineering today are directly related to the process of making pure mathematics applicable. The Norbert Wiener Center sees the *FFT* as the ideal venue to enhance this process in a constructive and creative way. Furthermore, we believe that our vision is shared by the scientific community, as shown by the steady growth of the *FFT* over the years.

The *FFT* is formatted as a two-day single-track meeting consisting of thirty-minute talks as well as the following:

- Norbert Wiener Distinguished Lecturer series
- General interest keynote address
- Norbert Wiener Colloquium
- Graduate and postdoctoral poster session

The talks are given by experts in applied and pure harmonic analysis, including academic researchers and invited scientists from industry and government agencies.

The Norbert Wiener Distinguished Lecture caps the technical talks of the first day. It is given by a senior harmonic analyst, whose vision and depth through the years have had profound impact on our field. In contrast to the highly technical day sessions, the keynote address is aimed at a general public audience and highlights the role of mathematics, in general, and harmonic analysis, in particular. Furthermore, this address can be seen as an opportunity for practitioners in a specific area to present mathematical problems that they encounter in their work. The concluding lecture of each *FFT*, our Norbert Wiener Colloquium, features a mathematical talk by a renowned applied or pure harmonic analyst. The objective of the Norbert Wiener Colloquium is to give an overview of a particular problem or a new challenge in the field. We include here a list of speakers for these three lectures:

Distinguished lecturer

- Peter Lax
- Richard Kadison
- Elias Stein
- Ronald Coifman
- Gilbert Strang

Keynote address

- Frederick Williams
- Steven Schiff
- Peter Carr
- Barry Cipra
- William Noel
- James Coddington
- Mario Livio

Colloquium

- Christopher Heil
- Margaret Cheney
- Victor Wickerhauser
- Robert Fefferman
- Charles Fefferman
- Peter Jones

The Norbert Wiener Center

The Norbert Wiener Center for Harmonic Analysis and Applications provides a national focus for the broad area of mathematical engineering. Applied harmonic analysis and its theoretical underpinnings form the technological basis for this area. It can be confidently asserted that mathematical engineering will be to today's mathematics departments what mathematical physics was to those of a century ago. At that time, mathematical physics provided the impetus for tremendous advances within mathematics departments, with particular impact in fields such as differential equations, operator theory, and numerical analysis. Tools developed in these fields were essential in the advances of applied physics, e.g., the development of the solid-state devices which now enable our information economy.

Mathematical engineering impels the study of fundamental harmonic analysis issues in the theories and applications of topics such as signal and image processing, machine learning, data mining, waveform design, and dimension reduction into mathematics departments. The results will advance the technologies of this millennium.

The golden age of mathematical engineering is upon us. The Norbert Wiener Center reflects the importance of integrating new mathematical technologies and algorithms in the context of current industrial and academic needs and problems. The Norbert Wiener Center has three goals:

- Research activities in harmonic analysis and applications
- Education—undergraduate to postdoctoral
- Interaction within the international harmonic analysis community

We belicve that educating the next generation of harmonic analysts, with a strong understanding of the foundations of the field and a grasp of the problems arising in applications, is important for a high-level and productive industrial, government, and academic workforce.

The Norbert Wiener Center web site: www.norbertwiener.umd.edu

The Structure of the Volumes

To some extent the eight parts of these two volumes are artificial placeholders for all the diverse chapters. It is an organizational convenience that reflects major areas in harmonic analysis and its applications, and it is also a means to highlight significant modern thrusts in harmonic analysis. Each of the following parts includes an introduction that describes the chapters therein:

Volume 1	Volume 2
I Sampling Theory	V Measure Theory
II Remote Sensing	VI Filtering
III Mathematics of Data Processing	VII Operator Theory
IV Applications of Data Processing	VIII Biomathematics

Acknowledgements

The Norbert Wiener Center gratefully acknowledges the indispensable support of the following groups: Australian Academy of Science, Birkhäuser, the IEEE Baltimore Section, MiMoCloud, Inc., Patton Electronics Co., Radyn, Inc., the SIAM Washington–Baltimore Section, and SR2 Group, LLC. One of the successes of the February Fourier Talks has been the dynamic participation of graduate student and postdoctoral engineers, mathematicians, and scientists. We have been fortunate to be able to provide travel and living expenses to this group due to continuing, significant grants from the National Science Foundation, which, along with the aforementioned organizations and companies, believes in and supports our vision of the FFT.

Acknowledgments

Contents

Part I
Sampling Theory

Sampling theory has a long history going at least as far back as Lagrange with explicit sampling formulas. The classical sampling theorem due to Cauchy in the 1840s is often associated with names such as Hadamard, E. T. Whittaker, J.M. Whittaker, Kotel'nikov, Wiener, Raabe, Someya, Shannon (see Chapter 1 of [1]). It is also a subject that has expanded significantly in recent years both with regard to compelling theoretical advances and modern applicability.

The first of our six chapters on sampling theory is by AKRAM ALDROUBI. Aldroubi gives an overview of the theory of nonlinear signal representations in terms of unions of subspaces. It is a theory that Aldroubi and his collaborators have spearheaded. It is formulated as a general model for data representations, which includes a variety of applications from compressive sensing and dimension reduction to motion and tracking in video. Frames, shift-invariant subspaces, and the so-called minimum subspace approximation property are key concepts in the theory, and useful algorithms are part and parcel of the presentation.

BERNHARD G. BODMANN, PETER G. CASAZZA, JESSE D. PETERSON, IHAR SMALYANAU, AND JANET C. TREMAIN provide original constructions of fusion frames with rigid geometric properties. Fusion frames can be viewed as a generalization of frames. By their definition it is tantalizing to try to invoke them in modeling environments in which data sets, e.g., communications data, overlap and where the goal is to provide efficient connectivity, e.g., good communications, over the union of the sets. Part of the authors' setting is based on Naimark's classical theorem relating frames and orthonormal bases, but their technology is both deep and beyond the original Naimark theorem. Their specific constructions are of equi-isoclinic Parseval fusion frames, motivated from a rich history and a setting for present-day applicability.

JENS G. CHRISTENSEN AND GESTUR OLAFSSON exposit their theory of the classical sampling theorem in the grand setting of commutative connected homogeneous spaces. Lie groups and homogeneous spaces are the starting point, leading to invariant differential operators on homogeneous spaces. There is a necessary and technical excursion dealing with oscillation estimates on Lie groups, as well as state-of-the-art results on the path from Gelfand pairs to Ruffino's recent topological isomorphism theorem for positive definite spherical functions. It is a virtuosic route to the definition of bandlimited functions for homogeneous spaces, and, then, to the authors' classical sampling theorem in this generality.

CHARLES FEFFERMAN began a program going back to fundamental questions, posed by Hassler Whitney in the 1930s, dealing with smooth extensions of functions. He led an effort, along with several others, solving some of these problems and addresing the comparably fundamental problems of devising best-possible and useful algorithms for interpolation of data. He and Klartay (2009) have best-possible algorithms, and Fefferman has shown the importance of constructing useful algorithms with regard to this problem. Fefferman's chapter is an important, readable introduction to Whitney's extension theorem, his problems, and to the exciting algorithmic questions that remain. It provides a marvelous setting for mathematicians of various stripes to transcend the pure power of the solutions to Whitney's problems to their relevance in current applicability and the importance of viable algorithms to effect this applicability.

JOSEPH D. LAKEY resurrects the classical vision of Henry J. Landau's poignant encapsulation in 1983 of the fundamental issues of sampling theory. Lakey brings to bear an imposing analytic overview and also integrates the most recent topics in sampling beginning with the fundamental work of Candès, Romberg, and Tao, as well as a potpourri of deep results ranging from those of Logan, Widom, Rokhlin, and others. It is a reassuring testament to Henry Landau's magisterial understanding of the uncertainty principle and classical sampling theory.

KABE MOEN, HRVOJE ŠIKIĆ, GUIDO WEISS, AND EDWARD WILSON go beyond the sampling theory of the previous chapter (by Lakey) by formulating the essential ideas of classical sampling theory in terms of technological notions such as convolution idempotents, Zak transforms, frames, and principal shift-invariant spaces. This is a state-of-the-art setting for formulating and proving sampling formulas.

Reference

1. Benedetto, J.J., Ferreira, P.J.S.G. (eds.): Modern Sampling Theory. Applied and Numerical Harmonic Analysis. Birkhäuser, Boston (2001)

Unions of Subspaces for Data Modeling and Subspace Clustering

Akram Aldroubi

Abstract Signals and data are often modeled as vectors belonging to a subspace of a Hilbert space (or Banach space) $\mathscr{H}$, e.g., bandlimited signals. However, nonlinear subsets can also serve as signal model, e.g., signals with finite rate of innovation or sparse signals. Discovering the model from data observations is a fundamental task. The goal of this chapter is to discuss the problem of learning a nonlinear model of the form $\mathscr{M} = \cup_{i=1}^{l} V_i$ from a set of observations $\mathbf{F} = \{f_1, \dots, f_m\}$, where the V_i are the unknown subspaces to be found. Learning this nonlinear model from observed data has applications to computer vision and signal processing and is connected to several areas of mathematics. In this chapter we give a brief description of the theoretical as well as the applied aspects related to this type of nonlinear modeling in terms of unions of subspaces.

Keywords Union of subspaces • Sparse signals • Subspace segmentation • Signal learning from observations • Spectral clustering • Non-linear signal models • Motion segmentation • Minimum subspace approximation property (MSAP) • Nearness to local subspaces

1 Introduction

This chapter is an overview of the problem of nonlinear signal representations in terms of unions of subspaces and its connection to other areas of mathematics and engineering. The point presented stems from recent research with my many collaborators [1–7].

A. Aldroubi (✉)
Department of Mathematics, Vanderbilt University, 1520 Stevenson Center, Nashville, TN 37240, USA
e-mail: akram.aldroubi@vanderbilt.edu

T.D. Andrews et al. (eds.), *Excursions in Harmonic Analysis, Volume 1*, Applied and Numerical Harmonic Analysis, DOI 10.1007/978-0-8176-8376-4_1,

A standard model for signals, images, or multivariate data is often a shift-invariant space, such as a set of bandlimited functions, a B-spline space, or the level zero subspace of a multiresolution of $L^2(\mathbb{R}^d)$. Given a class of signals, the appropriate model can be learned from signal observations. For example, we may observe a set of signals $\mathbf{F} = \{f_1, \ldots, f_m\} \subset L^2(\mathbb{R}^d)$ from a signal class of interest and then wish to use these observations to find a shift-invariant space that is consistent with these observations. Mathematically, this problem can be stated as follows. Let $\mathscr{V}_n$ be the set of all shift-invariant spaces with length at most n. That is, $V \in \mathscr{V}_n$ if

$$V = V(\Phi) := \overline{\operatorname{span}\{\varphi_i(\cdot - k) : i = 1, \ldots, s, k \in \mathbb{Z}^d\}} \tag{1}$$

for some set of generators $\Phi = \{\varphi_i \in L^2(\mathbb{R}^d) : i = 1, \ldots, s\}$ with $s \leq n$. The problem is to find a shift-invariant space that is nearest to the data $\mathbf{F}$ in the sense that

$$V^o = \operatorname{argmin}_{V' \in \mathscr{V}_n} \sum_{i=1}^{m} \|f_i - P_{V'} f_i\|^p, \quad 0 < p < \infty, \tag{2}$$

where $P_{V'}$ is the orthogonal projection on V' and $\|\cdot\|$ denotes the standard norm in $L^2(\mathbb{R}^d)$. The choice $p = 2$ is good for noisy data; however, the choice $0 < p \leq 1$ may be more advantageous when outliers are present [23]. Observe that the shift-invariant spaces in $\mathscr{V}_n$ have infinite dimensions, while there are finitely many data observations. However, the spaces in $\mathscr{V}_n$ are constrained (to be shift-invariant). Thus, if the number of generators $n < m$, it may not possible to find a space $V \in \mathscr{V}_n$ that contains $\mathbf{F}$. For $p = 2$, the existence of a minimizer V^o as in (2) that is nearest to the data $\mathbf{F}$ has been established in [1]. Moreover, a construction of a set $\{\varphi_i \in L^2(\mathbb{R}^d) : i = 1, \ldots, s\}$ such that $\{\varphi_i(x - k) : i = 1, \ldots, s, k \in \mathbb{Z}^d\}$ is a Parseval frame for V^o and a formula for the error $e(\mathbf{F}, V^o) = \sum_{i=1}^{m} \|f_i - P_{V^o} f_i\|^2$ have also been established in [1].

A general nonlinear model for signal or data representation in the finite- or infinite-dimensional cases can be stated as follows:

Problem 1 (Unions of Subspaces Model). Let $\mathscr{H}$ be a Hilbert space, $\mathbf{F} = \{f_1, \ldots, f_m\}$ a finite set of vectors in $\mathscr{H}$, $\mathscr{C}$ a family of closed subspaces of $\mathscr{H}$, and $\mathscr{V}$ the set of all sequences of elements in $\mathscr{C}$ of length l (i.e., $\mathscr{V} = \mathscr{V}(l) = \{\{V_1, \ldots, V_l\} : V_i \in \mathscr{C}, 1 \leq i \leq l\}$). Find a sequence of l subspaces $\mathbf{V}^o = \{V_1^o, \ldots, V_l^o\} \in \mathscr{V}$ (if it exists) such that

$$e(\mathbf{F}, \mathbf{V}^o) = \inf\{e(\mathbf{F}, \mathbf{V}) : \mathbf{V} \in \mathscr{V}\} = \sum_{f \in F} d^2(f, EV). \tag{3}$$

The problem above is that of finding the unions of subspaces model $\mathscr{M} = \cup_{i=1}^{l} V_i$ that is nearest to a set of data $\mathbf{F} = \{f_1, \ldots, f_m\}$ (see Fig. 1 for an illustration). For $l = 1$, this problem reduces to least squares problem. However, for $l > 1$ the problem becomes a nonlinear, generalized version of the least squares problem ($l = 1$), which has many applications in mathematics and engineering.

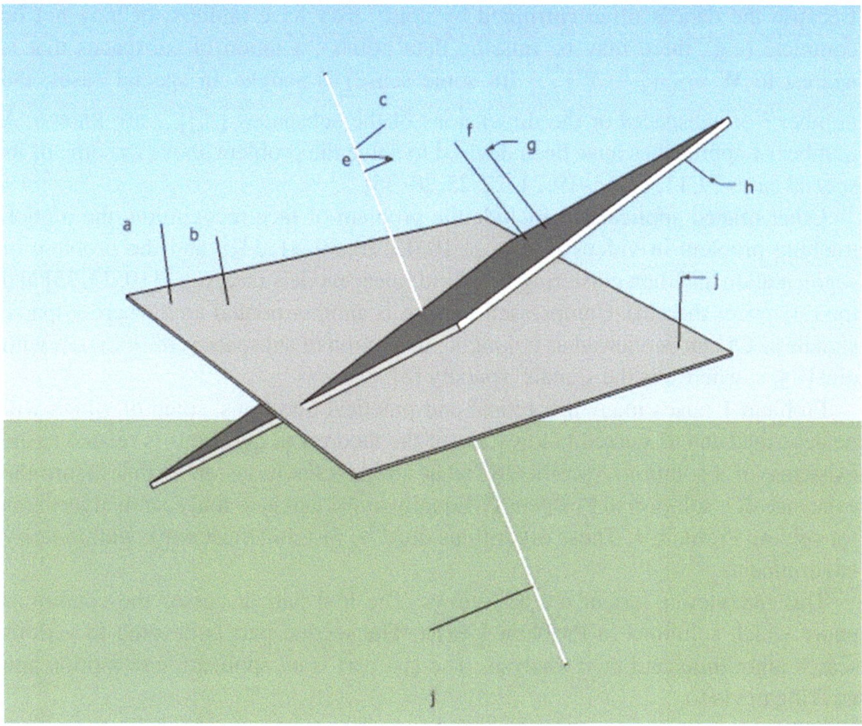

Fig. 1 An example illustrating a set of data points in $\mathbb{R}^3$ that we wish to model by a union of three subspaces, two of which are of dimension two and one of dimension one. The union drawn for this situation is clearly not the one nearest to the data points

Examples where $\mathscr{H}$ is infinite dimensional occur in signal modeling. The typical situation is $\mathscr{H} = L^2(\mathbb{R}^d)$, $\mathbf{F} \subset L^2(\mathbb{R}^d)$ is a finite set of signals, and $\mathscr{C}$ is the set of all finitely generated shift-invariant spaces of $L^2(\mathbb{R}^d)$ of length at most n [3,24,27]. The class of signals with finite rate of innovation is another example where the space $\mathscr{H}$ is infinite dimensional [26]. Applications where a union of subspaces underlies the signal model in infinite dimensions can be found in [2, 24, 26].

An important application in finite dimensions is the subspace segmentation problem in computer vision (see, e.g., [31] and the references therein). The subspace segmentation problem can be described as follows: Let $\mathscr{M} = \bigcup_{i=1}^{l} V_i$ where $\{V_i \subset \mathscr{H}\}_{i=1}^{l}$ is an unknown set of subspaces of a Hilbert space $\mathscr{H}$. Let $\mathbf{W} = \left\{w_j \in \mathscr{H}\right\}_{j=1}^{m}$ be a set of data points drawn from $\mathscr{M}$. Then,

1. Determine the number of subspaces l.
2. Determine the set of dimensions $\{d_i\}_{i=1}^{l}$.
3. Find an orthonormal basis for each subspace V_i.
4. Collect the data points belonging to the same subspace into the same cluster.

Because the data is often corrupted by noise, may have outliers, or may not be complete (e.g., there may be missing data points), a union of subspaces that is nearest to $\mathbf{W} = \{w_j \in \mathscr{H}\}_{j=1}^m$ (in some sense) is sought. In special cases, the number l of subspaces or the dimensions of the subspaces $\{d_i\}_{i=1}^l$ are known. A number of approaches have been devised to solve the problem above or some of its special cases [9, 11, 12, 15–19, 21, 22, 25, 28–35].

Other related applications include the problem of face recognition, the motion tracking problem in videos (see, e.g., [6, 13, 20, 30, 31, 33]), and the problem of segmentation and data clustering in hybrid linear models (see, e.g., [10, 23, 35] and the references therein). Compressed sensing is another related area where s-sparse signals in $\mathbb{C}^d$ can be viewed as belonging to a union of subspaces $\mathscr{M} = \cup_{i\in I} V_i$, with $\dim V_i \leq s$, where s is the signals' sparsity [8].

Problem 1 raises many theoretical and practical questions, some of which will be described and discussed below. One of the theoretical questions is related to the existence of a solution. Specifically, what are the conditions on $\mathscr{C}$ that assure the existence of a solution to Problem 1? Equally important is to find search algorithms for solving Problem 1. These algorithms must be fast and must work well in noisy environments.

This overview is organized as follows. The first part discusses the conditions under which solutions to Problem 1 exist. The second part is devoted to various search algorithms and their analysis. The last part is an application to motion and tracking in video.

2 Existence of Solutions and the Minimum Subspace Approximation Property

It has been shown that, given a family of closed subspaces $\mathscr{C}$, the existence of a minimizing sequence of subspaces $\mathbf{V}^o = \{V_1^o, \dots, V_l^o\}$ that solves Problem 1 is equivalent to the existence of a solution to the same problem but for $l = 1$ [2]:

Theorem 1. *Problem 1 has a minimizing set of subspaces for all finite sets of data and for any $l \geq 1$ if and only if it has a minimizing subspace for all finite sets of data and for $l = 1$.*

This suggests the following definition:

Definition 1. A set of closed subspaces $\mathscr{C}$ of a separable Hilbert space $\mathscr{H}$ has the minimum subspace approximation property (MSAP) if for every finite subset $\mathbf{F} \subset \mathscr{H}$, there exists an element $V \in \mathscr{C}$ that minimizes the expression

$$e(\mathbf{F}, V) = \sum_{f\in\mathbf{F}} d^2(f, V), \quad V \in \mathscr{C}. \tag{4}$$

We will say that $\mathscr{C}$ has MSAP(k) for some $k \in \mathbb{N}$ if the previous property holds for all subsets $\mathbf{F}$ of cardinality $j \leq k$.

Using this terminology, Problem 1 has a minimizing sequence of subspaces if and only if $\mathscr{C}$ satisfies the MSAP. It should be noted that MSAP($k+1$) is strictly stronger than MSAP(k). Obviously, MSAP is stronger than MSAP(k) for any $k \in \mathbb{N}$.

There are some cases for which it is known that the MSAP is satisfied. For example, if $\mathscr{H} = \mathbb{C}^d$ and $\mathscr{C} = \{V \subset \mathscr{H} : \dim V \leq s\}$, the Eckart–Young theorem [14] implies that $\mathscr{C}$ satisfies MSAP. Another example is the one described in the introduction when $\mathscr{H} = L^2(\mathbb{R}^d)$ and $\mathscr{C} = \mathscr{V}_n = \{V : V = \operatorname{span}\{\varphi_1(x-k), \ldots, \varphi_n(x-k) : k \in \mathbb{Z}^d\}\}$ is the set of all shift-invariant spaces of length at most n. For this last example, a result in [1] implies that $\mathscr{C}$ satisfies the MSAP.

The general approach for the existence of a minimizer has been recently considered in [7]. The family $\mathscr{C}$ is viewed as a set of projectors. Specifically, a subspace $V \in \mathscr{C}$ is identified with the orthogonal projector $Q = Q_V$ whose kernel is exactly V (i.e., $Q = I - P_V$ where P_V is the orthogonal projector on V). In this way, any set of closed subspaces $\mathscr{C}$ of $\mathscr{H}$ is identified with a set of projectors $\{Q \in \Pi(\mathscr{H}) : \ker(Q) \in \mathscr{C}\} \subset \Pi(\mathscr{H})$, where $\Pi(\mathscr{H}) \subset \mathscr{B}(\mathscr{H})$ denotes the set of orthogonal projectors. Using this identification, this set of projectors is denoted by $\mathscr{C}$ as well, and $e(\mathbf{F}, V)$ in (4) is expressed as

$$e(\mathbf{F}, V) = \Phi_{\mathbf{F}}(Q_V) = \sum_{f \in \mathbf{F}} \langle Q_V f, f \rangle. \tag{5}$$

The functional $\Phi_{\mathbf{F}}$ defined above is easily seen to be linear on the set $\mathscr{B}(\mathscr{H})$ of bounded linear operators on $\mathscr{H}$.

The right topology for the set of bounded linear operators $\mathscr{B}(\mathscr{H})$ in this case is the weak operator topology. Indeed, in this topology, $\Phi_{\mathbf{F}}$ is a continuous functional for each $\mathbf{F}$, and the set of projectors $\Pi(\mathscr{H})$ is pre-compact. Thus it is evident that if the set $\mathscr{C} \subset \Pi(\mathscr{H})$ is weakly closed in $\mathscr{B}(\mathscr{H})$, then $\Phi_{\mathbf{F}}$ attains its minimum (and maximum) for some $V^o \in \mathscr{C}$. This condition, however, is too strong as can be seen from this obvious example: let $\mathscr{H} = \mathbb{R}^3$ and consider the set $\mathscr{C} = \mathscr{C}_1 \cup \mathscr{C}_2$ which is the union of the plane $\mathscr{C}_1 = \operatorname{span}\{e_1, e_2\}$ and the set of lines $\mathscr{C}_2 = \cup_v \{\operatorname{span}\{v\} : v = e_3 + ce_2, \text{for some } c \in \mathbb{R}\}$. Then $\mathscr{C}$ (identified with a set of projectors as described earlier) is not closed (since $Q_{\operatorname{span}\{e_2\}} \notin \mathscr{C}$), but $\mathscr{C}$ satisfies the MSAP, since if the infimum is achieved by the missing line $\operatorname{span}\{e_2\}$, it is also achieved by the plane $\mathscr{C}_1$.

2.1 A Sufficient Geometric Condition for MSAP

Clearly, as the example above has shown, it is not necessary for $\mathscr{C}$ to be closed in order for $\mathscr{C}$ to have the MSAP even if $\mathscr{H}$ has finite dimension. It turns out that the right set to analyze is the set $\mathscr{C}^+$ consisting of $\mathscr{C}$ together with the positive semidefinite operators added to it. For example, the theorem [7] below gives a sufficient condition for MSAP in terms of the set $\mathscr{C}^+ = \mathscr{C} + \mathscr{P}^+(\mathscr{H})$ where $\mathscr{P}^+(\mathscr{H})$ is the set of positive semidefinite (self-adjoint operators). It also gives a sufficient condition for MSAP in terms of the convex hull $\mathrm{co}(\mathscr{C})$ of $\mathscr{C}$, i.e., the smallest convex set containing $\mathscr{C}$:

Theorem 2 ([7]).

$$(\mathscr{C} = \overline{\mathscr{C}}) \Rightarrow (\mathscr{C}^+ = \overline{\mathscr{C}^+}) \Rightarrow (\mathrm{co}(\mathscr{C}^+) = \mathrm{co}(\overline{\mathscr{C}^+})) \Rightarrow (\mathscr{C} \text{ satisfies MSAP}),$$

and the implications are strict in general.

The theorem above can be used to obtain another proof of the Eckart–Young theorem [14], or the fact that the set $\mathscr{C}(n) = \mathscr{V}_n = \{V : V = \overline{\mathrm{span}\{\varphi_1(\cdot - k), \dots, \varphi_n(\cdot - k) : k \in \mathbb{Z}^d\}} \subset L^2(\mathbb{R}^d)$ of all shift-invariant spaces of length at most n satisfies MSAP as proved in [1]. It can also be used to prove that the set $\mathscr{C}_m(n) = \{V : \overline{\mathrm{span}\{\varphi_1(\cdot - k), \dots, \varphi_n(\cdot - k) : k \in \mathbb{Z}^d\}}$ of all shift-invariant spaces of length at most n that are also $\frac{1}{m}\mathbb{Z}$ invariant satisfies MSAP as proved in [4]. Note that $\mathscr{C}_m \subset \mathscr{C}$. However, unlike the proofs in [1, 4], the use of Theorem 2 does not give constructive proofs.

2.2 *Characterization of MSAP in Finite Dimensions*

In finite dimensions, the last two implications of Theorem 2 can be reversed. Thus, the necessary and sufficient condition for MSAP is that $\mathscr{C}^+$ is closed. An equivalent characterization is that the convex hull $\mathrm{co}(\mathscr{C}^+)$ of $\mathscr{C}^+$ is equal to the convex hull $\mathrm{co}(\overline{\mathscr{C}^+})$ of its closure. A third characterization for a finite d-dimensional space $\mathscr{H}$ is that $\mathscr{C}$ satisfies the MSAP $(d-1)$ (see Definition 1). These characterizations are proved in [7], and we have:

Theorem 3. *Suppose $\mathscr{H}$ has dimension d. Then the following are equivalent:*

(i) $\mathscr{C}$ satisfies MSAP(d-1).
(ii) $\mathscr{C}$ satisfies MSAP.
(iii) $\mathscr{C}^+$ is closed.
(iv) $\mathrm{co}(\mathscr{C}^+) = \mathrm{co}(\overline{\mathscr{C}^+})$.

Unfortunately, the above equivalences are false in infinite dimensions. The difficulty is that although the operators of norm ≤ 1 form a compact subset for the weak operator topology, in infinite-dimensional spaces, the set of projectors is not closed in this topology. This creates most of the complications in infinite dimensions.

For example, to see why in infinite dimensions the last implication of Theorem 2 cannot be reversed, take $\mathscr{C}$ to be the set of all finite-dimensional subspaces (except the trivial vector space) of some infinite-dimensional space $\mathscr{H}$. It has MSAP since for all finite sets F in $\mathscr{H}$, one can find a finite-dimensional subspace containing it. On the other hand, the convex hull of $\mathscr{C}^+$ does not contain $\{0\}$, while the weak closure of $\mathscr{C}$ does.

An example that shows that, in infinite dimensions, the second implication of Theorem 2 is not an equivalence can be constructed as follows: let $\mathscr{H} = \ell^2$ and consider the sequence of vectors $v_n = e_1 + e_n$, which weakly converges to e_1, and the sequence $w_n = e_2 + e_{n+1}$, which weakly converges to e_2. For all $n \geq 3$, let P_n be the projector on the space spanned by v_n and w_n. One checks that the sequence $\{P_n\}$ converges weakly to $Q = (P_{E_1} + P_{E_2})/2$, where $E_1 = \text{span}\{e_1\}$ and $E_2 = \text{span}\{e_2\}$. Moreover, since $P_n = Q + (P_{E_n} + P_{E_{n+1}})/2$, where $E_n = \text{span}\{e_n\}$ and $E_{n+1} = \text{span}\{e_{n+1}\}$, we conclude that $Q < P_n$ for any n. Now define $\mathscr{C} = \{P_n,\ n \geq 3\} \cup \{P_{E_1}, P_{E_2}\}$. The closure of $\mathscr{C}$ consists of $\mathscr{C} \cup \{Q\}$. By the previous remark, Q does not belong to $\mathscr{C}^+$, so that $\mathscr{C}^+$ is not closed. But on the other hand, $Q \in \text{co}(\mathscr{C})$, hence $\overline{\mathscr{C}} \subset \text{co}(\mathscr{C})$ which implies that $\text{co}(\mathscr{C}) = \text{co}(\overline{\mathscr{C}})$. It follows that $\text{co}(\mathscr{C}^+) = \text{co}(\overline{\mathscr{C}^+})$.

2.3 *Characterization of MSAP in Infinite Dimensions*

For the infinite-dimensional case, the necessary and sufficient conditions are found in terms of the set of contact half-spaces. A contact half-space to a set E is a half-space containing E such that its boundary has a nontrivial intersection with E; a half-space is a closed set of the form $H_{\phi,a} = \{x \in \mathscr{B}(\mathscr{H}) : \phi(x) \geq a\}$, for some $a \in \mathbb{R}$ and ϕ an $\mathbb{R}$-linear functional on $\mathscr{B}(\mathscr{H})$. The boundary of such a half-space is the (affine) hyperplane given by $\phi(x) = a$. The set of contact half-spaces containing E is denoted by $\mathscr{T}(E)$. Using these concepts, the necessary and sufficient condition for MSAP in infinite dimensions is:

Theorem 4 ([7]). *Let $\mathscr{C}$ be a set of projectors in $\mathscr{B}(\mathscr{H})$. Then $\mathscr{C}$ has MSAP if and only if*

$$\mathscr{T}(\mathscr{C}^+) = \mathscr{T}\left(\overline{\mathscr{C}^+}\right).$$

3 Algorithms and Dimensionality Reduction

Search algorithms for finding solutions to Problem 1 are often iterative. For computational efficiency, dimensionality reduction is generally needed. Moreover, iterative algorithms often need a good initial approximation to the solution. A general, abstract algorithm of this kind is described in [2].

3.1 Algorithm for the Noise-Free Case

A fast algorithm for the noise-free case can be found using a simple reduced row echelon algorithm. Before getting into the details, we first need some definitions.

Definition 2. Let S be a linear subspace of $\mathbb{R}^D$ with dimension d. A set of data $\mathbf{F}$ drawn from $S \subset \mathbb{R}^D$ with dimension d is said to be *generic* if (i) $|\mathbf{F}| > d$, (ii) span $\mathbf{F} = S$, and (iii) every d vector from $\mathbf{F}$ forms a basis for S.

Definition 3. Matrix R is said to be the *binary reduced row echelon form* of matrix A if all non-pivot column vectors are converted to binary vectors by setting nonzero entries to one.

Theorem 5. *Let $\{S_i\}_{i=1}^k$ be a set of nontrivial linearly independent subspaces of $\mathbb{R}^D$ with corresponding dimensions $\{d_i\}_{i=1}^k$. Let $\mathbf{F} = [w_1 \cdots w_N] \in \mathbb{R}^{D \times N}$ be a matrix whose columns are drawn from $\bigcup_{i=1}^k S_i$. Assume the data is drawn from each subspace and that it is generic. Let Brref($\mathbf{F}$) be the binary reduced row echelon form of $\mathbf{F}$. Then:*

1. *The inner product $\langle e_i, b_j \rangle$ of a pivot column e_i and a non-pivot column b_j in Brref($\mathbf{F}$) is one, if and only if the corresponding column vectors $\{w_i, w_j\}$ in $\mathbf{F}$ belong to the same subspace S_l for some $l = 1, \ldots, k$.*
2. *Moreover, $\dim(S_l) = \|b_j\|_1$, where $\|b_j\|_1$ is the l_1-norm of b_j.*
3. *Finally, $w_p \in S_l$ if and only if $b_p = b_j$ or $\langle b_p, b_j \rangle = 1$.*

The theorem above provides an effective approach to cluster the data points. We simply find the reduced row echelon form of the data matrix W and cluster the data points according to the corresponding columns in the reduced row echelon form. For example, consider the following data matrix that contains data points drawn from the union of three linearly independent subspaces (i.e., from $\mathscr{M} = \bigcup \{V_i\}_{i=1}^3$) as column vectors.

$$W = \begin{bmatrix} 2872 & 138 & 342 & 263 & 1956 & 2016 & 1793 & 801 & 195 & 360 & 1076 & 1882 & 1918 & 2350 & 83 \\ 4041 & 249 & 467 & 516 & 129 & 288 & 2612 & 769 & 312 & 174 & 241 & 176 & 3019 & 3270 & 219 \\ 2906 & 4292 & 352 & 7240 & 2861 & 3072 & 1847 & 665 & 6968 & 646 & 1709 & 2794 & 2080 & 2366 & 1012 \\ 5803 & 1405 & 657 & 2498 & 549 & 864 & 3854 & 687 & 2158 & 390 & 629 & 628 & 4711 & 4654 & 545 \\ 5124 & 744 & 2092 & 1335 & 662 & 1056 & 2835 & 1116 & 1131 & 484 & 774 & 762 & 4867 & 4546 & 309 \\ 6701 & 3192 & 757 & 5420 & 775 & 1248 & 4502 & 578 & 5148 & 578 & 919 & 896 & 5638 & 5354 & 812 \\ 7102 & 1625 & 802 & 2862 & 888 & 1440 & 4793 & 522 & 2522 & 672 & 1064 & 1030 & 6059 & 5666 & 585 \\ 495 & 223 & 117 & 577 & 322 & 960 & 266 & 247 & 169 & 668 & 866 & 520 & 275 & 430 & 388 \\ 1184 & 2910 & 192 & 8282 & 435 & 1152 & 755 & 200 & 1482 & 762 & 1011 & 654 & 951 & 970 & 6320 \\ 2065 & 1117 & 287 & 3027 & 4040 & 4800 & 1376 & 159 & 715 & 1360 & 2920 & 4100 & 1797 & 1662 & 2172 \end{bmatrix}. \tag{6}$$

The computed (binary) reduced row echelon form of W is W_b:

$$W_b = \begin{bmatrix} 1&0&0&0&0&0&0&1&0&0&0&0&1&1&0 \\ 0&1&0&0&0&0&0&0&1&0&0&0&0&0&1 \\ 0&0&1&0&0&0&0&1&0&0&0&0&1&1&0 \\ 0&0&0&1&0&0&0&0&1&0&0&0&0&0&1 \\ 0&0&0&0&1&0&0&0&0&1&1&1&0&0&0 \\ 0&0&0&0&0&1&0&0&0&1&1&1&0&0&0 \\ 0&0&0&0&0&0&1&1&0&0&0&0&1&1&0 \\ 0&0&0&0&0&0&0&0&0&0&0&0&0&0&0 \\ 0&0&0&0&0&0&0&0&0&0&0&0&0&0&0 \\ 0&0&0&0&0&0&0&0&0&0&0&0&0&0&0 \end{bmatrix}. \tag{7}$$

Then, from the binary reduced echelon form, it is trivial to correctly clusters columns of W as $\{1,3,7,8,13,14\}$, $\{2,4,9,15\}$, and $\{5,6,10,11,12\}$.

Although an algorithm based on this method is fast and effective, it fails dramatically in a noisy environment, and a simple thresholding of small entry values is not enough to provide robust results in practice. A modification of this algorithm can be used to obtain an initial set of optimal subspaces in an iterative algorithm such as the one in [2] for solving Problem 1.

3.2 A Subspace Segmentation Algorithm for Subspaces of Equal and Known Dimensions

When the dimensions of the subspaces are equal and known, the algorithm described below uses local subspaces and spectral clustering technique to provide a very efficient and robust to noise [5] approach for solving the subspace clustering problem in the special case where the dimensions of the subspaces are equal and known.

Let W be an $n \times m$ data matrix whose columns are drawn from a union of subspaces of dimensions at most d, possibly perturbed by noise. The singular value decomposition (SVD) not only reduces the noise (assuming that the rank of W is estimated correctly) but it also projects the high-dimensional data to a lower dimension. We then can represent W as follows (using SVD):

$$W = U\Sigma V^t, \tag{8}$$

where $U = \begin{bmatrix} u_1 \ u_2 \ \cdots \ u_n \end{bmatrix}$ is an $n \times n$ matrix, $V = \begin{bmatrix} v_1 \ v_2 \ \cdots \ v_m \end{bmatrix}$ is an $m \times m$ matrix, and Σ is an $n \times m$ diagonal matrix with diagonal entries $\sigma_1, \ldots, \sigma_l$, $l = \min\{m,n\}$.

Let r be the rank (known or estimated) of W. Then $U_r\Sigma_r(V_r)^t$ is the best rank-r approximation of $W = U\Sigma V^t$, where U_r refers to a matrix that has the first r columns of U as its columns and $(V_r)^t$ refers to the first r rows of V^t. It can be shown that the columns of $(V_r)^t$ group (cluster) in the same way as the columns of W, and therefore $(V_r)^t$ can be used to replace the data matrix W [5]. Note that the data points in W are m dimensional columns, whereas they are r dimensional in $(V_r)^t$.

3.3 The Nearness to a Local Subspace Algorithm

The following algorithm assumes that the subspaces have dimension d, and it generates a subspace segmentation and data clustering:

- *Dimensionality Reduction and Normalization*: Compute the SVD of W and estimate the effective rank r of W using

$$r = \operatorname{argmin}_j \frac{\sigma_{j+1}^2}{\sum_{i=1}^{j} \sigma_i^2} + \kappa j, \tag{9}$$

where σ_j is the jth singular value and κ is a suitable constant. This allows us to replace the data matrix W with the matrix $(V_r)^t$ that consists of the first r rows of V^t (thereby reducing the dimensionality of data).

Another type of data reduction is to normalize it to lie on the (ℓ_2) unit sphere. Specifically, the columns of $(V_r)^t$ are normalized to lie on the unit sphere $\mathbb{S}^{r-1}$. This is because by projecting the subspace on the unit sphere, we effectively reduce the dimensionality of the data by one.

- *Local Subspace Estimation*: The data points (i.e., each column vector of $(V_r)^t$) that are close to each other are likely to belong to the same subspace. For this reason, we estimate a local subspace for each data point using its closest neighbors. Then we can find, for each point x_i, a d-dimensional subspace that is nearest (in the least square sense) to the data x_i and its k nearest neighbors.
- *Construction of a Binary Similarity Matrix*: So far, we have associated a local subspace S_i for each point x_i. Ideally, the points and only those points that belong to the same subspace as x_i should have zero distance from S_i. This suggests computing the distance of each point x_j to the local subspace S_i and forming a distance matrix H.

  A distance matrix H is then generated whose component d_{ij} is the average of the Euclidian distance of x_j to the subspace S_i associated with x_i and the distance of x_i to the subspace S_j associated with x_j. Note that as d_{ij} decreases, the probability of having x_j on the same subspace as x_i increases.

  Since we are assuming noisy data, a point x_j that belongs to the same subspace as x_i may have nonzero distance to S_i. However, this distance is likely to be small compared to the distance between x_j and S_k if x_j and x_k do not belong to the same subspace. Thus, we compute a threshold that will distinguish between these two cases and transform the distance matrix into a binary matrix in which a zero in the (i,j) entry means x_i and x_j are likely to belong to the same subspace, whereas (i,j) entry of one means x_i and x_j are not likely to belong to the same subspace.

  To do this, we convert the distance matrix $H = (d_{ij})_{m\times m}$ into a binary similarity matrix $S = (s_{ij})$ by applying a data-driven thresholding as described in [5].
- *Segmentation*: The last step is to use the similarity matrix S to segment the data. To do this, we first normalize the rows of S to obtain the normalized similarity matrix $\tilde{S}$. Observe that the initial data segmentation problem has now been converted to segmentation of n 1-dimensional subspaces from the rows of $\tilde{S}$. This is because, in the ideal case, from the construction of $\tilde{S}$, if x_i and x_j are in the same subspace, the ith and jth rows of $\tilde{S}$ are equal. Since there are l subspaces, there will be l 1-dimensional subspaces.

  Now, the problem is again a subspace segmentation problem, but this time the data matrix is $\tilde{S}$ with each row as a data point. Also, each subspace is 1-dimensional and there are l subspaces. Therefore, we can apply SVD again to obtain

$$\tilde{S}^t = U_l \Sigma_l (V_l)^t$$

and cluster the columns of $(V_l)^t$, that is, the rows of V_l. Note that V_l is $m \times l$, and as before, $(V_l)^t$ replaces $\tilde{S}^t$. Since the problem is only segmentation of subspaces of dimension 1, we can use any traditional segmentation algorithm such as k-means to cluster the data points.

4 Application to Motion Segmentation

Consider a moving affine camera that captures N frames of a scene that contains multiple moving objects. Let p be a point of one of these objects and let $x_i(p), y_i(p)$ be the coordinates of p in frame i. Define the *trajectory vector* of p as the vector $w(p) = (x_1(p), y_1(p), \ldots, x_N(p), y_N(p))^t$ in $\mathbb{R}^{2N}$. It can be shown that the trajectory vectors of all points of an object in a video belong to a vector subspace in $\mathbb{R}^{2N}$ of dimension no larger than 4 [21]. Thus, trajectory vectors in videos can be modeled by a union $\mathscr{M} = \cup_{i=1}^{l} V_i$ of l subspaces where l is the number of moving objects (background is itself a motion).

A precise description of motion tracking in video can be found in [6]. Finding the nearest unions of subspaces to a set of trajectory vectors as in Problem 1 allows for segmenting and tracking the moving objects. Techniques for motion tracking can be compared to state of the art methods on the Hopkins 155 dataset [30]. The Hopkins 155 dataset was created as a benchmark database to evaluate motion segmentation algorithms. It contains two and three motion sequences. Cornerness features that are extracted and tracked across the frames are provided along with the dataset. The ground truth segmentations are also provided for comparison. Figure 2 shows two samples from the dataset with the extracted features.

Fig. 2 Samples from the Hopkins 155 dataset

Table 1 Percentage classification errors for sequences with two motions

Checker (78)	GPCA	LSA	RANSAC	MSL	ALC	SSC-B	SSC-N	NLS
Average (%)	6.09	2.57	6.52	4.46	1.55	0.83	1.12	0.23
Median (%)	1.03	0.27	1.75	0.00	0.29	0.00	0.00	0.00
Traffic (31)	GPCA	LSA	RANSAC	MSL	ALC	SSC-B	SSC-N	NLS
Average (%)	1.41	5.43	2.55	2.23	1.59	0.23	0.02	1.40
Median (%)	0.00	1.48	0.21	0.00	1.17	0.00	0.00	0.00
Articulated (11)	GPCA	LSA	RANSAC	MSL	ALC	SSC-B	SSC-N	NLS
Average (%)	2.88	4.10	7.25	7.23	10.70	1.63	0.62	1.77
Median (%)	0.00	1.22	2.64	0.00	0.95	0.00	0.00	0.88
All (120 seq)	GPCA	LSA	RANSAC	MSL	ALC	SSC-B	SSC-N	NLS
Average (%)	4.59	3.45	5.56	4.14	2.40	0.75	0.82	0.57
Median (%)	0.38	0.59	1.18	0.00	0.43	0.00	0.00	0.00

4.1 Results

Tables 1–3 display some of the experimental results for the Hopkins 155 dataset. Our nearness to local subspace (NLS) approach has been compared with six (6) motion detection algorithms: (1) GPCA [31], (2) RANSAC [18], (3) local subspace affinity (LSA) [33], (4) MLS [19, 21], (5) agglomerative lossy compression (ALC) [32], and (6) sparse subspace clustering (SSC) [15]. An evaluation of those algorithms is presented in [15] with a minor error in the tabulated results for articulated three motion analysis of SSC-N. SSC-B and SSC-N correspond to Bernoulli and normal random projections, respectively [15]. Table 1 displays the misclassification rates for the two motions video sequences. The NLS algorithm outperforms all of the algorithms for the checkerboard sequences, which are linearly independent motions. The overall misclassification rate is 0.57% when three neighboring points are used for local subspace calculations. This is 24 % better than the next best algorithm. Table 2 shows the misclassification rates for the three motion sequences. NLS has 1.31% misclassification rate and performs 47% better than the next best algorithm (i.e., SSC-N). Table 3 presents the misclassification rates for all of the video sequences. NLS has 0.76% misclassification rate and performs 39% better than the next best algorithm (i.e., SSC-N). In general, the NLS algorithm outperforms SSC-N, which is given as the best algorithm for the two and three motion sequences together.

In conclusion, the NLS algorithm is reliable in the presence of noise, and applied to the Hopkins 155 dataset, it generates the best results to date for motion segmentation. The two motion, three motion, and overall segmentation rates for the video sequences are 99.43%, 98.69%, and 99.24%, respectively.

Acknowledgments This research is supported in part by NSF grants DMS-0807464 and DMS-0908239.

Table 2 Percentage classification errors for sequences with three motions

Checker (26)	GPCA	LSA	RANSAC	MSL	ALC	SSC-B	SSC-N	NLS
Average(%)	31.95	5.80	25.78	10.38	5.20	4.49	2.97	0.87
Traffic (7)	GPCA	LSA	RANSAC	MSL	ALC	SSC-B	SSC-N	NLS
Average (%)	19.83	25.07	12.83	1.80	7.75	0.61	0.58	1.86
Median (%)	19.55	23.79	11.45	0.00	0.49	0.00	0.00	1.53
Articulated (2)	GPCA	LSA	RANSAC	MSL	ALC	SSC-B	SSC-N	NLS
Average (%)	16.85	7.25	21.38	2.71	21.08	1.60	1.60	5.12
Median (%)	16.85	7.25	21.38	2.71	21.08	1.60	1.60	5.12
All (35 seq)	GPCA	LSA	RANSAC	MSL	ALC	SSC-B	SSC-N	NLS
Average (%)	28.66	9.73	22.94	8.23	6.69	3.55	2.45	1.31
Median (%)	28.26	2.33	22.03	1.76	0.67	0.25	0.20	0.45

Table 3 Percentage classification errors for all sequences

All (155 seq)	GPCA	LSA	RANSAC	MSL	ALC	SSC-B	SSC-N	NLS
Average (%)	10.34	4.94	9.76	5.03	3.56	1.45	1.24	0.76
Median (%)	2.54	0.90	3.21	0.00	0.50	0.00	0.00	0.20

References

1. Aldroubi, A., Cabrelli, C., Hardin, D., Molter, U.: Optimal shift-invariant spaces and their Parseval frame generators. Appl. Comput. Harmon. Anal. **23**(2), 273–283 (2007)
2. Aldroubi, A., Cabrelli, C., Molter, U.: Optimal nonlinear models for sparsity and sampling, J. Fourier Anal. Appl. **14**, 793–812 (2008)
3. Aldroubi, A., Gröchenig, K.: Non-uniform sampling in shift-invariant spaces. SIAM Rev. **43**, 585–620 (2001)
4. Aldroubi, A., Krishtal, I., Tessera, R., Wang, H.: Principle shift-invariant spaces with extra invariance nearest to observed data. Collectanea Mathematica **63**(3), 393–401 (2012)
5. Aldroubi, A., Sekmen, A.: Nearness to local subspace algorithm for subspace and motion segmentation. IEEE Signal Process. Letters, in press, 2012. (2011, submitted)
6. Aldroubi, A., Sekmen, A.: Reduced row echelon form and non-linear approximation for subspace segmentation and high dimensional data clustering. (2012, preprint)
7. Aldroubi, A., Tessera, R.: On the existence of optimal unions of subspaces for data modelling and clustering. Found. Comput. Math. **11**, 363–379 (2011) .
8. Candès, E., Romberg, J., Tao, T.: Robust uncertainty principles: exact signal reconstruction from highly incomplete frequency information. IEEE Trans. Inform. Theor. **52**, 489–509 (2006)
9. Chen, G., Atev, S., Lerman, G.: Kernel spectral curvature clustering (kscc). In: 4th International Workshop on Dynamical Vision (2009)
10. Chen, G., Lerman, G.: Foundations of a multi-way spectral clustering framework for hybrid linear modeling. Found. Comput. Math. **9**, 517–558 (2009)
11. Chen, G., Lerman, G.: Motion segmentation by SCC on the hopkins 155 database. In: 4th International Workshop on Dynamical Vision (2009)
12. Chen, G., Lerman, G.: Spectral curvature clustering (SCC). Int. J. Comput. Vis. **81**, 317–330 (2009)
13. Costeira, J., Kanade, T.: A multibody factorization method for independently moving objects. Int. J. Comput. Vis. **29**, 159–179 (1998)

14. Eckart, C., Young, G.: The approximation of one matrix by another of lower rank. Psychometrica **1**, 211–218 (1936)
15. Elhamifar, E., Vidal, R.: Sparse subspace clustering. In: IEEE Conference on Computer Vision and Pattern Recognition, pp. 2790–2797 (2009)
16. Elhamifar, E., Vidal, R.: Clustering disjoint subspaces via sparse representation. In: IEEE International Conference on Acoustics, Speech, and Signal Processing (2010)
17. Favaro, P., Vidal, R., Ravichandran, A.: A closed form solution to robust subspace estimation and clustering. In: IEEE Conference on Computer Vision and Pattern Recognition (2011)
18. Fischler, M., Bolles, R.: Random sample consensus: a paradigm for model fitting with appilcations to image analysis and automated cartography. Commun. ACM **24**(6), 381–395 (1981)
19. Gruber, A., Weiss, Y.: Multibody factorization with uncertainty and missing data using the em algorithm. In: International Conference on Computer Vision and Patern Recognition, pp. 707–714 (2004)
20. Ho, J., Yang, M., Lim, J., Lee, K., Kriegman, D.: Clustering appearances of objects under varying illumination conditions. In: Proceedings of International Conference on Computer Vision and Pattern Recognition, vol. 1, pp. 11–18 (2003)
21. Kanatani, Y. Sugaya, Multi-stage optimization for multi-body motion segmentation. IEICE Trans. Inf. Syst. **E87-D**, 335–349 (2003)
22. Lauer, F., Schnorr, C.: Spectral clustering of linear subspaces for motion segmentation. In: IEEE International Conference on Computer Vision (2009)
23. Lerman, G., Zhang, T.: Robust recovery of multiple subspaces by geometric ℓ_p minimization. Annals of Statistics, **9**,(5) 2686–2715 (2011)
24. Lu, Y., Do, M.N.: A theory for sampling signals from a union of subspaces. IEEE Trans. Signal Process. **56**, 2334–2345 (2008)
25. Ma, Y., Yang, A.Y., Derksen, H., Fossum, R.: Estimation of subspace arrangements with applications in modeling and segmenting mixed data. SIAM Rev. **50**, 413–458 (2008)
26. Maravic, I., Vetterli, M.: Sampling and reconstruction of signals with finite rate of innovation in the presence of noise. IEEE Trans. Signal Process. **53**, 2788–2805 (2005)
27. Smale, S., Zhou, D.X.: Shannon sampling II. Connections to learning theory. Appl. Comput. Harmon. Anal. **19**, 285–302 (2005)
28. Tron, R., Vidal, R.: A benchmark for the comparison of 3-d motion segmentation algorithms. In: Computer Vision and Pattern Recognition, pp. 1–8 (2007)
29. Tseng, P.: Nearest q-flat to m points. J. Optim. Theor. Appl. **105**(1), 249–252 (2000)
30. Vidal, R.: A tutorial on subspace clustering. IEEE Signal Process. Mag. **28**, 52–68, 2010
31. Vidal, R., Ma, Y., Sastry, S.: Generalized principal component analysis (GPCA). IEEE Trans. Pattern Anal. Mach. Intell. **27**(12), 1945–1959 (2005)
32. Vidal, R., Rao, S., Tron, R., Ma, Y.: Motion segmentation in the presence of outlying, incomplete, or corrupted trajectories, IEEE Trans. Pattern Anal. Mach. Intell. **32**(10), 1832–1845 (2010)
33. Yan, J., Pollefeys, M.: A general framework for motion segmentation: Independent, articulated, rigid, non-rigid, degenerate and nondegenerate, vol. 4, pp. 94–106. In: 9th European Conference on Computer Vision (2006)
34. Zelnik-Manor, L., Perona, P.: Self-tuning spectral clustering. In: Saul, L.K., Weiss, Y., Bottou, L. (eds.) Advances in Neural Information Processing Systems, vol. 17, pp. 1601–1608. MIT Press, Cambridge (2004)
35. Zhang, T., Szlam, A., Wang, Y., Lerman, G.: Randomized hybrid linear modeling by local best-fit flats. In: CVPR, San Francisco (2010)

Fusion Frames and Unbiased Basic Sequences

**Bernhard G. Bodmann, Peter G. Casazza, Jesse D. Peterson,
Ihar Smalyanau, and Janet C. Tremain**

Abstract The construction of Parseval frames with special, rigid geometric properties has left many open problems even after decades of efforts. The construction of similar types of fusion frames is even less developed. We construct a large family of equi-isoclinic Parseval fusion frames by taking the Naimark complement of the union of orthonormal bases. If these bases are chosen to be mutually unbiased, then the resulting fusion frame subspaces are spanned by mutually unbiased basic sequences. By giving an explicit representation for Naimark complements, we are able to construct concrete fusion frames in their respective Hilbert spaces.

Keywords Frames • Parseval frames • Tight frames • Fusion frames • Parseval fusion frames • Unbiased basic sequences • Equi-isoclinic subspaces • Naimark complement • Chordal distance • Mutually unbiased bases • Mutually unbiased basic sequences • Tight fusion frame • Naimark's theorem • Naimark complement fusion frame • Synthesis operator • Principal angles

1 Introduction

Hilbert space frame theory is a powerful tool for robust and stable representation of signals [19, 20]. It has found a broad range of applications to problems in wireless communication, time-frequency analysis, coding, sampling theory, and

B.G. Bodmann (✉)
Department of Mathematics, University of Houston, Houston, TX 77204, USA
e-mail: bgb@math.uh.edu

P.G. Casazza • J.D. Peterson • I. Smalyanau • J.C. Tremain
Department of Mathematics, University of Missouri, Columbia, MO 65211-4100, USA
e-mail: casazzap@missouri.edu; jdpq6c@mail.missouri.edu; ivsvhd@mizzou.edu;
J.Tremain@mchsi.com

T.D. Andrews et al. (eds.), *Excursions in Harmonic Analysis, Volume 1*,
Applied and Numerical Harmonic Analysis, DOI 10.1007/978-0-8176-8376-4_2,

much more [13]. A generalization of frames, called *fusion frames*, was developed in [10] (see also [28,30,31]) and quickly found application to problems in engineering including sensor networks, distributed processing, hierarchical data processing, packet encoding, and more [2,5,11]. In frame theory, we measure the amplitudes of the projections of a signal onto the frame vectors. A fusion frame is a collection of subspaces of a Hilbert space (see Sect. 3 for definitions) where the signal is represented by the projections onto the fusion frame subspaces. This provides a platform for performing hierarchical data processing. Fusion frames provide multilevel processing capabilities with resilience to noise and erasures [2,12,23,24] due to their redundancy. For applications, we generally need additional properties on the fusion frame, the most desired being the *Parseval* property. Apart from the use of group representations [20], and although a recent breakthrough called *spectral tetris* has allowed for large classes of Parseval fusion frames to be constructed [7,9], it has proven to be extremely difficult to construct Parseval fusion frames.

One of the main drawbacks of fusion frame theory is that until now, results were geared towards constructing tight fusion frames. In [26], there is a nearly complete classification of dimensions for the spaces, dimensions for the subspaces, number of subspaces, and weights for which tight fusion frames exist. In applications. however, it is possible that the fusion frame is given—such as the case of sensor networks—and we do not get to construct a tight fusion frame. Recently, in [4], the notion of *non-orthogonal fusion frames* was introduced to deal with this case. By using non-orthogonal projections, the authors are able to show that most fusion frames can be turned into tight fusion frames if we use non-orthogonal projections.

Apart from tightness, additional desirable properties in frame design are of geometric nature: frames whose vectors have *equal norm* or are *equiangular*. Among equal-norm Parseval frames there is a special class obtained by forming the union of orthonormal bases and scaling the vectors with a common factor. Within this class, using mutually unbiased bases provides a near match to equiangular Parseval frames which is desirable for applications [1], in particular in quantum state tomography [32]. The construction of large frames made of mutually unbiased bases is possible in prime power dimensions [33], but is otherwise similarly challenging as the construction of highly or minimally redundant equiangular Parseval frames.

We wish to consider the fusion frame analogue of these desired geometric properties in frame design. One could consider the fusion frame version of equiangular Parseval frames either to be Parseval fusion frames whose subspaces have equal chordal distances [14] or as the stricter requirement that the subspaces are equi-isoclinic [25]. In [23,24], it is shown that a tight fusion frame (see Sect. 3 for definitions) has maximal robustness against one subspace erasure when all subspaces have equal dimensions. They further show that a tight fusion frame consisting of equi-dimensional subspaces with equal chordal distances (called *equi-distance tight fusion frames*) gives the best robustness with respect to two and more subspace erasures when the performance is measured in terms of the mean-square error. Equi-distance Parseval fusion frames can be constructed, for example, from Hadamard matrices [22]. When the worst-case error is optimized instead, then *equi-isoclinic* fusion frames, if they exist, are found to be optimal for two erasures.

Optimality for larger numbers of erasures requires a more detailed analysis [2]. Lemmens and Seidel first introduced the notion of equi-isoclinic subspaces [25], which was further studied by Hoggar [21] and others [15–18].

In dimensions where equi-distance or equi-isoclinic Parseval fusion frames are not realizable, one still has a notion of optimality in the more general type of Grassmannian frames. In [6], the authors study the problem of packing equi-dimensional subspaces of $\mathbb{R}^M$ so they are as far apart as possible (called *optimal packings*). They give packings so that either all the principal angles are the same or have at most two different values.

In this chapter we will use the notion of Naimark complements to construct a large class of equi-isoclinic Parseval fusion frames. In addition, we construct the special type of equi-isoclinic Parseval fusion frame for which the subspaces are spanned by mutually unbiased basic sequences.

2 Preliminary Results

In this section we will introduce definitions and the main ideas studied in this chapter. We begin with Hilbert space frames.

Definition 1. A family of vectors $\{f_i\}_{i\in I}$ is a frame for a Hilbert space $\mathscr{H}$ if there are constants $0 < A \le B < \infty$, called a lower and upper frame bound, respectively, so that for all $f \in \mathscr{H}$ we have

$$A\|f\|^2 \le \sum_{i\in I} |\langle f, f_i\rangle|^2 \le B\|f\|^2.$$

If $A = B$, this is called an A-*tight* frame, and if $A = B = 1$, it is a *Parseval frame*. The *analysis operator* of the frame is $T : \mathscr{H} \to \ell_2(I)$ given by

$$Tf = \sum_{i\in I} \langle f, f_i\rangle e_i,$$

where $\{e_i\}_{i\in I}$ is the canonical orthonormal basis of $\ell_2(I)$. The *synthesis operator* of the frame is $T^* : \ell_2(I) \to \mathscr{H}$ given by

$$T^*\left(\sum_{i\in I} a_i e_i\right) = \sum_{i\in I} a_i f_i.$$

The *frame operator* is the positive, self-adjoint, and invertible operator $S = T^*T : \mathscr{H} \to \mathscr{H}$ given by

$$Sf = \sum_{i\in I} \langle f, f_i\rangle f_i, \text{for all} f \in \mathscr{H}.$$

It follows that $AI \le S \le BI$ and the frame is Parseval if and only if $S = I$.

Definition 2. Given two subspaces W_1, W_2 of a Hilbert space $\mathscr{H}$ with $\dim W_1 = k \leq \dim W_2 = \ell$, the *principal angles* $(\theta_1, \theta_2, \dots, \theta_k)$ between the subspaces are defined as follows: The first principal angle is

$$\theta_1 = \min\{\arccos|\langle f, g\rangle| : f \in S_{W_1}, g \in S_{W_2}\},$$

where $S_{W_i} = \{f \in W_i : \|f\| = 1\}$. Two vectors f_1, g_1 are called *principal vectors* if they give the minimum above.

The other principal angles and vectors are then defined recursively via

$$\theta_i = \min\{\arccos|\langle f, g\rangle| : f \in S_{W_1}, g \in S_{W_2}, \text{and} f \perp f_j, g \perp g_j, 1 \leq j \leq i-1\}.$$

Conway, Hardin, and Sloane introduced the *chordal distance* between subspaces of a Hilbert space [14].

Definition 3. If W_1, W_2 are subspaces of $\mathscr{H}$ of dimension k, the *chordal distance* $d_c(W_1, W_2)$ between the subspaces is given by

$$d_c^2(W_1, W_2) = k - tr[P_1 P_2] = k - \sum_{j=1}^{k} \cos^2\theta_j,$$

where P_i is the orthogonal projection onto W_i and $\{\theta_j\}_{j=1}^k$ are the principal angles for W_1, W_2.

If P_1, P_2 are the orthogonal projections onto the k-dimensional subspaces W_1, W_2, the nonzero singular values of $P_1 P_2$ are the squares of the cosines of the principal angles $\{\theta_j\}_{j=1}^k$ between W_1 and W_2. Hence,

$$tr[P_1 P_2] = \sum_{j=1}^{k} \cos^2\theta_j = k - d_c^2(W_1, W_2).$$

Definition 4. Two k-dimensional subspaces W_1, W_2 of a Hilbert space are isoclinic with parameter λ, if the angle θ between any $f \in W_1$ and its orthogonal projection Pf in W_2 is unique with $\cos^2\theta = \lambda$.

Multiple subspaces are equi-isoclinic if they are pairwise isoclinic with the same parameter λ.

An alternative definition is given in [16] where two subspaces are called isoclinic if the stationary values of the angles of two lines, one in each subspace, are equal. The geometric characterization given by Lemmens and Seidel is that when a sphere in one subspace is projected onto the other subspace, then it remains a sphere, although the radius may change [25]. This is all equivalent to the principal angles between the subspaces being identical.

Much work has been done on finding the maximum number of equi-isoclinic subspaces given the dimensions of the overall space and the subspaces (and often the parameter λ). Specifically, Seidel and Lemmens [25] give an upper bound on the number of real equi-isoclinic subspaces, and Hoggar [21] generalizes this to vector spaces over $\mathbb{R}$ and $\mathbb{C}$.

We will show that equi-isoclinic subspaces can be constructed with the help of the Naimark complement from a simple type of Parseval frame formed by taking the union of orthonormal bases and scaling the vectors by a common factor. A special case occurs when we form the union of *mutually unbiased bases*.

Definition 5. A family of orthonormal sequences $\{g_{ij}\}_{i=1}^{N}$, $j = 1,2,\ldots,L$ in an M-dimensional Hilbert space $\mathscr{H}_M$ is called mutually unbiased if there is a constant $0 < c$ so that

$$|\langle g_{ij}, g_{k\ell}\rangle| = c, \text{for all} j \neq \ell.$$

If $N = M$, then we say the family consists of mutually unbiased bases. In this case, c must equal $\sqrt{\frac{1}{N}}$.

3 Fusion Frames and Their Complements

Fusion frames were developed in [10] and quickly generated much literature (see www.fusionframe.org) because of their application to problems in engineering.

Definition 6. Given a Hilbert space $\mathscr{H}$ and a family of closed subspaces $\{W_i\}_{i\in I}$ with associated positive weights v_i, $i \in I$, a collection of weighted subspaces $\{(\mathscr{W}_i, v_i)\}_{i\in I}$ is a *fusion frame* for $\mathscr{H}$ if there exist constants $0 < A \leq B < \infty$ satisfying

$$A\|f\|^2 \leq \sum_{i\in I} v_i^2 \|P_i f\|^2 \leq B\|f\|^2 \qquad \text{for any} f \in \mathscr{H},$$

where P_i is the orthogonal projection onto $\mathscr{W}_i$.

The constants A and B are called *fusion frame bounds*. A fusion frame is called *tight* if A and B can be chosen to be equal, *Parseval* if $A = B = 1$, and *orthonormal* if

$$\mathscr{H} = \oplus_{i\in I} W_i.$$

The fusion frame operator is the positive, self-adjoint, and invertible operator $S_W : \mathscr{H} \to \mathscr{H}$ given by

$$S_W f = \sum_{i\in I} v_i^2 P_i f, \text{for all} f \in \mathscr{H}.$$

It is known [11] that $\{W_i, v_i\}_{i\in I}$ is a fusion frame with fusion frame bounds A, B if and only if $AI \leq S_W \leq BI$. Any signal $f \in \mathscr{H}$ can be reconstructed [11] from its fusion frame measurements $\{v_i P_i f\}_{i\in I}$ by performing

$$f = \sum_{i\in I} v_i S^{-1}(v_i P_i f).$$

A frame $\{f_i\}_{i\in I}$ can be thought of as a fusion frame of one-dimensional subspaces where $W_i = \text{span}\{f_i\}$ for all $i \in I$. The fusion frame is then $\{W_i, \|f_i\|\}$. A difference between frames and fusion frames is that for frames, an input signal $f \in \mathcal{H}$ is represented by a collection of scalar coefficients $\{\langle f, f_i\rangle\}_{i\in I}$ that measure the projection of the signal onto each frame vector f_i, while for fusion frames, an input signal $f \in \mathcal{H}$ is represented by a collection of *vector coefficients* $\{\Pi_{W_i}(f)\}_{i\in I}$ corresponding to projections onto each subspace W_i.

There is an important connection between fusion frame bounds and bounds from frames taken from each of the fusion frame's subspaces [10].

Theorem 1. *For each $i \in I$, let $v_i > 0$ and W_i be a closed subspace of $\mathcal{H}$, and let $\{f_{ij}\}_{j\in J_i}$ be a frame for W_i with frame bounds A_i, B_i. Assume that $0 < A = \inf_{i\in I} A_i \leq \sup_{i\in I} B_i = B < \infty$. Then the following conditions hold:*

1. *$\{W_i, v_i\}_{i\in I}$ is a fusion frame for $\mathcal{H}$.*
2. *$\{v_i f_{ij}\}_{i\in I, j\in J_i}$ is a frame for $\mathcal{H}$.*

In particular, if $\{W_i, v_i\}_{j\in J_i}\}_{i\in I}$ is a fusion frame for $\mathcal{H}$ with fusion frame bounds C, D, then $\{v_i f_{ij}\}_{i\in I, j\in J_i}$ is a frame for $\mathcal{H}$ with frame bounds AC, BD. Also, if $\{v_i f_{ij}\}_{i\in I, j\in J_i}$ is a frame for $\mathcal{H}$ with frame bounds C, D, then $\{W_i, v_i, \}_{j\in J_i}\}_{i\in I}$ is a fusion frame for $\mathcal{H}$ with fusion frame bounds $\frac{C}{B}, \frac{D}{A}$.

Corollary 1. *For each $i \in I$, let $v_i > 0$ and W_i be a closed subspace of $\mathcal{H}$. The following are equivalent:*

1. *$\{W_i, v_i\}_{i\in I}$ is a fusion frame for $\mathcal{H}$ with fusion frame bounds A, B.*
2. *For every orthonormal basis $\{e_{ij}\}_{j\in K_i}$ for W_i, the family $\{v_i e_{ij}\}_{i\in I, j\in K_i}$ is a frame for $\mathcal{H}$ with frame bounds A, B.*
3. *For every Parseval frame $\{f_{ij}\}_{i\in I, j\in J_i}$ for W_i, the family $\{v_i f_{ij}\}_{i\in I, j\in J_i}$ is a frame for $\mathcal{H}$ with frame bounds A, B.*

Corollary 2. *For each $i \in I$, let $v_i > 0$ and W_i be a closed subspace of $\mathcal{H}$. The following are equivalent:*

1. *$\{W_i, v_i\}_{i\in I}$ is a Parseval fusion frame for $\mathcal{H}$.*
2. *For every orthonormal basis $\{e_{ij}\}_{j\in K_i}$ for W_i, the family $\{v_i e_{ij}\}_{i\in I, j\in K_i}$ is a Parseval frame for $\mathcal{H}$.*
3. *For every Parseval frame $\{f_{ij}\}_{i\in I, j\in J_i}$ for W_i, the family $\{v_i f_{ij}\}_{i\in I, j\in J_i}$ is a Parseval frame for $\mathcal{H}$.*

We now turn our attention to complements of fusion frames.

3.1 Naimark Complements

A fundamental result in frame theory is *Naimark's theorem* (see, e.g., [8, 13, 20] for a generalization).

Theorem 2. *A family $\{f_i\}_{i=1}^N$ is a Parseval frame for a Hilbert space $\mathscr{H}$ if and only if there is a Hilbert space $\ell_2(N)$ with orthonormal basis $\{e_i\}_{i=1}^N$, $\mathscr{H} \subset \ell_2(N)$, and the orthogonal projection P of $\ell_2(N)$ onto $\mathscr{H}$ satisfies $Pe_i = f_i$, for all $i = 1,2,\ldots,N$.*

Definition 7. If $\{Pe_i\}_{i=1}^N$ as in Naimark's theorem is a Parseval frame for $\mathscr{H}_M$, then $\{(I-P)e_i\}_{i=1}^N$ is a Parseval frame for the subspace $\mathscr{H}_{N-M}$ of $\ell_2(N)$ called the Naimark complement of $\{Pe_i\}_{i=1}^N$.

For a Parseval fusion frame $\{W_i, v_i\}_{i=1}^M$ we also have a Naimark complement. For this, we choose orthonormal bases $\{g_{ij}\}_{i=1}^{n_j}$ for W_j, $j = 1,2,\ldots,M$. Then $\{\sqrt{v_j}g_{ij}\}_{i=1,j=1}^{n_j\ ,\ M}$ is a Parseval frame which we can denote as $\{Pe_{ij}\}_{i=1,j=1}^{n_j\ ,\ M}$. Since

$$\langle Pe_{ij}, Pe_{k\ell}\rangle = -\langle (I-P)e_{ij}, (I-P)e_{k\ell}\rangle, \qquad j \neq l,$$

it follows that $\{(I-P)e_{ij}\}_{i=1}^{n_j}$ is an orthogonal set. Letting

$$W_j' = \text{span}\ \{(I-P)e_{ij}\}_{i=1}^{n_j},$$

it can be shown that $\{W_j', \sqrt{1-v_j^2}\}_{j=1}^M$ is a Parseval fusion frame called the *Naimark complement fusion frame*. We note that the Naimark complement fusion frame actually depends on the choice of orthonormal bases for the fusion frame, but we will now check that all possible Naimark complements are unitarily equivalent.

Lemma 1. *Let $\mathscr{H}_1$, $\mathscr{H}_3$ be M-dimensional Hilbert spaces, $\mathscr{H}_2$, $\mathscr{H}_4$ - $(N-M)$-dimensional Hilbert spaces, and let $U : \mathscr{H}_1 \to \mathscr{H}_3$ be a unitary operator. If $\{f_i \oplus g_i\}_{i=1}^N \subseteq \mathscr{H}_1 \oplus \mathscr{H}_2$ and $\{Uf_i \oplus h_i\}_{i=1}^N \subseteq \mathscr{H}_3 \oplus \mathscr{H}_4$ are orthonormal sets, then $Vg_i = h_i$ is a unitary operator.*

Proof. For all families of scalars $\{a_i\}_{i=1}^N$ we have

$$\left\|\sum_{i=1}^N a_i(f_i \oplus g_i)\right\|^2 = \sum_{i=1}^N |a_i|^2 = \left\|\sum_{i=1}^N a_i(Uf_i \oplus h_i)\right\|^2.$$

Thus

$$\begin{aligned}\left\|\sum_{i=1}^N a_i f_i\right\|^2 + \left\|\sum_{i=1}^N a_i g_i\right\|^2 &= \left\|\sum_{i=1}^N a_i(f_i \oplus g_i)\right\|^2 = \left\|\sum_{i=1}^N a_i(Uf_i \oplus h_i)\right\|^2 \\ &= \left\|\sum_{i=1}^N a_i Uf_i\|^2 + \|\sum_{i=1}^N a_i h_i\right\|^2 = \left\|\sum_{i=1}^N a_i f_i\|^2 + \|\sum_{i=1}^N a_i h_i\right\|^2.\end{aligned}$$

Hence, $\|\sum_{i=1}^N a_i g_i\|^2 = \|\sum_{i=1}^N a_i h_i\|^2$. □

One consequence of the above lemma is that Naimark complements for Parseval fusion frames are independent of the choice of orthonormal bases for the subspaces up to the application of a unitary operator. For our constructions, this will allow us to explicitly identify our equi-isoclinic fusion frames in the Naimark complement of our original fusion frame. A detailed proof for the following corollary may be found in [8].

Corollary 3. *If $\{f_i\}_{i=1}^N$ is a Parseval frame for $\mathscr{H}_M$ and $\{g_i\}_{i=1}^N$ is a family of vectors in $\mathscr{H}_{N-M}$ so that $\{f_i \oplus g_i\}_{i=1}^N$ is an orthonormal basis for $\mathscr{H}_N$, then up to a unitary operator, $\{g_i\}_{i=1}^N$ is the Naimark complement of $\{f_i\}_{i=1}^N$.*

3.2 Identifying the Naimark Complement

Now we will give an explicit representation of the Naimark complements for the fusion frames we will be working with.

Proposition 1. *Let $A_1, \ldots, A_M$ be $N \times N$ unitary matrices, and let*

$$A = \begin{bmatrix} a_{11} & a_{12} & \cdots & a_{1M} \\ a_{21} & a_{22} & \cdots & a_{2M} \\ \vdots & \vdots & \cdots & \vdots \\ a_{M1} & a_{M2} & \cdots & a_{MM} \end{bmatrix}$$

be a unitary matrix. Then

$$\mathscr{A} = \begin{bmatrix} a_{11}A_1 & a_{12}A_2 & \cdots & a_{1M}A_M \\ a_{21}A_1 & a_{22}A_2 & \cdots & a_{2M}A_M \\ \vdots & \vdots & \cdots & \vdots \\ a_{M1}A_1 & a_{M2}A_2 & \cdots & a_{MM}A_M \end{bmatrix}$$

is a unitary matrix.

Proof. A direct calculation shows that $A^*A = Id$. □

Corollary 4. *Given orthonormal bases $\{g_{ij}\}_{i=1}^N$, $j = 1, 2, \ldots, M$ for $\mathscr{H}_N$, consider the $N \times MN$ matrix*

$$[A_1\ A_2\ \cdots\ A_M] = \begin{bmatrix} | & & | & | & & | & & | & & | \\ g_{11} & \cdots & g_{N1} & g_{12} & \cdots & g_{N2} & \cdots & g_{1M} & \cdots & g_{NM} \\ | & & | & | & & | & & | & & | \end{bmatrix}.$$

This is the synthesis operator for the M-tight frame $\{g_{ij}\}_{i=1,j=1}^{N,\ M}$. Choose a unitary matrix

$$A = \begin{bmatrix} a_{11} & a_{12} & \cdots & a_{1M} \\ a_{21} & a_{22} & \cdots & a_{2M} \\ \vdots & \vdots & \cdots & \vdots \\ a_{M1} & a_{M2} & \cdots & a_{MM} \end{bmatrix}.$$

Then the column vectors of the matrix

$$\mathscr{A}' = \sqrt{\frac{1}{M}} \begin{bmatrix} a_{21}A_1 & a_{22}A_2 & \cdots & a_{2M}A_M \\ \vdots & \vdots & \cdots & \vdots \\ a_{M1}A_1 & a_{M2}A_2 & \cdots & a_{MM}A_M \end{bmatrix}$$

are the Naimark complement for the Parseval frame $\left\{\sqrt{\frac{1}{M}} g_{ij}\right\}_{i=1,j=1}^{N,\,M}$.

Proof. This is immediate from Corollary 3 and Proposition 1. □

Corollary 5. *Letting W_j be the span of the column vectors of $a_{1j}A_j$, then $\{W_j, \sqrt{a_{1j}}\}_{j=1}^M$ is a Parseval fusion frame for $\mathscr{H}_N$. Letting W_j' be the span of the column vectors of*

$$\begin{bmatrix} a_{2j}A_j \\ a_{3j}A_j \\ \vdots \\ a_{Mj}A_j \end{bmatrix},$$

then $\{W_j', \sqrt{\sum_{i=2}^M a_{ij}}\}_{j=1}^M$ is a Parseval fusion frame which is the Naimark complement of the Parseval fusion frame $\{W_j, \sqrt{a_{1j}}\}_{j=1}^M$.

Now that we have identified our Parseval fusion frames, we find the principal angles and the chordal distance between our fusion frame subspaces by recalling results from [8]:

Theorem 3. *Let $\{W_i, v_i\}_{i=1}^K$ be a Parseval fusion frame for $\mathscr{H}_N$ with Naimark complement Parseval fusion frame $\{W_i', \sqrt{1-v_i^2}\}_{i=1}^K$. Let $\{\theta_{\ell,ij}\}_{\ell=1}^k$ be the principal angles for the subspaces W_i, W_j. Then the principal angles $\{\theta_{l,ij}'\}$ for the subspaces W_i', W_j' are*

$$\left\{ \arccos\left[\frac{v_i}{\sqrt{1-v_i^2}} \frac{v_j}{\sqrt{1-v_j^2}} \cos(\theta_{\ell,ij}) \right] \right\}_{\ell=1}^k.$$

Theorem 4. *Let $\{W_i, v_i\}_{i=1}^K$ be a Parseval fusion frame for $\mathscr{H}_N$. Then the Naimark complement Parseval fusion frame $\{W_i', \sqrt{1-v_i^2}\}_{i=1}^K$ satisfies*

$$d_c^2(W_i', W_j') = \left[1 - \frac{v_i^2}{1-v_i^2}\frac{v_j^2}{1-v_j^2}\right] k + \left[\frac{v_i^2}{1-v_i^2}\frac{v_j^2}{1-v_j^2}\right] d_c^2(W_i, W_j).$$

This construction creates equi-isoclinic fusion frame subspaces in our Naimark complement which we shall further discuss in Sect. 4. Now, however, we briefly leave our discussion of Naimark complements to consider the *spatial complement* of a fusion frame.

3.3 Spatial Complements

Definition 8. Let $\{(W_i, v_i)\}_{i\in I}$ be a fusion frame for $\mathscr{H}$. If the family $\{(W_i^\perp, v_i)\}_{i\in I}$, where $W_i^\perp$ is the orthogonal complement of W_i, is also a fusion frame, then we call $\{(W_i^\perp, v_i)\}_{i\in I}$ the orthogonal fusion frame to $\{(W_i, v_i)\}_{i\in I}$.

Proposition 2. *Let $\{W_i, v_i\}_{i\in I}$ be a Parseval fusion frame for $\mathscr{H}$. Then*

1. $\sum_{i\in I} v_i^2 \geq 1$
2. $\sum_{i\in I} v_i^2 = 1$ *if and only if* $W_i = \mathscr{H}$ *for all* $i \in I$

Proof. To prove the first statement, let P_i be the orthogonal projection of $\mathscr{H}$ onto W_i. Since the fusion frame is Parseval, for all $f \in \mathscr{H}$,

$$f = \sum_{i\in I} v_i^2 P_i(f).$$

Thus

$$\begin{aligned} \|f\| &= \|\sum_{i\in I} v_i^2 P_i(f)\| \\ &\leq \sum_{i\in I} \|v_i^2 P_i(f)\| \\ &= \sum_{i\in I} v_i^2 \|P_i(f)\| \\ &\leq \left(\sum_{i\in I} v_i^2\right) \|f\|. \end{aligned} \tag{1}$$

Then the second statement follows from considering when equality holds in inequality (1) for all $f \in \mathscr{H}$:

$$\begin{aligned} \sum_{i\in I} v_i^2 = 1 &\Leftrightarrow \sum_{i\in I} v_i^2 \|P_i(f)\| = \|f\| \\ &\Leftrightarrow P_i(f) = f \text{ for all } i \in I \\ &\Leftrightarrow W_i = \mathscr{H} \text{ for all } i \in I. \end{aligned}$$

□

We need a result from [5].

Theorem 5. *Let $\{(W_i, v_i)\}_{i\in I}$ be a fusion frame for $\mathscr{H}$ with optimal fusion frame bounds $0 < A \le B < \infty$ such that $\sum_{i\in I} v_i^2 < \infty$. Then the following conditions are equivalent:*

1. $\bigcap_{i\in I} W_i = \{0\}$.
2. $B < \sum_{i\in I} v_i^2$.
3. *The family $\{(W_i^{\perp}, v_i)\}_{i\in I}$ is a fusion frame for $\mathscr{H}$ with optimal fusion frame bounds $\sum_{i\in I} v_i^2 - B$ and $\sum_{i\in I} v_i^2 - A$.*

Corollary 6. *Let $\{(W_i, v_i)\}_{i\in I}$ be a Parseval fusion frame for $\mathscr{H}$ with $\sum_{i\in I} v_i^2 > 1$. Then $\{(W_i^{\perp}, v_i)\}_{i\in I}$ is a $\left(\sum_{j\in I} v_j^2 - 1\right)$ tight fusion frame, and*

$$\left\{W_i^{\perp}, \frac{v_i}{\sqrt{\sum_{j\in I} v_j^2 - 1}}\right\}$$

is a Parseval fusion frame.

Moreover, we have

$$d_c^2(W_i, W_j) = d_c^2(W_i^{\perp}, W_j^{\perp}).$$

Proof. The only part that needs a proof is the moreover part, but this is a well-known result from [27,29] where the authors show that the nonzero principal angles between two subspaces and the nonzero principal angles between their orthogonal complements are equal. □

4 Equi-isoclinic Fusion Frames and Mutually Unbiased Basic Sequences

Now we combine our results to find equi-isoclinic fusion frames, and we find a special case where the fusion frame subspaces are spanned by mutually unbiased basic sequences. We will need a result from [3].

Theorem 6. *Let P be the orthogonal projection of $\ell_2(M)$ onto $\mathscr{H}_N$. Given the Parseval frame $\{Pe_i\}_{i=1}^M$ for $\mathscr{H}_N$, and $J \subseteq \{1,2,\ldots,M\}$, the following are equivalent:*

1. *span*$\{(I-P)e_i\}_{i\in J} = (I-P)(\ell_2(M))$.
2. $\{Pe_i\}_{i\in J^c}$ *is linearly independent.*

Now we present the main results.

Theorem 7. *Let $\{g_{ij}\}_{i=1,j=1}^{N\ ,\ M}$ be the union of M orthonormal bases for a Hilbert space $\mathscr{H}_N$. Let $\left\{W_j, \sqrt{\frac{1}{M}}\right\}_{i=1}^{M}$ be the Parseval fusion frame for $\mathscr{H}_N$ where $W_j =$*

span $\{g_{ij}\}_{i=1}^{N} = \mathscr{H}_N$. *Let* $\{W_j', \sqrt{\frac{M-1}{M}}\}_{j=1}^{M}$ *be the Naimark complement, a Parseval fusion frame for* $\mathscr{H}_{(M-1)N}$. *Using Naimark's theorem, we can write* $g_{ij} = \sqrt{\frac{1}{M}} Pe_{ij}$ *where* P *is an orthogonal projection from* $\ell_2(MN)$ *onto* $\mathscr{H}_N$. *Then the following hold:*

1. *For any* $j \in \{1,2,\dots,M\}$,

$$span\{W_i'\}_{i \neq j} = \mathscr{H}_{(M-1)N}.$$

2. *The principal angles for* W_i', W_j' *are*

$$\theta_{l,ij} = \arccos\left[\frac{1}{M-1}\right],$$

so this is a equi-isoclinic Parseval fusion frame.

Proof. To prove (1), we use that $\{g_{ij}\}_{i=1}^{N}$ is, for any fixed j, a linearly independent set. This implies by the preceding theorem that selecting the complement of the index set, that is, all $j' \neq j$, provides a spanning subset $\left\{\sqrt{\frac{1}{M}}(I-P)e_{ij'} : j' \neq j\right\}$ for the Naimark complement. Since the subspaces W_j' are spanned by $\left\{\sqrt{\frac{1}{M}}(I-P)e_{ij'}\right\}$, the result follows.

To prove (2), we note that the principal angles for W_i, W_j are all $\arccos 1 = 0$. The conversion for principal angles gives

$$\arccos\left[\frac{\frac{1}{\sqrt{M}}}{\sqrt{1-(\frac{1}{\sqrt{M}})^2}}\frac{\frac{1}{\sqrt{M}}}{\sqrt{1-(\frac{1}{\sqrt{M}})^2}}\cos 0\right] = \arccos\left[\frac{1}{M-1}\right]. \qquad \square$$

The second item in the theorem has a converse.

Corollary 7. *Let* $\{W_i, v_i\}_{i=1}^{M}$ *be a Parseval fusion frame for* $\mathscr{H}_{N(M-1)}$ *with* $\dim W_i = N$ *for all* $i = 1,2,\dots,M$. *Then*

$$\sum_{i=1}^{M} v_i^2 = M-1.$$

If $v_i = \sqrt{\frac{M-1}{M}}$ *for all* $i = 1,2,\dots,M$, *then all principal angles between the subspaces* W_i *are equal to* $\arccos\left[\frac{1}{M-1}\right]$.

Proof. Since $\sum_{i=1}^{M} \dim W_i = NM$, the Naimark complement fusion frame $\{W_i', \sqrt{1-v_i^2}\}_{i=1}^{M}$ spans $\mathscr{H}_{NM-N(M-1)} = \mathscr{H}_N$. Since $\dim W_i = \dim W_i' = N$, $W_i' = \mathscr{H}_N$ for every i. The result is now immediate from the preceding theorem. $\square$

Additionally, in the special case where our bases are mutually unbiased, the resulting Naimark fusion frame subspaces are spanned by mutually unbiased basic sequences.

Corollary 8. *Let $\{g_{ij}\}_{i=1,j=1}^{N\ ,\ M}$ be a family of mutually unbiased bases for a Hilbert space $\mathscr{H}_N$. Let $\{W_j,\sqrt{\frac{1}{M}}\}_{i=1}^{M}$ be the Parseval fusion frame for $\mathscr{H}_N$ where $W_j =$ span $\{g_{ij}\}_{i=1}^{N}$. Let $\{W_j',\sqrt{\frac{M-1}{M}}\}_{j=1}^{M}$ be the Naimark complement Parseval fusion frame for $\mathscr{H}_{(M-1)N}$. Using Naimark's theorem, we can write $g_{ij} = \sqrt{\frac{1}{M}}Pe_{ij}$ where P is an orthogonal projection from $\ell_2(MN)$ onto $\mathscr{H}_N$. Then the family $\{\sqrt{\frac{M}{M-1}}(I-P)e_{ij}\}_{i=1}^{N}$, for $j=1,2,\ldots,M$ consists of mutually unbiased orthonormal sequences, and the subspaces*

$$W_j = span\{(I-P)e_{ij}\}_{i=1}^{N}$$

are equi-isoclinic.

Proof. The equi-isoclinic property is a consequence of the preceding theorem. The fact that the basic sequences spanning the subspaces are mutually unbiased comes from the fact that the Gram matrix of the Naimark complement differs on all off-diagonal elements only by an overall change of sign from the Gram matrix of the mutually unbiased bases. □

Maximal mutually unbiased bases are known to exist in all prime power dimensions p^r [33], so by the Naimark complement, there are Parseval fusion frames with $M \leq p^r + 1$ equi-isoclinic subspaces of dimension p^r, spanned by mutually unbiased basic sequences in a Hilbert space of dimension $N = (M-1)p^r$. The union of these unbiased basic sequences forms a tight frame of redundancy $M/(M-1)$, and thus we can achieve redundancies arbitrarily close to one if p^r is chosen sufficiently large.

Lastly, we now present a concrete example of a special case of our main theorem. In $\mathscr{H}_N$, consider M copies of an orthonormal basis $\{g_i\}_{i=1}^{N}$ as $\{g_{ij}\}_{i=1,j=1}^{N,M}$. Hence $\{\frac{1}{\sqrt{M}}g_{ij}\}_{i=1,j=1}^{N,M}$ is a Parseval frame. Let $T : \mathscr{H}_N \to \ell_2(MN)$ be the analysis operator of this Parseval frame, and let $W_j = \text{span}\{g_{ij}\}_{i=1}^{N}$. Then $\{W_j, \frac{1}{\sqrt{M}}\}_{j=1}^{M}$ is a Parseval fusion frame. Let P be the orthogonal projection of $\ell_2(MN)$ onto the range of T. Let $\{e_{ij}\}_{i=1,j=1}^{N,\ M}$ be the orthonormal basis for $\ell_2(MN)$ so that $Pe_{ij} = Tg_{ij}$. Write our Parseval fusion frame in the order $\frac{1}{\sqrt{M}}Pe_{11}, \frac{1}{\sqrt{M}}Pe_{12}, \ldots, \frac{1}{\sqrt{M}}Pe_{1M}, \frac{1}{\sqrt{M}}Pe_{21}, \ldots, \frac{1}{\sqrt{M}}Pe_{NM}$. Now,

$$T = \begin{pmatrix} Pe_{11} \\ \vdots \\ Pe_{1M} \\ Pe_{21} \\ \vdots \\ Pe_{2M} \\ \vdots \\ Pe_{N1} \\ \vdots \\ Pe_{NM} \end{pmatrix} = \begin{pmatrix} \frac{1}{M} & \cdots & \frac{1}{M} & 0 & \cdots & 0 & \cdots & 0 & \cdots & 0 \\ \vdots & & \vdots & \vdots & & \vdots & & \vdots & & \vdots \\ \frac{1}{M} & \cdots & \frac{1}{M} & 0 & \cdots & 0 & \cdots & 0 & \cdots & 0 \\ 0 & \cdots & 0 & \frac{1}{M} & \cdots & \frac{1}{M} & \cdots & 0 & \cdots & 0 \\ \vdots & & \vdots & \vdots & & \vdots & & \vdots & & \vdots \\ 0 & \cdots & 0 & \frac{1}{M} & \cdots & \frac{1}{M} & \cdots & 0 & \cdots & 0 \\ & & & & & & \ddots & & & \\ 0 & \cdots & 0 & 0 & \cdots & 0 & \cdots & \frac{1}{M} & \cdots & \frac{1}{M} \\ \vdots & & \vdots & \vdots & & \vdots & & \vdots & & \vdots \\ 0 & \cdots & 0 & 0 & \cdots & 0 & \cdots & \frac{1}{M} & \cdots & \frac{1}{M} \end{pmatrix}$$

and

$$\begin{pmatrix} (I-P)e_{11} \\ \vdots \\ (I-P)e_{1M} \\ (I-P)e_{21} \\ \vdots \\ (I-P)e_{2M} \\ \vdots \\ (I-P)e_{N1} \\ \vdots \\ (I-P)e_{NM} \end{pmatrix} = \begin{pmatrix} 1-\frac{1}{M} & \cdots & -\frac{1}{M} & 0 & \cdots & 0 & \cdots & 0 & \cdots & 0 \\ \vdots & & \vdots & \vdots & & \vdots & \cdots & \vdots & & \vdots \\ -\frac{1}{M} & \cdots & 1-\frac{1}{M} & 0 & \cdots & 0 & \cdots & 0 & \cdots & 0 \\ 0 & \cdots & 0 & 1-\frac{1}{M} & \cdots & -\frac{1}{M} & \cdots & 0 & \cdots & 0 \\ \vdots & & \vdots & \vdots & & \vdots & \cdots & \vdots & & \vdots \\ 0 & \cdots & 0 & -\frac{1}{M} & \cdots & 1-\frac{1}{M} & \cdots & 0 & \cdots & 0 \\ & & & & & & \ddots & & & \\ 0 & \cdots & 0 & 0 & \cdots & 0 & \cdots & 1-\frac{1}{M} & \cdots & -\frac{1}{M} \\ \vdots & & \vdots & \vdots & & \vdots & \cdots & \vdots & & \vdots \\ 0 & \cdots & 0 & 0 & \cdots & 0 & \cdots & -\frac{1}{M} & \cdots & 1-\frac{1}{M} \end{pmatrix}.$$

In this setting, W_1' is

$$\begin{pmatrix} 1-\frac{1}{M} & \cdots & -\frac{1}{M} & 0 & \cdots & 0 & \cdots & 0 & \cdots & 0 \\ \vdots & & \vdots & \vdots & & \vdots & \cdots & \vdots & & \vdots \\ 0 & \cdots & 0 & 0 & \cdots & 0 & \cdots & 0 & \cdots & 0 \\ 0 & \cdots & 0 & 1-\frac{1}{M} & \cdots & -\frac{1}{M} & \cdots & 0 & \cdots & 0 \\ \vdots & & \vdots & \vdots & & \vdots & \cdots & \vdots & & \vdots \\ 0 & \cdots & 0 & 0 & \cdots & 0 & \cdots & 0 & \cdots & 0 \\ & & & & & & \ddots & & & \\ 0 & \cdots & 0 & 0 & \cdots & 0 & \cdots & 1-\frac{1}{M} & \cdots & -\frac{1}{M} \\ \vdots & & \vdots & \vdots & & \vdots & \cdots & \vdots & & \vdots \\ 0 & \cdots & 0 & 0 & \cdots & 0 & \cdots & 0 & \cdots & 0 \end{pmatrix}.$$

The above represents an orthonormal basis for W_1'. Similarly we have W_j' for $j = 1,2,\ldots,M$. This leads to:

Proposition 3. *For the projections above we have:*

1. $W_j = \{Pe_{1j}, Pe_{2j}, \ldots, Pe_{Nj}\}$ *implies* $span\{W_i'\}_{i \neq j} = \mathscr{H}_{(M-1)N}$.
2. *The family* $\{W_i, v_i\}_{i=1}^M$, $v_i = \frac{1}{\sqrt{M}}$ *for* $i = 1,2,\ldots,M$ *is a Parseval fusion frame for* $\mathscr{H}_N$.
3. *The family* $\{W_i', \sqrt{1 - v_i^2} = \sqrt{\frac{M-1}{M}}\}_{i=1}^M$ *is a Parseval fusion frame for* $\mathscr{H}_{(M-1)N}$.
4. *Letting* $U_i = span\{(I-P)e_{ij}\}_{j=1}^M$, *then* $\{U_i, 1\}_{i=1}^N$ *is an orthonormal fusion frame with* $\dim U_i = M-1$.
5. *The principal angles for* W_i', W_j' *are*

$$\arccos\left[\frac{\frac{1}{\sqrt{M}}}{\sqrt{1-(\frac{1}{\sqrt{M}})^2}}\frac{\frac{1}{\sqrt{M}}}{\sqrt{1-(\frac{1}{\sqrt{M}})^2}}\cos 0\right] = \arccos\left[\frac{1}{M-1}\right],$$

and this is an equi-isoclinic Parseval fusion frame.

Acknowledgements Bernhard G. Bodmann was Supported by NSF DMS 0807399. Peter G. Casazza, Jesse D. Peterson, Ihar Smalyanau and Janet C. Tremain were Supported by NSF DMS 1008183, NSF ATD 1042701, and AFOSR FA9550-11-1-0245.

References

1. Balan, R., Bodmann, B.G., Casazza, P.G., Edidin, D.: Painless reconstruction from magnitudes of frame coefficients. J. Fourier Anal. Appl. **15**(4), 488–501 (2009)
2. Bodmann, B.G.: Optimal linear transmission by loss-insensitive packet encoding. Appl. Comput. Harmon. Anal. **22**, 274–285 (2007)
3. Bodmann, B.G., Casazza, P.G., Paulsen, V.I., Speegle, D.: Spanning and independence properties of frame partitions. Proc. Amer. Math. Soc. 140, 2193–2207 (2012).
4. Cahill, J., Casazza, P.G., Li, S.: Non-orthogonal fusion frames and the sparsity of the fusion frame operator. J. Fourier Anal. Appl. 18, 287–308 (2012)
5. Calderbank, A.R., Casazza, P.G., Heinecke, A., Kutyniok, G., Pezeshki, A.: Sparse fusion frames: existence and construction. Adv. Comput. Math. **35**(1), 1–31 (2011)
6. Calderbank, A.R., Hardin, R.H., Rains, E.M., Shore, P.W., Sloane, N.J.A.: A Group-Theoretic Framework for the Construction of Packings in Grassmannian Spaces. AT&T Labs, Florham Park (1997)
7. Casazza, P.G., Fickus, M., Heinecke, A., Wang, Y., Zhou, Z.: Spectral tetris fusion frame constructions. J. Fourier Anal. Appl. 18(4), 828–851 (2012)
8. Casazza, P.G., Fickus, M., Mixon, D.G., Peterson, J., Smalyanau, I.: Every Hilbert space frame has a Naimark complement. Preprint
9. Casazza, P.G., Fickus, M., Mixon, D.G., Wang, Y., Zhou, Z.: Constructing tight fusion frames. Appl. Comput. Harmon. Anal. 30, 175–187 (2011)
10. Casazza, P.G., Kutyniok, G.: Frames of subspaces. In: Wavelets, Frames and Operator Theory: College Park 2003. Contemp. Math. vol. 345, pp. 87–113 AMS, Providence RI (2004)

11. Casazza, P.G., Kutyniok, G., Li, S.: Fusion frames and distributed processing. Appl. Comput. Harmonic Anal. **25**, 114–132 (2008)
12. Casazza, P.G., Kutyniok, G.: Robustness of fusion frames under erasures of subspaces and of local frame vectors. In: Radon Transforms, Geometry, and Wavelets: New Orleans 2006. Contemp. Math. vol. 464, pp. 149–160, Amer. Math. Soc., Providence, RI (2008)
13. Christensen, O.: An Introduction to Frames and Riesz Bases. Birkhäuser, Boston (2003)
14. Conway, J.H., Hardin, R.H., Sloane, N.J.A.: Packing lines, planes, etc.: packings in Grassmannian spaces. Experiment. Math. **5**(2), 139–159 (1996)
15. Et-Taoui, B.: Equi-isoclinic planes of Euclidean spaces. Indag. Math. (N.S.) **17**(2), 205–219 (2006)
16. Et-Taoui, B.: Equi-isoclinic planes in Euclidean even dimensional spaces. Adv. Geom. **7**(3), 379–384 (2007)
17. Et-Taoui, B., Fruchard, A.: Sous-espaces qui-isoclins de l'espace euclidien. (French) [Equi isoclinic subspaces of Euclidean space] Adv. Geom. **9**(4), 471–515 (2009)
18. Godsil, C.D., Hensel, A.D.: Distance regular covers of the complete graph. J. Combin. Theor. Ser. B **56**(2), 205–238 (1992)
19. Gröchenig, K.: Foundations of Time-Frequency Analysis. Birkhaüser, Boston (2000)
20. Han, D., Larson, D.R.: Frames, bases and group representations. Memoirs AMS **697** (2000)
21. Hoggar, S.G.: New sets of equi-isoclinic n-planes from old. Proc. Edinburgh Math. Soc. **20**(2), 287–291 (1976/77)
22. King, E.J.: Grassmannian fusion frames. Preprint
23. Kutyniok, G., Pezeshki, A., Calderbank, A.R.: Fusion frames and Robust Dimension Reduction. In: 42nd Annual Conference on Information Sciences and Systems (CISS). Princeton University, Princeton, NJ (2008)
24. Kutyniok, G., Pezeshki, A., Calderbank, A.R., Liu, T.: Robust dimension reduction, fusion frames, and grassmannian packings. Appl. Comput. Harmonic Anal. **26**, 64–76 (2009)
25. Lemmens, P.W.H., Seidel, J.J.: Equi-isoclinic subspaces of Euclidean spaces. Nederl. Akad. Wetensch. Proc. Ser. A 76 Indag. Math. **35**, 98–107 (1973)
26. Masses, P., Ruiz, M., Stojanoff, D.: The structure of minimizers of the frame potential on fusion frames. J. Fourier Anal. Appl. J. Fourier Anal. Appl. 16(4), 514–543 (2010)
27. Miao, J., Ben-Israel, A.: On Principal Angles between Subspaces in R^n. Linear Algebra Appl. 171, 81–98 (1992)
28. Oswald, P.: Frames and space splittings in Hilbert spaces. Lecture notes. http://www.faculty.jacobs-university.de/poswald
29. Qiu, L., Zhang, Y., Li, C.-K.: Unitarily invariant metrics on the Grassmann space. SIAM J. Matrix Anal. Appl. 27(2), 507–531 (2005)
30. Sun, W.: G-frames and g-Riesz bases. J. Math. Anal. Appl. **322**(1), 437–452 (2006)
31. Sun, W.: Stability of g-frames. J. Math. Anal. Appl. **326**(2), 858–868 (2007)
32. Wootters, W.K.: Quantum mechanics without probability amplitudes. Found. Phys. **16**(4), 391–405 (1986)
33. Wootters, W.K., Fields, B.D.: Optimal state-determination by mutually unbiased measurements. Ann. Physics **191**(2), 363–381 (1989)

Sampling in Spaces of Bandlimited Functions on Commutative Spaces

Jens Gerlach Christensen and Gestur Ólafsson

Abstract A homogeneous space $\mathbf{X} = G/K$ is called commutative if G is a locally compact group, K is a compact subgroup, and the Banach $*$-algebra $L^1(\mathbf{X})^K$ of K-invariant integrable functions on $\mathbf{X}$ is commutative. In this chapter we introduce the space $L^2_\Omega(\mathbf{X})$ of Ω-bandlimited function on $\mathbf{X}$ by using the spectral decomposition of $L^2(\mathbf{X})$. We show that those spaces are reproducing kernel Hilbert spaces and determine the reproducing kernel. We then prove sampling results for those spaces using the smoothness of the elements in $L^2_\Omega(\mathbf{X})$. At the end we discuss the example of $\mathbb{R}^d$, the spheres S^d, compact symmetric spaces, and the Heisenberg group realized as the commutative space $\mathrm{U}(n) \ltimes \mathbb{H}_n/\mathrm{U}(n)$.

Keywords Sampling • Bandlimited functions • Reproducing kernel Hilbert spaces • Gelfand pairs • Commutative spaces • Representation theory • Abstract harmonic analysis

1 Introduction

Reconstruction or approximation of a function using a discrete set of values of the function, or a transformation of the function, has an old and prominent history. A well-known example is the reconstruction of a function using discrete sets of line integrals, a fundamental tool in computerized tomography. Stepping into the digital age has only made this more important. But *sampling theory* as independent

J.G. Christensen (✉)
Department of Mathematics, Tufts University, 503 Boston Avenue, Medford, MA 02155, USA
e-mail: jens.christensen@tufts.edu

G. Ólafsson
Department of Mathematics, Louisiana State University, Baton Rouge, LA 70803, USA
e-mail: olafsson@math.lsu.edu

T.D. Andrews et al. (eds.), *Excursions in Harmonic Analysis, Volume 1,* Applied and Numerical Harmonic Analysis, DOI 10.1007/978-0-8176-8376-4_3,

mathematical subject originates from the fundamental article [44]. We refer to [50], in particular the introduction, and [51] as good places to consult about the history of the subject.

Sampling theory is a field of interest to engineers, signal analysts, and mathematicians alike. It is concerned with the reconstruction of a function or signal from its values at a certain collection of points. Sampling theory is concerned with many questions:

1. Which classes of signals can we expect to reconstruct?
2. Which conditions do the sampling points have to satisfy?
3. How are the signals reconstructed from samples?
4. Reconstruction algorithms and error analysis.
5. The speed of the reconstruction.

The first, and arguably most famous, result is the Whittaker–Shannon–Kotelnikov sampling theorem which states that an audible (bandlimited) signal can be reconstructed from its values at equidistant sample points if the samples are taken at the Nyquist rate. If sampling takes place at a slower rate the signal cannot be reconstructed. With a higher sampling rate than the Nyquist rate (oversampling) the signal can be reconstructed by many different methods. Some of the developed methods also apply in the case when the samples are not equidistant but within the Nyquist rate (irregular sampling).

Frames [8, 33] generalize orthonormal bases and are relatively new additions to mathematics, yet they have become increasingly important for approximation theory, reconstruction in function spaces, time frequency analysis, and generalizations to shift (translation) invariant spaces on topological groups and homogeneous spaces. So there is no surprise that frames have also been widely used in sampling theory. We will not go into detail here, but would like to point out the article by Benedetto [3] and by Benedetto and Heller [4] as well as the fundamental work by Feichtinger, Gröchenig, and their coauthors [16, 17, 21–25, 29]. Again, we refer to [50], in particular Chapter 10, for a good overview.

The natural generalization of the spaces of bandlimited functions on $\mathbb{R}^n$ is similarly defined translation invariant spaces of functions on Lie groups and homogeneous spaces. Of particular interest are homogeneous Riemannian spaces, a subject closely related to the coorbit theory of Feichtinger and Gröchenig and more general reproducing kernel Hilbert spaces in connection with unitary representations of Lie groups [9–11, 18–20, 28]. Here we have some natural tools from analysis at our disposal, including an algebra of invariant differential operators and in particular the Laplace operator. Most commonly the spaces of bandlimited functions are defined in terms of the boundness of the eigenvalues of the Laplace operator. The bounded geometry of the space allows us then to derive the needed Sobolev and Bernstein inequalities. We point out [26, 27, 29, 30, 41] as important contributions to the subject.

In this chapter we follow an approach which allows for more general spectral sets than those coming from the Laplacian. In particular we find sampling theorems for square integrable functions on the homogeneous space $\mathbf{X} = G/K$ whose vector-valued Fourier transform (defined using spherical representations) is supported in a compact subset of the Gelfand spectrum of $L^1(\mathbf{X})^K$.

This chapter is organized as follows. In Sect. 2 we recall some standard notation for Lie groups G and homogeneous spaces $\mathbf{X}$. We introduce the algebra of invariant differential operators on homogeneous spaces and connect it with the algebra of invariant polynomials on $T_{x_o}(\mathbf{X})$, where x_o is a fixed base point. We then recall some basic fact about representations, and in particular, we introduce the space of smooth and analytic vectors. In Sect. 3 we discuss sampling in reproducing kernel Hilbert spaces on Lie groups. The main ideas are based on the work of H. Feichtinger and K. Gröchenig.

Section 4 deals with oscillation estimates on Lie groups. The exposition is based on [9] and uses smoothness of the functions to derive oscillation results and hence sampling theorems.

We introduce the notion of Gelfand pairs and commutative spaces in Sect. 5. Here we assume that the group is connected, which allows us to state that $\mathbf{X} = G/K$ is commutative if and only if the algebra $\mathbb{D}(\mathbf{X})$ of G-invariant differential operators is commutative. We review well-known facts about positive-definite functions and the spherical Fourier transform on $\mathbf{X}$. One of the main results in this section is a recent theorem of Ruffino [43] which implies that we identify the parameter set for the spherical Fourier transform with a subset of $\mathbb{C}^s$, where s is the number of generators for $\mathbb{D}(\mathbf{X})$. The result by Ruffino generalizes statements about Gelfand pairs related to the Heisenberg group by Benson, Jenkins, and Ratcliff [7]. In Sect. 6 we relate the positive-definite spherical functions to K-spherical representations of G and introduce the vector-valued Fourier transform on $\mathbf{X}$. This relates the representation theory of G to the harmonic analysis on $\mathbf{X}$.

In Sect. 7 we finally introduce the space of bandlimited functions on $\mathbf{X}$. The definition is based on the support of the Plancherel measure on $\mathbf{X}$ rather than the spectrum of the Laplacian on $\mathbf{X}$ as in [26, 27, 41]. We do not prove it, but in all examples the definitions of a bandlimited function are equivalent, though our definition allows for more general spectral sets. Another benefit of our approach is that one does not have to worry about injectivity radius for the exponential function nor about the construction of smooth partitions of unity (characteristic functions for a disjoint cover can be used just as well). Our sampling result is proved in Sect. 8 using a Bernstein inequality for the space of bandlimited functions. Finally, in Sect. 9, we give some examples of commutative spaces and their spherical harmonic analysis. Those examples include the spheres and, more generally, compact symmetric spaces and the Heisenberg group as a homogeneous space for the group $\mathrm{U}(n) \ltimes \mathbb{H}_n$.

2 Notation and Preliminaries

2.1 Locally Compact Groups

In the following G denotes a locally compact Hausdorff group with left-invariant Haar measure μ_G. Sometimes we write $\mathrm{d}x$ instead of $\mathrm{d}\mu_G(x)$. For $1 \le p < \infty$ we let $L^p(G)$ denote the space of equivalence classes of p-integrable functions on G with norm

$$\|f\|_{L^p} = \left(\int |f(x)|^p \,\mathrm{d}x \right)^{1/p}.$$

Further, let $L^\infty(G)$ denote the space of essentially bounded function on G with norm

$$\|f\|_\infty = \operatorname*{ess\,sup}_{x \in G} |f(x)|.$$

The spaces $L^p(G)$ are Banach spaces for $1 \le p \le \infty$, and $L^2(G)$ is a Hilbert space with inner product

$$(f,g) = \int f(x)\overline{g(x)} \,\mathrm{d}x.$$

When it makes sense (either the integrand is integrable or as a vector-valued integral) we define the convolution

$$f * g(x) = \int f(y) g(y^{-1}x) \,\mathrm{d}y.$$

Equipped with convolution the space $L^1(G)$ becomes a Banach algebra. For functions on G we denote the left and right translations by

$$\ell(a)f(x) = f(a^{-1}x) \qquad \text{and} \qquad \rho(a)f(x) = f(xa),$$

respectively. Now, let K be a compact subgroup of G with bi-invariant Haar measure μ_K. We always normalize μ_K so that $\mu_K(K) = 1$. The same convention applies to other compact groups and compact spaces.

If $\mathscr{A}$ is a set of functions on G we denote the left K-fixed subset as

$$\mathscr{A}^K = \{ f \in \mathscr{A} \mid \ell(k)f = f \},$$

and similarly the right K-fixed subset is denoted

$$\mathscr{A}^{\rho(K)} = \{ f \in \mathscr{A} \mid \rho(k)f = f \}.$$

Let $\mathbf{X} = G/K$ and $x_0 = eK$, and let $\kappa : G \to \mathbf{X}$ be the canonical map $g \mapsto gx_0$. The space $\mathbf{X}$ possesses a G-invariant measure $\mu_{\mathbf{X}}$, and we define the L^p-spaces

$$L^p(\mathbf{X}) = \{f \mid f \text{ is } \mu_{\mathbf{X}}\text{-measurable and } \int |f(x)|^p \, \mathrm{d}\mu_{\mathbf{X}} < \infty\}.$$

We will identify functions on $\mathbf{X}$ by K-invariant functions via $f \leftrightarrow f \circ \kappa$. Since K is compact the map $f \mapsto f \circ \kappa$ is an isometric isomorphism $L^p(\mathbf{X}) \simeq L^p(G)^{\rho(K)}$. In particular, $L^p(\mathbf{X})$ is a closed G-invariant subspace of $L^p(G)$. The projection $L^p(G) \to L^p(\mathbf{X})$ is

$$p_K(f)(x) = \int_K f(xk) \, \mathrm{d}\mu_K(k).$$

If $f \in L^1(G)$ and $g \in L^p(\mathbf{X})$, $1 \le p \le \infty$, then $f * g \in L^p(\mathbf{X})$ and $\|f * g\|_{L^p} \le \|f\|_{L^1} \|g\|_{L^p}$. If f is further assumed to be left K-invariant, then

$$\begin{aligned} f * g(ky) &= \int_G f(x) g(x^{-1}ky) \, \mathrm{d}\mu_G(x) \\ &= \int_G f(kx) g(x^{-1}y) \, \mathrm{d}\mu_G(x) \\ &= \int_G f(x) g(x^{-1}y) \, \mathrm{d}\mu_G(x) \\ &= f * g(y). \end{aligned}$$

Thus $f * g$ is also left K-invariant. Denote by m_G the modular function on G. Note that m_G is usually denoted by Δ or Δ_G, but we will need that notation for the Laplace operator on $\mathbf{X}$ respectively G. We have $m_G|_K = 1$ as K is compact. Hence m_G is K-bi-invariant. It follows that $L^1(\mathbf{X})^K$ is invariant under the anti-involutions $f^\vee(x) = m_G(x)^{-1} f(x^{-1})$ and $f^* = \overline{f^\vee}$. In particular, $L^1(\mathbf{X})^K$ is a closed Banach $*$-subalgebra of $L^1(G)$.

2.2 Lie Theory

Let G be a connected Lie group and K a compact subgroup. Most of the statements hold for non-connected groups, but some technical problems turn up as we start to deal with the Lie algebra and invariant differential operators. We will therefore for simplicity assume that G is connected from the beginning.

Denote by $\mathfrak{g}$ the Lie algebra of G and $\mathfrak{k}$ the Lie algebra of K. Fix a K-invariant inner product $\langle \, , \, \rangle$ on $\mathfrak{g}$. That is always possible as K is compact. Let $\mathfrak{s} := \mathfrak{k}^\perp$. Then $\mathfrak{s}$ is K-invariant and $\mathfrak{s} \simeq T_{x_o}(\mathbf{X})$ (as a K-module) via the map

$$X \mapsto D(X), \quad D(X)(f)(x_o) := \left.\frac{\mathrm{d}}{\mathrm{d}t}\right|_{t=0} f(\exp(tX)x_o).$$

Denote also by $\langle\, , \rangle$ the restriction of $\langle\, , \rangle$ to $\mathfrak{s}\times\mathfrak{s}$. As the tangent bundle on $T(\mathbf{X})$ is isomorphic to $G\times_K\mathfrak{s}$ as a G-bundle, it follows that the restriction of $\langle\, , \rangle$ to $\mathfrak{s}$ defines a G-invariant Riemannian structure on $\mathbf{X}$.

Let $D: C_c^\infty(\mathbf{X})\to C_c^\infty(\mathbf{X})$ be a differential operator. For $g\in G$ let $g\cdot D: C_c^\infty(\mathbf{X})\to C_c^\infty(\mathbf{X})$ be the differential operator

$$g\cdot D(f)(x) = D(\ell(g^{-1})f)(g^{-1}x).$$

D is said to be G-invariant if $g\cdot D = D$ for all $g\in G$. Thus D is G-invariant if and only if D commutes with left translation, $D(\ell(g)f) = \ell(g)D(f)$. Denote by $\mathbb{D}(\mathbf{X})$ the algebra of G-invariant differential operators on $\mathbf{X}$. The algebra $\mathbb{D}(\mathbf{X})$ has a simple description. For a polynomial function $p:\mathfrak{g}\to\mathbb{C}$ define a left-invariant differential operator $D_p: C_c^\infty(G)\to C_c^\infty(G)$ by

$$\begin{aligned} D_p(f)(g) &:= p\left(\frac{\partial}{\partial t_1},\dots,\frac{\partial}{\partial t_m}\right) f(g\exp(t_1X_1+\cdots+t_mX_m))|_{t_1=\cdots=t_m=0} \\ &= p\left(\frac{\partial}{\partial t_1},\dots,\frac{\partial}{\partial t_m}\right) f(g\exp(t_1X_1)\cdots\exp(t_mX_m))|_{t_1=\cdots=t_m=0}, \end{aligned}$$

where we have extended our basis of $\mathfrak{s}$ to a basis $X_1,\dots,X_m$ of $\mathfrak{g}$. If p is a K-invariant polynomial, then

$$\begin{aligned} D_p(f)(gk) &= p\left(\frac{\partial}{\partial t_1},\dots,\frac{\partial}{\partial t_m}\right) f(gk\exp(t_1X_1+\cdots+t_nX_m))|_{t_1=\cdots=t_m=0} \\ &= p\left(\frac{\partial}{\partial t_1},\dots,\frac{\partial}{\partial t_n}\right) f(g\exp(t_1X_1+\cdots+t_mX_m)k)|_{t_1=\cdots=t_m=0} \end{aligned}$$

for all $k\in K$. Hence, if p is K-invariant and f is right K-invariant, it is clear that D_p only depends on the polynomial $q = p|_{\mathfrak{s}}$, and $D_pf = D_qf$ is right K-invariant and defines a function on $\mathbf{X}$. Therefore D_q is a G-invariant differential operator on $\mathbf{X}$.

Denote by $S(\mathfrak{s})$ the symmetric algebra over $\mathfrak{s}$. Then $S(\mathfrak{s})$ is commutative and isomorphic to the algebra of polynomial functions.

Theorem 1. *The map $S(\mathfrak{s})^K\to\mathbb{D}(\mathbf{X})$ is bijective.*

Proof. This is Theorem 10 in [34]. □

Remark 1. If we take $p(X) = \|X\|^2$, then $D_p =: \Delta$ is the Laplace operator on $\mathbf{X}$.

Remark 2. The algebra $\mathbb{D}(\mathbf{X})$ is not commutative in general. Hence the above map is not necessarily an algebra homomorphism.

For a fixed basis $X_1,\dots,X_m$ for $\mathfrak{g}$ it will ease our notation to introduce the differential operator $D^\alpha : C_c^\infty(G)\to C_c^\infty(G)$ for a multi-index α of length k with entries between 1 and m:

$$D^\alpha f(x) = D(X_{\alpha(k)})D(X_{\alpha(k-1)})\cdots D(X_{\alpha(1)})f(x).$$

2.3 Representation Theory

Let π be a representation of the Lie group G on a Hilbert space $\mathscr{H}$. Then $u \in \mathscr{H}$ is called *smooth* respectively *analytic* if the $\mathscr{H}$-valued function $\pi_u(x) = \pi(x)u$ is smooth respectively analytic. Denote by $\mathscr{H}^\infty$, respectively $\mathscr{H}^\omega$, the space of smooth, respectively analytic, vectors in $\mathscr{H}$. For $u \in \mathscr{H}^\infty$ and $X \in \mathfrak{g}$ let

$$\pi^\infty(X)u := \lim_{t\to 0} \frac{\pi(\exp tX)u - u}{t}$$

and $\pi^\omega(X) := \pi^\infty(X)|_{\mathscr{H}^\omega}$. We have:

Lemma 1. *Let $(\pi, \mathscr{H})$ be a unitary representation of G. Then the following holds:*

1. *The space $\mathscr{H}^\infty$ is G-invariant.*
2. *$\pi^\infty(\mathfrak{g})\mathscr{H}^\infty \subseteq \mathscr{H}^\infty$ and $(\pi^\infty, \mathscr{H}^\infty)$ are a representation of $\mathfrak{g}$. In particular*

$$\pi^\infty([X,Y]) = \pi^\infty(X)\pi^\infty(Y) - \pi^\infty(Y)\pi^\infty(X).$$

3. *$\pi^\infty(\mathrm{Ad}(g)X) = \pi(g)\pi^\infty(X)\pi(g^{-1})$.*
4. *$\pi^\infty(X)^*|_{\mathscr{H}^\infty} = -\pi^\infty(X)$.*
5. *$\mathscr{H}^\infty$ is dense in $\mathscr{H}$.*

Corresponding statements are also true for $\mathscr{H}^\omega$. To show that $\mathscr{H}^\infty$ is dense in $\mathscr{H}$, let $f \in L^1(G)$. Define $\pi(f) : \mathscr{H} \to \mathscr{H}$ by

$$\pi(f)u = \int_G f(x)\pi(x)u \,\mathrm{d}\mu_G(x).$$

Then $\|\pi(f)\| \leq \|f\|_{L^1}$, $\pi(f * g) = \pi(f)\pi(g)$, and $\pi(f^*) = \pi(f)^*$. Thus, $\pi : L^1(G) \to B(\mathscr{H})$ is a continuous $*$-homomorphism. If $f \in C_c^\infty(G)$ then it is easy to see that $\pi(f)\mathscr{H} \subseteq \mathscr{H}^\infty$. The main step in the proof is to show that

$$\pi^\infty(X)\pi(f)u = \pi(\ell^\infty(X)f)u,$$

where

$$\ell^\infty(X)f(x) = \lim_{t\to 0} \frac{f(\exp(-tX)x) - f(x)}{t}.$$

Lemma 2. *If $\{U_j\}$ is a decreasing sequence of e-neighborhoods such that $\bigcap U_j = \{e\}$ and $f_j \in C_c^\infty(G)$ is so that $f_j \geq 0$, $\mathrm{supp} f_j \subset U_j$, and $\|f\|_{L^1} = 1$, then $\pi(f_j)u \to u$ for all $u \in \mathscr{H}$. In particular, $\mathscr{H}^\infty$ is dense in $\mathscr{H}$.*

3 Reconstruction in Reproducing Kernel Hilbert Spaces

We will be concerned with sampling in subspaces of the Hilbert space $L^2(G)$. We start with the definition of a frame due to Duffin and Schaeffer [14]. For further references see the introduction.

Definition 1. For a Hilbert space $\mathscr{H}$ a set of vectors $\{\phi_i\} \subseteq \mathscr{H}$ is a frame for $\mathscr{H}$ if there are constants $0 < A \leq B < \infty$ such that

$$A\|f\|_{\mathscr{H}}^2 \leq \sum_i |(f,\phi_i)|^2 \leq B\|f\|_{\mathscr{H}}^2$$

for all $f \in \mathscr{H}$.

These conditions ensure that the frame operator $S : \mathscr{H} \to \mathscr{H}$ given by

$$Sf = \sum_i (f,\phi_i)\phi_i$$

is invertible and that f can be reconstructed by

$$f = \sum_i (f,\phi_i)\psi_i,$$

where $\psi_i = S^{-1}\phi_i$. The sequence $\{\psi_i\}$ is also a frame called the dual frame. In general there are other ways to reconstruct f from the sequence $\{(f,\phi_i)\}$.

The inversion of S can be carried out via the Neumann series

$$S^{-1} = \frac{2}{A+B}\sum_{n=0}^{\infty}\left(I - \frac{2}{A+B}S\right)^n, \tag{1}$$

which has rate of convergence $\|I - \frac{2}{A+B}S\| \leq \frac{B-A}{A+B}$ (which is the best possible for optimal frame bounds [31, 37, 39]).

In order to obtain frames from samples it is natural to restrict ourselves to a class of functions with nice properties for point evaluation. A Hilbert space $\mathscr{H}$ of functions on G is called a *reproducing kernel Hilbert space* if point evaluation is continuous, i.e., if for every $x \in G$, there is a constant C_x such that for all $f \in \mathscr{H}$

$$|f(x)| \leq C_x\|f\|_{\mathscr{H}}.$$

A classical reference for reproducing kernel Hilbert spaces is [1]. For such spaces point evaluation is given by an inner product $f(x) = (f, g_x)$, where $g_x \in \mathscr{H}$ is called the *reproducing kernel* for $\mathscr{H}$. Our aim is to find a sequence of points $\{x_j\}$ such that $\{g_{x_j}\}$ is a frame.

Here are the main facts about closed reproducing kernel subspaces of $L^2(G)$ which is all what we will need here:

Proposition 1. *If $\mathscr{H}$ is a closed and left invariant reproducing kernel subspace of $L^2(G)$, then:*

1. *There is a $\phi \in \mathscr{H}$ such that $f = f * \phi$ for all $f \in \mathscr{H}$.*
2. *The functions in $\mathscr{H}$ are continuous.*
3. *The kernel ϕ satisfies $\overline{\phi(x^{-1})} = \phi(x)$ so $f(x) = f * \phi(x) = (f, \ell(x)\phi)$.*
4. *The mapping $f \mapsto f * \phi$ is a continuous projection from $L^2(G)$ onto $\mathscr{H}$. In particular $\mathscr{H} = \{f \in L^2(G) \mid f * \phi = f\}$.*

Proof. Here are the main ideas of the proof. By Riesz' representation theorem there is a $g_x \in H$ such that

$$f(x) = \int f(y)\overline{g_x(y)}\,\mathrm{d}y.$$

Let $g(x) := g_e(x)$. The left invariance of $\mathscr{H}$ ensures that

$$f(x) = [\ell(x^{-1})f](e) = \int f(xy)\overline{g_e(y)}\,\mathrm{d}y = \int f(y)\overline{g(x^{-1}y)}\,\mathrm{d}y = (f, \ell(x)g).$$

Hence $g_x(y) = g(x^{-1}y)$. We also have

$$g(x^{-1}y) = (g_x, g_y) = \overline{(g_y, g_x)} = \overline{g(y^{-1}x)}.$$

Thus, if we set $\phi(x) = \overline{g(x^{-1})}$, which agrees with g^* in case G is unimodular, we get $f = f * \phi$, which in particular implies that $\mathscr{H} \subseteq C(G)$ as claimed.

Assume that $f * \phi = f$ and that $f \perp \mathscr{H}$. Then $f(x) = (f, g_x) = 0$ as $g_x \in \mathscr{H}$. Hence $f = 0$ and $\mathscr{H} = L^2(G) * \phi = \{f \in L^2(G) \mid f * \phi = f\}$. □

Remark 3. It should be noted that several functions $\phi \in L^2(G)$ could satisfy $f = f * \phi$ for $f \in \mathscr{H}$. Just take an arbitrary function η such that $\overline{\eta^\vee} \in \mathscr{H}^\perp$. Then $f * (\phi + \eta) = f * \phi$. For example, as we will see later, if $\mathscr{H}$ is a space of bandlimited functions, then ϕ could be a sinc function with larger bandwidth than the functions in $\mathscr{H}$. However, the restriction that $\phi \in \mathscr{H}$ ensures uniqueness of ϕ.

The sampling theory of H. Feichtinger, K. Gröchenig, and H. Führ (see the introduction for references) builds on estimating the variation of a function under small right translations. The local oscillations were therefore introduced as follows: For a compact neighborhood U of the identity define

$$\operatorname{osc}_U(f) = \sup_{u \in U} |f(x) - f(xu^{-1})|.$$

Before stating the next result we need to introduce a reasonable collection of points at which to sample: For a compact neighborhood U of the identity, the points x_i are called U-relatively separated if the x_iU cover G and there is an N such that each x_iU intersects at most N other x_jU. This in particular implies that each $x \in G$ belongs to at most N of the x_iU and that

$$\mathbf{1}_G \leq \sum_i \mathbf{1}_{x_iU} \leq N\mathbf{1}_G.$$

Lemma 3. *Let $\mathscr{H}$ be a reproducing kernel subspace $L^2(G)$ with reproducing convolution kernel ϕ. Assume that for any compact neighborhood U of the identity there is a constant C_U such that for any $f \in \mathscr{H}$ the estimate $\|\mathrm{osc}_U(f)\|_{L^2} \leq C_U \|f\|_{\mathscr{H}}$ holds. If we can choose U such that $C_U < 1$, then for any U-relatively separated points $\{x_i\}$ the norms $\|\{f(x_i)\}\|_{\ell^2}$ and $\|f\|_{L^2}$ are equivalent, and $\{\ell(x_i)\phi\}$ is a frame for $\mathscr{H}$.*

Proof.

$$
\begin{aligned}
\|\{f(x_i)\}\|_{\ell^2}^2 &= |U|^{-1} \Big\| \sum_i |f(x_i)|^2 \mathbf{1}_{x_i U} \Big\|_{L^1} \\
&\leq |U|^{-1} \Big\| \sum_i |f(x_i)| \mathbf{1}_{x_i U} \Big\|_{L^2}^2 \\
&\leq |U|^{-1} \Big(\Big\| \sum_i |f(x_i) - f| \mathbf{1}_{x_i U} \Big\|_{L^2} + \Big\| \sum_i |f| \mathbf{1}_{x_i U} \Big\|_{L^2} \Big)^2 \\
&\leq |U|^{-1} \Big(\Big\| \sum_i |\mathrm{osc}_U(f)| \mathbf{1}_{x_i U} \Big\|_{L^2} + \Big\| \sum_i |f| \mathbf{1}_{x_i U} \Big\|_{L^2} \Big)^2 \\
&\leq |U|^{-1} N^2 (\|\mathrm{osc}_U(f)\|_{L^2} + \|f\|_{L^2})^2 \\
&\leq |U|^{-1} N^2 (1 + C_U)^2 \|f\|_{L^2}^2 .
\end{aligned}
$$

Here N is the maximal number of overlaps between the $x_i U$'s. To get the other inequality we let ψ_i be a bounded partition of unity such that $0 \leq \psi_i \leq \mathbf{1}_{x_i} U$ and $\sum_i \psi_i = 1$. Then,

$$
\begin{aligned}
\|f\|_{L^2} &\leq \Big\| f - \sum_i f(x_i) \psi_i \Big\|_{L^2} + \Big\| \sum_i f(x_i) \psi_i \Big\|_{L^2} \\
&\leq \Big\| \sum_i \mathrm{osc}_U(f) \psi_i \Big\|_{L^2} + \Big\| \sum_i |f(x_i)| \mathbf{1}_{x_i U} \Big\|_{L^2} \\
&\leq \|\mathrm{osc}_U(f)\|_{L^2} + N|U| \|f(x_i)\|_{\ell^2} \\
&\leq C_U \|f\|_{L^2} + N|U| \|f(x_i)\|_{\ell^2}
\end{aligned}
$$

If $C_U < 1$ then we get

$$
(1 - C_U) \|f\|_{L^2} \leq N|U| \|f(x_i)\|_{\ell^2} .
$$

This concludes the proof. □

Remark 4. From the proof of the lemma follows that the norm equivalence becomes

$$\left(\frac{1-C_U}{|U|N}\right)^2 \|f\|_{L^2}^2 \leq \|\{f(x_i)\}\|_{\ell^2}^2 \leq \left(N\frac{1+C_U}{|U|}\right)^2 \|f\|_{L^2}^2, \tag{2}$$

and thus the frame constants A and B can be chosen to be

$$A = \left(\frac{1-C_U}{|U|N}\right)^2 \quad \text{and} \quad B = \left(N\frac{1+C_U}{|U|}\right)^2.$$

It follows that the rate of convergence for the Neumann series (1) can be estimated by

$$\frac{B-A}{B+A} = \frac{N^2(1+C_U)^2-(1-C_U)^2/N^2}{N^2(1+C_U)^2+(1-C_U)^2/N^2} \to \frac{N^4-1}{N^4+1} \quad \text{as } C_U \to 0.$$

This shows that as the sampling points x_i are chosen closer (U gets smaller) the rate of convergence can be very slow (assuming that we can choose the overlaps of the x_iU's bounded by a certain N even if U gets smaller). We therefore have very little control of the rate of convergence in this case.

To obtain operators with faster-decaying Neumann series, H. Feichtinger and K. Gröchenig introduced new sampling operators. An example is the sampling operator $T : \mathscr{H} \to \mathscr{H}$ defined as

$$Tf = \sum_i f(x_i)\psi_i * \phi.$$

Using oscillations it is possible to estimate the norm of $I - T$ by C_U:

$$\|f - Tf\|_{L^2} = \left\|\left(\sum_i |f - f(x_i)|\psi_i\right) * \phi\right\|_{L^2} \leq \|\mathrm{osc}_U f\|_{L^2} \leq C_U\|f\|_{L^2}.$$

Thus T is invertible on $\mathscr{H}$ if $C_U < 1$, and the rate of convergence of the Neumann series is governed directly by C_U. By increasing the rate of sampling (decreasing U and thereby C_U) fewer iterations are necessary in order to obtain good approximation. This was not the case for the frame inversion above.

4 Oscillation Estimates on Lie Groups

In this section we will show how oscillation estimates can be obtained for functions on Lie groups.

First we set up the notation. As before we let G be a Lie group with Lie algebra $\mathfrak{g}$. Fix a basis $\{X_i\}_{i=1}^{\dim(G)}$ for $\mathfrak{g}$. Denote by U_ε the set

$$U_\varepsilon := \left\{\exp(t_1X_1)\cdots\exp(t_nX_n) \,\middle|\, -\varepsilon \leq t_k \leq \varepsilon, 1 \leq k \leq n\right\}.$$

Remark 5. Note that U_ε depends on the choice of basis as well as the ordering of the vectors. It would therefore be more natural to use sets of the form $V_\varepsilon := \exp\{X \in \mathfrak{g} \mid \|X\| \le \varepsilon\}$ or even $W_\varepsilon := \exp\{X \in \mathfrak{s} \mid \|X\| \le \varepsilon\}\exp\{X \in \mathfrak{k} \mid \|X\| \le \varepsilon\}$. Both of those sets are invariant under conjugation by elements in K. The reason to use U_ε as defined above is that this is the definition that works best for the proofs! But it should be noted that $V_\varepsilon, W_\varepsilon \subseteq U_\varepsilon$. Hence the local oscillation using either V_ε or W_ε is controlled by the local oscillation using U_ε.

Set

$$\mathrm{osc}_\varepsilon(f) = \mathrm{osc}_{U_\varepsilon}(f).$$

By δ we denote an n-tuple $\delta = (\delta_1, \ldots, \delta_n)$ with $\delta_i \in \{0,1\}$. The length $|\delta|$ of δ is the number of nonzero entries $|\delta| = \delta_1 + \cdots + \delta_n$. Further, define the function $\tau_\delta : (-\varepsilon, \varepsilon)^n \to G$ by

$$\tau_\delta(t_1, \ldots, t_n) = \exp(\delta_1 t_n X_1) \cdots \exp(\delta_n t_1 X_n).$$

Lemma 4. *If f is right differentiable of order $n = \dim(G)$ then there is a constant C_ε such that*

$$\mathrm{osc}_\varepsilon(f)(x) \le C_\varepsilon \sum_{1 \le |\alpha| \le n} \sum_{|\delta| = |\alpha|} \underbrace{\int_{-\varepsilon}^{\varepsilon} \cdots \int_{-\varepsilon}^{\varepsilon}}_{|\delta| \text{ integrals}} |D^\alpha f(x\tau_\delta(t_1, \ldots, t_n)^{-1})| (\mathrm{d}t_1)^{\delta_1} \cdots (\mathrm{d}t_n)^{\delta_n}.$$

For $\varepsilon' \le \varepsilon$ w e have $C_{\varepsilon'} \le C_\varepsilon$.

Proof. We refer to [9] for a full proof. Instead we restrict ourselves to a proof in 2 dimension that easily carries over to arbitrary dimensions. We will sometimes write e^X instead of $\exp X$.

For $y \in U_\varepsilon$ there are $s_1, s_2 \in [-\varepsilon, \varepsilon]$ such that $y^{-1} = \mathrm{e}^{-s_2X_2}\mathrm{e}^{-s_1X_1}$. Hence

$$\begin{aligned}
&|f(x) - f(xy^{-1})| \\
&\quad = |f(x) - f(x\mathrm{e}^{-s_2X_2}\mathrm{e}^{-s_1X_1})| \\
&\quad \le |f(x) - f(x\mathrm{e}^{-s_2X_2})| + |f(x\mathrm{e}^{-s_2X_2}) - f(x\mathrm{e}^{-s_2X_2}\mathrm{e}^{-s_1X_1})| \\
&\quad = \left| \int_0^{s_2} \frac{\mathrm{d}}{\mathrm{d}t_2} f(x\mathrm{e}^{-t_2X_2})\,\mathrm{d}t_2 \right| + \left| \int_0^{s_1} \frac{\mathrm{d}}{\mathrm{d}t_1} f(x\mathrm{e}^{-s_2X_2}\mathrm{e}^{-t_1X_1})\,\mathrm{d}t_1 \right| \\
&\quad \le \int_{-\varepsilon}^{\varepsilon} \left| D(X_2) f(x\mathrm{e}^{-t_2X_2}) \right| \mathrm{d}t_2 \\
&\qquad + \int_{-\varepsilon}^{\varepsilon} \left| D(X_1) f(x\mathrm{e}^{-s_2X_2}\mathrm{e}^{-t_1X_1}) \right| \mathrm{d}t_1.
\end{aligned} \tag{3}$$

Since

$$\mathrm{e}^{-s_2X_2}\mathrm{e}^{-t_1X_1} = \mathrm{e}^{-t_1X_1}\mathrm{e}^{-s_2X(t_1)} \quad \text{with} \quad X(t) = \mathrm{Ad}(\exp(tX_1))X_2$$

the term $|D(X_1)f(xe^{-s_2X_2}e^{-t_1X_1})|$ can be estimated by

$$
\begin{aligned}
&|D(X_1)f(xe^{-s_2X_2}e^{-t_1X_1})| \\
&\quad = |D(X_1)f(xe^{-t_1X_1}e^{-s_2X(t_1)})| \\
&\quad \leq |D(X_1)f(xe^{-t_1X_1}e^{-s_2X(t_1)}) - D(X_1)f(xe^{-t_1X_1})| + |D(X_1)f(xe^{-t_1X_1})| \\
&\quad = \left| \int_0^{s_1} \frac{\mathrm{d}}{\mathrm{d}t_2} D(X_1)f(xe^{-t_1X_1}e^{-t_2X(t_1)})\,\mathrm{d}t_2 \right| + |D(X_1)f(xe^{-t_1X_1})| \\
&\quad = \left| \int_0^{s_1} D(X(t_1))D(X_1)f(xe^{-t_1X_1}e^{-t_2X(t_1)})\,\mathrm{d}t_2 \right| + |X_1f(xe^{-t_1X_1})| \\
&\quad \leq C_\varepsilon \int_{-\varepsilon}^{\varepsilon} \left|D(X_2)D(X_1)f(xe^{-t_2X_2}e^{-t_1X_1})\right| \mathrm{d}t_2 \\
&\qquad + \int_{-\varepsilon}^{\varepsilon} \left|D(X_1)D(X_1)f(xe^{-t_2X_2}e^{-t_1X_1})\right| \mathrm{d}t_2 \\
&\qquad + |D(X_1)f(xe^{-t_1X_1})|.
\end{aligned} \tag{4}
$$

The last inequality follows since $D(X(t_2)) = a(t_1)D(X_1) + b(t_1)D(X_2)$ is a differential operator with coefficients a and b depending continuously, in fact analytically, on all variables. Together (3) and (4) provide the desired estimate. □

Since right translation is continuous on $L^2(G)$ and $\sup_{u\in U} \|r_uf\|_{L^2} \leq C_U\|f\|_{L^2}$ for compact U [42, Theorem 3.29] gives

$$
\begin{aligned}
\|\mathrm{osc}_\varepsilon(f)\|_{L^2} &\leq \sum_{1\leq|\alpha|\leq n} \sum_{|\delta|=|\alpha|} \underbrace{\int_{-\varepsilon}^{\varepsilon}\cdots\int_{-\varepsilon}^{\varepsilon}}_{|\delta|\text{ integrals}} \|r_{\tau_\delta(t_1,\ldots,t_n)^{-1}} D^\alpha f\|_{L^2} (\mathrm{d}t_1)^{\delta_1}\cdots(\mathrm{d}t_n)^{\delta_n} \\
&\leq C_{U_\varepsilon} \sum_{1\leq|\alpha|\leq n} \sum_{|\delta|=|\alpha|} \underbrace{\int_{-\varepsilon}^{\varepsilon}\cdots\int_{-\varepsilon}^{\varepsilon}}_{|\delta|\text{ integrals}} \|D^\alpha f\|_{L^2} (\mathrm{d}t_1)^{\delta_1}\cdots(\mathrm{d}t_n)^{\delta_n} \\
&\leq C_{U_\varepsilon} \sum_{1\leq|\alpha|\leq n} \binom{n}{|\alpha|} \varepsilon^{|\alpha|} \|D^\alpha f\|_{L^2}.
\end{aligned}
$$

To sum up we get

Theorem 2. *If $D^\alpha f \in L^2(G)$ for all $|\alpha| \leq n$, then*

$$\|\mathrm{osc}_\varepsilon(f)\|_{L^2} \leq C_\varepsilon \sum_{1\leq|\alpha|\leq n} \|D^\alpha f\|_{L^2},$$

where $C_\varepsilon \to 0$ as $\varepsilon \to 0$.

We will need the following fact later when we obtain a Bernstein-type inequality for bandlimited functions on a commutative space. If $\langle X,Y\rangle$ defines an inner product on $\mathfrak{g}$ and $X_1,\dots,X_n$ is an orthonormal basis, then the associated Laplace operator has the form $\Delta_G = D(X_1)^2+\cdots+D(X_n)^2$. We have:

Lemma 5. *Let the notation be as above. Then*

$$\sum_{1\le|\alpha|\le n} \|D^\alpha f\|_{L^2} \le C\|(I-\Delta_G)^{n/2} f\|_{L^2}.$$

Proof. According to Theorem 4 in [48] the Sobolev norm on the left can be estimated by the Bessel norm, defined in [45], on the right. □

5 Gelfand Pairs and Commutative Spaces

In this section we introduce the basic notation for *Gelfand pairs* and *commutative spaces*. Our standard references are [12], Chapter 22, [13,35], Chapter IV, and [49]. We give several examples in Sect. 9.

Theorem 3. *Suppose that G is a connected Lie group and K a compact subgroup. Then the following are equivalent:*

1. *The Banach $*$-algebra $L^1(\mathbf{X})^K$ is commutative.*
2. *The algebra $C_c^\infty(\mathbf{X})^K$ is commutative.*
3. *The algebra $\mathbb{D}(\mathbf{X})$ is commutative.*

Definition 2. Let G be a connected Lie group and K a compact subgroup. (G,K) is called a Gelfand pair if one, and hence all, of the conditions in Theorem 3 holds. In that case $\mathbf{X}$ is called a *commutative space*.

If G is abelian, then $(G,\{e\})$ is a Gelfand pair. Similarly, if K is a compact group that acts on the abelian group G by group homomorphisms, i.e., $k\cdot(xy)=(k\cdot x)(k\cdot y)$ then $(G\rtimes K,K)$ is a Gelfand pair. One of the standard ways to decide if a given space is commutative is the following lemma:

Lemma 6. *Assume there exists a continuous involution $\tau: G\to G$ such that $\tau(x)\in Kx^{-1}K$ for all $x\in G$. Then $\mathbf{X}=G/K$ is commutative.*

Proof. As $x\mapsto x^{-1}$ is an antihomomorphism it follows that $f\mapsto f^\vee$ is an antihomomorphism on $L^1(\mathbf{X})^K$. On the other hand if we define $f^\tau(x):=f(\tau(x))$ then $f\mapsto f^\tau$ is a homomorphism. But as $\tau(x)=k_1x^{-1}k_2$ it follows that $f^\vee=f^\tau$ for all $f\in L^1(\mathbf{X})^K$, and hence $L^1(\mathbf{X})^K$ is abelian. □

Example 1. Let $G=\mathrm{SO}(d+1)$ and $K=\mathrm{SO}(d)$ is the group of rotations around the e_1-axis

$$K=\left\{\begin{pmatrix}1&0\\0&A\end{pmatrix}\,\middle|\,A\in\mathrm{SO}(d)\right\}.$$

Then $K = \{k \in G \mid k(e_1) = e_1\}$. For $a \in G$ write $a = [a_1, \ldots, a_{d+1}]$ where a_j are the row vectors in the matrix a. Then $a \cdot e_1 = a_1$. If $x \in S^d$ set $a_1 = x$ and extend a_1 to a positively oriented orthonormal basis $a_1, \ldots, a_{d+1}$ and set $a = [a_1, \ldots, a_{d+1}] \in G$. Then $a \cdot e_1 = x$, so the action of G is transitive on S^d. This also shows that the stabilizer of e_1 is the group

$$K = \left\{ \begin{pmatrix} 1 & 0 \\ 0 & k \end{pmatrix} \middle| \, k \in \mathrm{SO}(d) \right\} \simeq \mathrm{SO}(d).$$

Hence $S^d = G/K$. Let

$$A := \left\{ a_t = \begin{pmatrix} \cos(t) & -\sin(t) & 0 \\ \sin(t) & \cos(t) & 0 \\ 0 & 0 & I_{d-1} \end{pmatrix} \middle| \, t \in \mathbb{R} \right\}.$$

Then every element $g \in G$ can be written as $k_1 a k_2$ with $k_1, k_2 \in K$ and $a \in A$. Define

$$\tau(a) = \begin{pmatrix} -1 & 0 \\ 0 & I_d \end{pmatrix} a \begin{pmatrix} -1 & 0 \\ 0 & I_d \end{pmatrix}.$$

Then $\tau|_K = \mathrm{id}$ and $\tau(a) = a^{-1}$ if $a \in A$. Hence $\tau(x) \in Kx^{-1}K$ which implies that S^d is a commutative space.

Instead of working with the group it is better to work directly with the sphere. Think of S^{d-1} as a subset of S^d by $v \mapsto (0v)$. If $u \in S^d$ then there is a t and $v \in S^{d-1}$ such that

$$u = \cos(t)e_1 + \sin(t)v = k_v a_t e_1,$$

where k_v is a rotation in K. The involution τ is now simply

$$u \mapsto \cos(t)e_1 - \sin(t)v = k_v a_t^{-1} e_1,$$

which can be rotated, using an element from K, back to u.

From now on (G, K) will always—if nothing else is stated—denote a Gelfand pair and $\mathbf{X}$ will stand for a commutative space. We start with a simple lemma (see [13], p. 75).

Lemma 7. *Assume that (G, K) is a Gelfand pair. Then G is unimodular.*

Recall that a function $\varphi : G \to \mathbb{C}$ is *positive definite* if φ is continuous, and for all $N \in \mathbb{N}$, all $c_j \in \mathbb{C}$, and all $x_j \in G$, $j = 1, \ldots, N$, we have

$$\sum_{i,j=1}^{N} c_i \overline{c_j} \varphi(x_i^{-1} x_j) \geq 0.$$

The following gives different characterizations of positive-definite spherical functions. In particular, they arise as the $*$-homomorphisms of the commutative Banach $*$-algebra $L^1(\mathbf{X})^K$ and as positive-definite normalized eigenfunctions of $\mathbb{D}(\mathbf{X})$. Recall that we are always assuming that G and hence also $\mathbf{X}$ are connected.

Theorem 4. *Let $\varphi \in L^\infty(\mathbf{X})$. Then the following assertions are equivalent:*

1. *φ is K-bi-invariant and $L^1(\mathbf{X})^K \to \mathbb{C}$, $f \mapsto \widehat{f}(\varphi) := \int_G f(x)\overline{\varphi(x)}\,\mathrm{d}\mu_G(x)$, is a homomorphism.*
2. *φ is continuous and for all $x, y \in G$ we have*
$$\int_G \varphi(xky)\,\mathrm{d}\mu_K(k) = \varphi(x)\varphi(y).$$
3. *φ is K-bi-invariant, analytic, $\varphi(e) = 1$, and there exists a homomorphism $\chi_\varphi : \mathbb{D}(\mathbf{X}) \to \mathbb{C}$ such that*
$$D\varphi = \chi_\varphi(D)\varphi \tag{5}$$
for all $D \in \mathbb{D}(X)$.

The homomorphism in (a) is a $*$-homomorphism, if and only if φ is positive definite.

Remark 6. We note that (5) implies that φ is analytic because $\Delta\varphi = \chi_\varphi(\Delta)\varphi$ and Δ is elliptic.

Definition 3. $\varphi \in L^\infty(\mathbf{X})^K$ is called a *spherical function* if it satisfies the conditions in Theorem 4.

Denote by $\mathscr{P}_{\mathrm{sp}}(\mathbf{X})$ the space of positive-definite spherical functions on $\mathbf{X}$. It is a locally compact Hausdorff topological vector space in the topology of uniform convergence on compact sets. The *spherical Fourier transform* $\mathscr{S} : L^1(\mathbf{X})^K \to \mathscr{C}(\mathscr{P}_{\mathrm{sp}}(\mathbf{X}))$ is the map

$$\mathscr{S}(f)(\varphi) = \widehat{f}(\varphi) := \int_G f(x)\overline{\varphi(x)}\,\mathrm{d}\mu_G(x) = \int_G f(x)\varphi(x^{-1})\,\mathrm{d}\mu_G(x).$$

The last equality follows from the fact that $\overline{\varphi(x)} = \varphi(x^{-1})$ if φ is positive definite. We note that $\widehat{f * g} = \widehat{f}\widehat{g}$.

Theorem 5. *There exists a unique measure $\mu_{\mathscr{P}}$ on $\mathscr{P}_{\mathrm{sp}}(\mathbf{X})$ such that the following holds:*

1. *If $f \in L^1(\mathbf{X})^K \cap L^2(\mathbf{X})$ then $\|f\|_{L^2} = \|\widehat{f}\|_{L^2}$.*
2. *The spherical Fourier transform extends to a unitary isomorphism*
$$L^2(\mathbf{X})^K \to L^2(\mathscr{P}_{\mathrm{sp}}(\mathbf{X}), \mathrm{d}\mu_{\mathscr{P}})$$
with inverse

$$f = \int_{\mathscr{P}_{\mathrm{sp}}(\mathbf{X})} \widehat{f}(\varphi)\varphi \, \mathrm{d}\mu_{\mathscr{P}}(\varphi), \tag{6}$$

where the integral is understood in L^2-sense.

3. *If $f \in L^1(\mathbf{X})^K \cap L^2(\mathbf{X})$ and $\widehat{f} \in L^1(\mathscr{P}_{\mathrm{sp}}(\mathbf{X}), \mathrm{d}\mu_{\mathscr{P}})$ then (6) holds pointwise.*

At this point we have not said much about the set $\mathscr{P}_{\mathrm{sp}}(\mathbf{X})$. However, it was proved in [43] that $\mathscr{P}_{\mathrm{sp}}(\mathbf{X})$ can always be identified with a subset of $\mathbb{C}^s$ for some $s \in \mathbb{N}$ in a very simple way.

Lemma 8. *The algebra $\mathbf{D}(\mathbf{X})$ is finitely generated.*

Proof. This is the corollary on p. 269 in [34]. □

Let $D_1, \ldots, D_s$ be a set of generators and define a map

$$\Phi : \mathscr{P}_{\mathrm{sp}}(\mathbf{X}) \to \mathbb{C}^s, \quad \varphi \mapsto (D_1\varphi(e), \ldots, D_s\varphi(e)).$$

Let $\Lambda_1 := \Phi(\mathscr{P}_{\mathrm{sp}}(\mathbf{X}))$ with the topology induced from $\mathbb{C}^s$, $\Lambda := \Phi(\mathrm{supp}\mu_{\mathscr{P}})$ and $\widehat{\mu} := \Phi^*(\mu_{\mathscr{P}})$ is the push-forward of the measure on $\mathscr{P}_{\mathrm{sp}}(\mathbf{X})$.

Theorem 6 (Ruffino [43]). *The map $\Phi : \mathscr{P}_{\mathrm{sp}}(\mathbf{X}) \to \Lambda$ is a topological isomorphism.*

Remark 7. In [43] the statement is for the set of bounded spherical functions. But $\mathscr{P}_{\mathrm{sp}}(\mathbf{X})$ is a closed subset of the set of bounded spherical functions, so the statement holds for $\mathscr{P}_{\mathrm{sp}}(\mathbf{X})$. Furthermore, we can choose the generators D_j such that $\overline{D_j} = D_j$, i.e., D_j has real coefficients. If $\varphi \in \mathscr{P}_{\mathrm{sp}}(\mathbf{X})$, then $\overline{\varphi} \in \mathscr{P}_{\mathrm{sp}}(\mathbf{X})$ and it follows that $\overline{\Lambda_1} = \Lambda_1$. We will always assume that this is the case.

For $\lambda \in \Lambda_1$ we let $\varphi_\lambda := \Phi^{-1}(\lambda)$. We view the spherical Fourier transform of $f \in L^1(\mathbf{X})^K \cap L^2(\mathbf{X})^K$ as a function on Λ given by $\widehat{f}(\lambda) := \widehat{f}(\varphi_\lambda)$.

6 Spherical Functions and Representations

To extend Theorem 5 to all of $L^2(\mathbf{X})$ one needs to connect the theory of spherical functions to representation theory. In this section (G,K) will always denote a Gelfand pair. A unitary representation $(\pi, \mathscr{H})$ of G is called *spherical* if the space of K-fixed vectors

$$\mathscr{H}^K := \{u \in \mathscr{H} \mid (\forall k \in K)\ \pi(k)u = u\}$$

is nonzero.

Lemma 9. *Let $f \in L^1(\mathbf{X})^K$. Then $\pi(f)\mathscr{H} \subseteq \mathscr{H}^K$.*

Proof. We have for $k \in K$:

$$\pi(k)\pi(f)v = \int_G f(x)\pi(kx)v \, \mathrm{d}x = \int_G f(k^{-1}x)\pi(x)v \, \mathrm{d}x = \pi(f)v.$$

□

For the following statement, see for example Proposition 6.3.1 in [13].

Lemma 10. *If (G,K) is a Gelfand pair and $\mathcal{H}$ is an irreducible unitary representation of G, then* $\dim \mathcal{H}^K \leq 1$.

Corollary 1. *Let $(\pi, \mathcal{H})$ be an irreducible unitary representation of G such that $\mathcal{H}^K \neq \{0\}$. Then there exists a $*$-homomorphism $\chi_\pi : L^1(\mathbf{X})^K \to \mathbb{C}$ such that*

$$\pi(f)u = \chi_\pi(f)u$$

for all $u \in \mathcal{H}^K$.

Proof. Let $e_\pi \in \mathcal{H}^K$ be a unit vector. As $\dim \mathcal{H}^K = 1$ it follows that $\mathcal{H}^K = \mathbb{C}e_\pi$. It follows from Lemma 2 that $\pi(f)e_\pi = (\pi(f)e_\pi, e_\pi)e_\pi$. The lemma follows now by defining $\chi_\pi(f) := (\pi(f)e_\pi, e_\pi)$. □

Using the heat kernel one can show that $\mathcal{H}^K \subset \mathcal{H}^\omega$, but the following is enough for us.

Theorem 7. $\mathcal{H}^K \subseteq \mathcal{H}^\infty$.

Proof. It is enough to show that $e_\pi \in \mathcal{H}^\infty$. By Lemma 2 it is possible to choose $f \in C_c^\infty(\mathbf{X})$ so that $(\pi(f)e_\pi, e_\pi) \neq 0$. Let

$$h(x) = \int_K f(kx)\, \mathrm{d}\mu_K(k),$$

then

$$\chi_\pi(f) = (\pi(h)e_\pi, e_\pi) = (\pi(f)e_\pi, e_\pi) \neq 0.$$

Hence

$$e_\pi = \frac{1}{\chi_\pi(f)} \pi(h)e_\pi \in \mathcal{H}^K \cap \mathcal{H}^\infty.$$

□

Theorem 8. *Let $(\pi, \mathcal{H})$ be an irreducible spherical representation of G and $e_\pi \in \mathcal{H}^K$ a unit vector. Then the function*

$$\varphi_\pi(x) := (e_\pi, \pi_\pi(x)e_\pi)$$

is a positive-definite spherical function. If φ is a positive-definite spherical function on G, then there exists an irreducible unitary representation $(\pi, \mathcal{H})$ of G such that $\dim \mathcal{H}^K = 1$ *and $\varphi = \varphi_\pi$.*

Proof. Here are the main ideas of the proof. First we note that

$$\int_K \pi(ky)e_\pi\, \mathrm{d}\mu_K(k) = (\pi(y)e_\pi, e_\pi)e_\pi.$$

Hence

$$\begin{aligned}\int_K \varphi_\pi(xky)\,\mathrm{d}\mu_K(k) &= (\pi(x^{-1})e_\pi, \int_K \pi(ky)e_\pi)\,\mathrm{d}\mu_K(k)\\ &= (\pi(x^{-1})e_\pi, (\pi(y)e_\pi, e_\pi)e_\pi)\\ &= (e_\pi, \pi(x)e_\pi)(e_\pi, \pi(y)e_\pi)\\ &= \varphi_\pi(x)\varphi_\pi(y).\end{aligned}$$

Hence φ_π is a spherical function. It is positive definite because

$$\sum_{i,j=1}^{N} c_i\overline{c_j}\varphi_\pi(x_i^{-1}x_j) = \left\|\sum_{i=1}^{N} c_i\pi(x_i)e_\pi\right\|^2 \geq 0.$$

□

Theorem 9. *Let $\varphi : G \to \mathbb{C}$ be a positive-definite function. Then $\varphi \in \mathscr{P}_{\mathrm{sp}}(\mathbf{X})$ if and only if there exists an irreducible spherical unitary representation $(\pi, \mathscr{H})$ and $e_\pi \in \mathscr{H}^K$, $\|e_\pi\| = 1$ such that*

$$\varphi(g) = (e_\pi, \pi(g)e_\pi).$$

Proof. We have already seen one direction. The other direction follows by the classical Gelfand-Naimark-Segal construction. Assume that φ is a positive-definite function. Let $\mathscr{H}$ denote the space of functions generated by linear combinations of $\ell(x)\varphi$, $x \in G$. Define

$$\left(\sum_{j=0}^{N} c_j\ell(x_j)\varphi, \sum_{j=0}^{N} d_j\ell(y_j)\varphi\right)_0 := \sum_{i,j} c_i\overline{d_j}\varphi(x_i^{-1}y_j).$$

(By adding zeros we can always assume that the sum is taken over the same set of indices.) Then $(\, , \,)_0$ is a positive semidefinite Hermitian form on $\mathscr{H}_0$. Let $\mathscr{N} := \{\psi \in \mathscr{H}_0 \mid \|\psi\|_0 = 0\}$. Then $\mathscr{N}$ is G-invariant under left translations and G acts on $\mathscr{H}_0/\mathscr{N}$ by left translation. The form $(\, , \,)_0$ defines an inner product on $\mathscr{H}/\mathscr{N}$ by $(f+\mathscr{N}, g+\mathscr{N}) := (f,g)_0$. Let $\mathscr{H}$ be the completion of $\mathscr{H}_0/\mathscr{N}$ with respect with the metric given by $(\, , \,)$. Then $\mathscr{H}$ is a Hilbert space and the left translation on $\mathscr{H}_0$ induces a unitary representation π_φ on $\mathscr{H}$. If e is the equivalence class of $\varphi \in \mathscr{H}$, then, since φ is K-invariant and $\|\varphi\|_0 = 1$, we get $e \in \mathscr{H}^K \setminus \{0\}$ and $(e, \pi_\varphi(x)e) = \varphi(x)$. □

For $\lambda \in \Lambda_1$ and $\varphi = \varphi_\lambda$ we denote the corresponding representation by $(\pi_\lambda, \mathscr{H}_\lambda)$. We fix once and for all a unit vector $e_\lambda \in \mathscr{H}_\lambda^K$. Let $\mathrm{pr}_\lambda = \int_K \pi_\lambda(k)\,\mathrm{d}k$. Then pr_λ is the orthogonal projection $\mathscr{H}_\lambda \to \mathscr{H}_\lambda^K$, $\mathrm{pr}_\lambda(u) = (u, e_\lambda)e_\lambda$. Let $f \in L^1(\mathbf{X})$. Then, as f is right K-invariant, we get $\pi_\lambda(f) = \pi_\lambda(f) \circ \mathrm{pr}_\lambda$. It therefore makes sense to define a vector-valued Fourier transform by

$$\widetilde{f}(\lambda) := \pi_\lambda(f)e_\lambda$$

(see [40]). We note that if f is K-invariant, then $\widehat{f}(\lambda) = (\widetilde{f}(\lambda), e_\lambda)$.

If $f \in L^1(\mathbf{X}) \cap L^2(\mathbf{X})$, then

$$\widetilde{\ell(x)f} = \pi_\lambda(x)\widetilde{f}(\lambda) \quad \text{and} \quad \mathrm{Tr}(\pi_\lambda(f)) = (\pi_\lambda(f)e_\lambda, e_\lambda).$$

Let $g = \int f^* * f(kx)\, \mathrm{d}\mu_K$. Then g is K-bi-invariant and the function

$$\widehat{g}(\lambda) = (\pi_\lambda(f^*)\pi_\lambda(f)e_\lambda, e_\lambda) = \|\widetilde{f}(\lambda)\|^2$$

is integrable on Λ. Finally

$$\|f\|^2 = f^* * f(e) = g(e) = \int_\lambda \widehat{g}(\lambda)\, \mathrm{d}\widehat{\mu}(\lambda) = \int_\lambda \|\widetilde{f}(\lambda)\|^2\, \mathrm{d}\widehat{\mu}(\lambda).$$

Furthermore, if $\lambda \mapsto (\widetilde{f}(\lambda), e_\lambda)$ is integrable, then by the same argument as above

$$f(x) = \ell(x^{-1})f(e) = \int_\lambda (\pi_\lambda(x^{-1})\widetilde{f}(\lambda), e_\lambda)\, \mathrm{d}\widehat{\mu}(\lambda) = \int_\lambda (\widetilde{f}(\lambda), \pi_\lambda(x)e_\lambda)\, \mathrm{d}\widehat{\mu}(\lambda).$$

Thus we have proved the following theorem:

Theorem 10. *The vector-valued Fourier transform defines a unitary G-isomorphism*

$$L^2(\mathbf{X}) \simeq \int_\lambda^\oplus (\pi_\lambda, \mathscr{H}_\lambda)\, \mathrm{d}\widehat{\mu}.$$

If $f \in L^2(\mathbf{X})$ is so that the function $\lambda \mapsto \|\widetilde{f}(\lambda)\|$ is integrable, then

$$f(x) = \int_\lambda (\widetilde{f}(\lambda), \pi_\lambda(x)e_\lambda)\, \mathrm{d}\widehat{\mu}.$$

7 The Space of Bandlimited Functions

As before (G,K) denotes a Gelfand pair with G connected and $\mathbf{X} = G/K$ the corresponding commutative space. In this section we introduce the bandlimited functions and prove a sampling theorem for the space $L^2_\Omega(\mathbf{X})$ of Ω-bandlimited functions on $\mathbf{X}$.

Definition 4. Suppose $\Omega \subset \Lambda$ is compact. We say that $f \in L^2(\mathbf{X})$ is *Ω-bandlimited* if $\mathrm{supp}\widetilde{f} \subseteq \Omega$. A function f in $L^2(\mathbf{X})$ is bandlimited if there is a compact $\Omega \subseteq \Lambda$ such that f is Ω-bandlimited.

We denote by $L^2_\Omega(\mathbf{X})$ the space of Ω-bandlimited functions. As Ω will be fixed, we just say that f is bandlimited if $f \in L^2_\Omega(\mathbf{X})$. Let $\phi = \phi_\Omega$ be such that $\widetilde{\phi}(\lambda) = \mathbf{1}_\Omega e_\lambda$. Since Ω is compact it follows that $\phi \in L^2_\Omega(\mathbf{X})$. However, ϕ is generally not integrable, because $\lambda \mapsto \mathbf{1}_\Omega(\lambda)e_\lambda$ is not necessarily continuous.

Lemma 11. *We have that*

$$\phi_\Omega(x) = \int_\Omega \varphi_\lambda(x)\,\mathrm{d}\widehat{\mu}(\lambda)$$

is K-invariant and positive definite. In particular, $\phi_\Omega^* = \phi_\Omega$.

Proof. Since $|(e_\lambda, \pi_\lambda(x)e_\lambda)_\lambda|\mathbf{1}_\Omega(\lambda) \le \mathbf{1}_\Omega(\lambda)$, we observe that the function $\lambda \mapsto (\mathbf{1}_\Omega(\lambda)e_\lambda, \pi_\lambda(x)e_\lambda)_\lambda$ is integrable. Therefore Theorem 5 implies that

$$\phi_\Omega(x) = \int_\Omega (e_\lambda, \pi_\lambda(x)e_\lambda)_\lambda\,\mathrm{d}\widehat{\mu}(\lambda) = \int_\Omega \varphi_\lambda(x)\,\mathrm{d}\widehat{\mu}(\lambda).$$

We have

$$\sum_{i,j} c_i\overline{c_j}\phi(x_i^{-1}x_j) = \int_\Omega \sum_{i,j} c_i\overline{c_j}\varphi_\lambda(x_i^{-1}x_j)\,\mathrm{d}\widehat{\mu}(\lambda) \ge 0$$

as the spherical functions φ_λ, $\lambda \in \Omega$, are positive definite. □

Theorem 11. *$L^2_\Omega(\mathbf{X})$ is a reproducing kernel Hilbert space with reproducing kernel $K(x,y) = \phi_\Omega(y^{-1}x)$. Furthermore, the orthogonal projection $L^2(G) \to L^2_\Omega(\mathbf{X})$ is given by $f \mapsto f * \phi_\Omega$.*

Proof. We have for $f \in L^2_\Omega(\mathbf{X})$

$$\left|\int_\lambda (\widetilde{f}(\lambda), \pi_\lambda(x)e_\lambda)_\lambda\,\mathrm{d}\widehat{\mu}(\lambda)\right| \le \int_\Omega \|\widetilde{f}(\lambda)\|_\lambda\,\mathrm{d}\widehat{\mu}(\lambda) \le |\Omega|^{1/2}\|\widetilde{f}\|_{L^2},$$

where $|\Omega|$ denotes the volume $\int_\Omega \mathrm{d}\widehat{\mu}$ of Ω which is finite as Ω is compact. It follows that

$$\begin{aligned}
f(x) &= \int_\Omega (\widetilde{f}(\lambda), \pi_\lambda(x)e_\lambda)_\lambda\,\mathrm{d}\widehat{\mu}(\lambda) \\
&= \int_\lambda (\widetilde{f}(\lambda), \mathbf{1}_\Omega(\lambda)\pi_\lambda(x)e_\lambda)_\lambda\,\mathrm{d}\widehat{\mu}(\lambda) \\
&= \int_{\mathbf{X}} f(y)\overline{\ell(x)\phi_\Omega(y)}\,\mathrm{d}y \\
&= \int_{\mathbf{X}} f(y)\phi_\Omega(y^{-1}x)\,\mathrm{d}y \\
&= f * \varphi_\Omega(x).
\end{aligned}$$

Thus $L^2_\Omega(\mathbf{X})$ is a reproducing kernel Hilbert space with reproducing kernel $K(x,y) = \phi_\Omega(y^{-1}x)$. The rest follows now from Proposition 1. □

Let us point out the following consequence of Proposition 1:

Corollary 2. *Let $f \in L^2(G)$. Then $f \in L^2_\Omega(\mathbf{X})$ if and only if $f * \phi_\Omega = f$.*

8 The Bernstein Inequality and Sampling of Bandlimited Functions

The definition of the topology on Λ inspired by [43] ensures that the eigenvalues c_λ for the Laplacian on e_λ are bounded when λ is in a compact set Ω. This enables us to obtain

Lemma 12. *For a compact set $\Omega \subseteq \Lambda$ the functions in $L^2_\Omega(\mathbf{X})$ are smooth, and there is a constant $c(\Omega)$ such that the following Bernstein inequality holds:*

$$\|\Delta^k f\|_{L^2} \leq c(\Omega)^k \|f\|_{L^2}.$$

Proof. As we have seen, each $f \in L^2_\Omega(\mathbf{X})$ can be written

$$f(x) = \int_\Omega (\widehat{f}(\lambda), \pi_\lambda(x)e_\lambda)_\lambda \, \mathrm{d}\widehat{\mu}(\lambda).$$

For fixed λ the function

$$t \mapsto (\widetilde{f}(\lambda), \pi_\lambda(x\mathrm{e}^{tX_i})e_\lambda)_\lambda$$

is differentiable as $e_\lambda \in \mathscr{H}^\infty_\lambda$. Thus, there exists a t_λ between zero and t such that

$$\left(\widetilde{f}(\lambda), \frac{\pi_\lambda(x\mathrm{e}^{tX_i})e_\lambda - \pi_\lambda(x)e_\lambda}{t}\right)_\lambda = (\widetilde{f}(\lambda), \pi_\lambda(x\mathrm{e}^{t_\lambda X_i})\pi_\lambda(X_i)e_\lambda)_\lambda.$$

It follows that

$$\begin{aligned}
\frac{f(x\mathrm{e}^{tX_i}) - f(x)}{t} &= \int_\Omega \left(\widetilde{f}(\lambda), \frac{\pi_\lambda(x\mathrm{e}^{tX_i})e_\lambda - \pi_\lambda(x)e_\lambda}{t}\right)_\lambda \mathrm{d}\widehat{\mu}(\lambda) \\
&= \int_\Omega (\widetilde{f}(\lambda), \pi_\lambda(x)\pi_\lambda(\mathrm{e}^{t_\lambda X_i})\pi_\lambda(X_i)e_\lambda)_\lambda \, \mathrm{d}\widehat{\mu}(\lambda) \\
&\leq \int_\Omega \|\widetilde{f}\|_\lambda \|\pi_\lambda(x)\pi_\lambda(\mathrm{e}^{t_\lambda X_i})\pi_\lambda(X_i)e_\lambda\|_\lambda \, \mathrm{d}\widehat{\mu}(\lambda) \\
&\leq \int_\Omega \|\widetilde{f}\|_\lambda \|\pi_\lambda(X_i)e_\lambda\|_\lambda \, \mathrm{d}\widehat{\mu}(\lambda).
\end{aligned}$$

Here we have used that e_λ is a smooth vector for π_λ and the unitarity of π_λ. Now

$$\|\pi_\lambda(X_i)e_\lambda\|^2_\lambda \leq |\left(e_\lambda, \sum_i \pi_\lambda(X_i)\pi_\lambda(X_i)e_\lambda\right)_\lambda| = c_\lambda\|e_\lambda\|^2_\lambda.$$

Therefore the Lebesgue-dominated convergence theorem ensures that

$$\lim_{t\to 0}\frac{f(xe^{tX_i})-f(x)}{t}=\int_{\Omega}(\widetilde{f}(\lambda),\pi_{\lambda}(x)\pi_{\lambda}(X_i)e_{\lambda})_{\lambda}\,\mathrm{d}\widehat{\mu}(\lambda),$$

which shows that f is differentiable. Repeat the argument to show that f is smooth and notice that

$$\Delta^k f(x)=\int_{\Omega}c_{\lambda}^{k}(\widetilde{f}(\lambda),\pi_{\lambda}(x)e_{\lambda})_{\lambda}\,\mathrm{d}\widehat{\mu}(\lambda).$$

It then finally follows that

$$\|\Delta^k f(x)\|_{L^2}^2=\int_{\Omega}|c_{\lambda}|^{2k}\|\widetilde{f}\|_{\lambda}^2\,\mathrm{d}\mu(\lambda)\leq c(\Omega)^{2k}\int_{\Omega}\|\widetilde{f}\|_{\lambda}^2\,\mathrm{d}\widehat{\mu}(\lambda).$$

We have thus proved the Bernstein inequality. □

Corollary 3. *Let $\Omega\subseteq\Lambda$ be a compact set and define the neighborhoods U_{ε} by*

$$U_{\varepsilon}=\{\exp(t_1X_1)\dots\exp(t_nX_n)\mid(t_1,\dots,t_n)\in[-\varepsilon,\varepsilon]^n\}.$$

It is possible to choose ε small enough that for any U_{ε}-relatively separated family $\{x_i\}$, the functions $\{\ell(x_i)\phi\}$ provide a frame for L^2_{Ω}.

Corollary 4. *Let $\Omega\subseteq\Lambda$ be a compact set and define the neighborhoods U_{ε} by*

$$U_{\varepsilon}=\{\exp(t_1X_1)\dots\exp(t_nX_n)\mid(t_1,\dots,t_n)\in[-\varepsilon,\varepsilon]^n\}.$$

It is possible to choose ε small enough that, for any U_{ε}-relatively separated family x_i and any partition of unity $0\leq\psi_i\leq 1_{x_iU_{\varepsilon}}$, the operator

$$Tf=\sum_i f(x_i)\psi_i*\phi$$

is invertible on L^2_{Ω}. If the functions $\{\ell(x_i)\phi\}$ also form a frame for L^2_{Ω}, the functions $\{T^{-1}(\psi_i*\phi)\}$ provide a dual frame.

Proof. Due to the expansion

$$f=\sum_i f(x_i)T^{-1}(\psi_i*\phi)=\sum_i(f,\ell_{x_i}\phi)T^{-1}(\psi_i*\phi),$$

Proposition 2.4 in [38] tells us it is enough to check that $T^{-1}(\psi_i*\phi)$ is a Bessel sequence, i.e., it satisfies $\sum_i|(f,T^{-1}(\psi_i*\phi))|^2\leq C\|f\|_{L^2}^2$. Since, T is a bounded invertible operator on L^2_{Ω}, the operator $(T^{-1})^*$ is also bounded on L^2_{Ω}. Then, since convolution with ϕ is self-adjoint and $(T^{-1})^*f=((T^{-1})^*f)*\phi$, we get

$$(f,T^{-1}(\psi_i*\phi))=((T^{-1})^*f,\psi_i*\phi)=((T^{-1})^*f,\psi_i)=\int_G(T^{-1})^*f(x)\psi_i(x)\,\mathrm{d}x.$$

Therefore

$$\begin{aligned}|(f,T^{-1}(\psi_i*\phi))|^2 &\leq \left(\int_G |(T^{-1})^*f(x)|\psi_i(x)\,\mathrm{d}x\right)^2\\ &\leq \int_G |(T^{-1})^*f(x)|^2\psi_i(x)\,\mathrm{d}x\int\psi_i(x)\,\mathrm{d}x\\ &\leq |U|\int_G |(T^{-1})^*f(x)|^2\psi_i(x)\,\mathrm{d}x\end{aligned}$$

and finally

$$\sum_i |(f,T^{-1}(\psi_i*\phi))|^2 \leq |U|\|(T^{-1})^*f\|_{L^2}^2 \leq |U|\|(T^{-1})^*\|^2\|f\|_{L^2}^2. \qquad \square$$

9 Examples of Commutative Spaces

In this section we give some examples of the theory developed in the previous section. We do not discuss the Riemannian symmetric spaces of the noncompact type as those can be found in [41].

9.1 The Space $\mathbb{R}^d$

The simplest example of a Gelfand pair is $(\mathbb{R}^d,\{0\})$. The algebra of invariant differential operators is $\mathbb{D}(\mathbb{R}^d)=\mathbb{C}[\partial_1,\ldots,\partial_d]$, the polynomials in the partial derivatives $\partial_j=\partial/\partial x_j$. The positive-definite spherical functions are the exponentials $\varphi_\lambda(x)=\mathrm{e}^{i\lambda\cdot x}$, $\lambda\in\mathbb{R}^d$. Using $\partial_1,\ldots,\partial_d$ as generators for $\mathbb{D}(\mathbb{R}^d)$ Theorem 6 identifies Λ with $i\mathbb{R}^d$ via the map $\varphi_\lambda\mapsto i(\lambda_1,\ldots,\lambda_d)$. Note the slight difference from our previous notation for φ_λ.

We can also consider $\mathbb{R}^d$ as the commutative space corresponding to the connected Euclidean motion group $G=\mathrm{SO}(d)\ltimes\mathbb{R}^d$ with $K=\mathrm{SO}(d)$. The K-invariant functions are now the radial functions $f(x)=F_f(\|x\|)$, where F_f is a function of one variable. We have $\mathbb{D}(\mathbb{R}^d)=\mathbb{C}[-\Delta]$ and Theorem 6 now identifies the spectrum Λ with $\mathbb{R}^+$. For $\lambda\in\mathbb{R}$ we denote by φ_λ the spherical function with $-\Delta\varphi_\lambda=\lambda^2\varphi_\lambda$.

Denote by J_ν the Bessel function

$$J_\nu(r)=\frac{(r/2)^\nu}{\Gamma(1/2)\Gamma(\nu+1/2)}\int_{-1}^{1}\cos(tr)(1-t^2)^{\nu-1/2}\,\mathrm{d}t$$

(see [36], p. 144).

Lemma 13. *The positive-definite spherical functions for the Euclidean motion group are given by*

$$\begin{aligned}\varphi_\lambda(x) &= \frac{2^{\frac{d-2}{2}}\Gamma\left(\frac{d}{2}\right)}{(\lambda\|x\|)^{\frac{d-2}{2}}} J_{(d-2)/2}(\lambda\|x\|)\\ &= \frac{\Gamma\left(\frac{d}{2}\right)}{\sqrt{\pi}\Gamma\left(\frac{d-1}{2}\right)} \int_{-1}^{1} \cos(\lambda\|x\|t)(1-t^2)^{\frac{d-3}{2}}\,\mathrm{d}t.\end{aligned}$$

Proof. Denote for the moment the right-hand side by ψ_λ. Then ψ_λ is analytic as cos is even. It is also a radial eigenfunction of $-\Delta$ with eigenvalue λ^2 and $\psi_\lambda(0) = 1$. Now Theorem 4 implies that $\varphi_\lambda = \psi_\lambda$. □

Remark 8. We note that we can write

$$\varphi_\lambda(x) = \int_{S^{d-1}} \mathrm{e}^{-i\lambda(\omega,x)}\,\mathrm{d}\sigma(\omega)$$

where $\mathrm{d}\sigma$ is the normalized rotational invariant measure on the sphere.

It is easy to describe the representation $(\pi_\lambda, \mathscr{H}_\lambda)$ associated to φ_λ. For $\lambda \in \mathbb{R}^*$ set $\mathscr{H}_\lambda = L^2(S^{d-1}, \mathrm{d}\sigma) = L^2(S^{d-1})$ and define

$$\pi_\lambda((k,x))u(\omega) := \mathrm{e}^{i\lambda(\omega,x)}u(k^{-1}(\omega)).$$

We take the constant function $\omega \mapsto 1$ as normalized K-invariant vector e_λ. Then

$$(e_\lambda, \pi_\lambda((k,x))e_\lambda) = \int_{S^{d-1}} \overline{\mathrm{e}^{i\lambda(\omega,x)}}\,\mathrm{d}\sigma(\omega) = \varphi_\lambda(x).$$

We refer to [40] for more information.

9.2 The Sphere S^d

Let $S^d = \{x \in \mathbb{R}^{d+1} \mid \|x\| = 1\}$ be the unit sphere in $\mathbb{R}^{d+1}$. We refer to Chapter 9 of [15] and Chapter III in [47] for more detailed discussion on harmonic analysis and representation theory related to the sphere. In particular, most of the proofs can be found there. Recall from Example 1 that $S^d = G/K$ where $G = \mathrm{SO}(d+1)$ and $K = \mathrm{SO}(d)$ and that S^d is a commutative space.

For $d = 1$ we have $S^1 = \mathbb{T} = \{z \in \mathbb{C} \mid |z| = 1\}$ is an abelian group, and the spherical functions are just the usual characters $z \mapsto z^n$ (or if we view $\mathbb{T} = \mathbb{R}/2\pi\mathbb{Z}$, the functions $\theta \mapsto e^{in\theta}$). We therefore assume that $d \geq 2$, but we would also like to point out another special case. $\mathrm{SO}(3) \simeq \mathrm{SU}(2)$ and $\mathrm{SO}_o(4) = \mathrm{SU}(2) \times \mathrm{SU}(2)$. The group $K = \mathrm{SU}(2)$ is embedded as the diagonal in $\mathrm{SU}(2) \times \mathrm{SU}(2)$. Then

$$S^3 = \mathrm{SO}(4)/\mathrm{SO}(3) \simeq \mathrm{SU}(2) \simeq \{z \in \mathbb{H} \mid |z| = 1\}$$

and the K-invariant functions on S^3 corresponds to the central functions on $\mathrm{SU}(2)$, i.e., $f(kuk^{-1}) = f(u)$. Hence Λ, the set of spherical representations, is just $\widehat{\mathrm{SU}(2)}$; the set of equivalence classes of irreducible representations of $\mathrm{SU}(2)$ and the spherical functions are

$$\varphi_\pi = \frac{1}{d(\pi)}\mathrm{Tr}\pi =: \frac{1}{d(\pi)}\chi_\pi,$$

where $d(\pi)$ denotes the dimension of V_π. We will come back to this example later.

Denote by $\mathfrak{g} = \mathfrak{so}(d+1) = \{X \in M_{d+1}(\mathbb{R}) \mid X^T = -X\}$ the Lie algebra of G. We can take $\langle X, Y\rangle = -\mathrm{Tr}(XY)$ as a K-invariant inner product on $\mathfrak{g}$. Then

$$\mathfrak{k} = \left\{ \begin{pmatrix} 0 & 0 \\ 0 & Y \end{pmatrix} \,\middle|\, Y \in \mathfrak{so}(d) \right\} \simeq \mathfrak{so}(d)$$

and $\mathfrak{s} = \mathfrak{k}^\perp$ is given by

$$\mathfrak{s} = \left\{ X(v) = \begin{pmatrix} 0 & -v^T \\ v & 0 \end{pmatrix} \,\middle|\, v \in \mathbb{R}^d \right\} \simeq \mathbb{R}^d.$$

A simple matrix multiplication shows that $kX(v)k^{-1} = X(k(v))$ where we have identified $k \in \mathrm{SO}(d)$ with its image in K. It follows that the only invariant polynomials on $\mathfrak{s}$ are those of the form $p(X(v)) = q(\|v\|^2)$ where q is a polynomial of one variable. It follows that $\mathbb{D}(S^d) = \mathbb{C}[\Delta]$ where Δ now denotes the Laplace operator on S^{d-1}. Thus $\mathbb{D}(S^d)$ is abelian and this shows again that $S^d = \mathrm{SO}(d+1)/\mathrm{SO}(d)$ is a commutative space.

Recall that a polynomial $p(x)$ on $\mathbb{R}^{d+1}$ is homogeneous of degree n if $p(\lambda x) = \lambda^n p(x)$ for all $\lambda \in \mathbb{R}$ and p is harmonic if $\Delta_{\mathbb{R}^{d+1}} p = 0$. Denote by $\mathscr{H}_n$ the space of harmonic polynomials that are homogeneous of degree n and set

$$\mathscr{Y}_n := \mathscr{H}_n|_{S^d} = \{p|_{S^d} \mid p \in \mathscr{H}_n\}. \tag{7}$$

As the action of G on $\mathbb{R}^{d+1}$ commutes with $\Delta_{\mathbb{R}^{d+1}}$ it follows that each of the spaces $\mathscr{Y}_n$ is G-invariant. Denote the corresponding representation by π_n, then $\pi_n(a)p(x) = p(a^{-1}x)$ for $p \in \mathscr{Y}_n$.

Theorem 12. *The following holds:*

1. $(\pi_n, \mathscr{Y}_n)$ *is an irreducible spherical representation of* $\mathrm{SO}_o(d+1)$.
2. *If* (π, V) *is an irreducible spherical representation of* G *then there exists an* n *such that* $(\pi, V) \simeq (\pi_n, \mathscr{Y}_n)$.
3. $\dim \mathscr{Y}_n = (2n+d-1)\frac{(d+n-2)!}{(d-1)!n!} =: d(n)$.
4. $-\Delta|_{\mathscr{Y}_n} = n(d+n-1)$.
5. $L^2(S^d) \simeq_G \bigoplus_{n=0}^{\infty} \mathscr{Y}_n$. *In particular, every* $f \in L^2(S^d)$ *can be approximated by harmonic polynomials.*

The last part of the above theorem implies that $\Lambda = \mathbb{N} = \{0, 1, \ldots\}$. We use this natural parametrization of Λ rather than the one given in Sect. 6.

For $\Omega \in \mathbb{N}$ the Paley–Wiener space for Ω is

$$L^2_\Omega(S^d) = \{p|_{S^d} \mid p \text{ is a harmonic polynomial of degree } \leq \Omega\}.$$

It is noted that $\dim L^2_\Omega(S^d) < \infty$ which is also the case in the more general case of compact Gelfand pairs.

The group $\mathrm{SO}(d)$ acts transitively on spheres in $\mathbb{R}^d$. Hence every $v \in S^d$ is K-conjugate to a vector of the form $(\cos(\theta), \sin(\theta), 0, \ldots, 0)^T$ and a function f is K-invariant if and only if there exists a function F_f of one variable such that

$$f(v) = F_f(\cos(\theta)) = F_f((v, e_1)) = F_f(v_1).$$

In particular, this holds for the spherical function $\varphi_n(x)$ corresponding to the representation π_n as well as the reproducing kernel ϕ of the space $L^2_\Omega(S^d)$. In fact, for $d \geq 2$, the spherical functions are determined by the Jacobi polynomials or normalized Gegenbauer polynomials in the following manner: $F_{\varphi_n}(t) = \Phi_n(t)$ or $\varphi_n(x) = \Phi_n((x, e_1)) = \Phi_n(\cos(\theta))$, where

$$\begin{aligned}
\Phi_n(\cos(\theta)) &= {}_2F_1\left(n+d-1, -n, \frac{d}{2}; \sin^2(\theta/2)\right) \\
&= {}_2F_1\left(n+d-1, -n, \frac{d}{2}; \frac{1-\cos(\theta)}{2}\right) \\
&= \frac{n!(d-2)!}{(n+d-2)!} C_n^{(d-1)/2}(\cos(\theta)).
\end{aligned}$$

As the polynomials $\varphi_n(t)$ are real valued we can write the spherical Fourier transform as

$$\widehat{f}(n) = \int_{S^d} f(x)\varphi_n(x)\,\mathrm{d}\sigma(x) = \frac{\Gamma\left(\frac{d+1}{2}\right)}{\sqrt{\pi}\Gamma\left(\frac{d}{2}\right)} \int_{-1}^{1} F_f(t)\Phi_n(t)(1-t^2)^{\frac{d}{2}-1}\,\mathrm{d}t$$

with inversion formula

$$f(x) = \sum_{n=0}^{\infty} d(n)\widehat{f}(n)\varphi_n(x) = \sum_{n=0}^{\infty} d(n)\widehat{f}(n)\Phi_n((x,e_1)).$$

In particular the sinc-type function is given by

$$\phi_\Omega(x) = F_{\phi_\Omega}((x,e_1)) = \sum_{n=0}^{\omega} d(n)\Phi_n((x,e_1)). \tag{8}$$

Note also that we can write the convolution kernel $\phi_\Omega(a^{-1}b)$, $a,b \in \mathrm{SO}(d+1)$ as $F_{\phi_\Omega}((x,y))$ where $x = be_1$ and $y = ue_1$.

For $d = 1$ the sphere is the torus $\mathbf{T} = \{z \in \mathbb{C} \mid |z| = 1\}$ and $\varphi_n(z) = z^n$. Hence

$$\phi_\Omega(\mathrm{e}^{it}) = \sum_{n=-\Omega}^{\omega} \mathrm{e}^{nit} = \frac{\sin((\Omega+1/2)t)}{\sin(t/2)}$$

is the Dirichlet kernel D_Ω. In the higher dimensional cases the kernel ϕ_Ω behaves very similar to the Dirichlet kernel. Here are some of its properties:

Lemma 14. *Let the notation be as above. Then the following holds:*

1. $\phi_\Omega(e_1) = \sum_{n=0}^{\omega} d(n) = \dim L^2_\Omega(S^d) \nearrow \infty$ *as* $\Omega \to \infty$.
2. $\int_{S^d} \phi_\Omega(x)\,\mathrm{d}\sigma(x) = 1$.
3. $\|\phi_\Omega\|^2_{L^2} = \sum_{n=0}^{\omega} d(n) \to \infty$ *as* $\Omega \to \infty$.
4. *If* $f \in L^2(S^d)$, *then* $f * \phi_\Omega = \int_{S^{d-1}} f(x)F_{\phi_\Omega}((\cdot,x))\,\mathrm{d}\sigma(x) \underset{\Omega\to\infty}{\longrightarrow} f$ *in* $L^2(S^d)$.

Let $N(\Omega) = \dim L^2_\Omega(S^d) = 1 + d(1) + \cdots + d(\Omega)$. Then every set of points $\{\omega_j \in S^d \mid j = 1,\ldots,N(\Omega)\}$ such that the functions $F_{\phi_\Omega}((\cdot,\omega_j))$ are linearly independent will give us a basis (and hence a frame) for $L^2_\Omega(S^d)$. Further $N(\Omega)$ is the minimal number of points so that the sampling will determine an arbitrary function $f \in L^2_\Omega(S^d)$. If $n > N(\Omega)$, then the functions $\{F_{\phi_\Omega}((\cdot,\omega_j))\}_{j=1}^n$ will form a frame if and only if the set is generating.

Let us go back to the special case $S^3 \simeq \mathrm{SU}(2)$. The set $\Lambda \simeq \widehat{\mathrm{SU}(2)}$ is isomorphic to $\mathbb{N}$ in such a way that $d(n) = d(\pi_n) = n+1$. Every element in $\mathrm{SU}(2)$ is conjugate to a matrix of the form

$$u(\theta) = \begin{pmatrix} \mathrm{e}^{i\theta} & 0 \\ 0 & \mathrm{e}^{-i\theta} \end{pmatrix}.$$

Thus the positive-definite spherical functions are given by

$$\varphi_n(u(\theta)) = \frac{1}{n+1}\chi_{\pi_n}(u(\theta)) = \frac{1}{n+1}\frac{\sin((n+1)\theta)}{\sin(\theta)}.$$

It follows that the reproducing kernel ϕ_Ω is

$$\begin{aligned}
\phi_\Omega(u(\theta)) &= \frac{1}{\sin(\theta)}\sum_{n=1}^{\Omega+1}\sin(n\theta)\\
&= \frac{1}{2i\sin(\theta)}\left(\sum_{n=1}^{\Omega+1}\left(e^{i\theta}\right)^n - \sum_{n=1}^{\Omega+1}\left(e^{-i\theta}\right)^n\right)\\
&= \frac{\sin((\Omega+2)\theta/2)\sin((\Omega+1)\theta/2)}{\sin(\theta)\sin(\theta/2)}\,.
\end{aligned}$$

9.3 Symmetric Spaces of the Compact Type

To avoid introducing too much new notation we will not go into much detail about general symmetric spaces $\mathbf{X} = G/K$ of the compact type. The general case follows very much the same line as the special case of the sphere. Recall that "symmetric space of the compact type" means that the group G is compact and there exists an involution $\tau : G \to G$ such that with $G^\tau = \{u \in G \mid \tau(u) = u\}$ we have

$$(G^\tau)_o \subseteq K \subseteq G^\tau\,.$$

An example is the sphere S^d where as in the last section $G = \mathrm{SO}_o(d+1)$ and the involution τ is given by

$$u \mapsto \begin{pmatrix} -1 & 0\\ 0 & I_d\end{pmatrix} u \begin{pmatrix} -1 & 0\\ 0 & I_d\end{pmatrix} = \begin{pmatrix} u_{11} & -v^t\\ -v & k\end{pmatrix}$$

as in Example 1 and the previous section. All symmetric spaces of the compact type are commutative. The spectral set Λ is well understood (see Theorem 4.1, p. 535 in [35]). In particular Λ is discrete. Each representation $(\pi_\lambda, \mathcal{H}_\lambda)$ occurs with multiplicity one in $L^2(\mathbf{X})$. Let

$$\mathcal{Y}_\lambda := \{(u, \pi_\lambda(\cdot)e_\pi) \mid u \in V_\lambda\}.$$

Then, if compact Ω is given, there exists a finite set $\Lambda(\Omega) \subset \Lambda$ such that

$$L^2_\Omega(\mathbf{X}) = \bigoplus_{\lambda\in\Lambda(\Omega)} \mathcal{Y}_\lambda$$

and $N(\Omega) = \dim L^2_\Omega(\mathbf{X})$ is finite. In particular, only finitely many points are needed to determine the elements in $L^2_\Omega(\mathbf{X})$.

The spherical functions are well understood. They are given by the generalized hypergeometric functions (and Jacobi polynomials) of Heckman and Opdam [32]. Again the function

$$\phi_\Omega = \sum_{\lambda \in \Lambda(\Omega)} d(\pi_\lambda)\varphi_\lambda \in C^\omega(\mathbf{X})$$

generalizes the Dirichlet kernel. Furthermore, Lemma 14 holds true in the general case.

9.4 Gelfand Pairs for the Heisenberg Group

For the details in the following discussion we refer to [6,46]. We let $\mathbb{H}_n = \mathbb{C}^n \times \mathbb{R}$ denote the $2n+1$-dimensional Heisenberg group with group composition

$$(z,t)(z',t') = \left(z+z',t+t'+\frac{1}{2}\mathrm{Im}(\bar{z}z')\right).$$

Denoting $z = x+iy$ the Heisenberg group is equipped with the left and right Haar measure $\mathrm{d}x\mathrm{d}y\mathrm{d}t$ where $\mathrm{d}x, \mathrm{d}y, \mathrm{d}t$ are Lebesgue measures on $\mathbb{R}^n$,$\mathbb{R}^n$, and $\mathbb{R}$, respectively. The group $K = \mathrm{U}(n)$ acts on $\mathbb{H}_n$ by the group homomorphism given by

$$k \cdot (z,t) = (kz,t).$$

Let $G = K \ltimes \mathbb{H}_n$. It follows that $L^p(G/K) \simeq L^p(\mathbb{H}_n)$ and $L^p(G/K)^K \simeq L^p(\mathbb{H}_n)^K = L^p(\mathbb{H}_n)_{\mathrm{rad}}$. It is known that the algebra $L^1(\mathbb{H}_n)^K$ of integrable radial functions on $\mathbb{H}_n$ is commutative and thus (G,K) is a Gelfand pair. This is also the case for several other subgroups of $\mathrm{U}(n)$ as shown in [5].

9.4.1 Representation Theory for $G = \mathrm{U}(n) \ltimes \mathbb{H}_n$

A collection of important representations for the Heisenberg group is the representations for $\lambda > 0$ given by

$$\pi_\lambda(z,t)f(w) = \mathrm{e}^{i\lambda t - \lambda \mathrm{Im}(w\bar{z})/2 - \lambda|z|^2/4} f(w+z).$$

They act irreducibly on the Fock space $\mathscr{F}_\lambda$ of entire functions on $\mathbb{C}^n$ with norm defined by

$$\|F\|_\lambda^2 = \left(\frac{\lambda}{2\pi}\right)^n \int_{\mathbb{C}^n} |F(z)|^2 \mathrm{e}^{-\lambda|z|^2/2}\,\mathrm{d}z < \infty.$$

For $\lambda < 0$ define the representations

$$\pi_\lambda(z,t)f(w) := \pi_{-\lambda}(\bar{z},t)f(w)$$

on the anti-holomorphic functions $\mathscr{F}_\lambda := \overline{\mathscr{F}_{-\lambda}}$. These representations are irreducible, and the left regular representation of $\mathbb{H}_n$ on $L^2(\mathbb{H}_n)$ decomposes as

$$(\ell_{\mathbb{H}_n}, L^2(\mathbb{H}_n)) \simeq \int_{\mathbb{R}^*}^{\oplus} (\pi_\lambda, \mathscr{F}_\lambda)\, |\lambda|^n \, \mathrm{d}\lambda.$$

We should note that there are more irreducible representations than the π_λ, but they are one dimensional and do not show up in the Plancherel formula (they are of Plancherel measure 0).

Let us now turn to the regular representation of $G = \mathrm{U}(n) \ltimes \mathbb{H}_n$ on $L^2(\mathbb{H}_n)$ which is

$$\ell_G(k,z,t)f(z',t') = f\left(k^{-1}(z'-z), t'-t-\frac{1}{2}\mathrm{Im}(z'\bar{z})\right).$$

Notice that $\mathrm{U}(n)$ acts only on the z-variable, and for fixed $k \in \mathrm{U}(n)$ the elements $G_k = \{(kz,t) \mid (z,t) \in \mathbb{H}_n\}$ form a group isomorphic to $\mathbb{H}_n$. For fixed k the left regular representation of $\mathbb{H}_n$ on $L^2(G_k)$ can be decomposed using the representations π_λ. We get

$$(\ell_{\mathbb{H}_n}, L^2(G_k)) = \int_{\mathbb{R}^*}^{\oplus} (\pi_\lambda^k, \mathscr{F}_\lambda)\, |\lambda|^n \, \mathrm{d}\lambda,$$

where $\pi_\lambda^k(z,t) = \pi_\lambda(kz,t)$. Note that for $\lambda > 0$ (the case of $\lambda < 0$ is handled similarly) we have, with $\nu(k)f(w) = f(k^{-1}w)$,

$$\pi_\lambda^k(z,t) = \nu(k)\pi_\lambda(z,t)\nu(k)^{-1}.$$

Denote by $\nu_{\lambda,m}$ the representation ν restricted to the homogeneous polynomials $P_{\lambda,m}$ of degree m with inner product from $\mathscr{F}_\lambda$. Then $(\nu, \mathscr{F}_\lambda)$ decomposes into

$$\bigoplus_{m=0}^{\infty} (\nu_{\lambda,m}, P_{\lambda,m}).$$

Note that $\dim(P_{\lambda,m}) = 2m+n$. Let $H_{\lambda,m}$ be the Hilbert space spanned by $\pi_\lambda(\mathbb{H}_n)u_{\lambda,m}$ with $u_{\lambda,m}$ in $P_{\lambda,m}$. The representations $\pi_{\lambda,m}$ of G on $H_{\lambda,m}$ thus obtained are irreducible [46], $\dim(H_{\lambda,m}^K) = 1$, and they provide a decomposition of the left regular representation of G on $L^2(\mathbb{H}_n)$:

$$(\ell_G, L^2(\mathbb{H}_n)) = \bigoplus_{m=0}^{\infty} \int_{\mathbb{R}^*}^{\oplus} (\pi_{\lambda,m}, H_{\lambda,m})\, |\lambda|^n \, \mathrm{d}\lambda.$$

9.4.2 Spherical Functions

The bounded $\mathrm{U}(n)$-spherical functions in this case are

$$\phi_{\lambda,m}(z,t) = \int_{\mathrm{U}(n)} (\pi_\lambda^k(z,t)u_\lambda, u_\lambda)_{\mathscr{F}_\lambda}\, \mathrm{d}k = \mathrm{e}^{i\lambda t} L_m^{(n-1)}\left(|\lambda||z|^2/2\right) \mathrm{e}^{-|\lambda||z|^2/4}$$

for $\lambda \in \mathbb{R}\setminus\{0\}$ and $m = 0,1,2,\ldots$. Here $L_m^{(n-1)}$ is the Laguerre polynomial of degree m and order $n-1$

$$L_m^{(n-1)}(x) = \binom{m+n-1}{m}^{-1} \sum_{k=0}^{m} (-1)^k \binom{m+n-1}{m-k} \frac{x^k}{k!}.$$

In this fashion the spectrum for $L^1(\mathbb{H}_n)^K$ can be identified with the Heisenberg fan [46]

$$\mathfrak{F}_n = \{((2m+n)|\lambda|,\lambda) \mid m = 0,1,\ldots;\lambda \neq 0\} \cup \mathbb{R}_+$$

with Plancherel measure supported on $\Lambda = \{((2m+n)|\lambda|,\lambda) \mid m = 0,1,\ldots;\lambda \neq 0\}$ and given explicitly as

$$\int_\Lambda F(\phi)\,\mathrm{d}\mu(\phi) = \int_{\mathbb{R}^*} \sum_{m=0}^{\infty} (2m+n)F(\phi_{\lambda,m})|\lambda|^n\,\mathrm{d}\lambda.$$

As shown by [7] and more generally in [43] the topologies on $\mathfrak{F}_n$ and Λ are the topologies inherited from $\mathbb{R}^2$.

9.4.3 Sampling and Oversampling of Bandlimited Functions

Let $L^2_\Omega(\mathbb{H}_n)$ be the space of functions in $L^2(\mathbb{H}_n)$ with Fourier transform supported in

$$\Omega = ((2m+n)|\lambda|,\lambda) \mid m = 0,\ldots,M; 0 < |\lambda| \leq R\}.$$

In this case the sinc-type function is given by the expression

$$\begin{aligned}\phi(z,t) &= \sum_{m=0}^{M} \Big[\int_0^R \mathrm{e}^{i\lambda t} L_m^{(n-1)}\Big(|\lambda||z|^2/2\Big)\mathrm{e}^{-|\lambda||z|^2/4}\,\mathrm{d}\lambda \\ &\quad + \int_0^R \mathrm{e}^{-i\lambda t} L_m^{(n-1)}\Big(|\lambda||z|^2/2\Big)\mathrm{e}^{-|\lambda||z|^2/4}\,\mathrm{d}\lambda \Big] \\ &= \sum_{m=0}^{M} \int_0^R 2\cos(\lambda t) L_m^{(n-1)}\Big(|\lambda||z|^2/2\Big)\mathrm{e}^{-|\lambda||z|^2/4}\,\mathrm{d}\lambda.\end{aligned}$$

Let x_iU with $x_i \in G$ be a cover of the group $G = \mathbb{H}_n \ltimes K$, then x_iKUK covers the Heisenberg group $\mathbb{H}_n$. Let ψ_i be a bounded partition of unity, which could for example be characteristic functions for disjoint sets $U_i \in x_iU$. The operator T then has the form

$$Tf = \sum_i f(x_iK)\psi_i * \phi,$$

where ϕ is given above. Choosing x_i close enough we can invert T to obtain

$$f = \sum_i f(x_i K) T^{-1}(\psi_i * \phi).$$

One typical problem with the above reconstruction is that the functions $T^{-1}(\psi_i * \phi)$ have slow decay. We therefore lose locality of the reconstruction which is undesirable in applications. This can be solved by expressing the inversion using functions g and h with faster decay but with the expense of having to oversample. Let $\widehat{g}$ be the restriction to $\mathfrak{F}_n$ of a compactly supported Schwartz function on $\mathbb{R}^2$ for which $\widehat{g}|_\Omega = 1$. Let us say the support of $\widehat{g}$ is the compact set Ω_g. Let $\widehat{h}$ be the restriction to $\mathfrak{F}_n$ of another Schwartz function which is one on Ω_g and has support in the compact set Ω_h. According to Theorem 1.1 in [2] there are K-invariant Schwartz functions g and h on $\mathbb{H}_n$ such that their Fourier transforms are equal to $\widehat{g}$ and $\widehat{h}$, respectively. The functions g and h are therefore integrable and bandlimited, and $f \in L^2_\Omega$ can be reconstructed by the following algorithm if the sampling points x_i are close enough. Denote by S the spline approximation

$$Sf = \sum_i f(x_i)\psi_i.$$

Let

$$f_0 = Sf$$

and define

$$f_{k+1} = f_k * h - S(f_k * h).$$

Then, as argued in Theorem 3.1 from [22], we can reconstruct f in L^2_Ω by

$$f = \left(\sum_{k=0}^{\infty} f_k\right) * g.$$

Remark 9. This oversampling situation is not possible for symmetric spaces of the noncompact type. The reason is that there are no integrable bandlimited functions with Fourier transform constant on a set with limit point.

Acknowledgements The research of J.G. Christensen was partially supported by NSF grant DMS-0801010, and ONR grants NAVY.N0001409103, NAVY.N000140910324. The research of G. Ólafsson was supported by NSF Grant DMS-0801010 and DMS-1101337.

References

1. Aronszajn, N.: Theory of reproducing kernels. Trans. Am. Math. Soc. **68**, 337–404 (1950)
2. Astengo, F., Di Blasio, B., Ricci, F.: Gelfand pairs on the Heisenberg group and Schwartz functions. J. Funct. Anal. **256**, 1565–1587 (2009)

3. Benedetto, J.: Irregular sampling and frames. In: Chui, C. (ed.) Wavelets - A Tutorial in Theory and Applications, pp. 1–63. Academic, New York (1991)
4. Benedetto, J., Heller, W.: Irregular sampling and the theory of frames, part I. Note Math. **X**(1), 103–125 (1990)
5. Benson, C., Jenkins, J., Ratcliff, G.: On Gelfand pairs associated with solvable Lie groups. Trans. Am. Math. Soc. **321**, 85–116 (1990)
6. Benson, C., Jenkins, J., Ratcliff, G.: Bounded K-spherical functions on Heisenberg groups. J. Funct. Anal. **105**(2), 409–443(1992)
7. Benson, C., Jenkins, J., Ratcliff, G.: Spectra for Gelfand pairs associated with the Heisenberg group. Colloq. Math. **71**, 305–328 (1996)
8. Christensen, O.: An Introduction to Frames and Riesz Bases, Applied and Numerical Harmonic Analysis. Birkhäuser, Basel (2003)
9. Christensen, J.G.: Sampling in reproducing kernel Banach spaces on Lie groups. J. Approx. Theor. **164**(1), 179–203 (2012)
10. Christensen, J.G., Ólafsson, G.: Examples of coorbit spaces for dual pairs. Acta Appl. Math. **107**, 25–48 (2009)
11. Christensen, J.G., Ólafsson, G.: Coorbit spaces for dual pairs. Appl. Comp. Harmonic Anal. **31**, 303–324 (2011)
12. Dieudonneé, J.: Grundzüge der Modernen Analysis. Band 5/6. Vieweg (1979)
13. van Dijk, G.: Introduction to Harmonic Analysis and Generalized Gelfand Pairs. Studies in Mathematics **36**, de Gruyter, 2009.
14. Duffin, R.J., Schaeffer, A.C.: A class of nonharmonic Fourier series. Trans. Am. Math. Soc. **72**(2), 341–366 (1952) (American Mathematical Society)
15. Faraut, J.: Analysis on lie groups, an introduction. In: Cambridge Studies in Advanced Mathematics, vol. 110. Cambridge University Press, Cambridge (2008)
16. Feichtinger, H.G.: Coherent frames and irregular sampling. In: Byrnes, J.S. (ed.) Recent Advances in Fourier Analysis and its Applications, pp. 427–440. Kluwer Academic Publishers, Boston (1990) (NATO Adv. Sci. Inst. Ser. C Math. Phys. Sci. **315**)
17. Feichtinger, H.G.: Discretization of convolution and reconstruction of band-limited functions from irregular sampling. In: Progress in Approximation Theory, pp. 333–345. Academic, Boston (1991)
18. Feichtinger, H.G., Grčhenig, K.H.: A unified approach to atomic decompositions via integrable group representations. In: Function Spaces and Applications, Lund, 1986. Lecture Notes in Math., vol. 1302, pp. 52–73. Springer, Berlin (1988)
19. Feichtinger, H.G., Gröchenig, K.H.: Banach spaces related to integrable group representations and their atomic decompositions, I. J. Funct. Anal. **86**, 307–340 (1989)
20. Feichtinger, H.G., Gröchenig, K.H.: Banach spaces related to integrable group representations and their atomic decompositions, II. Monatsh. Math. **108**, 129–148 (1989)
21. Feichtinger, H.G., Gröchenig, K.H.: Multidimensional irregular sampling of band-limited functions in L^p-spaces. In: Multivariate Approximation Theory, IV, Oberwolfach, 1989. Internat. Ser. Numer. Math., vol. 90, pp. 135–142. Birkhäuser, Basel (1989)
22. Feichtinger, H.G., Gröchenig, K.H.: Iterative reconstruction of multivariate band-limited functions from irregular sampling values. SIAM J. Math. Anal. **23**, 244–261 (1992)
23. Feichtinger, H.G., Gröchenig, K.H.: Irregular sampling theorems and series expansions of band-limited functions. J. Math. Anal. Appl. **167**, 530–556 (1992)
24. Feichtinger, H.G., Gröchenig, K.H.: Theory and practice of irregular sampling. In: Benedetto, J., Fraxier, M.W. (eds.) Wavelets: Mathematics and Applications, pp. 305–363, Stud. Adv. Math., CRC, Boca Raton (1994)
25. Feichtinger, H.G., Pandey, S.S.: Error estimates for irregular sampling of band-limited functions on a locally compact abelian group. J. Math. Anal. Appl. **279**, 380–397 (2003)
26. Feichtinger, H.G., Pesenson, I.A.: Recovery of band-limited functions on manifolds by an iterative algorithm. In: Heil, C., Jorgensen, P.E.T., Larson D.R. (eds.) Wavelets, Frames and Operator Theory. Contemp. Math. **345**, pp. 137–152. American Mathematical Society (2004)

27. Feichtinger, H.G., Pesenson, I.A.: A reconstruction method for band-limited signals on the hyperbolic plane. Sampl. Theor. Signal Image Process. **4**, 107–119 (2005)
28. Führ, H.: Abstract Harmonic Analysis of Continuous Wavelet Transforms. Lecture Notes in Mathematics, **1863**. Springer, Berlin (2005)
29. Führ, H., Gröchenig, K.: Sampling theorems on locally compact groups from oscillation estimates. Math. Z. **255**, 177–194 (2007)
30. Geller, D., Pesenson, I.Z.: Band-limited localized Parseval frames and Besov spaces on compact homogeneous manifolds. J. Geom. Anal. **21**, 334–371 (2011)
31. Gröchenig, K.: Acceleration of the frame algorithm. IEEE Trans. Signal Process. **41**(12), 3331–3340 (1993)
32. Heckman, G., Schlichtkrull, H.: Harmonic Analysis and Special Functions on Symmetric Spaces. Perspectives in Mathematics, vol. 16. Academic, New York (1994)
33. Heil, C.: A Basis Theory Primer: Expanded Edition, Applied and Numerical Harmonic Analysis. Birkhäuser/Springer, Boston (2011)
34. Helgason, S.: Differential operators on homogeneous spaces. Acta Math. **102**, 249–299 (1959)
35. Helgason, S.: Groups and Geometric Analysis. Academic, Orlando (1984)
36. Lebedev, N.N.: Special Functions and Their Applications. Dover, New York (1972)
37. Li, S.: The theory of multiresolution analysis frames and applications. Ph.D. dissertation, University of Maryland Baltimore County, United States (1993)
38. Li, S.: On general frame decompositions. Numer. Funct. Anal. Optim. **16**, 9–10, 1181–1191 (1995)
39. Lim, J.K.: Neumann series expansion of the inverse of a frame operator. Commun. Korean Math. Soc. **13**(4), 791–800, 1225–1763 (1998)
40. Ólafsson, G., Schlichtkrull, H.: Representation theory, radon transform and the heat equation on a Riemannian symmetric space. In: Group Representations, Ergodic Theory, and Mathematical Physics; A Tribute to George W. Mackey. Contemp. Math. **449**, 315–344 (2008)
41. Pesenson, I.: A discrete Helgason-Fourier transform for Sobolev and Besov functions on noncompact symmetric spaces. Radon transforms, geometry, and wavelets. In: Ólafsson, G., Grinberg, E.L., Larson, D., Palle, E.T., Jorgensen, P.R., Massopust, E.T., Quinto, and B. Rubin (eds.) Radon Transforms, Geometry, and Wavelets. Contemporary Mathematics **464**, 231–247. American Mathematical Society (2008)
42. Rudin, W.: Functional Analysis. In: International Series in Pure and Applied Mathematics, 2nd edn. McGraw-Hill, New York (1991)
43. Ruffino, F.F.: The Topology of the Spectrum for Gelfand Pairs on Lie Groups. Bullettino U.M.I. **10-B**(8), 569–579 (2007)
44. Shannon, C.E.: A mathematical theory of communication. Bell System Tech. J. **27**, 379–423, 623–656 (1948)
45. Strichartz, R.S.: Analysis of the Laplacian on the complete Riemannian manifold. J. Funct. Anal. **521**, 48–79 (1983)
46. Strichartz, R.S.: L^p harmonic analysis and Radon transforms on the Heisenberg group. J. Funct. Anal. **96**, 350–406 (1991)
47. Takeuchi, M.: Modern Spherical Functions. In: Transactions of Mathematical Monographs, vol. 135, AMS, Providence, RI (1994)
48. Triebel, H.: Spaces of Besov-Hardy-Sobolev type on complete Riemannian manifolds. Ark. Mat. **24**, 299–337 (1986)
49. Wolf, J.A.: Harmonic Analysis on Commutative Spaces. Mathematical Surveys and Monographs, vol. 142. American Mathematical Society, Providence, RI (2007)
50. Zayed, A.I.: Advances in Shannon's Sampling Theory. CRC, Boca Raton (1993)
51. Zayed, A.I.: A prelude to sampling, wavelets, and tomography. In: Benedetto, J.J., Zayed, A.I. (eds.) Sampling, Wavelets, and Tomography, pp. 1–32. Applied Numeric Harmonic Analysis. Birkhäuser, Basel (2004)

Smooth Interpolation of Data by Efficient Algorithms

C. Fefferman

Abstract In 1934, Whitney (Trans. Am. Math. Soc. 36:63–89, 1934; Trans. Am. Math. Soc. 36:369–389, 1934; Ann. Math. 35:482–485, 1934) posed several basic questions on smooth extension of functions. Those questions have been answered in the last few years, thanks to the work of Bierstone et al. (Inventiones Math. 151(2):329–352, 2003), Brudnyi and Shvartsman (Int. Math. Res. Notices 3:129–139, 1994; J. Geomet. Anal. 7(4):515–574, 1997), Fefferman (Ann. Math. 161:509–577, 2005; Ann. Math. 164(1):313–359, 2006; Ann. Math. 166(3):779–835, 2007) and Glaeser (J. d' Analyse Math. 6:1–124, 1958). The solution of Whitney's problems has led to a new algorithm for interpolation of data, due to Fefferman and Klartag (Ann. Math. 169:315–346, 2009; Rev. Mat. Iberoam. 25:49–273, 2009). The new algorithm is theoretically best possible, but far from practical. We hope it can be modified to apply to practical problems. In this expository chapter, we briefly review Whitney's problems, then formulate carefully the problem of interpolation of data. Next, we state the main results of Fefferman and Klartag (Ann. Math. 169:315–346, 2009; Rev. Mat. Iberoam. 25:49–273, 2009) on efficient interpolation. Finally, we present some of the ideas in the proofs.

Let us set up notation. We fix positive integers m, n throughout the chapter. We work in $C^m(\mathbb{R}^n)$, the space of m times continuously differentiable functions $F : \mathbb{R}^n \to \mathbb{R}$ for which the norm

$$\|F\| = \sup_{x \in \mathbb{R}^n} \max_{|\alpha| \leq m} |\partial^\alpha F(x)|$$

is finite. (We would like to be able to treat Sobolev spaces also. Work on Sobolev interpolation of data is just beginning; see A. Israel, G. Luli, C. Fefferman, and

C. Fefferman (✉)
Department of Mathematics, Princeton University, Princeton, NJ
e-mail: cf@math.princeton.edu

T.D. Andrews et al. (eds.), *Excursions in Harmonic Analysis, Volume 1*,
Applied and Numerical Harmonic Analysis, DOI 10.1007/978-0-8176-8376-4_4,

P. Shvartsman, to appear.)

Let $F \in C^m(\mathbb{R}^n)$ and $x \in \mathbb{R}^n$ be given. We write $J_x(F)$ (the "jet"of F at x) to denote the mth order Taylor polynomial of F at x:

$$[J_x(F)](y) = \sum_{|\alpha| \le m} \frac{1}{\alpha!}(\partial^\alpha F(x)) \cdot (y-x)^\alpha.$$

Thus, $J_x(F)$ belongs to $\mathscr{P}$, the vector space of all real-valued polynomials of degree at most m on $\mathbb{R}^n$. The jet $J_x(F)$ encodes the values at x of F and its derivatives through order m.

Whitney's classic problems are as follows. Suppose we are given a subset $E \subset \mathbb{R}^n$ and a function $f : E \to \mathbb{R}$. We make no assumptions on the set E.

Question 1: How can we tell whether f extends to a C^m function on $\mathbb{R}^n$? That is, how can we tell whether there exists $F \in C^m(\mathbb{R}^n)$ such that $F = f$ on E?

If such an F exists, then we ask:

Question 2: How small can we take $\|F\|$?

Question 3: What can we say about $J_x(F)$ for a given point x lying in or near E?

Question 4: Can we take F to depend linearly on f?

Whitney himself settled these questions in the one-dimensional case $(n = 1)$. He also proved the Whitney extension theorem, answering an easier version of Questions 1–4. Here is

Theorem 1 (Whitney's Extension Theorem [16]:). *Let $E \subset \mathbb{R}^n$ be a closed set, and let $(P^x)_{x \in E}$ be a family of polynomials $P^x \in \mathscr{P}$, indexed by the points of E.*

Then the following are equivalent:

(A) There exists $F \in C^m(\mathbb{R}^n)$ such that $J_x(F) = P^x$ for each $x \in E$.
(B) There exists a real number $M > 0$ such that:

- $|\partial^\alpha P^x(x)| \le M$ *for* $|\alpha| \le m, \ x \in E$.
- $|\partial^\alpha (P^x - P^y)(y)| \le M|x-y|^{m-|\alpha|}$ *for* $|\alpha| \le m-1, \ x, y \in E$.
- $|\partial^\alpha (P^x - P^y)(y)|/|x-y|^{m-|\alpha|}$ *tends to zero as* $|x-y| \to 0, \ |\alpha| \le m, \ x, y \in E$.

Moreover, the least possible M in (B) is comparable to the least possible (inf)$\|F\|$ *in (A). That is, $cM \le \inf\|F\| < CM$, for constants c, C depending only on m, n.*

Given a family of jets $(P^x)_{x \in E}$ satisfying *(B)*, Whitney's proof exhibits an $F \in C^m(\mathbb{R}^n)$ as in (A) such that $\|F\| \le CM$. This F is given by an explicit formula. In particular, it depends linearly on the family $(P^x)_{x \in E}$.

The proof of Whitney's extension theorem is a milestone in our understanding of analysis. Along with the work of Marcinkiewicz [14], it is an early appearance of the idea of a Calderón–Zygmund decomposition.

In Questions 1–4, we are given merely a function value $f(x)$ at each point $x \in E$, rather than a jet P^x as in the Whitney extension theorem. It is our responsibility to compute (or guess) the derivatives of F up to order m at the points of E.

Progress on Whitney's problems occurred over several decades. In 1958, Glaeser [12] answered Questions 1 and 3 in the case of $C^1(\mathbb{R}^n)$ $(m = 1)$. Glaeser's solution was based on a geometric construction, which he called the "iterated paratangent space."

In the 1970s, Brudnyi and Shvartsman [3,4] discovered the key idea of "finiteness principles." The simplest finiteness principle is as follows:

Theorem 2. *Let $E \subset \mathbb{R}^2$ be finite, and let $f : E \to \mathbb{R}$. For every subset $S \subset E$ with at most 6 points, suppose there exists $F^S \in C^2(\mathbb{R}^2)$ with norm at most 1, such that $F^S = f$ on S. Then there exists $F \in C^2(\mathbb{R}^2)$ with norm less than a universal constant, such that $F = f$ on E.*

Brudnyi and Shvartsman also proved the analogous result for $C^2(\mathbb{R}^n)$; the number of points in the subset S is at most $k^\# \equiv 3 \cdot 2^{n-1}$ (= 6 in two dimensions); Brudnyi–Shvartsman showed that this $k^\#$ is best possible. For several related results and conjectures, we refer the reader to [3,4].

In 2002, Bierstone, Milman, and Pawłuski [2] found an analogue of Glaeser's iterated paratangent space relevant to $C^m(\mathbb{R}^n)$. They answered a close relative of Questions 1 and 3 for the case in which $E \subset \mathbb{R}^n$ is a subanalytic set.

Finally, the papers [6–8] proved a finiteness principle for $C^m(\mathbb{R}^n)$, modified slightly the iterated paratangent space of Bierstone–Milman–Pawłuski, and gave complete answers to Questions 1–4. For further details, we refer the reader to the expository paper[9].

Our purpose here is to discuss the problem of interpolation of data. Again, we fix $m,n \geq 1$, and we work in $C^m(\mathbb{R}^n)$. This time, we are given $f : E \to \mathbb{R}$ with E *finite*; say, $\#(E) = N$, where $\#(E)$ denotes the number of points of E. We may not choose to fit f perfectly. Therefore, we suppose we are given a "tolerance" $\sigma : E \to [0,\infty)$.

We want to compute a function $F \in C^m(\mathbb{R}^n)$ and a real number $M \geq 0$ such that

$$\|F\| \leq M, \text{ and } |F(x) - f(x)| \leq M\sigma(x) \text{ for all } x \in E. \tag{1}$$

We would like to take M as small as possible. In the special case $\sigma \equiv 0$, we demand that $F = f$ on E.[1]

Let us define $\|f\|_{E,\sigma}$ to be the infimum of M over all pairs (F,M) that satisfy (1). Elementary examples show that this infimum need not be attained. Therefore, for a constant $C > 1$, we define a "*C-optimal interpolant*" for (E,f,σ) to be a function $F \in C^m(\mathbb{R}^n)$ such that for some $M \geq 0$, we have

$$\|F\| \leq M; \; |F(x) - f(x)| \leq M\sigma(x) \text{ for all } x \in E; \text{ and } M \leq C\|f\|_{(E,\sigma)}. \tag{2}$$

[1] It is perhaps natural to use two different positive numbers M_1, M_2 in the two inequalities in (1). However, since we are free to multiply $\sigma(x)$ by our favorite positive constant, we lose no generality in taking a single M.

We can now give a crude statement of the problem of interpolating data. Given m, n, E, f, σ, we pose:

Problem 1: Compute a C–optimal interpolant with C not too big.
Problem 2: Compute the order of magnitude of $||f||_{(E,\sigma)}$.

Our solution will consist of an algorithm, to be run on an (idealized) computer.

To make the above problems precise, we owe the reader several clarifications. We must explain:

- What it means to say that C is "not too big"
- What we mean by the "order of magnitude"
- What is an (idealized) computer
- What it means to "compute a function"
- Efficient vs. wasteful computation

The explanations are as follows:

A "not-too-big"constant is simply a constant depending only on m and n. (Recall, we are working in $C^m(\mathbb{R}^n)$.) We denote such constants by c, C, C', etc. These symbols may denote different constants in different occurrences.

To "compute the order of magnitude"of a real number $X \geq 0$ is to compute some real number Y for which we guarantee that

$$cX \leq Y \leq CX.$$

Our idealized "computer" has standard von Neumann architecture. Unlike a real computer, our idealized machine is assumed here to deal with exact real numbers, without roundoff error. Thus, we assume that an arbitrary real number may be stored at a single memory address and that the registers perform arithmetic operations (including powers and logarithms) to perfect accuracy.[2]

Our idealized computer can deal with only finitely many real numbers. What does it mean to "compute a function" $F \in C^m(\mathbb{R}^n)$?

We have in mind the following dialogue with the computer:

First, we enter the data (m, n, E, f, σ). The computer performs *one-time work*, then responds to our *queries*.

A *query* consists of a point $\underline{x} \in \mathbb{R}^n$. When we enter the coordinates of a query point $\underline{x}$, the computer responds by producing the list of numbers $(\partial^\alpha F(\underline{x}))_{|\alpha| \leq m}$.

Since $F \in C^m(\mathbb{R}^n)$, this is the most we can expect.

The above notion of "computing a function" is clearly too restrictive for many purposes. It demands in particular that $F(\underline{x})$ can be computed exactly from $\underline{x}$ by performing finitely many arithmetic operations. Thus, Bessel functions cannot be computed according to our stringent definition. Since our main theorem will

---

[2]This unrealistic model of computation is subject to serious criticism[15]. In Fefferman–Klartag[10, 11], we make a rigorous analysis of the roundoff error. That analysis is omitted in this expository chapter for the sake of simplicity.

assert that the desired F can be computed, we are undisturbed by the overly strict definition.

Finally, we owe the reader a discussion of "efficient" vs. "wasteful" computation. The resources used to compute $F \in C^m(\mathbb{R}^n)$ (as explained above) are as follows:

- The number of computer operations used to perform the one-time work.
- The number of operations used to respond to a query.
- The number of memory cells in the RAM. (Recall that each memory cell can hold a single real number.).

We refer to these, respectively, as the "one-time work," the "query work," and the "storage." For an "efficient" algorithm, all the resources used are as small as possible.

We note a few trivial lower bounds for the resources needed to compute a C-optimal interpolant for (m,n,E,f,σ). First of all, any interpolation algorithm must at least read the data and reproduce $f(\underline{x})$ perfectly for $\underline{x} \in E$ if we take $\sigma \equiv 0$.

Since E consists of N points, it follows that any interpolation algorithm entails at least N operations of one-time work, and at least N memory cells of storage.

Similarly, to respond to a query $\underline{x}$, any interpolation algorithm must at least read the query and print out a response. Thus, the query work for any interpolation algorithm is at least 1.

We point out also that any algorithm that computes the order of magnitude of $\|f\|_{(E,\sigma)}$ requires at least N operations, since at least it looks at the data.

We are now ready to state our main results. We are given positive integers m,n; a finite set $E \subset \mathbb{R}^n$ consisting of N points; and functions $f : E \to \mathbb{R}$, $\sigma : E \to [0,\infty)$.

Theorem 3. *The algorithm given in [11] computes a C_1-optimal interpolant, using one-time work at most $C_2 N \log N$, storage at most $C_3 N$, and query work at most $C_4 \log N$.*

Theorem 4. *The algorithm given in [10] computes the order of magnitude of $\|f\|_{(E,\sigma)}$ using work at most $C_5 N \log N$ and storage at most $C_6 N$.*

Recall that $C_1 \cdots C_6$ denote constants depending only on m and n.

The computer resources indicated by Theorems 3 and 4 differ only by a factor $\log N$ from the trivial lower bounds we pointed out above. Very likely, Theorems 3 and 4 are best possible. Nevertheless, the algorithms in [10, 11] are not of practical use, because they compute a C_1-optimal interpolant for a very large constant C_1. (The constants $C_2,\ldots,C_6$ are not nearly so bad.) We hope that an improved version of our algorithm may (someday) yield practical results.

Note that our algorithms apply to arbitrary interpolation problems (m,n,E,f,σ). We have made no assumptions on the geometry of the set E. Such assumptions greatly simplify the task of interpolating data.

The rest of this chapter gives some ideas from the proofs of Theorems 3 and 4. We will define certain basic convex sets $\Gamma(x,M)$, compute their approximate size and shape, and use them to compute a C-optimal F. We begin with the definition. Let m,n,E,f,σ be given. For each $x \in \mathbb{R}^n$ and $M > 0$, we define $\Gamma(x,M)$ to consist of the Taylor polynomials $J_x(F)$ of all functions $F \in C^m(\mathbb{R}^n)$ that satisfy

$$\|F\| \le M \text{ and } |F - f| \le M\sigma \text{ on } E. \tag{3}$$

Immediately from the definition, we see that $\Gamma(x,M)$ is a (possibly empty) convex subset of the vector space $\mathscr{P}$ of mth degree polynomials. (In particular, $\Gamma(x,M)$ is empty if we take M too small, since there are then no functions F satisfying (3).)

The convex sets $\Gamma(x,M)$ are a key tool in computing a C-optimal interpolant. Moreover, we hope to convince the reader that they are interesting in their own right. To see this, consider a trivial one-dimensional example: Suppose I take a car trip. My position is a function of time. Say $y = F(t)$. I don't know the function F, but I know that its second derivative satisfies $|F''(t)| \le M$ for some explicit M, because my car can't accelerate very fast. At particular times $t_1, t_2, \ldots, t_N$, I look out the window and observe my approximate position. This tells me that

$$|F(t) - f(t)| \le M\sigma(t) \quad \text{for } t \in E \equiv \{t_1,\, t_2, \ldots,\, t_N\},$$

where f and σ arise from my observations. Given the above, what can we say about my position, velocity and acceleration at some given time t_0? This question amounts to asking us to compute $\Gamma(t_0, M)$ in a one-dimensional case.

We now explain how to compute the approximate size and shape of $\Gamma(x,M)$. Our goal is to compute convex sets

$$\Gamma_*(x,M) \subset \mathscr{P} \text{ (possibly empty), such that}$$
$$\Gamma_*(x,cM) \subset \Gamma(x,M) \subset \Gamma_*(x,M).$$

We explain how to exhibit such $\Gamma_*(x,M)$. In this expository chapter, we restrict attention to $x \in E$, even though $\Gamma_*(x,M)$ can be computed for general $x \in \mathbb{R}^n$.

By induction on $\ell \ge 0$, we will define (possibly empty) convex sets $\Gamma_\ell(x,M)$ (all $x \in E$). These Γ_ℓ will satisfy

$$\Gamma_\ell(x,M) \supset \Gamma(x,M), \text{ and}$$
$$\Gamma_\ell(x,M) \supset \Gamma_{\ell+1}(x,M), \text{ for each } x \in E \text{ and } \ell \ge 0.$$

We will then set $\Gamma_*(x,M) \equiv \Gamma_{\ell_*}(x,M)$ for a large enough integer constant ℓ_* depending only on m and n.

The inductive definition of the $\Gamma_\ell(x,M)$ proceeds as follows:

In the base case $\ell = 0$, we simply define

$$\Gamma_0(x,M) = \{P \in \mathscr{P} : |\partial^\alpha P(x)| \le M \quad \text{for } |\alpha| \le m, \text{ and } |P(x) - f(x)| \le M\sigma(x)\}$$

for each $x \in E$. Note that $\Gamma_0(x,M)$ is a (possibly empty) convex subset of $\mathscr{P}$ and that $\Gamma_0(x,M) \supset \Gamma(x,M)$.

The set $\Gamma_0(x,M)$ is defined trivially, ignoring the information available at points of E other than x.

For the induction step, we fix $\ell \geq 0$, and we suppose that $\Gamma_\ell(x,M)$ has already been defined for all $x \in E$. We suppose that each $\Gamma_\ell(x,M)$ is a (possibly empty) convex subset of $\mathscr{P}$ and that $\Gamma_\ell(x,M) \supset \Gamma(x,M)$ for each $x \in E$.

Our task is to define a (possibly empty) convex set $\Gamma_{\ell+1}(x,M)$ for each $x \in E$ and check that

$$\Gamma_\ell(x,M) \supset \Gamma_{\ell+1}(x,M) \supset \Gamma(x,M).$$

This will complete our induction on ℓ.

To define $\Gamma_{\ell+1}(x,M)$, we just use Taylor's theorem, which can be stated in the following form:

$$\text{Let } F \in C^m(\mathbb{R}^n) \quad \text{with } \|F\| \leq M, \quad \text{and suppose } x,y \in \mathbb{R}^n.$$
$$\text{Set } P = J_x(F) \quad \text{and} \quad P' = J_y(F). \text{ Then}$$
$$|\partial^\alpha(P-P')(x)| \leq M|x-y|^{m-|\alpha|} \quad \text{for all } |\alpha| \leq m.$$

As a corollary, we obtain a basic property of the sets $\Gamma(x,M)$.

Proposition 1. *Let $x,y \in E$. Given $P \in \Gamma(x,M)$, there exists $P' \in \Gamma(y,M)$ such that*

$$|\partial^\alpha(P-P')(x)| \leq M|x-y|^{m-|\alpha|} \quad \textit{for } |\alpha| \leq m.$$

The above proposition motivates our definition of $\Gamma_{\ell+1}(x,M)$. For each $x \in E$, we take $\Gamma_{\ell+1}(x,M)$ to consist of all $P \in \Gamma_\ell(x,M)$ such that for each $y \in E$, there exists $P' \in \Gamma_\ell(y,M)$ for which we have

$$|\partial^\alpha(P-P')(x)| \leq M|x-y|^{m-|\alpha|} \quad (\text{all } |\alpha| \leq m).$$

Note that $\Gamma_{\ell+1}(x,M)$ is a (possibly empty) convex subset of $\mathscr{P}$ and that $\Gamma_\ell(x,M) \supset \Gamma_{\ell+1}(x,M)$.

Moreover, $\Gamma_{\ell+1}(x,M) \supset \Gamma(x,M)$, thanks to the above proposition.

This completes our induction on ℓ. We have succeeded in defining the $\Gamma_\ell(x,M)$ for all $x \in E$, $\ell \geq 0$.

Note that our definition of $\Gamma_{\ell+1}(x,M)$ for a given $x \in E$ involves $\Gamma_\ell(y,M)$ for all $y \in E$. We will return to this point soon.

We know that each $\Gamma_\ell(x,M)$ is a (possibly empty) convex subset of P, and that

$$\Gamma_\ell(x,M) \supset \Gamma(x,M) \quad \text{for each } \ell, \text{ and } \Gamma_\ell(x,M) \supset \Gamma_{\ell+1}(x,M).$$

The basic mathematical result on the above Γ_ℓ is as follows:

Theorem 5. *For a large enough integer constant ℓ_* (depending only on m and n), we have*

$$\Gamma_{\ell_*}(x,cM) \subset \Gamma(x,M) \subset \Gamma_{\ell_*}(x,M) \text{ for all } x \in E \text{ and } M > 0.$$

Thus, we have succeeded in computing the approximate size and shape of the $\Gamma(x,M)$. Unfortunately, the above computation is too expensive. Recall that each $\Gamma_{\ell+1}(x,M)$ is defined using all the $\Gamma_\ell(y,M)(y \in E)$. Since each y talks to each x, the work required to compute all the $\Gamma_{\ell+1}(x,M)(x \in E)$ from all the $\Gamma_\ell(y,M)(y \in E)$ contains a factor N^2. In Theorems 3 and 4, we promised algorithms that do only $N \log N$ work.

Thus, we cannot use the above Γ_ℓ's to compute the approximate size and shape of the $\Gamma(x,M)$.

The remedy is to change the definition of the Γ_ℓ without losing their usefulness. To do so, we bring in an idea from computer science, namely, the *well-separated pairs decomposition* (WSPD), due to Callahan–Kosaraju [5]; see also Har-Peled and Mendel [13]. (We also use the closely related *balanced box decomposition tree* of Arya–Mount–Netanyahu–Silverman–Wu [1], but we suppress that discussion here.)

To understand how the WSPD can overcome the need to do N^2 work, we next discuss a problem much easier than the interpolation problems considered above.

Let $E \subset \mathbb{R}^n$ be a finite set, consisting of N points. Let $f : E \to \mathbb{R}$. We want to compute the Lipschitz constant of f, given by

$$\|f\|_{\mathrm{Lip}} = \max_{x',x'' \in E \text{ distinct}} \frac{|f(x') - f(x'')|}{|x' - x''|}.$$

The obvious computation of $\|f\|_{\mathrm{Lip}}$ requires work $\sim N^2$. However, a clever method computes $\|f\|_{\mathrm{Lip}}$ to within (say) a 1 % error by using $O(N \log N)$ operations. The idea is that for certain E', $E'' \subset E$, we can compute the restricted maximum

$$\Lambda \equiv \left[\max_{(x',x'') \in E' \times E''} \frac{|f(x') - f(x'')|}{|x' - x''|} \right]$$

much faster than the obvious way.

We will take E' and E'' *well separated*, i.e., we suppose that

$$\text{distance}(E',E'') > 10^3 \cdot [\text{diameter } (E') + \text{ diameter } (E'')].$$

Here, of course,

$$\text{distance } (E',E'') = \min\{|x' - x''| : x' \in E',\ x'' \in E''\},$$

$$\text{diameter } (E') = \max\{|x' - y'| : x',\ y' \in E'\}, \text{ and similarly for diameter } (E'').$$

Let us see how to compute $\wedge$ to within a 1 % error for such E', E'', faster than the obvious method that takes work $\sim \#(E') \cdot \#(E'')$.

First of all, as (x', x'') varies over $E' \times E''$, the distance $|x' - x''|$ is essentially constant. Therefore, to compute $\wedge$, it is enough to compute

$$\max\{|f(x') - f(x'')| : x' \in E',\ x'' \in E''\}.$$

To achieve the above maximum, either

Case 1: We maximize $f(x')$ over all $x' \in E'$, and minimize $f(x'')$ over all $x'' \in E''$;

or else

Case 2: We maximize $f(x'')$ over all $x'' \in E''$, and minimize $f(x')$ over all $x' \in E'$.

It follows easily that $\wedge$ can be computed (up to a 1 % error) with work $\sim \#(E') + \#(E'')$.

We have succeeded in beating the trivial algorithm to compute $\wedge$.

The above discussion motivates the following:

Theorem 6. *Let $E \subset \mathbb{R}^n$ be a finite set, consisting of N elements. Then $\{(x,y) \in E \times E : x \neq y\}$ can be partitioned into Cartesian products $E_1' \times E_1''$, $E_2' \times E_2''$, $\ldots$, $E_L' \times E_L''$, with $L \leq CN$, such that each pair (E_ℓ', E_ℓ'') is well separated. Moreover, an efficient algorithm computes the above decomposition [5].*

We omit a careful discussion of the meaning of the preceding sentence. Returning to the problem of computing the Lipschitz constant, we deliver the coup de grâce.

Let $E_1' \times E_1''$, $E_2' \times E_2''$, $\ldots$, $E_L' \times E_L''$ be as in Theorem 6. For each $\ell = 1, \ldots, L$, let $(x_\ell', x_\ell'') \in E_\ell' \times E_\ell''$ be given. Then the Lipschitz constant of f differs by at most 1 % from the quantity

$$\max\left\{\left|f(x_\ell') - f(x_\ell'')\right| \ / \ \left|x_\ell' - x_\ell''\right| : \ \ell = 1, \ldots, L\right\}. \tag{4}$$

(We will prove this in a moment).

Consequently, once we find the "representatives" (x_ℓ', x_ℓ''), $\ell = 1, \ldots, L$, we can then compute the Lipschitz constant of f with work at most $C \cdot L \leq C'N$. To compute the representatives by the algorithm of Callahan–Kosaraju requires work $O(N \log N)$. Thus, computing a Lipschitz constant, a task that seems to require N^2 operations, can actually be done in $O(N \log N)$ operations.

We pause to explain in detail why the Lipschitz constant of f is given by the restricted maximum (4) up to a 1 % error. The point is as follows.

Proposition 2. *Suppose $\left|f(x_\ell') - f(x_\ell'')\right| \leq \left|x_\ell' - x_\ell''\right|$ for $\ell = 1, \ldots, L$. Then $|f(x') - f(x'')| \leq (1.01)|x' - x''|$ for all $x', x'' \in E$.*

Proof. Suppose not. Let $(x',x'') \in E \times E$ be a counterexample with $|x'-x''|$ as small as possible. Thus,

$$\left|f(x') - f(x'')\right| > (1.01)|x'-x''| \quad \text{(strict inequality).} \tag{5}$$

In view of the strict inequality, x' and x'' are distinct. Consequently, we have $(x',x'') \in E'_\ell \times E''_\ell$ for some ℓ. Fix that ℓ. Since E'_ℓ and E''_ℓ are well separated and since also $(x'_\ell, x''_\ell) \in E'_\ell \times E''_\ell$, it follows that

$$|x'-x'_\ell| + |x''-x''_\ell| \leq \text{diam}\,(E'_\ell) + \text{diam}\,(E''_\ell) \leq 10^{-3}\,\text{dist}\,(E',E''),$$

and therefore

$$|x'-x'_\ell| + |x''-x''_\ell| \leq 10^{-3}|x'_\ell - x''_\ell|. \tag{6}$$

Thanks to our choice of x' and x'' to minimize $|x'-x''|$, it follows from (6) that

$$|f(x')-f(x'_\ell)| \leq (1.01)|x'-x'_\ell| \quad \text{and} \quad |f(x'')-f(x''_\ell)| \leq (1.01)|x''-x''_\ell|.$$

Consequently,

$$\begin{aligned}|f(x')-f(x'')| &\leq |f(x')-f(x'_\ell)| + |f(x'_\ell)-f(x''_\ell)| + |f(x''_\ell)-f(x'')| \\ &\leq (1.01)|x'-x'_\ell| + |x'_\ell - x''_\ell| + (1.01)|x''-x''_\ell| \\ &\leq (1.01)\cdot 10^{-3}|x'_\ell - x''_\ell| + |x'_\ell - x''_\ell| + (1.01)\cdot 10^{-3}|x'_\ell - x''_\ell| \\ &\leq (1.003)\cdot|x'_\ell - x''_\ell| \leq (1.01)|x'-x''|,\end{aligned}$$

where the last inequality follows from (6).
Thus, $|f(x')-f(x'')| \leq (1.01)|x'-x''|$, contradicting (5). □

This concludes our discussion of the Lipschitz constant. Returning to the interpolation problem, we now change the definition of the convex sets $\Gamma_\ell(x,M)$ by using the WSPD. Our new Γ_ℓ's can be computed with $N\log N$ work, and they have the following key properties in common with the old, expensive Γ_ℓ's:

Property 0: $\Gamma_0(x,M) = \{P \in \mathscr{P} : |\partial^\alpha P(x)| \leq M \quad (\text{all } \alpha)$ and $|P(x) - f(x)| \leq M\}$.

Property 1: $\Gamma_\ell(x,M)$ is a (possibly empty) convex subset of $\mathscr{P}$.

Property 2: $\Gamma_\ell(x,M) \supset \Gamma(x,M)$ for each $x \in E$, $\ell \geq 0$.

Property 3: $\Gamma_\ell(x,M) \supset \Gamma_{\ell+1}(x,M)$ for each $x \in E$, $\ell \geq 0$.

Property 4: Let $x,y \in E$, and let $P \in \Gamma_{\ell+1}(x,M)$.

Then there exists $P' \in \Gamma_\ell(y,M)$ such that

$$|\partial^\alpha(P-P')(x)| \leq M|x-y|^{m-|\alpha|} \quad (\text{all } |\alpha| \leq m).$$

In our earlier discussion, we essentially took Property 4 to be the definition of $\Gamma_{\ell+1}(x,M)$.

There are additional key properties enjoyed by the new $\Gamma_\ell(x,M)$, but we omit them here. Indeed, we omit the definition of our new Γ_ℓ, which requires additional structure of the WSPD not discussed in this article.

The proof of Theorem 5 uses only the key properties on the above list, not the precise definition of the Γ_ℓ. Therefore, thanks to Theorem 5 (in its generalized form), our new, cheaper Γ_ℓ allow us to compute the approximate size and shape of the convex sets $\Gamma(x,M)$.[3] The computation, running over all $x \in E$, takes at most $O(N \log N)$ operations, as promised in Theorems 3 and 4.

This concludes our discussion of the computation of the approximate size and shape of the $\Gamma(x,M)$.

At last it is time to explain how to prove Theorems 3 and 4, once we have computed the approximate size and shape of the $\Gamma(x,M)$ (by computing our cheap $\Gamma_\ell(x,M)$). These results reduce quickly to Theorem 5. In fact, by definition, the set $\Gamma(x,M)$ is nonempty if and only if $M \geq \|f\|_{(E,\sigma)}$.

By computing the approximate size and shape of the $\Gamma(x,M)$, we have in particular computed the order of magnitude of $\|f\|_{(E,\sigma)}$. Thus, Theorem 4 reduces easily to Theorem 5. Moreover, the proof of Theorem 5 is constructive. Given $P \in \Gamma_{\ell_*}(x,cM)$, we prove that $P \in \Gamma(x,M)$ by constructing an explicit interpolant F, satisfying $\|F\| \leq M$, $|F-f| \leq M\sigma$ on E, and $J_x(F)=P$. If we take M as small as possible with $\Gamma_{\ell_*}(x,cM)$ nonempty, then the interpolant F is C-optimal. The proof of Theorem 3 amounts to an efficient implementation of the proof of Theorem 5.

Thus, everything comes down to Theorem 5, which (after a trivial localization using a partition of unity) in turn amounts to solving the following local interpolation problem:

LIP (Q,x_0,P_0): Suppose we are given a cube Q, a point $x_0 \in E \cap Q$, and a polynomial $P_0 \in \Gamma_\ell(x_0,M)$ for a suitable ℓ. Produce a function $F_Q \in C^m(Q)$ such that:

(*1) The mth derivatives of F are bounded by CM on Q.
(*2) $|F_Q(x) - f(x)| \leq CM\sigma(x)$ for all $x \in E \cap Q$.
(*3) $J_{x_0}(F_Q) = P_0$.

Our task is to solve LIP (Q,x_0,P_0) for a cube Q of sidelength 1. To carry out this task, we will also consider LIP (Q,x_0,P_0) for smaller cubes Q.

In the explanation below, we sacrifice accuracy for ease of understanding.

A local interpolation problem carries a "label"$\mathscr{A}$, to be explained later. For the moment, we just remark that the label $\mathscr{A}$ tells us certain information on the geometry of the convex set $\Gamma_\ell(x_0,M)$ in LIP (Q,x_0,P_0). When we "attach" the label $\mathscr{A}$ to the problem LIP (Q,x_0,P_0), we guarantee in advance that the geometric

[3]For the rest of this chapter, whenever we refer to Theorem 5, we mean the generalized version in which the Γ_ℓ are not specified, but merely assumed to satisfy a list of key properties.

conditions indicated by $\mathscr{A}$ hold for the convex set $\Gamma_\ell(x_0,M)$. This information may help us in constructing functions F_Q that satisfy (*1)–(*3) above.

Thus LIP (Q,x_0,P_0) may or may not carry a given label $\mathscr{A}$, but if it does, then we have extra information that may help us solve the problem LIP (Q,x_0,P_0).

One particular label plays a special role; it is the empty set $\varnothing$, which provides no information whatever on the geometry of $\Gamma_\ell(x_0,M)$. Every LIP (Q,x_0,P_0) carries the label $\varnothing$.

There is a natural order relation $<$ on the set of all labels. If $\mathscr{A}' < \mathscr{A}$, then an interpolation problem that carries the label $\mathscr{A}'$ is easier than a problem that carries the label $\mathscr{A}$. In particular, the label $\varnothing$ corresponds to the hardest version of LIP (Q,x_0,P_0), in which we are given no additional information to work with.

There are only finitely many possible labels; indeed, the number of labels is a constant depending only on m and n.

We will solve a local interpolation problem with a given label $\mathscr{A}$ by reducing it to local interpolation problems with labels $\mathscr{A}' < \mathscr{A}$. Thus, we proceed by induction on the label. Let us now describe that induction.

In the base case, our local interpolation problem LIP (Q,x_0,P_0) carries the easiest possible label (the label called $\mathscr{M}$ later on). We may then simply take $F_Q = P_0$, and one checks without difficulty that F_Q satisfies (*1)–(*3). Thus, we can easily solve LIP (Q,x_0,P_0) in the base case.

For the induction step, we fix a label $\mathscr{A}$, and suppose that we can solve any local interpolation problem that carries a label $\mathscr{A}' < \mathscr{A}$. We must solve a LIP (Q,x_0,P_0) that carries the label $\mathscr{A}$. To do so, we partition Q into finitely many subcubes Q_ν and introduce a "representative point" $x_\nu \in E\cap Q_\nu$ for each ν. The construction of the partition involves the label $\mathscr{A}$ and the geometry of the $\Gamma_{\ell-1}(x,M)$ for all the points $x\in E\cap Q$. For each x_ν, we invoke the key property called Property 4, with ℓ in Property 4 replaced by our present $\ell-1$. Thus, for each ν, we obtain a polynomial

$$P_\nu \in \Gamma_{\ell-1}(x_\nu,M), \text{ such that } |\partial^\alpha(P_\nu - P_0)(x_0)| \le M|x_\nu - x_0|^{m-|\alpha|} \quad \text{for } |\alpha|\le m. \tag{7}$$

We now have a cube Q_ν, a point $x_\nu \in E\cap Q_\nu$, and a polynomial $P_\nu \in \Gamma_{\ell-1}(x_\nu,M)$. Thus, we can pose the local interpolation problem

$$\text{LIP }(\nu) \equiv \text{LIP}(Q_\nu,x_\nu,P_\nu) \quad \text{for each } \nu.$$

Our partition $\{Q_\nu\}$ was constructed to guarantee that each of the above problems LIP(ν) carries a label $\mathscr{A}'_\nu < \mathscr{A}$. Therefore, by our induction hypothesis, we can solve each LIP(ν), to produce a "local interpolant" $F_\nu \in C^m(Q_\nu)$ that satisfies:

(*1)$_\nu$ The mth derivatives of F_ν are bounded by CM on Q_ν.
(*2)$_\nu$ $|F_\nu(x) - f(x)| \le CM\sigma(x)$ for all $x\in E\cap Q_\nu$.
(*3)$_\nu$ $J_{x_\nu}(F_\nu) = P_\nu$.

By using a partition of unity adapted to the partition $\{Q_\nu\}$, we can then patch together the F_ν into a function $F_Q \in C^m(Q)$.

If we are careful, then our F_Q will satisfy (*1)–(*3) and thus solve LIP (Q, x_0, P_0). This will complete our induction on the label $\mathscr{A}$, solve all local interpolation problems, and complete the proof of Theorem 5.

To make the above construction work, we have to take one main precaution. We must pick the polynomials P_ν in (7) to satisfy the consistency condition

$$|\partial^\alpha (P_\nu - P_{\nu'})(x_\nu)| \le CM\,|x_\nu - x_{\nu'}|^{m-|\alpha|} \qquad (\text{all } |\alpha| \le m) \tag{8}$$

whenever the cubes Q_ν and $Q_{\nu'}$ touch. This is much stronger than the defining condition (7), since Q_ν and $Q_{\nu'}$ may be much smaller than Q, and therefore $|x_\nu - x_{\nu'}|$ may be much smaller than $|x_0 - x_\nu|$. If (8) fails, then we have no chance to control the mth derivatives of F as in (*1).

Thus, it is essential to pick the P_ν to satisfy (8) in addition to (7). Achieving this extra consistency is the most delicate point in our proof of Theorem 5. We postpone for a few paragraphs our brief remarks on how to achieve (8).

We owe the reader an explanation of a local interpolation problem LIP (Q, x_0, P_0) with a "label" $\mathscr{A}$. In fact, a label $\mathscr{A}$ is simply a subset of the set $\mathscr{M}$ of all multi-indices $\alpha = (\alpha_1, \alpha_2, \cdots, \alpha_n) \in \{0,1,2,\cdots\}^n$ of order $|\alpha| = \alpha_1 + \cdots + \alpha_n \le m-1$. The problem LIP (Q, x_0, P_0) carries the label $\mathscr{A}$ if there exist polynomials $P_\alpha \in P$ indexed by $\alpha \in \mathscr{A}$, such that the following conditions are satisfied, where δ_Q denotes the sidelength of Q and $\delta_{\beta\alpha}$ denotes the Kronecker delta:

- $P_0 + M\delta_Q^{m-|\alpha|} P_\alpha \in \Gamma_\ell(x_0, CM)$ for each $\alpha \in \mathscr{A}$.
- $\partial^\beta P_\alpha(x_0) = \delta_{\beta\alpha}$ for $\beta, \alpha \in \mathscr{A}$.
- $|\partial^\beta P_\alpha(x_0)| \le C\delta_Q^{|\alpha|-|\beta|}$ for $\beta \in \mathscr{M}$, $\alpha \in \mathscr{A}$.

These conditions assert that the convex set $\Gamma_\ell(x_0, CM) \subset \mathscr{P}$ is "fat enough "in certain directions corresponding to the $P_\alpha \in P$.

Note that the above conditions hold vacuously when $\mathscr{A}$ is the empty set $\varnothing$. Thus, as promised, the label $\varnothing$ provides no extra information.

However, when $\mathscr{A}$ is nonempty, the above conditions provide us with some room to maneuver—we can change P_0 and stay inside $\Gamma_\ell(x_0, CM)$. In the proof of Theorem 5 sketched above, we exploit this freedom of maneuver for each of the local problems LIP(Q_ν, x_ν, P_ν). Initially, the P_ν satisfy the consistency condition in (7), but not the strong consistency condition (8). However, by exploiting our freedom of maneuver, we can modify slightly each P_ν, to achieve (8) without sacrificing (7).

This completes our sketch of the proof of Theorem 5. We again warn the reader that it is not completely accurate. See [10, 11] for the correct version.

Acknowledgements C. Fefferman was Supported by NSF Grant No. DMS-09-01-040 and ONR Grant No. N00014-08-1-0678. The author is grateful to Frances Wroblewski for TeXing this chapter.

References

1. Arya, S., Mount, D., Netanyahu, N., Silverman, R., Wu, A.: An optimal algorithm for approximate nearest neighbor searching in fixed dimensions. J. Assoc. Comput. Mach. **45**(6), 891–923 (1998)
2. Bierstone, E., Milman, P., Pawłucki, W.: Differentiable functions defined on closed sets, A problem of Whitney. Inventiones Math. **151** (2), 329–352 (2003)
3. Brudnyi, Y., Shvartsman, P.: Generalizations of Whitney's extension theorem. Int. Math. Res. Notices **3**, 129–139 (1994)
4. Brudnyi, Y., Shvartsman, P.: The Whitney problem of existence of a linear extension operator. J. Geomet. Anal. **7**(4), 515–574 (1997)
5. Callahan, P.B., Kosaraju, S.R.: A decomposition of multi-dimensional point sets with applications to k-nearest neighbors and n-body potential fields. J. Assoc. Comput. Mach. **42**(1), 67–90 (1995)
6. Fefferman, C.: A sharp form of Whitney's extension theorem. Ann. Math. **161**, 509–577 (2005)
7. Fefferman, C.: Whitney's extension problem for C^m. Ann. Math. **164**(1), 313–359 (2006)
8. Fefferman, C.: C^m extension by linear operators. Ann. Math. **166**(3), 779–835 (2007)
9. Fefferman, C.: Whitney's extension problems and interpolation of data. Bull. Am. Math. Soc. **46**(2), 207–220 (2009)
10. Fefferman, C., Klartag, B.: Fitting a C^m-smooth function to data I. Ann. Math. **169**(1), 315–346 (2009)
11. Fefferman, C., Klartag, B.: Fitting a C^m-smooth function to data II. Rev. Mat. Iberoam. **25**(1), 49–273 (2009)
12. Glaeser, G.: Etudes de quelques algebres tayloriennes. J. d'Analyse Math. **6**, 1–124 (1958)
13. Har-Peled, S., Mendel, M.: Fast construction of nets in low-dimensional metrics and their applications. SIAM J. Comput. **35**(5), 1148–1184 (2006)
14. Marcinkiewicz, J.: Sur les series de Fourier. Fund. Math. **27**, 38–69 (1936)
15. Schönhage, A.: On the power of random access machines. In: Proceedings 6th International Colloquium, on Automata, Languages and Programming. Lecture Notes in Computer Science, vol. 71, pp. 520–529. Springer, London (1979)
16. Whitney, H.: Analytic extensions of differentiable functions defined in closed sets. Trans. Am. Math. Soc. **36**, 63–89 (1934)

An Overview of Time and Multiband Limiting

Joseph D. Lakey

Abstract The purpose of this chapter is to provide an up-to-date overview of time and multiband limiting somewhat parallel to Landau's (Fourier Techniques and Applications (Kensington, 1983), pp. 201–220. Plenum, New York, 1985) overview. Particular focus is given to the theory of time and frequency limiting of multiband signals and to time-localized sampling approximations of Shannon type for band-limited signals.

Keywords Time-limiting • Band-limiting • Shannon sampling • Multiband • Prolate spheroidal wave function • Prolate discrete prolate spheroidal sequence • Uncertainty principle • Legendre polynomial

Dedicated to the memory of Dennis Healy

1 Introduction

This exposition provides an overview of certain developments in the theory of time and multiband limiting made since Landau's [21] overview. Two principal developments are outlined here. The first addresses behavior of eigenvalues. For the case of time limiting of signals band limited to an interval, the results reviewed are by now classical. In the multiband case, there are two essential regimes, one involving mildly disconnected time and frequency supports—this will be called the *Landau–Widom* regime—which allows for some number of independent signals that are well localized in time and frequency and the other involving

J.D. Lakey (✉)
Department of Mathematical Sciences, New Mexico State University,
Las Cruces, NM 88003–8001, USA
e-mail: jlakey@nmsu.edu

T.D. Andrews et al. (eds.), *Excursions in Harmonic Analysis, Volume 1*,
Applied and Numerical Harmonic Analysis, DOI 10.1007/978-0-8176-8376-4_5,

highly disconnected time and frequency supports—this will be called the *Candès–Romberg–Tao* regime—which rarely allows for any signals that are jointly well localized. The second development involves signals band limited to an interval and addresses the numerical approximation of a time- and band-limited signal on the one hand and the ability to interpolate such a signal locally from samples on the other.

Given a compact set $S \subset \mathbb{R}$ with indicator function $\mathbb{1}_S(t) = 1$ if $t \in S$ and $\mathbb{1}_S(t) = 0$, otherwise, one defines the *time-limiting operator* $Q_S(f)(t) = f(t)\,\mathbb{1}_S(t)$. Similarly, given a compact subset $\Sigma \subset \mathbb{R}$, one defines the *band-limiting operator* $P_\Sigma = \mathscr{F}^{-1} Q_\Sigma \mathscr{F}$, where $\mathscr{F}(f)(\omega) = \int_{-\infty}^{\infty} f(t)\,\mathrm{e}^{-2\pi i t\omega}\,\mathrm{d}t$ is the Fourier transform of f. We will abuse notation freely in abbreviating $Q_T = Q_{[-T,T]}$ with $Q = Q_1$ and $P_\Omega = P_{[-\Omega/2,\Omega/2]}$ with $P = P_1$. The area of the product $S \times \Sigma$ will be denoted by $a = a(S,\Sigma) = |S||\Sigma|$, where $|S|$ is the Lebesgue measure of S. When $a(S,\Sigma) < \infty$, the kernel of the operator $P_\Sigma Q_S$ is $K = K_{S,\Sigma}$ with $K(s,t) = (\mathbb{1}_\Sigma)^\vee(s-t)\,\mathbb{1}_S(t)$. Its trace and Hilbert–Schmidt norm are, respectively,

$$\mathrm{tr}\,(P_\Sigma Q_S) = \int K(s,s)\,\mathrm{d}s = (\mathbb{1}_\Sigma)^\vee(0)\int \mathbb{1}_S = |S||\Sigma| = a(S,\Sigma) \quad \text{and}$$

$$\|P_\Sigma Q_S\|_{\mathrm{HS}}^2 = \|K_{S,\Sigma}\|_{L^2(S\times S)}^2 = \int_S\int_S |(\mathbb{1}_\Sigma)^\vee(t-s)|^2\,\mathrm{d}s\,\mathrm{d}t. \tag{1}$$

In the following section we will review properties of the (discrete) spectrum of $P_\Sigma Q_S$, emphasizing the special case in which Σ and S are intervals. We also discuss those properties of the eigenfunctions that depend only on the operator $P_\Sigma Q_S$. In the subsequent section we will discuss further properties of the eigenfunctions of PQ_T that depend on Sturm–Liouville properties of these functions.

2 Eigenfunctions of Time and Band Limiting to an Interval

The unitary dilation $(D_\alpha f)(t) = \sqrt{\alpha}\, f(\alpha t)$ satisfies $P = D_{1/\Omega} P_\Omega D_\Omega$ and $Q_T = D_{1/T} Q D_T$ (see Table 1). Thus $P_\Omega Q_T$ is unitarily equivalent to $PQ_{\Omega T}$, since

$$PQ_{\Omega T} = D_{1/\Omega} P_\Omega D_\Omega Q_{\Omega T} = D_{1/\Omega} P_\Omega D_\Omega D_{1/\Omega} Q_T D_\Omega = D_{1/\Omega}(P_\Omega Q_T) D_\Omega.$$

2.1 Eigenfunction Consequences of the Prolate Operator

2.1.1 Commutation of PQ_T with the Prolate Differential Operator

One of the fundamental observations made by Slepian and Pollak in [27] is that the differential operator $\mathscr{P}$ in (3) commutes with the time-localization operator Q and the frequency-localization operator P_a, for a an appropriate fixed multiple

Table 1 Scaling relations for time and band limiting

Operation	Formula	Relations
D_α	$(D_\alpha f)(t) = \sqrt{\alpha}\, f(\alpha t)$	$D_{\alpha_1} D_{\alpha_1} = D_{\alpha_1 \alpha_2}$
$\mathscr{F}$	$(\mathscr{F} f)(s) = \int_{-\infty}^{\infty} f(t)\,\mathrm{e}^{-2\pi i s t}\,\mathrm{d}t$	$(\mathscr{F}^2 f)(t) = f(-t);\;\; D_\alpha \mathscr{F} = \mathscr{F} D_{1/\alpha}$
Q_β	$(Q_\beta f)(t) = f(t)\,\mathbb{1}_{[-\beta,\beta]}(t)$	$Q_\beta D_\alpha = D_\alpha Q_{\alpha\beta}$
P_γ	$(P_\gamma f)(t) = (\mathscr{F}^{-1} Q_{\gamma/2} \mathscr{F} f)(t)$	$P_{\alpha\gamma} D_\alpha = D_\alpha P_\gamma$
		$\mathscr{F} P_\gamma Q_\beta = Q_{\gamma/2} P_{2\beta} \mathscr{F}^{-1}$

of c. Slepian and Pollak simply observed that this "lucky accident," as Slepian [26] called it, followed from general properties relating differential operators to corresponding integral operators. Considerably later, Walter [28] viewed this accident as a characterization of certain differential operators that commute with multiplication by the indicator functions of intervals. Walter's result can be phrased as follows.

Theorem 1. *Suppose that $P(t,\mathrm{d}/\mathrm{d}t)$ is a differential operator of the form $\rho_0(t)\frac{\mathrm{d}^2}{\mathrm{d}t^2} + \rho_1 \frac{\mathrm{d}}{\mathrm{d}t} + \rho_2$ with quadratic coefficients ρ_i such that $P(t,\mathrm{d}/\mathrm{d}t)$ commutes with multiplication by the characteristic function of $[-T,T]$ and such that $\mathscr{F}(P(t,\mathrm{d}/\mathrm{d}t))$ commutes with multiplication by the characteristic function of $[-\Omega/2,\Omega/2]$. Then there exist constants a and b such that*

$$\rho_0(t) = a(t^2 - T^2) \quad \text{and} \quad \rho_1(t) = 2at \quad \text{while} \quad \rho_2(t) = \pi^2 a \Omega^2 t^2 + b. \tag{2}$$

The *prolate differential operator*, defined by

$$\mathscr{P} = \mathscr{P}_c = \frac{\mathrm{d}}{\mathrm{d}t}(t^2-1)\frac{\mathrm{d}}{\mathrm{d}t} + c^2 t^2 = (t^2-1)\frac{\mathrm{d}^2}{\mathrm{d}t^2} + 2t\frac{\mathrm{d}}{\mathrm{d}t} + c^2 t^2, \tag{3}$$

arises in solving the wave equation in *prolate spheroidal coordinates* by means of separation of variables. It corresponds to the particular case of (2) in which $a = 1$, $T = 1$, $c = \pi\Omega$, and $b = 0$. When the time–frequency area a equals 2Ω, one has $a = 2c/\pi$. Theorem 1 states that any second-order differential operator with quadratic coefficients that the operator commutes, not the coefficients with the time- and band-limiting operators is a multiple of a rescaling of the differential operator $\mathscr{P}$, plus a multiple of the identity. Because of this lucky accident, the eigenfunctions of $P_\Omega Q_T$ are, up to a dilation factor, *prolate spheroidal wave functions* (PSWFs).

2.1.2 Completeness of the Eigenfunctions of PQ_T in $L^2[-T,T]$

Completeness follows from Sturm–Liouville theory. It also follows from the spectral theorem for compact self-adjoint operators: the eigenfunctions of PQ_T are complete in PW and from Parseval's theorem and Proposition 1, completeness in $L^2[-T,T]$ follows.

2.1.3 PQ_T Has a Simple Spectrum

That $P_\Omega Q_T$ has a simple spectrum follows from the fact that the operator $\mathscr{P}_c$ has a simple spectrum; see [27]. To establish the latter requires somewhat advanced Sturm–Liouville theory. We denote by $\lambda_0 > \lambda_1 > \dots$ the eigenvalues of PQ_T and by φ_n the $L^2(\mathbb{R})$-normalized prolate spheroidal eigenfunction corresponding to λ_n.

2.2 *Eigenfunction Consequences of Time and Band Limiting*

2.2.1 Fourier Covariance of Eigenfunctions of PQ_T

A PSWF eigenfunction ψ of PQ_T is real analytic and has a Fourier transform supported in $[-1/2, 1/2]$.

Proposition 1. *If $\psi = \varphi_n$ is a λ-eigenfunction of PQ_T then*

$$\widehat{\psi}\left(\frac{\xi}{2T}\right) = (-\mathrm{i}^n)/\sqrt{\lambda}\mathbb{1}_{[-T,T]}\,\psi(\xi).$$

2.2.2 The PSWF Parameter c and the Time–Bandwidth Product a

As mentioned above, the parameter c in the operator $\mathscr{P}_c$ in (3) is related to the time–bandwidth product $a(T,\Omega) = 2\Omega T$ of the operator $P_{\Omega T}Q$ by $a = 2c/\pi$.

2.2.3 Double Orthogonality of Eigenfunctions of PQ_T

Since PQ_T has a simple spectrum, the PSWFs are orthogonal in $L^2(\mathbb{R})$. In $L^2[-T,T]$ one has $\int_{-T}^{T} \varphi_n(s)\,\varphi_m(s)\,\mathrm{d}s = \lambda_n\,\delta_{nm}$ as well. This double orthogonality extends to the case in which P is replaced by P_Σ such that $\Sigma = -\Sigma$; see [27].

2.2.4 A "Square Root" for PQ_T

Up to a dilation factor, one can regard $\mathscr{F}Q_T$ as *half* of a time- and band-limiting operation: $(D_a\mathscr{F}Q)^2 = \pm P_{2a}Q$ when acting, respectively, on real-valued even or odd functions, expressed as follows.

Proposition 2. *If ψ is an eigenfunction of $P_{a/2}Q$ with eigenvalue $\lambda = \lambda_n(a)$ then ψ is an eigenfunction of F_a with $\mu = \mu_n(a) = 2\mathrm{i}^n\sqrt{\lambda_n(a)/a}$, where*[1]

$$F_a(f)(t) = \int_{-1}^{1} \mathrm{e}^{\frac{\pi a}{2}\mathrm{i}st} f(s)\,\mathrm{d}s = \frac{2}{\sqrt{a}}(D_{a/4}\mathscr{F}^{-1}Q)(f)(t)\,.$$

One has $P_{a/2}Q = \frac{a}{4}F_a^*F_a$ so that $\lambda_n(a) = \frac{a}{4}|\mu_n(a)|^2$.

3 Eigenvalues of Time and Band Limiting

PQ_T has a simple, discrete spectrum [27]. An eigenfunction having an *eigenvalue close to one* is band limited and, approximately, time limited. Having a concrete definition of "eigenvalue close to one" would allow one to define the dimension of the space of signals band limited to $[-1/2,1/2]$ that are approximately time limited to $[-T,T]$. This makes particularly good sense when there is a sharply defined transition from eigenvalues close to one down to eigenvalues close to zero.

The eigenvalue λ_n of $P_\Omega Q_T$ depends only on the area $a = 2\Omega T$. We will write $A_a = PQ_{a/2}$. Landau [22] provided an intuitive but somewhat imprecise estimate of the width of the *plunge region* $\{n : 1-\alpha > \lambda_n(a) > \alpha\}$ for a and $\alpha > 0$ fixed in terms of $\mathrm{tr}(A_a) = \sum\lambda_n$ and $\|A_a\|_{\mathrm{HS}} = \sum\lambda_n^2$.

3.1 The Number of Eigenvalues of PQ_T Between α and $1-\alpha$

Fix the time–frequency area $a = 2T$ and consider the number of eigenvalues of A_{2T} that are neither close to one nor close to zero. By (1),

$$\sum_{n=0}^{\infty}\lambda_n = \mathrm{tr}(A_{2T}) = 2T.$$

With $\mathrm{sinc}\,(t) = \sin\pi t/\pi t$, the kernel of A_{2T} is $K(t,s) = \mathbb{1}_{[-T,T]}\,\mathrm{sinc}\,(t-s)$ and

$$\sum_{n=0}^{\infty}\lambda_n = \mathrm{tr}(A_{2T}) = \int K(t,t)\,\mathrm{d}t = \int_{-T}^{T} 1\,\mathrm{d}s = 2T,$$

while the Hilbert–Schmidt norm of PQ_T satisfies

$$\sum_{n=0}^{\infty}\lambda_n^2 = \|PQ_T\|_{\mathrm{HS}}^2 = \int_{-T}^{T}\int_{-T}^{T}\mathrm{sinc}^2(t-s)\,\mathrm{d}s\,\mathrm{d}t = \int_{-T}^{T}\int_{-T-s}^{T-s}\mathrm{sinc}^2 t\,\mathrm{d}t\,\mathrm{d}s\,.$$

[1] The operator given by integration against the kernel $\mathrm{e}^{-\mathrm{i}cst}\mathbb{1}_{[-1,1]}(s)$ is often denoted by F_c. One has $F_c = F_a$ when $c = a\pi/2$ as we assume here.

An integration by parts of the *outer* integral over s and an application of the fundamental theorem of calculus yield

$$\sum_{n=0}^{\infty}\lambda_n^2 = 2T\int_0^{2T}\operatorname{sinc}^2 t\,\mathrm{d}t + 2\int_0^{2T}(T-t)\operatorname{sinc}^2 t\,\mathrm{d}t \ge 2T - M_1\log(2T) - M_2.$$

Here, M_1 and M_2 are independent of T since $\int_0^\infty \operatorname{sinc}^2 t\,\mathrm{d}t = 1/2$, $\int_{2T}^\infty \operatorname{sinc}^2(t)\,\mathrm{d}t < 1/(2\pi^2 T)$, and $\int_0^{2T} t\operatorname{sinc}^2(t)\,\mathrm{d}t$ is comparable to $\log(2T)$. Subtracting the Hilbert–Schmidt estimate from the trace identity yields

$$\sum_{n=0}^{\infty}\lambda_n(1-\lambda_n) \le M_1\log(2T) + M_2\,.$$

Thus, for any fixed $\alpha\in(0,1/2)$, one has

$$\sum_{\alpha<\lambda_n<1-\alpha}\lambda_n(1-\lambda_n) \ge \alpha(1-\alpha)\#\{n:\alpha<\lambda_n<1-\alpha\} \quad \text{or}$$

$$\#\{n:\alpha<\lambda_n<1-\alpha\} \le \frac{\sum_{\alpha<\lambda_n<1-\alpha}\lambda_n(1-\lambda_n)}{\alpha(1-\alpha)} \le \frac{M_1\log 2T + M_2}{\alpha(1-\alpha)}. \tag{4}$$

Generalizing to any pair Ω and T, one sees that several eigenvalues of $P_\Omega Q_T$ are close to one, followed by a *plunge region* of width proportional to $\log 2\Omega T$ over which the eigenvalues transition from being close to one to being close to zero. The remaining eigenvalues, as it happens, decay to zero superexponentially, e.g., [30]. We will see how many eigenvalues of $P_\Omega Q_T$ are close to one momentarily.

3.2 The Multiband Case: Plunge Width Proportional to the Number of Intervals

In [19], Landau used much the same method as above to estimate the decay of eigenvalues of the time–frequency localization operator $P_\Sigma Q_S$ when S and Σ are finite unions of intervals. In the particular case $S=[-T,T]$, $\widehat{\varphi}_n$ is an eigenfunction of $P_{2T}Q_\Sigma$ with kernel $K(\xi,\eta)=\mathbb{1}_\Sigma(\eta)\sin 2\pi T(\xi-\eta)/(\pi(\xi-\eta))$. Arguing along the same lines as before, using the fact that $\int_I \operatorname{sinc}^2(t)\,\mathrm{d}t \le 2\int_0^{|I|/2}\operatorname{sinc}^2(t)\,\mathrm{d}t$ tells us that when $\Sigma=\cup_{\nu=1}^M I_\nu$ with pairwise disjoint intervals I_ν of length $|I_\nu|=\ell_\nu$,

$$\begin{aligned}\sum_n \lambda_n^2(2T,\Sigma) &= \int_\Sigma\int_\Sigma\left|\frac{\sin 2\pi T(\xi-\eta)}{\pi(\xi-\eta)}\right|^2 \mathrm{d}\xi\,\mathrm{d}\eta\\ &\ge \sum_{\nu=1}^M\Big(2T\ell_\nu - \frac{1}{\pi^2}\log^+ 2T\ell_\nu - 1\Big) = 2T|\Sigma| - A\log^+(2T) - M,\end{aligned}$$

where $\log^+(x) = \max(\log(x),0)$. In particular, A depends only on the linear distribution of Σ and $A \geq N_\Sigma/\pi^2$ when each of the N_Σ intervals comprising A has the same length. Combining the estimates for $\sum \lambda_n$ and $\sum \lambda_n^2$ one obtains

$$\sum_n \lambda_n(1-\lambda_n) \leq A\log^+(2T) - M. \tag{5}$$

When Σ is a fixed finite union of intervals and $A \geq N_\Sigma/\pi^2$,

$$\#\{n : \alpha < \lambda_n < 1-\alpha\} \leq \frac{\sum_{\alpha<\lambda_n<1-\alpha} \lambda_n(1-\lambda_n)}{\alpha(1-\alpha)} \leq \frac{A\log 2T + M}{\alpha(1-\alpha)}.$$

Suppose that $T = 1$ and Σ is a finite, pairwise disjoint union of a frequency intervals $I_1, \ldots, I_a$ each of unit length. Then $P_\Sigma Q_{1/2}$ will be a complicated operator, but it should have on the order of a eigenvalues of magnitude *at least* $1/2$. Consider now the limiting case in which the frequency intervals become *separated at infinity*. Any function ψ_j that is concentrated in frequency on I_j will be almost orthogonal over $[-1/2, 1/2]$, in the separation limit, to any function ψ_k that is frequency-concentrated on I_k when $j \neq k$. To see this, write $\psi_j(t) = \mathrm{e}^{2\pi i m_j t}\varphi_j(t)$ where m_j is the midpoint of I_j and $\widehat{\varphi}_j$ is essentially concentrated on $[-1/2, 1/2]$. Then

$$\int_{-1/2}^{1/2} \mathrm{e}^{2\pi\mathrm{i}(m_j-m_k)t}\varphi_j(t)\overline{\varphi_k}(t)\,\mathrm{d}t = \widehat{\varphi}_j * \widehat{\overline{\varphi_k}} * \mathrm{sinc}\,(m_j - m_k) = O(1/|m_j - m_k|)$$

as $|m_j - m_k| \to \infty$. This almost orthogonality prevents eigenvalues from the separate interval operators $P_I Q_{1/2}$ from *combining* into large eigenvalues of $P_\Sigma Q_{1/2}$. Consequently, $P_\Sigma Q_{1/2}$ will have on the order of a eigenvalues of size *approximately equal to* $\lambda_0(a = 1)$ in the separation limit, while the remaining eigenvalues will not be much larger than $\lambda_1(a=1) < 1/2$. The operator $PQ_{1/2}$, corresponding to single time and frequency intervals of unit length, has norm $\lambda_0(a = 1) \geq \|\mathrm{sinc}\,\mathbb{1}_{[-1/2,1/2]}\| > 0.88$. The trace of $PQ_{1/2}$ is equal to $a = 1$ on the one hand and to $\sum \lambda_n$ on the other, so $\lambda_1(a=1) \leq 1 - \lambda_0(a=1) < 1/2$. Incidentally, similar reasoning shows that $P_\Sigma Q$ cannot have *any* eigenvalues larger than $1/2$ when Σ is a union of a large number of short, mutually distant intervals, even if $a = |\Sigma| > 1$.

3.3 Landau and Widom's $2\Omega T$ Theorem

Landau and Widom [23] applied advanced techniques from spectral theory in order to estimate precisely, albeit asymptotically, the logarithmic term that gets subtracted from the time–bandwidth product to estimate the number of eigenvalues of $P_\Omega Q_T$ that are close to one. Theorem 2 was stated and proved by Landau and Widom in an equivalent, slightly different manner. As before, $A_a = PQ_{a/2}$.

Theorem 2 ([23]).

(i) The number $N(A_a,\alpha)$ of eigenvalues of A_a larger than α satisfies

$$N(A_a,\alpha)=a+\frac{1}{\pi^2}\log\Big(\frac{1-\alpha}{\alpha}\Big)\log a+o(\log a)\quad\text{as}\quad a\to\infty. \tag{6}$$

(ii) Let S and Σ be finite pairwise disjoint unions of N_S and N_Σ intervals, respectively, with $|S|=|\Sigma|=1$. Set $B_a=B_a(S,\Sigma)=P_{a\Sigma}Q_SP_{a\Sigma}$ where $a\Sigma=\{a\xi:\xi\in\Sigma\}$. Then the number $N(B_a,\alpha)$ of eigenvalues of B_a larger than α satisfies

$$N(B_a,\alpha)=a+\frac{N_SN_\Sigma}{\pi^2}\log\Big(\frac{1-\alpha}{\alpha}\Big)\log a+o(\log a),\qquad a\to\infty. \tag{7}$$

The estimate (6) boils down to estimating the polynomial moments of the discrete measure $\mathrm{d}_t[-N(A_a,t)]$ such that

$$N(A_a,\alpha)=\int_\alpha^1 \mathrm{d}_t[-N(A_a,t)]\,.$$

The result then follows from approximating $\mathbb{1}_{[\alpha,1]}$ by polynomials. The spectral measure estimate requires a variant of Szegő's eigenvalue distribution theorem [17] —a statement that the number of eigenvalues of a Toeplitz operator grows like the measure of the set on which the Fourier transform of the kernel exceeds α—also due to Landau [20], that is asymptotic in a.

The case of finitely many time and frequency intervals involves a reduction to the single interval case, which also requires asymptotic separation of the intervals. Although the factor N_SN_Σ disappears when $\alpha=1/2$, it appears prominently for other $\alpha\in(0,1)$.

3.4 Further Asymptotic Behavior of the Eigenvalues

In the following, $\phi_{n-1}^{(c)}$ is the nth eigenfunction of $F_{2c/\pi}$ with norm one in $L^2[-1,1]$.

Lemma 1 ([24]). *Let $c>0$. Then*

$$\mu_n\Big(\frac{c}{2\pi}\Big)=\frac{\mathrm{i}^n\sqrt{\pi}c^n(n!)^2}{(2n)!\Gamma(n+3/2)}\exp\left(\int_0^c\Big(\frac{2\phi_n^{(\tau)}(1)^2-1}{2\tau}-\frac{n}{\tau}\Big)\,\mathrm{d}\tau\right).$$

Consequently,

$$\Big|\mu_n\Big(\frac{c}{2\pi}\Big)\Big|\leq\frac{\sqrt{\pi}(n!)^2c^n}{\Gamma(n+3/2)(2n)!}.$$

The lemma proves that, for any fixed value of c, the eigenvalues decay *super-exponentially* as $n\to\infty$ by Stirling's approximation, cf. Widom [30].

3.5 *The Number of Eigenvalues of A_a Larger than 1/2*

Theorem 2 deals with asymptotic behavior of eigenvalues. Theorem 3, proved by Landau in [22], (cf. [18]), shows that there are, in essence, a eigenvalues of A_a larger than $1/2$: the "$o(\log a)$" term in (6) disappears when $\alpha = 1/2$.

In [22], (cf. [18]) Landau proved the following:

Theorem 3. *The eigenvalues of A_a satisfy*

$$\lambda_{\lfloor a\rfloor -1} \geq 1/2 \geq \lambda_{\lceil a\rceil}.$$

In the theorem, $\lfloor x\rfloor$ and $\lceil x\rceil$ denote the greatest integer less than or equal to x and least integer greater than or equal to x respectively. The Weyl–Courant minimax characterization of the singular values $\lambda_0 \geq \lambda_1 \geq \dots$ of $P_\Sigma Q_S P_\Sigma$ can be stated as

$$\lambda_n = \begin{cases} \min_{\mathscr{S}_n} \max\{\|Q_S f\|^2 : f \in \mathrm{PW}_\Sigma, \quad \|f\| = 1, f \perp \mathscr{S}_n\} \\ \max_{\mathscr{S}_{n+1}} \min\{\|Q_S f\|^2 : f \in \mathrm{PW}_\Sigma,\ \|f\| = 1, f \in \mathscr{S}_{n+1}\}. \end{cases}$$

Here, $\mathscr{S}_n$ ranges over all n-dimensional subspaces including, notably, the subspace spanned by the first n eigenfunctions of $P_\Sigma Q_S P_\Sigma$. In his 1965 work [18], Landau identified a convolver h such that if $f \in \mathrm{PW}$ and $f * h(m)$ vanishes at any integer in $[-a/2, a/2]$ then $\|Q_T f\|^2 \leq 0.6$. A sharper bound with 0.6 replaced by 0.5 was attributed to B.F. Logan in [18], but the sharp bound was never published until Landau's 1993 work [22] in which the convolver $h(t) = \sqrt{2}\cos \pi t \mathbb{1}_{[-1/2,1/2]}(t)$ was used. It satisfies $\|h\|^2 = 1$ and $\widehat{h}(\xi) \geq 1/\sqrt{2}$ whenever $|\xi| \leq 1/2$. We refer to [14] for the details of Landau's estimates. The theorem can also be viewed as a corollary of the multiple interval case Proposition 3, whose proof fundamentally relies on Landau's method.

3.5.1 Extension of Theorem 3 to Multiple Intervals

Landau's technique can be extended to the case in which S is a finite union of intervals (but still $\Sigma = [-1/2, 1/2]$). In [18], Landau observed that if S is a union of m intervals then

$$\begin{aligned} |S| - 2m &\leq \#\{k : (k - 1/2, k + 1/2) \subset S\} \\ &\leq \#\{k : (k - 1/2, k + 1/2) \cap S \neq \emptyset\} \leq |S| + 2m. \end{aligned}$$

The two numbers quantify dimensions of *vanishing subspaces* that can be used in the Weyl–Courant lemma. In Izu's dissertation [15], one can find a proof of an equivalent version of the following.

Proposition 3. *Let* $\Sigma = [-1/2, 1/2]$ *and let* S *be a finite union of* m *pairwise disjoint intervals. Denote by*

$$\nu = \max_{\alpha} \#\{k \in \mathbb{Z} : (k - 1/2, k + 1/2) \subset S + \alpha\} \quad \text{and}$$
$$\mu = \min_{\beta} \#\{\ell \in \mathbb{Z} : (\ell - 1/2, \ell + 1/2) \cap S + \beta \neq \emptyset\}.$$

Then the eigenvalues λ_n *of* $Q_S P$ *satisfy*

$$\lambda_{\nu-1} \geq 1/2 \geq \lambda_{\mu}. \tag{8}$$

In particular, for $|S| \geq 1$, $\lfloor |S| \rfloor - 2m + 2 \leq \nu \leq \mu \leq \lceil |S| \rceil + 2m - 2$ *so that*

$$\lambda_{\lfloor |S| \rfloor - 2m + 1} \geq 1/2 \geq \lambda_{\lceil |S| \rceil + 2m - 2}. \tag{9}$$

3.5.2 Discussion

Landau's estimate, effectively (9), represents a worst case in terms of the distribution of intervals in S. Both ν and μ will be closer to $|S|$ when the intervals are close to being aligned along a grid. If each of the intervals comprising S has the form $[k - 1/2, k + 1/2)$ then $\nu = \mu = |S|$ and one recovers the bounds $\lambda_{\lfloor |S| \rfloor - 1} \geq 1/2 \geq \lambda_{\lceil |S| \rceil}$, even though S can be disconnected (see [15]).

4 Discrete Theory

4.1 Finite Discrete Prolate Spheroidal Sequences

The discrete theory of index and band limiting (eigensequences are *discrete prolate spheroidal* (DPS) sequences on $\mathbb{Z}$) was developed by Slepian [25]. There is a parallel theory for the finite (discrete) Fourier transform on $\mathbb{Z}_N$ as well. Xu and Chamzas [32] referred to the corresponding eigenvectors as *periodic discrete prolate spheroidal sequences* (P-DPSS) and regarded them as periodic sequences. We will call them *finite discrete prolate spheroidal sequences* (FDPS), thinking of time and frequency localization as operations on functions $\mathbf{x} : \mathbb{Z}_N \to \mathbb{C}$. FDP sequences were also used in the work of Jain and Ranganath [16]. Grünbaum [10, 11] addressed the analogue of Theorem 1 for finite matrices. For K fixed such that $2K + 1 \leq N$, define the Toeplitz matrix $A = A^K$:

$$A_{k\ell} = a_{k-\ell} = \frac{\sin((2K+1)(k-\ell)\pi/N)}{N \sin((k-\ell)\pi/N)}, \quad (k, \ell = 0, \ldots, N-1). \tag{10}$$

A vector in the image of A is said to be *K-band limited*, since its N-point discrete Fourier transform (DFT) vanishes at any index m such that $m \mod N > K$. The dimension of the range of A is $2K+1$.

In contrast to the infinite case, the finite matrix $A_{k\ell}$ can have degenerate eigenvalues one and zero for time and band limiting. Denote by $A_M = A_M^K$ the $M \times M$ principal minor of A. Multiplication of a vector (a function on $\mathbb{Z}_N$) by A_M plays the analogous role to integrating against the sinc kernel over $[-T,T]$. Assuming that $M \geq 2K+1$, Xu and Chamzas [32] proved that, when ordered in nonincreasing order, the eigenvalues λ_n of A_M satisfy $\lambda_0 = \cdots = \lambda_{M+2K-N} = 1$ if $M+2K \geq N$, while $\lambda_{2K+1} = \cdots = \lambda_{M-1} = 0$ in this case. The intermediate eigenvalues are simple and have values strictly between zero and one.

The eigenvectors can be computed by noting that A_M commutes with a symmetric tridiagonal matrix T, e.g., Grünbaum [10, 11]. Eigenvectors $\mathbf{s}_n = (s_n(0),\ldots,s_n(M-1))^T$ ($n = 0,\ldots,\min(M-1,2K)$) of A_M corresponding to nonzero eigenvalues have K-band-limited extensions $\widetilde{\mathbf{s}}_n$ to $\mathbb{C}^N$ obtained by applying the matrix A in (10) to the vectors $(s_n(0),\ldots,s_n(M-1),0,\ldots,0)^T$. The $\widetilde{\mathbf{s}}_n$ are called *FDP* sequences. They are *doubly orthogonal*, just as in the case of the PSWFs in $L^2(\mathbb{R})$. If $M < 2K+1$ then the restrictions of the $\widetilde{\mathbf{s}}_n$ to their first M coordinates form a basis for $\mathbb{C}^M$, but the $\mathbb{Z}_N$ periodic extensions of $\mathbf{s}_0,\ldots,\mathbf{s}_{M-1}$ are not complete in the space of K-band-limited sequences. On the other hand, if $M > 2K+1$ then these periodic extensions are complete in the space of K-band-limited sequences, but the restrictions of $\mathbf{s}_0,\ldots,\mathbf{s}_{2K}$ to $\mathbb{C}^M$ are not complete [32].

4.2 Finite Fourier Uncertainty Inequalities

That A_M can have unit eigenvalues is reminiscent of the fact that if M divides N then the indicator vector $\mathbb{1}_{\ell+M\mathbb{Z}_N}$ of the coset $\mathbb{Z}_N/\mathbb{Z}_M$ has a DFT that is a modulation of $\mathbb{1}_{(N/M)\mathbb{Z}_N}$, e.g., [13, Chap. 4]. In particular, time and band limiting to such cosets gives rise to operators having some number of eigenvalues equal to one. Donoho and Stark [7] observed that these *picket fence* signals $\mathbf{x}$ on $\mathbb{C}^N$ are minimizers of the quantity $|\mathrm{supp}\,\mathbf{x}||\mathrm{supp}\,\widehat{\mathbf{x}}|$, which is always at least N if $\mathbf{x} \neq \mathbf{0}$. If sparse subsets S and Σ are chosen at random—e.g., from the uniform distribution of subsets of $\mathbb{Z}_N$ having a fixed size—then the probability that the corresponding time- and band-limiting operator will have a large norm—greater than $1/2$, say—is very small.

4.3 Quantitative Robust Uncertainty Principles

Candès, Romberg, and Tao [3–6] found that norm estimates on time- and band-limiting operators corresponding to sparse time and frequency supports could be useful in signal recovery problems. They considered the problem of finding a bound on the norm—and hence on the largest eigenvalue—of the discrete version of the operator $P_\Sigma Q_S P_\Sigma$ when the time–frequency area is small. In the finite case, the normalized area is $a = |S||\Sigma|/N$, where $|S|$ is the counting measure of S. Denote by $A_{S\Sigma}$ the operator with matrix $D_S \mathscr{F}_N^{-1} D_\Sigma \mathscr{F}_N$ where $\mathscr{F}_N$ is the matrix of the N-point DFT and D_S is the diagonal matrix with $D_S(j,j) = 1$ if $j \in S$ and $D_S(j,k) = 0$ otherwise, corresponding to multiplication by the discrete indicator function $\mathbb{1}_S$. Then

$$\begin{aligned} A_{S\Sigma} A_{S\Sigma}^* &= D_S \mathscr{F}_N^{-1} D_\Sigma \mathscr{F}_N (D_S \mathscr{F}_N^{-1} D_\Sigma \mathscr{F}_N)^* \\ &= D_S \mathscr{F}_N^{-1} D_\Sigma \mathscr{F}_N \mathscr{F}_N^* D_\Sigma \mathscr{F}_N D_S = D_S \mathscr{F}_N^{-1} D_\Sigma \mathscr{F}_N D_S, \end{aligned} \tag{11}$$

since $\mathscr{F}_N^* = \mathscr{F}_N^{-1}$ and D_Σ is idempotent.

Candès et al. were motivated by applications to compressed sensing in which a signal $\mathbf{x}$ could be recovered (via optimization techniques) from its values on S provided that its Fourier transform $\widehat{\mathbf{x}}$ vanishes outside Σ. This recovery is contingent upon invertibility of $I - A_{S\Sigma}$, which holds if $\|A_{S\Sigma}\| \ll 1$. Candès et al. were able to obtain such bounds in a probabilistic sense using a technical array of probabilistic and combinatorial methods. For technical reasons, Candès et al. assumed that $N \geq 512$. In order to discuss the rate at which the probability that $\|A_{S\Sigma} A_{S\Sigma}^*\| > 1/2$ decays (in problem size N) they fixed a parameter $1 \leq \beta \leq (3/8) \log N$, so β is at most a fraction of $\log N$. A sparse vector defined on $\mathbb{Z}_N$ should be supported in a set of size at most a fraction of N. To quantify sparsity, set

$$M(N,\beta) = \frac{N}{\sqrt{(\beta+1)\log N}} \left(\frac{1}{\sqrt{6}} + o(1) \right).$$

The "$o(1)$" term arises in technical estimates for the proof of the following theorem.

Theorem 4 ([5, 6]). *Fix $S \subset \mathbb{Z}_N$ of size smaller than $M(N,\beta)$, and let $\Sigma \subset \mathbb{Z}_N$ be randomly generated from the uniform distribution of subsets of $\mathbb{Z}_N$ of given size $|\Sigma|$, chosen so that $|S| + |\Sigma| \leq M(N,\beta)$. Then, with probability at least $1 - O((\log N)^{1/2}/N^\beta)$, every signal $\mathbf{x}$ supported in S satisfies*

$$\|\widehat{\mathbf{x}} \mathbb{1}_\Sigma\|^2 \leq \frac{1}{2} \|\mathbf{x}\|^2,$$

while every signal $\mathbf{x}$ frequency supported in Σ satisfies

$$\|\mathbf{x} \mathbb{1}_S\|^2 \leq \frac{1}{2} \|\mathbf{x}\|^2.$$

The second inequality says that $\|A_{S\Sigma}\|^2 \leq 1/2$. The arithmetic–geometric inequality implies that the normalized time–frequency area satisfies

$$a = \frac{|S||\Sigma|}{N} \leq \frac{1}{4N} M(N,\beta)^2 \leq \frac{N}{24(\beta+1)\log N}(1 + o(1)),$$

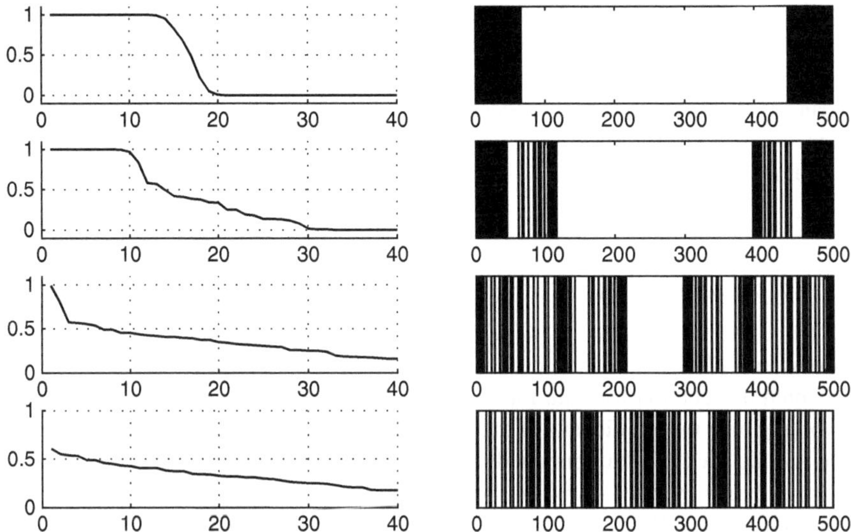

Fig. 1 Eigenvalues for disconnected Fourier supports. Eigenvalue scree plots (*left*) are shown with corresponding Fourier supports (*right*) for a 512-point DFT. In each case, the time support is a discrete interval of length 64 points, and the Fourier support has 128 points. The normalized area is $a = 16$. As the Fourier support becomes more disconnected, the eigenvalues become more evenly spread out

suggesting that even *small N* can support time–frequency set pairs of normalized area substantially larger than one on which no signal is mostly localized. The probability is defined with respect to the uniform distribution among all sets Σ of fixed size $|\Sigma|$. The proof involves complicated combinatorial estimates for Fourier sums arising in traces of powers of $A_{S\Sigma}$.

4.4 *Discussion*

Figure 1 illustrates the dispersion of eigenvalues of $A_{S\Sigma} = \mathscr{F}^{-1}D_\Sigma\mathscr{F}D_S$ as the frequency support becomes disconnected. Here, the normalized area is held fixed at $a = 16$ and the time support is a fixed interval. Only the bottom row of Fig. 1 corresponds to a "typical" randomly generated frequency support.

5 Further Analytical and Numerical Properties

5.1 *The Case* $c = 0$*: The Legendre Polynomials*

The nth Legendre polynomial $P_n(t)$ is a solution of

$$\frac{\mathrm{d}}{\mathrm{d}t}(t^2-1)\frac{\mathrm{d}P_n}{\mathrm{d}t} = \chi_n(0)P_n(t) \quad \text{or} \quad (t^2-1)\frac{\mathrm{d}^2P_n}{\mathrm{d}t^2} + 2t\frac{\mathrm{d}P_n}{\mathrm{d}t} - \chi_n(0)P_n = 0, \tag{12}$$

in which $\chi_n(0) = n(n+1)$. This equation is the $c = 0$ limit of $\mathscr{P}\varphi = \chi_n(c)\,\varphi$, with $\mathscr{P}$ as in (3). The constant functions are solutions with $\chi_0(0) = 0$, and the function $P_1(t) = t$ with $\chi_1(0) = 2$. The Legendre polynomials can be defined iteratively by means of the Gram–Schmidt process in order that the polynomials are orthogonal over $[-1,1]$ with respect to $\langle f, g\rangle = \int_{-1}^{1} f(t)\,g(t)\,\mathrm{d}t$. It is standard to normalize P_n so that $P_n(1) = 1$.

5.2 Approximations of Prolates in Legendre Series

Bouwkamp [1] first suggested a method to express PSWFs in Legendre series in which a recursion formula for Legendre polynomials gives rise to a recursion involving both the coefficients of the Legendre series expansion and the eigenvalue. Bouwkamp then used a continued fractions approach to estimate the eigenvalues $\chi_n(c)$ and then the coefficients of the Legendre series. A more up-to-date approach was outlined by Boyd [2]. Let $\bar{P}_m = \sqrt{m+1/2}\,P_m$ $(m \geq 0)$ be the $L^2[-1,1]$-normalized basis of Legendre polynomials. As in the Bouwkamp method, each prolate $\phi_n = \phi_n^{(c)}$ is expressed as its Legendre expansion

$$\phi_n = \sum_{m=0}^{\infty} \beta_{nm}\bar{P}_m, \tag{13}$$

with the goal of estimating the coefficients β_{nm} assuming that the eigenvalues of the prolate differential operator $\mathscr{P}$ are known. Applying $\mathscr{P}$ to both sides of this equation gives rise to an equation for the coefficients. Let $A = \{a_{mk}\}_{m,k=0}^{\infty}$, the doubly infinite tridiagonal matrix with nonzero elements

$$a_{mk} = \begin{cases} \dfrac{c^2 m(m-1)}{(2m-1)\sqrt{(2m-3)(2m+1)}} & \text{if } m \geq 2,\ k = m-2 \\ m(m+1) + \dfrac{c^2(2m^2+2m-1)}{(2m+3)(2m-1)} & \text{if } m = k \geq 0 \\ \dfrac{c^2(m+2)(m+1)}{(2m+3)\sqrt{(2m+5)(2m-1)}} & \text{if } m \geq 0,\ k = m+2 \\ 0 & \text{else.} \end{cases}$$

Then $\mathscr{P}\bar{P}_m = \sum_{k=0}^{\infty} A_{mk}\bar{P}_k$. Applying $\mathscr{P}$ to both sides of (13) gives

$$\chi_n\phi_n = \sum_{k=0}^{\infty}\left(\sum_{m=0}^{\infty}\beta_{nm}A_{mk}\right)\bar{P}_k. \tag{14}$$

For fixed $n \geq 0$, let $\mathbf{b}_n = (\beta_{n0}, \beta_{n1}, \beta_{n2}, \ldots)$. Equating coefficients in (14) and (13) gives the matrix equations

$$A^T \mathbf{b}_n = \chi_n \mathbf{b}_n, \tag{15}$$

that is, the vector $\mathbf{b}_n$ (whose entries are the L^2-normalized Legendre coefficients of ϕ_n) is an eigenvector of A^T with eigenvalue χ_n.

Truncating the sum in (13) after N_{tr} terms and following the procedure outlined above yield a finite matrix eigenvalue problem

$$(A^{\mathrm{tr}})^T \mathbf{b}_n^{\mathrm{tr}} = \chi_n^{\mathrm{tr}} \mathbf{b}_n^{\mathrm{tr}} \tag{16}$$

with $\mathbf{b}_n^{\mathrm{tr}} \in \mathbb{R}^{N_{\mathrm{tr}}}$ and A^{tr} the top left $N_{\mathrm{tr}} \times N_{\mathrm{tr}}$ submatrix of the matrix A above. Because of the structure of A^{tr} (whereby the only nonzero entries are of the form $a_{m,m-2}$, a_{mm}, and $a_{m,m+2}$), equation (16) may be written as a pair of uncoupled eigenproblems for even and odd values of n, respectively:

$$(A_{\mathrm{e}}^{\mathrm{tr}})^T \mathbf{b}_{\mathrm{e}}^{(2n)} = \chi_{2n}^{\mathrm{tr}} \mathbf{b}_{\mathrm{e}}^{(2n)}; \qquad (A_{\mathrm{o}}^{\mathrm{tr}})^T \mathbf{b}_{\mathrm{o}}^{(2n+1)} = \chi_{2n+1}^{\mathrm{tr}} \mathbf{b}_{\mathrm{o}}^{(2n+1)} \tag{17}$$

with $A_{\mathrm{o}}, A_{\mathrm{e}}$ both $(N_{\mathrm{tr}}/2) \times (N_{\mathrm{tr}}/2)$ tridiagonal matrices (assuming N_{tr} even) given by $(A_{\mathrm{e}})_{mk} = a_{2m,2k}$ and $(A_{\mathrm{o}})_{mk} = a_{2m+1,2k+1}$. Once solved, the eigenproblems (17) yield approximations of the Legendre coefficients β_{nm} of the prolates $\phi_0^{(c)}, \phi_1^{(c)}, \ldots, \phi_{N-1}^{(c)}$ as in (13). Boyd [2] reported that for all N and c, the worst approximated prolate is that of highest order $\phi_{N-1}^{(c)}$ so that if the truncation N_{tr} is chosen large enough so that $\phi_{N-1}^{(c)}$ is computed with sufficient accuracy, then so too will $\phi_n^{(c)}$ with $0 \leq n \leq N-2$. Numerical evidence was given to suggest that if $N_{\mathrm{tr}} \approx 30 + 2N$, then $\beta_{N-1,N_{\mathrm{tr}}} < 10^{-20}$, and it was claimed that if $N_{\mathrm{tr}} > 30 + 2N$ then the approximations to $\{\phi_n^{(c)}\}_{n=0}^{N-1}$ are accurate as long as $c \leq c_*(N) = \pi(N+1/2)/2$.

5.3 Another Look Legendre Series of PSWFs

Assume now that Legendre coefficients can be estimated accurately and consider, in turn, the accuracy of approximation of PSWFs over $[-1,1]$ by truncated partial sums of Legendre series. Proposition 4 is due to Xiao et al. [31]. As before, let $\phi_n^{(c)}$ denote the nth $L^2[-1,1]$-normalized PSWF eigenfunction of

$$F_{\frac{2c}{\pi}}(f)(t) = \int_{-1}^{1} \mathrm{e}^{icxt} f(x)\, \mathrm{d}x. \tag{18}$$

Let P_n denote the nth Legendre polynomial on $[-1,1]$ normalized such that $P(1) = 1$, so that $\int_{-1}^{1} P_n^2 = 1/(n+1/2)$. Since $\int_{-1}^{1} x^m P_n(x)\, \mathrm{d}x = 0$ if $m < n$, expanding the exponential in its power series gives

$$\int_{-1}^{1} \mathrm{e}^{\mathrm{i}cxt} P_n(x)\,\mathrm{d}x = \sum_{\nu=[n/2]}^{\infty} \int_{-1}^{1} (a_\nu x^{2\nu} + \mathrm{i} b_\nu x^{2\nu+1}) P_n(x)\,\mathrm{d}x$$

$$a_\nu(t) = \frac{(-1)^\nu (ct)^{2\nu}}{(2\nu)!}; \qquad b_\nu(t) = \frac{(-1)^\nu (ct)^{2\nu+1}}{(2\nu+1)!}.$$

The following *remainder estimate* is based on [31, Lemma 3.3].

Proposition 4. *(i) If* $n > (2[|\mathrm{e}tc|]+1)$ *then*

$$|F_{\frac{2c}{\pi}}(P_n)(t)| < \frac{C_n}{2^n\sqrt{2n+1}} \qquad (C_n \to 0 \text{ as } n \to \infty). \tag{19}$$

(ii) If ϕ_m^c *denotes the mth eigenfunction of* $F_{\frac{2c}{\pi}}$ *with eigenvalue* μ_m^c *then, for* $n > 2([|\mathrm{e}tc|]+1)$,

$$\left|\int_{-1}^{1} \phi_m^c(x) P_n(x)\,\mathrm{d}x\right| \le \frac{C}{2^n\sqrt{n+1/2}|\mu_m^c|}.$$

With $\alpha_{mn} = (n+1/2)\beta_{mn}$ as in (13), one has

$$\phi_m^c(t) = \sum_{n=0}^{\infty} \alpha_{mn} P_n(t); \quad \alpha_{mn} = \left(n+\frac{1}{2}\right)\int_{-1}^{1} \phi_m^c P_n. \tag{20}$$

Proposition4 provides an estimate of the error of approximating any value $\phi_m^c(t)$ for t inside a bounded interval by applying $F_{\frac{2c}{\pi}}$ to a partial sum of the Legendre expansion of ϕ_m^c, since

$$\left|F_{\frac{2c}{\pi}}(\phi_m)(t) - F_{\frac{2c}{\pi}}\left(\sum_{n=0}^{N} \alpha_{mn} P_n\right)(t)\right| \le \sum_{N+1}^{\infty} \left|F_{\frac{2c}{\pi}}(\alpha_{mn} P_n)\right|$$
$$\le \sum_{N+1}^{\infty} \frac{C\sqrt{n+1/2}}{2^n|\mu_m^c|}\left|F_{\frac{2c}{\pi}}(P_n)\right| \le \sum_{N+1}^{\infty} \frac{C^2}{4^n|\mu_m^c|}$$

if $N > 2[|\mathrm{e}tc|]+1$. In particular, the integer values of $\phi_m^c(k)$ can be approximated by applying $F_{2c/\pi}$ to the Nth partial sum of the Legendre expansion of ϕ_m^c on $[-1,1]$. This approximation will be effective for any k such that $N > 2[|\mathrm{e}kc|]+1$, with an error controlled by $1/(4^N|\mu_m^c|)$. In fact, the estimates just considered can be used to justify effective approximations of the values $\phi_n^c(k)$ for larger k as well. However, in what follows we are interested in approximating integer samples for k essentially within a fixed multiple of the time-limiting interval.

5.4 Approximation of Samples of Prolates

5.4.1 Integer Values of φ_m^T

Let φ^T again denote an $L^2(\mathbb{R})$-normalized eigenfunction of PQ_T, that is, φ^T is band limited to $[-1/2, 1/2]$ and approximately time limited to $[-T, T]$. Then, by Proposition 2, $\psi(t) = D_T\varphi^T$ is an eigenfunction of F_{2T} with F_a as in (18), that is,

$$F_{2T}(\psi)(t) = \int_{-1}^{1} \mathrm{e}^{\pi \mathrm{i} T st}\,\psi(s)\,\mathrm{d}s = \mu\,\psi(t)$$

for an appropriate $\mu \in \mathbb{C}$. Equivalently, if φ^T is an eigenfunction of PQ_T then $D_T\varphi^T$ will satisfy (see Table 1)

$$(D_T(\varphi^T))(t) = \frac{1}{\mu}\int_{-1}^{1} \mathrm{e}^{\pi \mathrm{i} T st}(D_T\varphi^T)(s)\,\mathrm{d}s\,.$$

Therefore, with α_{mn} as in (20) and φ_m^T the mth eigenfunction of PQ_T,

$$\begin{aligned}
\varphi_m^T(k) &= \frac{1}{\sqrt{T}}(D_T(\varphi_m^T))\Big(\frac{k}{T}\Big) = \frac{1}{\sqrt{T}\mu_m}\int_{-1}^{1} \mathrm{e}^{\pi \mathrm{i} sk}(D_T\varphi_m^T)(s)\,\mathrm{d}s \\
&\approx \frac{1}{\sqrt{T}\mu_m}\sum_{n=0}^{N(T)}(n+1/2)\langle D_T\varphi_m^T, P_n\rangle \int_{-1}^{1} \mathrm{e}^{\pi \mathrm{i} sk}P_n(s)\,\mathrm{d}s \\
&\approx \frac{1}{\sqrt{T}\mu_m}\sum_{n=0}^{N(T)}\alpha_{mn}(2T)\begin{cases}(-\mathrm{i})^n J_{n+1/2}(\frac{\pi}{?}k)\sqrt{k}, & k\neq 0\\ \delta_{0,n}, & k=0\,.\end{cases}
\end{aligned}$$

The formula for the Fourier coefficients of P_n is classical and can be found in the Bateman project manuscripts [9, p. 122] or [8, vol. II, p. 213].

6 Sampling Formulas for Prolates

Our goal in this section is to show that if φ_n is the nth eigenfunction of PQ_T with $n \ll 2T$ so that $\lambda_n \approx 1$, then $Q_T\varphi_n$ can be approximated accurately by sinc-interpolating its samples near $[-T, T]$. Since the methods of the previous section already provide accurate approximations of prolates via Legendre expansions, one might wonder what is the purpose of this alternative method. The purpose is to be able to construct an approximation of $Q_T f$ when $f \in \mathrm{PW}$ directly from the integer samples of f in a suitable neighborhood of $[-T, T]$. The primary goal is to obtain approximate coefficients $c_n(f) \approx \langle f, \varphi_n\rangle$, $(n = 0, \ldots, N(T))$ from which the expansion $f \approx \sum c_n(f)\,\varphi_n$ can be computed by whatever means of approximation of φ_n is available. Since the eigenfunctions of PQ_T form an orthonormal basis for PW, and since the sinc function provides a reproducing kernel for PW, one has the *Mercer expansion*:

$$\operatorname{sinc}(t-s) = \sum_{n=0}^{\infty} \langle \operatorname{sinc}(\cdot - s),\, \varphi_n(\cdot)\rangle \varphi_n(t) = \sum_{n=0}^{\infty} \varphi_n(s)\varphi_n(t)\,.$$

Evaluating at $s = k \in \mathbb{Z}$ yields

$$\operatorname{sinc}(t-k) = \sum_{n=0}^{\infty} \psi_n(\lambda)\varphi_n(t)\,.$$

Walter and Shen [29] observed that this identity can be coupled with the Shannon sampling theorem to provide the following proposition, to which we will refer as the *first Walter–Shen sampling formula*.

Proposition 5. *Fix $T > 0$. Let $\{\varphi_n\}$ denote the $2T$-concentrated PSWFs frequency supported in $[-1/2, 1/2]$. For any $f \in \mathrm{PW}$, one has*

$$f(t) = \sum_{n=0}^{\infty} \sum_{k=-\infty}^{\infty} f(k)\, \varphi_n(k)\, \varphi_n(t)\,.$$

The proposition is also a direct consequence of the fact that the shifted sinc functions form an orthonormal basis for PW and the inner product on PW is the same as the ℓ^2-inner product of the sequences of integer samples.

The Shen–Walter formula depends implicitly on the duration parameter T defining the PSWFs, but the eigenvalues of PQ_T do not appear. It is natural to ask whether some variation of the first Walter–Shen sampling formula can lead to an efficient approximation of a nearly time- and band-limited signal f in terms of its integer samples in or close to the time concentration interval. In particular, if $f \in \mathrm{PW}$ is concentrated in $[-T, T]$, how is this concentration reflected in the samples $\varphi_n(k)$ on the one hand and, if $f \in \operatorname{span}\{\varphi_0, \dots, \varphi_N\}$, how is this concentration reflected in the samples of f, on the other?

A function $f \in \mathrm{PW}$ is essentially concentrated in $[-T, T]$ if it nearly belongs to $\mathrm{PSWF}_N = \operatorname{span}\{\varphi_0, \dots, \varphi_N\}$ in the sense that $\sum_{n \geq N} |\langle f, \varphi_n\rangle|^2$ is small. A function whose samples $f(k)$ satisfy $\sum_{|k|>MT} |f(k)|^2$ is also, in a sense, approximately concentrated around $[-T, T]$. It is natural to ask for a sense in which these two conditions are effectively the same. To answer this question one must bound the quantities $\sum_{n \geq N} |\langle f, \varphi_n\rangle|^2$ and $\sum_{|k|>M} |f(k)|^2$ in terms of one another for specific values of N and M, when $f \in \mathrm{PW}$ is assumed either nearly to belong to PSWF_N or to have rapidly decaying samples away from $[-T, T]$.

6.1 Samples of PSWFs

The sample sequences $\{\varphi_n(k)\}$ of the PSWFs are orthogonal to one another:

$$\begin{aligned}\delta_{nm} = \langle \varphi_n, \varphi_m \rangle &= \int \big(\sum_k \varphi_n(k)\,\mathrm{sinc}\,(t-k)\big)\big(\sum_\ell \varphi_m(\ell)\,\mathrm{sinc}\,(t-\ell)\big)\,\mathrm{d}t \\ &= \sum_k \sum_\ell \varphi_n(k)\varphi_m(\ell) \int \mathrm{sinc}\,(t-k)\mathrm{sinc}\,(t-\ell)\,\mathrm{d}t \\ &= \sum_k \sum_\ell \varphi_n(k)\varphi_m(\ell)\,\delta_{k\ell} = \sum_k \varphi_n(k)\varphi_m(k). \qquad (21)\end{aligned}$$

The sequences $\{\varphi_n(k)\}$ are also complete in $\ell^2(\mathbb{Z})$ since they can be used to construct the sequence of samples of any $f \in \mathrm{PW}$.

Since the PSWFs are orthogonal on $[-T,T]$, one also has

$$\lambda_n \delta_{nm} = \langle Q_T \varphi_n, \varphi_m \rangle = \sum_k \sum_\ell \varphi_n(k)\,\varphi_m(\ell) \int_{-T}^{T} \mathrm{sinc}\,(t-k)\,\mathrm{sinc}\,(t-\ell)\,\mathrm{d}t\,.$$

Consider the matrix $A : \mathbb{Z} \times \mathbb{Z} \to \mathbb{R}$ defined by

$$A_{k\ell} = \int_{-T}^{T} \mathrm{sinc}\,(t-k)\,\mathrm{sinc}\,(t-\ell)\,\mathrm{d}t\,. \qquad (22)$$

The equation above can be written

$$\lambda_n \delta_{nm} = \sum_k \varphi_n(k) \sum_\ell A_{k\ell}\,\varphi_m(\ell). \qquad (23)$$

Since the sample sequences of the PSWFs form a complete orthonormal basis for $\ell^2(\mathbb{Z})$, this implies that these sequences are eigenvectors of A.

Proposition 6. *The matrix A defined in (22) has the same eigenvalues as PQ_T. Additionally, the eigenvector for λ_n is the sample sequence $\{\varphi_n(k)\}$ of the eigenfunction φ_n of PQ_T.*

An immediate corollary is the *second Walter–Shen sampling formula.*

Corollary 1. *If $f \in \mathrm{PW}$, then*

$$PQ_T f(t) = \sum_{n=0}^{\infty} \lambda_n \sum_{k=-\infty}^{\infty} f(k)\,\varphi_n(k)\,\varphi_n(t)\,.$$

6.2 Quadratic Decay for PSWF Samples

Shen and Walter [29] used the mean value theorem for differentiable functions of a real variable to show that the integer samples of the nth eigenfunction of PQ_T satisfies the quadratic decay estimate $\sum_{|k|>T}(\varphi_n(k))^2 \le CT\sqrt{1-\lambda_n}$. An alternative approach was taken in [12] in which the mean value theorem for analytic functions was employed to obtain a quadratic decay estimate on the samples. However, in this case the *tail* of the sample sequence has to be taken with regard to samples not just outside $[-T,T]$ but, instead, having some distance from $[-T,T]$.

Proposition 7. *Let $0 \le N < 2T$ and let $M(T) = (\pi^2+1)(1+\log^\gamma(T))T$ for some $\gamma > 1$. Then there is a $C > 0$ such that, for any $n \le 2T$, one has*

$$\sum_{|k|>M(T)} \varphi_n^2(k) \le C(1-\lambda_n).$$

Remark. The quantity $M(T)$ arises from the use of a Fourier bump function in order to obtain suitable mean-value inequalities. In particular, the $\log^\gamma(T)$ term arises from limitations on the best known decay in the time domain of such a mollifier. We conjecture that the estimate of Proposition 7 remains true with $M(T) = (\pi^2+1)T$.

6.3 Approximate Time-Localized Projections

In this section we consider the problem of approximating the projection of $f \in \mathrm{PW}$ onto PSWF_N using a collection of samples near $[-T,T]$. As above, φ_n will be the nth PSWF eigenfunction of PQ_T. We will let

$$f_{N,K} = \sum_{|k|\le K} f(k) \sum_{n=0}^{N} \varphi_n(k)\,\varphi_n(t) \quad \text{and} \quad f_N = \lim_{K\to\infty} f_{N,K}\,. \tag{24}$$

In [29], Walter and Shen proved that $\sum_{|k|>T} \varphi_n(k)^2 \le C(n,T)(1-\lambda_n)^{1/2}$ so that if $f \in \mathrm{PSWF}_N$, $N \le 2T$, then

$$\int_{-T}^{T} \left| f(t) - f_{N,[T]} \right|^2 \le \|f\|^2 \sum_{n=0}^{N} \lambda_n (1-\lambda_n)^{1/2} C(n,T).$$

Proposition 7 suggests that an approximation by $f_{N,M(T)}$ rather than by $f_{N,T}$ can provide an estimate of f whose squared error decays like $(1-\lambda_N)$ rather than $(1-\lambda_n)^{1/2}$. Since $f_N(t) = \sum_{n=0}^{N} \left(\sum_k f(k)\varphi_n(k)\right)\varphi_n(t)$ and since $\|f\|^2 = \sum (f(k))^2$,

$$|\langle f_N - f_{N,M(T)}, \varphi_n\rangle|^2 \le \sum_{n=0}^{N} |\langle f, \varphi_n\rangle|^2 \sum_{|k|>M(T)} \varphi_n^2(k) \le C\|f\|^2(1-\lambda_n)\,. \tag{25}$$

Using orthogonality of the φ_n over $[-T,T]$ one then has the following.

Proposition 8. *If $f \in \mathrm{PW}$ then $f_N \in \mathrm{PSWF}_N$ is the orthogonal projection of f onto the span of $\{\varphi_0,\dots,\varphi_N\}$ and, with $M(T)$ as in Proposition 7,*

$$\|Q_T(f_N - f_{N,M(T)})\|^2 \le \sum_{n=0}^{N} \lambda_n |\langle f_N - f_{N,M(T)},\, \varphi_n\rangle|^2 \le C\|f\|^2 \sum_{n=0}^{N} \lambda_n(1-\lambda_n)\,.$$

Proof. The orthogonality of the functions φ_n over $[-T,T]$ implies that

$$\|Q_T(f_N - f_{N,M(T)})\|^2 = \sum_{n=0}^{N} \lambda_n |\langle f_N - f_{N,M(T)},\, \varphi_n\rangle|^2,$$

and the proposition then follows from (25). □

Slepian and Pollak observed that $\|Q_T(f-f_N)\|^2 = \sum_{n=N+1}^{\infty} \lambda_n |\langle f,\, \varphi_n\rangle|^2$ (see [27, (17), p. 52]). Combining this with (25) and recalling that $\lambda_n < 1/2$ if $n > 2T+1$, the triangle inequality yields the following.

Corollary 2. *If $f \in \mathrm{PW}$ then for $T > 0$ fixed,*

$$\|Q_T(f - f_{N,M(T)})\|^2 \le C \sum_{n=0}^{\infty} |\langle f,\, \varphi_n\rangle|^2 \lambda_n(1-\lambda_n)\,.$$

Acknowledgments Several of the ideas presented here developed while the author was on sabbatical at Washington University in St. Louis. He is very grateful to the Department of Mathematics at WUSTL for its hospitality. Special thanks go to Guido Weiss and Ed Wilson. Jeff Hogan and Scott Izu contributed substantially to the ideas presented here.

References

1. Bouwkamp, C.J.: On spheroidal wave functions of order zero. J. Math. Phys. Mass. Inst. Tech. **26**, 79–92 (1947)
2. Boyd, J.P.: Algorithm 840: computation of grid points, quadrature weights and derivatives for spectral element methods using prolate spheroidal wave functions—prolate elements. ACM Trans. Math. Software **31**, 149–165 (2005)
3. Candès, E.J., Romberg, J.: Quantitative robust uncertainty principles and optimally sparse decompositions. Found. Comput. Math. **6**, 227–254 (2006)
4. Candès, E.J., Romberg, J.: Sparsity and incoherence in compressive sampling. Inverse Probl. **23**, 969–985 (2007)
5. Candès, E.J., Romberg, J.K., Tao, T.: Robust uncertainty principles: exact signal reconstruction from highly incomplete frequency information. IEEE Trans. Inform. Theory **52**, 489–509 (2006)
6. Candès, E.J., Romberg, J.K., Tao, T.: Stable signal recovery from incomplete and inaccurate measurements. Comm. Pure Appl. Math. **59**, 1207–1223 (2006)
7. Donoho, D.L., Stark, P.B.: Uncertainty principles and signal recovery. SIAM J. Appl. Math. **49**, 906–931 (1989)

8. Erdélyi, A., Magnus, W., Oberhettinger, F., Tricomi, F.G.: Higher Transcendental Functions, vols. I, II. McGraw-Hill, New York (1953)
9. Erdélyi, A., Magnus, W., Oberhettinger, F., Tricomi, F.G.: Tables of Integral Transforms, vol. I. McGraw-Hill, New York (1954)
10. Grünbaum, F.A.: Eigenvectors of a Toeplitz matrix: discrete version of the prolate spheroidal wave functions. SIAM J. Algebraic discrete methods **2**, 136–141 (1981)
11. Grünbaum, F.A.: Toeplitz matrices commuting with tridiagonal matrices. Linear Algebra Appl. **40**, 25–36 (1981)
12. Hogan, J.A., Izu, S., Lakey, J.D.: Sampling approximations for time- and bandlimiting. Sampling Theor. Signal Image Process. **9**, 91–118 (2010)
13. Hogan, J.A., Lakey, J.D.: Time–Frequency and Time–Scale Methods. Birkhäuser, Boston (2005)
14. Hogan, J.A., Lakey, J.D.: Duration and Bandwidth Limiting. Birkhäuser, Boston (2012)
15. Izu, S.: Sampling and Time–Frequency Localization in Paley–Wiener Spaces. PhD thesis, New Mexico State University (2009)
16. Jain, A.K., Ranganath, S.: Extrapolation algorithms for discrete signals with application in spectral estimation. IEEE Trans. Acoust. Speech Signal Process. **29**, 830–845 (1981)
17. Kac, M., Murdock, W.L., Szegő, G.: On the eigenvalues of certain Hermitian forms. J. Rational Mech. Anal. **2**, 767–800 (1953)
18. Landau, H.J.: The eigenvalue behavior of certain convolution equations. Trans. Am. Math. Soc. **115**, 242–256 (1965)
19. Landau, H.J.: Sampling, data transmission, and the Nyquist rate. Proc. IEEE **55**, 1701–1706 (1967)
20. Landau, H.J.: On Szegő's eigenvalue distribution theorem and non-Hermitian kernels. J. Analyse Math. **28**, 335–357 (1975)
21. Landau, H.J.: An overview of time and frequency limiting. In: Fourier Techniques and Applications (Kensington, 1983), pp. 201–220. Plenum, New York (1985)
22. Landau, H.J.: On the density of phase-space expansions. IEEE Trans. Inform. Theor. **39**, 1152–1156 (1993)
23. Landau, H.J., Widom, H.: Eigenvalue distribution of time and frequency limiting. J. Math. Anal. Appl. **77**, 469–481 (1980)
24. Rokhlin, V., Xiao, H.: Approximate formulae for certain prolate spheroidal wave functions valid for large values of both order and band-limit. Appl. Comput. Harmon. Anal. **22**, 105–123 (2007)
25. Slepian, D.: Prolate spheroidal wave functions, Fourier analysis, and uncertainty. V - The discrete case. ATT Tech. J. **57**, 1371–1430 (1978)
26. Slepian, D.: Some comments on Fourier analysis, uncertainty and modeling. SIAM Rev. **25**, 379–393 (1983)
27. Slepian, D., Pollak, H.O.: Prolate spheroidal wave functions, Fourier analysis and uncertainty, I. Bell System Tech. J **40**, 43–63 (1961)
28. Walter, G.G.: Differential operators which commute with characteristic functions with applications to a lucky accident. Complex Variables Theor. Appl. **18**, 7–12 (1992)
29. Walter, G.G., Shen, X.A.: Sampling with prolate spheroidal wave functions. Sampl. Theor. Signal Image Process. **2**, 25–52 (2003)
30. Widom, H.: Asymptotic behavior of the eigenvalues of certain integral equations, II. Arch. Rational Mech. Anal. **17**, 215–229 (1964)
31. Xiao, H., Rokhlin, V., Yarvin, N.: Prolate spheroidal wavefunctions, quadrature and interpolation. Inverse Probl. **17**, 805–838 (2001)
32. Xu, W.Y., Chamzas, C.: On the periodic discrete prolate spheroidal sequences. SIAM J. Appl. Math. **44**, 1210–1217 (1984)

A Panorama of Sampling Theory

Kabe Moen, Hrvoje Šikić, Guido Weiss, and Edward Wilson

Abstract By a sampling function we mean a member φ of a vector space V of, preferably, continuous, $\mathbf{C}$-valued functions on a topological space X for which there is an orbit $G \cdot x_0$ of a countable abelian group G acting continuously on X, and each $f \in V$ is the sum of the terms $f(k \cdot x_0)\varphi(k \cdot x)$, $k \in G$. Such a recovery formula generalizes the well-known Shannon sampling formula. This chapter presents a general discussion of sampling theory and introduces several new classes of sampling functions $\varphi : \mathbf{R} \to \mathbf{C}$ for sampling sets of the form $\mathbf{Z} + x_0$. In Sect. 2 we discuss the very close connection between general convolution idempotents and sampling functions. In Sect. 3 we review the properties of the Zak transform and use it to construct a large family of continuous sampling functions $\varphi \in L^2(\mathbf{R})$ where $\{T_k\varphi : k \in \mathbf{Z}\}$ is a frame for the principal shift-invariant space $V_\varphi = \langle \varphi \rangle$ generated by φ. This family includes all band-limited sampling functions as well as all continuous sampling functions $\varphi \in V_\psi$, $\psi \in C_c(\mathbf{R})$. In Sect. 4 we look at a class of continuous functions ψ which do not generate (via the Z-transform) any square-integrable sampling functions and use the Laurent transform (or Z-transform) to show how ψ generates a possibly infinite family of non-square-integrable sampling functions. In Sect. 5 we sketch the manner in which purely algebraic tools lead to construction of a very large class of convolution idempotents and associated sampling functions that cannot be obtained by Zak or Laurent transform methods.

K. Moen (✉)
University of Alabama, Tuscaloosa, AL-35487, USA
e-mail: kmoen@as.ua.edu

H. Šikić
Univervisty of Zagreb, Bijenička 30, HR-1000, Zagreb, Croatia
e-mail: hsikic@math.hr

G. Weiss • E. Wilson
Washington University in St. Louis, St. Louis, MO-63130, USA
e-mail: guido@math.wustl.edu; enwilson@math.wustl.edu

T.D. Andrews et al. (eds.), *Excursions in Harmonic Analysis, Volume 1*,
Applied and Numerical Harmonic Analysis, DOI 10.1007/978-0-8176-8376-4_6,

Keywords Sampling functions • Sampling spaces • Exotic sampling • Convolution idempotent • Zak transform • Laurent transform

1 Introduction

We present a rather general description of sampling theory. We begin by describing this subject in a simple setting that will allow us to extend it to a very general setting and to treat many different aspects of it using several algebraic and analytic tools.

We begin by considering complex-valued functions on $\mathbf{R}$. Suppose $f : \mathbf{R} \to \mathbf{C}$ and there exists a (comparatively) small subset $S \subset \mathbf{R}$ such that all the values of f on $\mathbf{R}$ are completely determined by the values f has on S. We say that S is a *sampling subset* for f on $\mathbf{R}$. A well-known result of sampling is the following: if $f : \mathbf{R} \to \mathbf{C}$ is square integrable and satisfies the "band-limited" condition supp $\hat{f} \subset [-\frac{1}{2}, \frac{1}{2}]$, then all the values of f are determined by the values of f on the sampling set $\mathbf{Z}$ (the integers). This is expressed by the equality

$$f(x) = \sum_{k \in \mathbf{Z}} f(k) \operatorname{sinc}(x-k), \tag{1}$$

where $\operatorname{sinc} x = \frac{\sin \pi x}{\pi x}$, $x \in \mathbf{R}$. This is often called the *Shannon sampling theorem.* The proof of formula (1) is not hard and uses the very strong band-limited assumption that f is a very smooth function which vanishes at infinity and that the series in (1) converges absolutely and uniformly (it also converges in $L^p(\mathbf{R})$ for $p \geq 2$). In fact, the sinc function is the Fourier transform of the characteristic function $\chi_{[-\frac{1}{2},\frac{1}{2}]}$ (as is easily calculated).

Let us first make some observations about this result that will lead us to several extensions. The series (1) has the form

$$f(x) = \sum_{k \in \mathbf{Z}} c(k) \varphi(x-k), \tag{2}$$

where $\varphi \in L^2(\mathbf{R})$ and $c = \{c(k)\} = \{c(k) : k \in \mathbf{Z}\}$ is a complex valued sequence. In the case that $\varphi = \operatorname{sinc}$, the family $\mathscr{B} = \{\varphi_k(x) = (T_k\varphi)(x) = \varphi(x-k) : k \in \mathbf{Z}\}$ is an orthonormal family in $L^2(\mathbf{R})$ ($T_k\varphi$ is the inverse Fourier transform of $e^{-2\pi i k\xi}\chi_{[-\frac{1}{2},\frac{1}{2}]}(\xi)$). Moreover, the function $\varphi = \operatorname{sinc}$ satisfies the *Nyquist condition*

$$\varphi(k) = \begin{cases} 1 \text{ if } k = 0 \\ 0 \text{ if } k \neq 0 \end{cases}$$

when $k \in \mathbf{Z}$; from this, it follows that $c_k = f(k)$. From these observations, it follows easily that we can construct many different functions φ that produce "sampling formulae" of this type. For example, let $\varphi : \mathbf{R} \to \mathbf{C}$ be a compactly supported complex-valued function satisfying the Nyquist condition $\varphi(0) = 1$ and $\varphi(k) = 0$

if $k \neq 0$ for $k \in \mathbf{Z}$, and let V be the vector space of all linear combinations of the form (2) with $c = \{c(k)\}$ a complex sequence. When we let $x = k \in \mathbf{Z}$ it follows that $c(k) = f(k)$ and, thus, (2) becomes

$$f(x) = \sum_{k \in \mathbf{Z}} f(k)\varphi(x-k), \tag{3}$$

which is a sampling formula of the type we have been considering. Since we are assuming that φ is compactly supported, the sum in (3) is finite. We will show in Sect. 2 that there are many functions φ that are not compactly supported and associated vector spaces V for which (3) is well defined; moreover, φ need not satisfy the Nyquist condition.

Another observation about these vector spaces V associated with sampling functions is that they are *shift invariant*. This means that if $f \in V$, then the translation operators, T_k, satisfying $(T_k f)(x) = f(x-k)$, $k \in \mathbf{Z}$, map V into V. The reader can check that if f satisfies (3), so does $T_k f$ for any $k \in \mathbf{Z}$. This observation indicates that the *principal shift-invariant* subspaces V_φ generated by $\varphi \in L^2(\mathbf{R})$ must have some association with the notion of sampling we are considering. We have studied shift-invariant spaces, and their extensions, in the papers [3, 4]. Our point of view was to study the properties of the family $\mathscr{B} = \{\varphi_k = T_k\varphi : k \in \mathbf{Z}\}$ generating these spaces and how much they tell us about the functions in them. The definition of V_φ, also denoted $\langle \varphi \rangle$, for $\varphi \in L^2(\mathbf{R})$ is $\overline{\operatorname{span}\mathscr{B}}$, the closure in $L^2(\mathbf{R})$ of the collection of all finite linear combinations of the φ_k's. As we shall see, the "analysis" that inspired part of our sampling results involves some of the results we obtained in our study of principal shift-invariant spaces.

2 The Role of Convolution Idempotents for Obtaining Sampling Formulae

We have shown how the simple Nyquist condition on a function $\varphi : \mathbf{R} \to \mathbf{C}$ produces a shift-invariant space $V = V_\varphi$ such that each $f \in V$ satisfies a sampling formula (3). We shall now present a property involving the coefficients in the expansions of the spanning set $\mathscr{B} = \{T_k\varphi : k \in \mathbf{Z}\}$ that provides such sampling equalities. The integers, $\mathbf{Z}$, are a group with respect to addition; thus, we can consider convolutions $c * d$ of two sequences in $\mathbf{C}^{\mathbf{Z}}$:

$$(c*d)(j) = \sum_{k \in \mathbf{Z}} c(k)d(j-k).$$

We will always assume that the sums defining the convolutions are unconditionally convergent; moreover, we shall always discuss the meaning of these sums. Suppose we consider those sequences that are the coefficients of the Fourier series of 1-periodic functions in $L^2([0,1))$. We know that the convolution of the sequences $\{\hat{f}(j) : j \in \mathbf{Z}\}$ and $\{\hat{g}(j) : j \in \mathbf{Z}\}$ of $f, g \in L^2([0,1))$ are the Fourier coefficients of the product fg. This product may no longer be in $L^2([0,1))$, but it is in $L^1([0,1))$ and its Fourier series is well defined. Hence, $(\hat{f} * \hat{g})(j) = (fg)\hat{}(j)$ for all $j \in \mathbf{Z}$. Suppose, now, that Ω is a measurable subset of $[0,1)$ and $f = g = \chi_\Omega$. Then $fg = \chi_\Omega^2 = \chi_\Omega$ since $f(x) = 1$ or 0 (depending whether $x \in \Omega$ or $x \notin \Omega$). It follows that the sequence $\{c(k)\}$ defined by $c(k) = \widehat{\chi_\Omega}(k)$ satisfies

$$(c * c)(j) = \sum_{k \in \mathbf{Z}} c(k)c(j-k) = c(j) \tag{4}$$

for all $j \in \mathbf{Z}$. When a sequence $c = \{c(k)\}$ satisfies (4), we call it a *convolution idempotent* sequence. We have just shown that the collection of convolution idempotent sequences is infinite. Let us observe that the 1-periodic extension of $\chi_{[-\frac{1}{2},\frac{1}{2}]}$ has Fourier coefficients that are the trivial convolution idempotent sequence $c(k) = \delta_{0k}$ (the Kronecker delta). Of course, the values of $\varphi = \chi_{[-\frac{1}{2},\frac{1}{2}]}$ at the integers show that φ satisfies the Nyquist condition ($\varphi(k) = \delta_{0k}, k \in \mathbf{Z}$).

Suppose that $\varphi : \mathbf{R} \to \mathbf{C}$. For each $x \in [0,1)$ let φ_x be the $\mathbf{C}$-valued sequence on $\mathbf{Z}$ satisfying $\varphi_x(j) = \varphi(x+j)$. Thus, each $\varphi : \mathbf{R} \to \mathbf{C}$ produces a family of $\mathbf{C}$-sequences $\{\varphi_x : x \in [0,1)\}$. Similarly, if we have a family $\{c_x : x \in [0,1)\}$ of $\mathbf{C}$-sequences on $\mathbf{Z}$, indexed by $x \in [0,1)$, we obtain a function $\varphi : \mathbf{R} \to \mathbf{C}$ by setting $\varphi(x+j) = c_x(j)$ at $x + j \in \mathbf{R}$.

We can now use the convolution operation, introduced at the beginning of this section, to write the sampling formula (3) in the form

$$f_x = f_0 * \varphi_x \quad x \in [0,1). \tag{5}$$

That is,

$$f(x+j) = \sum_{k \in \mathbf{Z}} f(k)\varphi(x+j-k)$$

for each $x \in [0,1)$ and $j \in \mathbf{Z}$, which is precisely (3) with $x + j$ replacing x.

We are now ready to use the convolution idempotent sequences to construct sampling formulae.

Theorem 1. *Suppose c is a convolution idempotent on $\mathbf{Z}$. Then there exists a complex-valued function φ on $\mathbf{R}$ such that $\varphi_0 = c$ and a shift-invariant linear space of functions, V, such that*

$$f(x) = \sum_{k \in \mathbf{Z}} f(k)\varphi(x-k) \qquad f \in V \tag{6}$$

and the series converges absolutely for each $x \in \mathbf{R}$.

Proof. Let $\psi : \mathbf{R} \to \mathbf{C}$ be any compactly supported Nyquist function. Thus $\psi_0(k) = \delta_{0k}$ and ψ_x has at most finitely many nonzero terms for each $x \in [0,1)$. Let

$$\varphi(x) = \sum_{k \in \mathbf{Z}} c(k)\psi(x-k).$$

Thus $\varphi_x = c * \psi_x$ and we have $c = c * \psi_0 = \varphi_0$. Consequently,

$$\varphi_x = c * \psi_x = (c * c) * \psi_x = c * (c * \psi_x) = \varphi_0 * (c * \psi_x) = \varphi_0 * \varphi_x, \tag{7}$$

where the use of the associativity property is justified since it is easy to check that each of the sums in (7) converges absolutely. We thus have the equality $\varphi_x = \varphi_0 * \varphi_x$, from which we will obtain the desired sampling formula. Let $\mathbf{C}_c^{\mathbf{Z}}$ be the set of all sequences $b = \{b(k)\}$ such that $b(k) \neq 0$ for, at most, finitely many $k \in \mathbf{Z}$, and let V be the linear space of all f of the form

$$f(x) = \sum_k b(k)\varphi(x-k)$$

for some $b = \{b(k)\} \in \mathbf{C}_c^{\mathbf{Z}}$. Then, for $l \in \mathbf{Z}$, we have

$$f_x(l) = \sum_{k\in\mathbf{Z}} b(k)\varphi_x(l-k) = (b * \varphi_x)(l)$$

which, by (7), equals

$$[b * (\varphi_0 * \varphi_x)](l) = [(b * \varphi_0) * \varphi_x](l).$$

On the other hand, $(b * \varphi_0)(j) = \sum_{k\in\mathbf{Z}} b(k)\varphi(j-k) = f(j) = f_0(j)$. Thus $f_x(l) = [(b * \varphi_0) * \varphi_x](l) = (f_0 * \varphi_x)(l)$ which is equivalent to $f(x+l) = \sum_{k\in\mathbf{Z}} f(k)\varphi(x+l-k)$ for $x \in [0,1)$ and $l \in \mathbf{Z}$. This last inequality for $y = x+l \in \mathbf{R}$ is the desired sampling formula (6). □

Remark 1. (1) The sampling function φ we constructed in this proof can be lacking any smoothness: just let ψ be any "weird" nonmeasurable function. φ can also be a very smooth square-integrable function: let ψ be a C^∞ function supported in $[-1/4, 1/4]$ and c a convolution idempotent in $\ell^2(\mathbf{Z})$. We only required the absolute convergence of the series (7). Many who looked at this manuscript felt that "smoothness" is a natural property for the functions involved in sampling formulae.

(2) We stress that we showed that there are infinitely many sampling functions that are *not* of Nyquist type. There are several interesting questions that arise naturally from Theorem 1. We postpone many of these questions for the moment. Let us, however, describe now a much more general algebraic setting for what we have done that also applies to other results we will develop.

Let X be a nondiscrete topological space and $\mathbf{C}^X$ the vector space of all functions mapping X into $\mathbf{C}$. We assume that our sampling points have an algebraic structure arising from a finitely generated abelian group $\mathscr{G} = (\mathscr{G}, +)$ with identity 0, acting continuously on X.[1] The group action is denoted by $(l,x) \mapsto l \cdot x$ and satisfies $(k+l) \cdot x = k \cdot (l \cdot x)$ and $0 \cdot x = x$ for all $x \in X$. We assume that this action is "free" in the sense that if $l \cdot x = x$ for some $x \in X$, then $l = 0$ (the only element of $\mathscr{G}$ that has a fixed point is the identity). The spaces $X = \mathbf{R}^n$ and $\mathscr{G} = \mathbf{Z}^n$ with $k \cdot x = k + x$ are the most basic example. Given $x_0 \in X$, $\mathscr{G} \cdot x_0$ denotes the orbit of x_0 under $\mathscr{G}$.

We can now state what is meant by a sampling space and the analog of Theorem 1 in this setting. We say that a topological vector space V which is a vector subspace of $\mathbf{C}^X$ is a *sampling space* if and only if there exists a triple $(\mathscr{G}, \varphi, \mathscr{C})$ satisfying the following properties:

1. $\mathscr{G}$ is a finitely generated abelian group with a continuous free action on X.
2. $\varphi \in V$ and the functions $(T_k\varphi)(x) = \varphi((-k) \cdot x)$ are in V for each $k \in \mathscr{G}$.
3. $\mathscr{C}$ is a shift-invariant vector subspace of $\mathbf{C}^{\mathscr{G}} = \{c : \mathscr{G} \to \mathbf{C}\} = \{c = \{c(k)\} : c(k) \in \mathbf{C}, k \in \mathscr{G}\}$, and there exists $x_0 \in X$ such that for $f \in V$, $c_f = \{f(k \cdot x_0)\}_{k \in \mathscr{G}} \in \mathscr{C}$.
4. For each $c \in \mathscr{C}$, the sum
$$\sum_{k \in \mathscr{G}} c(k)(T_k\varphi)(x)$$
converges absolutely for each $x \in X$ to a function $f_c \in V$, and, for each $f \in V$, $f = f_c = f_{c_f}$:
$$f(x) = \sum_{k \in \mathscr{G}} f(k \cdot x_0)(T_k\varphi)(x), \tag{8}$$
$x \in X$.

We emphasize: the function φ is called the *sampling function* for V, $\mathscr{C}$ is the *coefficient space*, $\mathscr{G} \cdot x_0$ is the sampling set, and (8) is the sampling formula for V. Observe that the set $\{T_k\varphi : k \in \mathscr{G}\}$ is a set of uniqueness for V in the sense that the map $f \mapsto c_f$ is a bijection from V to $\mathscr{C}$ and provides an explicit recovery of each $f \in V$ via (8).

Let us finish this section with the following easy example: Let $\mathscr{G} = \mathbf{Z}^n$ acting on $X = \{x = (x_1, \ldots, x_n) \in \mathbf{R}^n : x_i \neq 0, i = 1, \ldots, n\}$ with the group action

$$k \cdot x = \begin{pmatrix} e^{c_1k_1} & 0 & \cdots & 0 \\ 0 & e^{c_2k_2} & \cdots & 0 \\ \vdots & & \ddots & 0 \\ 0 & 0 & \cdots & e^{c_nk_n} \end{pmatrix} \begin{pmatrix} x_1 \\ x_2 \\ \vdots \\ x_n \end{pmatrix} = (e^{c_1k_1}x_1, \ldots, e^{c_nk_n}x_n),$$

where $k = (k_1, \ldots, k_n) \in \mathbf{Z}^n$ and $c_1, \ldots, c_n$ are fixed non-zero real numbers. It is routine to construct a large collection of more complicated examples where the action of $\mathscr{G}$ is not simultaneously differentiable and $\mathscr{G}$ may have a non-trivial finite subgroup.

---

[1] Think of $\mathbf{Z}$ acting on $\mathbf{R}$ or more generally $\mathbf{Z}^n$ acting on $\mathbf{R}^n$.

3 The Use of the Zak Transform for Obtaining Sampling Formulae

We have seen that the notion of sampling requires that the functions we deal with have well-defined pointwise values. Thus, we consider that many of the functions we deal with have some form of continuity. In particular, we introduce the following definition:

Definition 1. A function $\psi : \mathbf{R} \to \mathbf{C}$ is $\ell^2(\mathbf{Z})$-*continuous* if and only if the map $x \mapsto \psi_x = \{\psi(x+k)\}_{k\in\mathbf{Z}}$ is continuous from $\mathbf{R}$ into $\ell^2(\mathbf{Z})$. Specifically, this means that given $\varepsilon > 0$ and $x \in \mathbf{R}$ there exists a $\delta > 0$ such that $|x-y| < \delta$ implies

$$\left(\sum_{k\in\mathbf{Z}} |\psi(x+k) - \psi(y+k)|^2\right)^{1/2} = \|\psi_x - \psi_y\|_{\ell^2(\mathbf{Z})} < \varepsilon.$$

There is an equivalent definition of this notion. Suppose $\psi : \mathbf{R} \to \mathbf{C}$ is $\ell^2(\mathbf{Z})$-continuous. Then the function

$$a(x,\xi) = \sum_{k\in\mathbf{Z}} \psi(x+k)\mathrm{e}^{-2\pi ik\xi}$$

satisfies

$$\int_0^1 |a(x,\xi)|^2 \mathrm{d}\xi = \|\psi_x\|^2_{\ell^2(\mathbf{Z})} = \sum_{k\in\mathbf{Z}} |\psi(x+k)|^2 < \infty,$$

and, moreover, $a(x,\xi) - a(y,\xi) = \sum_{k\in\mathbf{Z}}[\psi(x+k) - \psi(y+k)]\mathrm{e}^{-2\pi ik\xi}$. Thus

$$\int_0^1 |a(x,\xi) - a(y,\xi)|^2 \mathrm{d}\xi = \|\psi_x - \psi_y\|^2_{\ell^2(\mathbf{Z})},$$

and since ψ is $\ell^2(\mathbf{Z})$-continuous, the last expression is less than ε^2 if $|x-y| < \delta$. This means that the map

$$\varphi(x) \mapsto a(x,\cdot) \in L^2(\mathbf{T}) = \left\{a(x,\cdot) : \int_0^1 |a(x,\xi)|^2 \mathrm{d}\xi = \|a(x,\cdot)\|^2_{L^2(\mathbf{T})} < \infty\right\}$$

satisfies the "$L^2(\mathbf{T})$-continuity condition":

$$\|a(x,\cdot) - a(y,\cdot)\|_{L^2(\mathbf{T})} < \varepsilon \text{ if } |x-y| < \delta. \tag{9}$$

We will make use of the Zak transform, as defined and developed in [5] (see also [6]), to show how the functions ψ we just described can be used to obtain sampling formulae. Formally, the Zak transform is the operator Z acting on $L^2(\mathbf{R})$ by the equality

$$(Z\psi)(x,\xi) = \sum_{k\in\mathbf{Z}} \psi(x+k)\mathrm{e}^{-2\pi ik\xi} \equiv a(x,\xi) \tag{10}$$

for each $\psi \in L^2(\mathbf{T})$. Since

$$\int_0^1 \sum_{k\in\mathbf{Z}} |\psi(x+k)|^2 \,\mathrm{d}x = \sum_{k\in\mathbf{Z}} \int_k^{k+1} |\psi(x)|^2 \,\mathrm{d}x = \int_{-\infty}^{\infty} |\psi(x)|^2 \,\mathrm{d}x,$$

we have $\sum_{k\in\mathbf{Z}} |\psi(x+k)|^2 < \infty$ a.e. This allows us to consider (for a.e. $x \in \mathbf{R}$) the Fourier series of the function $a(x,\cdot) \in L^2(\mathbf{T}) = L^2([0,1))$ with Fourier coefficients $\psi_x(-k) = \psi(x-k)$, $k \in \mathbf{Z}$. As is shown in [5] the functions $a(x,\xi)$ (the images of Z) are 1-periodic in ξ and satisfy

$$a(x+l,\xi) = \mathrm{e}^{2\pi i l\xi} a(x,\xi) \tag{11}$$

for each $l \in \mathbf{Z}$, $x \in \mathbf{R}$. Moreover,

$$\int\int_{\mathbf{T}^2} |(Z\psi)(x,\xi)|^2 \,\mathrm{d}x\mathrm{d}\xi = \int_0^1 \int_0^1 |a(x,\xi)|^2 \,\mathrm{d}x\mathrm{d}\xi = \|\psi\|^2_{L^2(\mathbf{R})} \tag{12}$$

for each $\psi \in L^2(\mathbf{R})$. In fact, if $a(x,\xi)$ is any square-integrable function on $\mathbf{T}^2 = [0,1) \times [0,1)$, then there exists $\psi \in L^2(\mathbf{R})$ such that $(Z\psi)(x,\xi) = a(x,\xi)$. If we extend $a(x,\xi)$ to all $(x,\xi) \in \mathbf{R}^2$ by 1-periodicity in ξ and by (11) in x we obtain a space $\mathscr{M}$ of functions on $a : \mathbf{R}^2 \to \mathbf{C}$, with norm

$$\|a\|_{\mathscr{M}} = \|a\|_{L^2(\mathbf{T}^2)} = \left(\int_0^1 \int_0^1 |a(x,\xi)|^2 \,\mathrm{d}x\mathrm{d}\xi\right)^{1/2}$$

that is the isometric image of $L^2(\mathbf{R})$ under the operator Z; that is, $\|Z\psi\|_{\mathscr{M}} = \|\psi\|_{L^2(\mathbf{R})}$ for all $\psi \in L^2(\mathbf{R})$.

There exists a "companion" space to $\mathscr{M}$, the space $\tilde{\mathscr{M}}$. The definition of $\tilde{\mathscr{M}}$ is very much like the definition of $\mathscr{M}$; essentially, it involves the "interchange" of the roles played by x and ξ. We begin by the definition of the operator $\tilde{Z}$ on $L^2(\mathbf{R})$. If $\theta \in L^2(\mathbf{R})$ we define $\tilde{Z}\theta$ by letting

$$(\tilde{Z}\theta)(x,\xi) = \sum_{k\in\mathbf{Z}} \theta(\xi+k)\mathrm{e}^{2\pi i kx} = \tilde{a}(x,\xi).$$

The function $\tilde{a}(x,\xi)$ is 1-periodic in x, and, for almost every ξ and $l \in \mathbf{Z}$,

$$\tilde{a}(x,\xi+l) = \mathrm{e}^{-2\pi i lx} \tilde{a}(x,\xi) \tag{13}$$

[compare with (11)].

Let us extend any function $\tilde{a}(x,\xi) \in L^2(\mathbf{T}^2)$ to a function $\tilde{a}$ defined on $\mathbf{R}^2$ which is 1-periodic in x and satisfies (13) in ξ, to obtain a member of the space $\tilde{\mathscr{M}}$; furthermore let the norm of this space be $\|\tilde{a}\|_{\tilde{\mathscr{M}}} = \|\tilde{a}\|_{L^2(\mathbf{T}^2)}$. It is easy to check (and this is done in [5]) that $\tilde{Z}$ maps $L^2(\mathbf{R})$ isometrically onto $\tilde{\mathscr{M}}$. Furthermore, $\tilde{\mathscr{M}} = U\mathscr{M}$, where U is the unitary operator defined by $(Ua)(x,\xi) = \mathrm{e}^{-2\pi i x\xi} a(x,\xi)$, for $a \in \mathscr{M}$. It follows that Z^{-1} and $\tilde{Z}^{-1}$ must exist and are, simply,

$$(Z^{-1}a)(x) = \int_0^1 a(x,\xi)\,\mathrm{d}\xi, \quad (\tilde{Z}^{-1}a)(\xi) = \int_0^1 \tilde{a}(x,\xi)\,\mathrm{d}x. \tag{14}$$

Another simple calculation gives us

$$(UZ\psi)(x,\xi) = \sum_{k\in\mathbf{Z}} \psi(x+k)\mathrm{e}^{-2\pi\mathrm{i}(k+x)\xi} = \tilde{a}(x,\xi). \tag{15}$$

Combining Eqs. (14) and (15), therefore, we obtain

$$(\tilde{Z}^{-1}UZ\psi)(\xi) = \int_0^1 \sum_{k\in\mathbf{Z}} \psi(x+k)\mathrm{e}^{-2\pi\mathrm{i}(x+k)\xi}\,\mathrm{d}x = \int_{-\infty}^{\infty} \psi(x)\mathrm{e}^{-2\pi\mathrm{i}x\xi}\,\mathrm{d}x = \hat{\psi}(\xi).$$

This shows:

Theorem 2. *$\mathscr{F} = \tilde{Z}^{-1}UZ$ is the unique extension of the Fourier transform on $L^1(\mathbf{R})\cap L^2(\mathbf{R})$ to a unitary operator on $L^2(\mathbf{R})$.*

All of these assertions are proved rigorously in [5] (the appropriate indications are when an equality should be interpreted in the a.e. sense). We remind our reader(s) that the continuity assumptions on the functions ψ we are using permit us to "ignore the a.e. sense."

In [4] (and, more generally, in [3]) various properties of the generating system $\mathscr{B}_\psi = \{\psi(\cdot - k) = T_k\psi : k \in \mathbf{Z}\}$ of $\langle\psi\rangle = \overline{\operatorname{span}\mathscr{B}}$ (the closure in the $L^2(\mathbf{R})$-norm)[2] are shown to be equivalent to properties of the "weight" $p_\psi(\xi) = \sum_{k\in\mathbf{Z}} |\hat{\psi}(\xi+k)|^2$ for the space

$$L^2(\mathbf{T},p_\psi) = \left\{1\text{ -periodic functions}, m(\xi), \text{ satisfying} \int_0^1 |m(\xi)|^2 p_\psi(\xi)\,\mathrm{d}\xi < \infty\right\}.$$

For example, we showed that $\mathscr{B}_\psi$ is a frame for $\langle\psi\rangle$ if and only if there exist constants $0 < A \le B < \infty$ such that

$$A\chi_\Omega(\xi) \le p_\psi(\xi) \le B\chi_\Omega(\xi) \quad \text{a.e.} \tag{16}$$

for all $\xi \in [0,1)$, where $\Omega = \operatorname{supp} p_\psi$. It is an easy consequence of (15) and Theorem 2 that $|(UZ\psi)(x,\xi)| = |(\tilde{Z}\hat{\psi})(x,\xi)|$. Thus

$$p_\psi(\xi) = \int_0^1 |(Z\psi)(x,\xi)|^2\,\mathrm{d}x. \tag{17}$$

Let us now turn our attention to the use of the Zak transform for the construction of a large family of $\ell^2(\mathbf{Z})$-continuous sampling functions on translation invariant Hilbert spaces of continuous, square-integrable functions with sample sets of

[2] $\langle\psi\rangle = V_\psi$ is the principal shift-invariant space introduced at the end of Sect. 1.

the form $\mathbf{Z}+x_0$, $x_0 \in \mathbf{R}$ (for short, $(\mathbf{Z}+x_0)$-sampling functions). We begin by establishing the basic properties of $\ell^2(\mathbf{Z})$-continuous functions $\psi : \mathbf{R} \to \mathbf{C}$:

(1) Since $x \mapsto \|\psi_x\|_{\ell^2(\mathbf{Z})}$ is continuous and 1-periodic, $M_\psi = \max_{x\in[0,1)} \|\psi_x\|_{\ell^2(\mathbf{Z})} < \infty$. Hence, $\psi \in L^2(\mathbf{R})$ with $\|\psi\|_{L^2(\mathbf{R})} \le M_\psi$, and, by (12),

$$M_\psi = \max_{x\in[0,1)} \|(Z\psi)(x,\cdot)\|_{L^2(\mathbf{T})}.$$

(2) For $b \in \ell^2(\mathbf{Z})$, let $B_\psi b = b *_{\mathbf{Z}} \psi$ be the function on $\mathbf{R}$ defined by

$$(B_\psi b)(x) = (b * \psi_x)(0) = \sum_{j\in\mathbf{Z}} b(j) T_j \psi(x). \tag{18}$$

$B_\psi b$ is bounded and continuous since for $x, y \in \mathbf{R}$, the Schwartz inequality for $\ell^2(\mathbf{Z})$ implies $|(B_\psi b)(x)| \le M_\psi \|b\|_{\ell^2(\mathbf{Z})}$ and

$$|(B_\psi b)(x) - |(B_\psi b)(y)| \le \|b\|_{\ell^2(\mathbf{Z})} \|\psi_x - \psi_y\|_{\ell^2(\mathbf{Z})}.$$

It follows that (18) implies that $W_\psi = B_\psi(\ell^2(\mathbf{Z}))$ is a translation-invariant linear subspace of the Banach space $(BC(\mathbf{R}), \|\cdot\|_\infty)$ and M_ψ is the operator norm of the bounded linear operator B_ψ from $(\ell^2(\mathbf{Z}), \|\cdot\|_{\ell^2(\mathbf{Z})})$ onto $(W_\psi, \|\cdot\|_\infty)$.

(3) If $m_b = \sum_{k\in\mathbf{Z}} b(k) e_{-k}$, then (10) implies that

$$Z(B_\psi b)(x,\cdot) = m_b \cdot (Z\psi)(x,\cdot). \tag{19}$$

When $\|p_\psi\|_\infty < \infty$, it follows from (19) and (12) that $B_\psi b \in L^2(\mathbf{R})$ and B_ψ is a bounded linear operator from $(\ell^2(\mathbf{Z}), \|\cdot\|_{\ell^2(\mathbf{Z})})$ onto $(W_\psi, \|\cdot\|_{L^2(\mathbf{R})}) \subset (BC(\mathbf{R}) \cap L^2(\mathbf{R}), \|\cdot\|_{L^2(\mathbf{R})})$ with operator norm $\|p_\psi\|_\infty$ and kernel $K_\psi = \{b \in \ell^2(\mathbf{Z}) : m_b p_\psi = 0 \text{ a.e.}\}$. With $\Omega = \operatorname{supp} p_\psi$ and $K_\psi^\perp$ the orthogonal complement of K_ψ in $\ell^2(\mathbf{Z})$, the one to one map $B_\psi|_{K_\psi^\perp}$ from $(K_\psi^\perp, \|\cdot\|_{\ell^2(\mathbf{Z})})$ onto $(W_\psi, \|\cdot\|_{L^2(\mathbf{R})})$ has a bounded inverse precisely when

$$0 < \operatorname*{essinf}_{\xi\in\Omega} p_\psi(\xi) = \frac{1}{\|1/p_\psi\|_{L^\infty(\Omega)}}.$$

Consider this condition $0 < \operatorname{essinf}_\Omega p_\psi \le \|p_\psi\|_\infty < \infty$ together with (16); we see that, in this case, we have that $\mathscr{B}_\psi = \{T_k\psi : k \in \mathbf{Z}\}$ is a frame for $V_\psi = \langle\psi\rangle$. Thus, W_ψ is a Hilbert subspace of $(BC(\mathbf{R}) \cap L^2(\mathbf{R}), \|\cdot\|_{L^2(\mathbf{R})})$, $\mathscr{B}_\psi$ is a frame for $(W_\psi, \|\cdot\|_{L^2(\mathbf{R})})$, and this last space is a continuous version of $V_\psi = \langle\psi\rangle$. (The correspondence between $f \in W_\psi$ and the a.e. equivalence class determined by $f \in L^2(\mathbf{R})$ gives us an isometry between $(W_\psi, \|\cdot\|_{L^2(\mathbf{R})})$ and $(V_\psi, \|\cdot\|_{L^2(\mathbf{R})})$).

Theorem 3. *Suppose $\psi : \mathbf{R} \to \mathbf{C}$ is $\ell^2(\mathbf{Z})$-continuous, $\mathscr{B}_\psi$ is a frame for V_ψ, $\Omega = \operatorname{supp} p_\psi$, and for some $x_0 \in \mathbf{R}$, both $\|Z\psi(x_0,\cdot)\|_\infty$ and $\left\|\frac{\chi_\Omega}{Z\psi(x_0,\cdot)}\right\|_\infty$ are finite. Then, with $m = \chi_\Omega / Z\psi(x_0,\cdot)$, the well-defined pointwise function $\varphi = Z^{-1}(m \cdot (Z\psi))$ is $\ell^2(\mathbf{Z})$-continuous, B_φ is a frame for $W_\varphi = W_\psi$, and φ is a $(x_0+\mathbf{Z})$-sampling function for the sample space W_φ.*

Proof. For $x, y \in \mathbf{R}$, we have

$$\begin{aligned}\|Z\varphi(x,\cdot) - Z\varphi(y,\cdot)\|_{L^2(\mathbf{T})} &= \|m\,(Z\psi(x,\cdot) - Z\psi(y,\cdot))\|_{L^2(\mathbf{T})} \\ &\le \|m\|_\infty \|Z\psi(x,\cdot) - Z\psi(y,\cdot)\|_{L^2(\mathbf{T})}.\end{aligned}$$

Since $\|m\|_\infty < \infty$ and $Z\psi$ is $L^2(\mathbf{T})$-continuous, $Z\varphi$ is $L^2(\mathbf{T})$-continuous, and, hence, φ is $\ell^2(\mathbf{Z})$-continuous. Because both m and p_ψ are essentially bounded on $\mathbf{T}$ and essentially bounded away from 0 on Ω, the same properties hold for $p_\varphi = m p_\psi$. By (16), B_φ is a frame for $W_\varphi = W_\psi$ as well as a frame for $V_\varphi = V_\psi$. Finally, $Z\varphi(x_0,\cdot) = m(\cdot) Z\psi(x_0,\cdot) = \chi_\Omega$ a.e.; thus, φ_{x_0} is a convolution idempotent in $\mathbf{C}^{\mathbf{Z}}$ with $b * \varphi_{x_0} = b$ for each $b \in K_\varphi^\perp = K_\psi^\perp = \{b \in \ell^2(\mathbf{Z}) : m_b = m_b \chi_\Omega \text{ a.e.}\}$. Since $W_\varphi = \{b *_{\mathbf{Z}} \varphi : b \in K_\varphi^\perp\}$, it follows that φ is a $(x_0 + \mathbf{Z})$-sampling function for the Hilbert space $W_\varphi \subseteq BC(\mathbf{R}) \cap L^2(\mathbf{R})$. □

Remark. Much of the extensive literature on $\mathbf{Z} + x_0$ sampling functions focuses on band-limited sampling functions and sampling functions in V_ψ where $\psi \in C_c(\mathbf{R})$. It is easy to see that every band-limited $\psi = \check{F}$ ($F \in L^2(\mathbf{R})$ with compact support) is $\ell^2(\mathbf{Z})$-continuous and the band-limited function $\psi^\circ = Z^{-1}\left(\frac{\chi_\Omega F}{\sqrt{p_\psi}}\right)$ yields a Parseval frame for V_ψ. In this way Theorem 3 recaptures the classification in [8] of band-limited sampling functions. When $\psi \in C_c(\mathbf{R})$, $Z\psi(x,\cdot)$ and p_ψ are trigonometric polynomials with $\|p_\psi\|_\infty = \max |p_\psi| < \infty$, and B_ψ is a frame for V_ψ when p_ψ is nonvanishing. As observed in [2,7], V_ψ contains a $(\mathbf{Z} + x_0)$-sampling function when $Z\psi(x_0,\cdot)$ is nonvanishing; obviously these sampling functions are special cases of those described in Theorem 3.

The class of $\ell^2(\mathbf{Z})$-continuous functions is much larger than the space of linear combinations of a band-limited function and one in $C_c(\mathbf{R})$; in particular, it includes all continuous functions $\psi : \mathbf{R} \to \mathbf{C}$ satisfying the mild decay condition

$$|\psi(x)| \le O(|x|^{-\frac{1}{2}-\varepsilon}) \quad \text{as } |x| \to \infty$$

for some $\varepsilon > 0$. For a general $\ell^2(\mathbf{Z})$-continuous function ψ, however, it is difficult to determine whether or not the Parseval frame generator $\psi^\circ = Z^{-1}\left(\frac{\chi_\Omega Z\psi}{\sqrt{p_\psi}}\right)$ for V_ψ is also $\ell^2(\mathbf{Z})$-continuous and whether or not $Z\psi(x_0,\cdot)$ satisfies the hypotheses of Theorem 3. Matters improve considerably by passing to functions which, in an obvious sense, are $\ell^1(\mathbf{Z})$-continuous (e.g., continuous and $|\psi(x)| = O(|x|^{-1-\varepsilon})$ for some $\varepsilon > 0$ as $|x| \to \infty$) and, hence, also $\ell^2(\mathbf{Z})$-continuous. Then $Z\psi(x,\xi)$ is jointly continuous in (x,ξ) and, as observed above, $\|p_\psi\|_\infty$, and Theorem 3 applies when we have some x_0 for which $Z\psi(x_0,\cdot)$ is nonvanishing. As we have emphasized, the rolê of Theorem 3 is to provide a large class of square-integrable sampling functions φ which are not only continuous, but $\mathscr{B}_\varphi$ is a frame for the continuous version of V_ψ or V_φ.

Several authors [1] have investigated measure theoretic sampling functions arising from a generator ψ for a principle shift-invariant space V_ψ by letting $\varphi = Z^{-1}\left(\frac{\chi_\Omega Z\psi}{Z\psi(x_0,\cdot)}\right)$ with $\chi_\Omega / Z\psi(x_0,\cdot) \in L^2(\mathbf{T})$ and ψ modified on a set of measure 0 in order to make φ pointwise well defined with $\varphi_x \in \ell^2(\mathbf{Z})$ for all x. But, without additional assumptions on ψ, one cannot make a.e. modifications of ψ in order that φ is continuous and φ is only certain to be a sampling function for $\operatorname{span}\{T_k\varphi : k \in \mathbf{Z}\}$.

4 Sampling via the Laurent Transform

As observed in [2], there are continuous compactly supported functions $\psi : \mathbf{R} \to \mathbf{C}$ such that $(Z\psi)(x_0,\cdot)$ has at least one zero for each $x_0 \in \mathbf{R}$; as a result, the Zak transform method for producing sampling functions φ from such a ψ fails. In this section, we will show that an appropriate use of the Laurent transform shows that such ψ's can produce a countable family of non-square-integrable sampling functions.

We will construct sampling functions that are very different from the ones we have obtained. In particular, some of these sampling functions may have exponential growth at infinity. In order to do this, we recall some facts about Laurent series and their coefficients.

Definition 2. Given a sequence $c = \{c(k) : k \in \mathbf{Z}\}$ set,

$$R_+(c) = \limsup_{\substack{k\to\infty \\ k\geq 0}} |c(k)|^{1/k} \quad \text{and} \quad R_-(c) = \limsup_{\substack{k\to\infty \\ k\geq 0}} |c(-k)|^{1/k}.$$

When $R_+(c)R_-(c) < 1$ we say that c is a *Laurent series coefficient sequence*. We denote this by $c \in \mathscr{L}$.

It is often useful to view a sequence $c \in \mathscr{L}$ as two sequences c_+ and c_- where $c_+ = (c(k))_{k\geq 0}$ and $c_- = (c(k))_{k<0}$ so that $R_+(c)$ and $R_-(c)$ depend only on c_+ and c_- respectively. Given $c \in \mathscr{L}$ we define the Laurent series

$$m_c(z) = \sum_{k\in\mathbf{Z}} c(k) z^k \quad z \in A_c,$$

where A_c denotes the annulus

$$A_c = A\left(0, R_-(c), \frac{1}{R_+(c)}\right) = \left\{z \in \mathbf{C} : R_-(c) < |z| < \frac{1}{R_+(c)}\right\}.$$

A_c is the maximum open annulus about zero for which the Laurent series for $m_{\mathbf{c}}$ converges to a holomorphic function. If $0 \le r < R \le \infty$ and m is a holomorphic function on $A = A(0,r,R)$, then there is a unique $c \in \mathscr{L}$ for which $A \subseteq A_c$ and $m = m_c$ on A. Moreover, for each $S \in (r,R)$ and $k \in \mathbf{Z}$, we have, using Cauchy's integral formula,

$$c(k) = \frac{1}{2\pi i} \int_{|z|=S} \frac{m(z)}{z^{k+1}}\,\mathrm{d}z = S^{-k} \int_0^1 m(Se^{2\pi i\xi})\mathrm{e}^{-2\pi \mathrm{i}k\xi}\mathrm{d}\xi. \tag{20}$$

This shows that, given $0 \le r < R \le \infty$, the map $c \mapsto m_c$ is a vector space isomorphism from

$$\mathscr{L}_{r,R} = \{c \in \mathscr{L} : R_-(c) \le r, R_+(c) \le 1/R\}.$$

onto

$$\mathrm{Hol}(A(0,r,R)) = \{m : m \text{ is holomorphic on } A(0,r,R)\}.$$

For $a, b \in \mathscr{L}_{r,R}$ we have $m_a(z)m_b(z) = m_c(z)$, where $c = a * b$ and

$$c(k) = \sum_{j \in \mathbf{Z}} a(j)b(k-j), \qquad k \in \mathbf{Z};$$

this last sum converges unconditionally. Thus, the isomorphism $c \mapsto m_c$ carries the algebraic structure of convolution to pointwise multiplication.

We are now ready to introduce our main tool: the Laurent transform (also called the Z-transform). Given a function $\psi : \mathbf{R} \to \mathbf{C}$ and $x \in [0,1)$, if $\psi_x \in \mathscr{L}$, we define

$$(L\psi)(x,z) = \sum_{k \in \mathbf{Z}} \psi(x+k)z^k \qquad R_-(\psi_x) < |z| < \frac{1}{R_+(\psi_x)}. \tag{21}$$

We will be concerned with functions $\psi \in C(\mathbf{R})$ such that $\psi_x \in \mathscr{L}_{r,R}$ for some $0 \le r < R \le \infty$ and for all $x \in [0,1)$. Moreover, we will assume that $L\psi : \mathbf{R} \times A(0,r,R) \to \mathbf{C}$ is a continuous function. We will refer to such functions ψ as class $\mathscr{L}_{r,R}$. Let us make a few useful calculations. Suppose, first, that ψ is of class $\mathscr{L}_{r,R}$ and $c \in \mathscr{L}_{r,R}$; then

$$\varphi(x+k) = \sum_{k \in \mathbf{Z}} c(j)\psi(x+k-j) = (c * \psi_x)(k)$$

is well defined (absolutely convergent) for each $(x,k) \in [0,1) \times \mathbf{Z}$ since it is the kth coefficient of the Laurent series expansion for the function $m_c(z)(L\varphi)(x,z)$, $(x,k) \in [0,1) \times \mathbf{Z}$. That is,

$$(L\varphi)(x,z) = m_c(z)(L\psi)(x,z). \tag{22}$$

Moreover, if ψ is of class $\mathscr{L}_{r,R}$, with $r < 1 < R$, then for $|z| = 1$, $(L\psi)(x,z) = (Z\psi)(x,\xi)$, where $z = \mathrm{e}^{-2\pi \mathrm{i}\xi}$.

It is clear that if ψ has a Laurent transform at 0, we have that $(L\psi)(0,\cdot) \equiv 1$ if and only if ψ satisfies the Nyquist condition. This observation allows us to construct sampling functions that are different from those constructed by the Zak transform. In particular, we are still able to find sampling functions when the Zak transform methods fail. In order to state the main result of this section, we make the following four comments (1–4) below.

When ψ is of type $\mathscr{L}_{r,R}$ and $c \in \mathscr{L}_{r',R'}$ for $r \le r' < R' \le R$ we define $c *_{\mathbf{Z}} \psi$ to be the function φ of type $\mathscr{L}_{r',R'}$ for which $\varphi_x = c * \psi_x$ for each $x \in [0,1)$. Thus, $(L\varphi)(x,z) = m_c(z)(L\psi)(x,z)$, and

$$\mathscr{L}_{r',R'} *_{\mathbf{Z}} \psi = \{c *_{\mathbf{Z}} \psi : c \in \mathscr{L}_{r',R'}\}$$

is the shift-invariant space generated by ψ and the convolution algebra $\mathscr{L}_{r',R'} \supset \mathscr{L}_{r,R} \supset \mathscr{L}_{0,\infty} \supset \mathbf{C}_c^{\mathbf{Z}}$:

1. If $(L\psi)(0,\cdot)$ is nonvanishing on the annulus $A = A(0,r,R)$ and $c \in \mathscr{L}_{r,R}$ for which $m_c(z) = \frac{1}{(L\psi)(0,z)}$ for each $z \in A$, $\varphi \in c *_{\mathbf{Z}} \psi$ is a sampling function for $\mathscr{L}_{r,R} *_{\mathbf{Z}} \psi$ since $(L\varphi)(0,\cdot) \equiv 1$ on A; this means that $\varphi_0 = \delta_0$ and φ is of Nyquist type. Moreover, $\mathscr{L}_{r,R} *_{\mathbf{Z}} \varphi = \mathscr{L}_{r,R} *_{\mathbf{Z}} \psi$.
2. Similarly, when $(L\psi)(0,\cdot)$ is nonvanishing on a subannulus $A' = A(0,r',R')$ of A, we obtain a sampling function $\varphi_{A'}$ for $\mathscr{L}_{r',R'} *_{\mathbf{Z}} \psi$ with $\mathscr{L}_{r',R'} *_{\mathbf{Z}} \varphi_{A'} = \mathscr{L}_{r',R'} *_{\mathbf{Z}} \psi$.
3. If $(L\psi)(0,\cdot)$ is not identically zero (i.e., $\psi_0 \neq 0$), then $(L\psi)(0,\cdot)$ has only finitely many zeros on compact subsets of A. We then have a countable (possibly finite) collection of mutually disjoint subannuli, $A_i = A(0,r_i,R_i)$, $i \in I$, of A for which $(L\psi)(0,\cdot)$ is nonvanishing on A_i but has at least one zero on
$$\partial A_i \cap A = \{z \in A : |z| = r_i \text{ or } |z| = R_i\}.$$
Also, $\overline{A} = \bigcup_{i\in I} \overline{A_i}$. We then obtain a family $(\varphi_i)_{i\in I} = (\varphi_{A_i})_{i\in I}$ of sampling functions φ_i for $\mathscr{L}_{r_i,R_i} *_{\mathbf{Z}} \psi = \mathscr{L}_{r_i,R_i} *_{\mathbf{Z}} \varphi_i \supset \mathscr{L}_{r,R} *_{\mathbf{Z}} \psi$. Note that the φ_i's are distinct since $\varphi_i = c_i *_{\mathbf{Z}} \psi$ where
$$m_{c_i} = \frac{\chi_{A_i}}{(L\psi)(0,\cdot)}$$
has at least one pole on ∂A_i. For $i \neq j$ $c_j * c_i^{-1}$ is not defined. We can pass, however, from φ_i to φ_j in two steps by $\varphi_j = c_j *_{\mathbf{Z}} \psi = c_j *_{\mathbf{Z}} (c_i^{-1} *_{\mathbf{Z}} \varphi_i)$.
4. We can replace 0 in (1), (2), or (3) by any $x_0 \in (0,1)$ to obtain, from ψ, various sampling functions for the sample set $x_0 + \mathbf{Z}$ when $\psi_{x_0} \neq 0$. Observe that if ψ is of type $\mathscr{L}_{r,R}$, with $r < 1 < R$, then continuity of $(L\psi)$ on $\mathbf{R} \times A(0,r,R)$ implies that ψ is $\ell^2(\mathbf{Z})$ continuous. Clearly, for $z = \mathrm{e}^{-2\pi i\xi}$ on the unit circle about 0, $(L\psi)(x,z) = (Z\psi)(x,\xi)$. In this sense, $L\psi$ is an analytic continuation of $Z\psi$. Trivially, when $\psi \in C_c(\mathbf{R})$, ψ is of type $\mathscr{L}_{0,\infty}$.

Theorem 4. *Suppose ψ is of type $\mathscr{L}_{r,R}$ with $0 \le r < 1 < R \le \infty$ and $(Z\psi)(x_0,\cdot)$, for each $x_0 \in \mathbf{R}$, has at least one zero in the other variable. Then, for each x_0 such that $\psi_{x_0} \neq 0$, we have annuli $A^+ = A(0,1,R')$ and $A^- = A(0,r',1)$ with $r \le r' < 1 < R' \le$*

R defining sampling functions φ^+ *and* φ^- *for the sample set* $x_0+\mathbf{Z}$ *with sampling spaces* $V^+ = \mathscr{L}_{1,R'} *_{\mathbf{Z}} \psi = \mathscr{L}_{1,R'} *_{\mathbf{Z}} \varphi^+$ *and* $V^- = \mathscr{L}_{r',1} *_{\mathbf{Z}} \psi = \mathscr{L}_{r',1} *_{\mathbf{Z}} \varphi^-$.

Proof. We only need to apply (2) above with r' and R' for which $(L\psi)(x_0,\cdot)$ has no zeros on the annuli A^+ and A^-. □

We point out to the reader that this theorem represents only one of the many variations that follow by similar arguments.

5 Sparse Subsets of Abelian Groups and Exotic Idempotents

As noted in Sect. 2, sampling formulae depend on convolution idempotents, that is, sequences $\{c(k)\}_{k\in\mathbf{Z}}$ that satisfy

$$c(k) = \sum_{k\in\mathbf{Z}} c(j)c(k-j), \qquad k\in\mathbf{Z},$$

where the above sum converges absolutely. There are exactly two sequences in $\ell^1(\mathbf{Z})$ (which is a convolution algebra) that are convolution idempotents: $c=0$ and $c=\delta_0$, the sequence that is 1 at the origin and zero everywhere else. For sequences in $\ell^p(\mathbf{Z})$ for $1<p\le 2$, as observed in Sect. 2, there are several convolution idempotents. In fact, if $1<p\le 2$ and $c\in\ell^p(\mathbf{Z})\subset\ell^2(\mathbf{Z})$, then c is a convolution idempotent if and only if $c(k)=\widehat{\chi_\Omega}(k)$ for some measurable subset Ω of $[0,1]$. We do not know if there are convolution idempotents that belong to $\ell^p(\mathbf{Z})\backslash\ell^2(\mathbf{Z})$ for some $p>2$. Below we present some exotic convolution idempotents in the setting of finitely generated abelian groups.

As in Sect. 1, let $\mathscr{G}=(\mathscr{G},+)$ be a finitely generated additive group whose maximal finite subgroup $F=\{l\in\mathscr{G}: m\cdot l=0, \text{for some } m\in\mathbf{N}\}$ is a proper subgroup of $\mathscr{G}$. Thus, for some n, $\mathscr{G}/F\cong\mathbf{Z}^n$, and each choice of generators for $\mathscr{G}$ leads to a direct sum decomposition $\mathscr{G}=\mathscr{G}_1+F$ with $\mathscr{G}_1\cong\mathbf{Z}^n$. A subset $S\subseteq\mathscr{G}$ is said to be *sparse* if, for each $l\in\mathscr{G}$, the set

$$\{(i,j)\in S\times S: i+j=l\}$$

is a finite set. When S is sparse and $a,b\in\mathbf{C}^{\mathscr{G}}$ are supported in S, that is, vanish off S, the convolution

$$(a*b)(l) = \sum_{i+j=l} a(i)b(j) \qquad l\in\mathscr{G}$$

is well defined (absolutely convergent) since the sum only involves finitely many nonzero terms. In fact, the same statement applies when we replace $\mathbf{C}$ by any other field. Subsets of F are trivially sparse, and idempotents in $\mathbf{C}^{\mathscr{G}}$ supported in F are uninteresting. We say an idempotent $c\in\mathbf{C}^{\mathscr{G}}$ is *exotic* if

$$\operatorname{supp} c = \{l\in\mathscr{G}: c(l)\neq 0\}$$

is an infinite sparse set.

The first examples of exotic idempotents were produced at Washington University in Saint Louis by Nik Weaver (2010, personal communication). They are constructed as follows. Let $\{z_i\}_{i\geq 0}$ be *any* nonzero sequence of complex numbers. Weaver devised an inductive algorithm producing an exotic idempotent $c \in \mathbf{C}^{\mathbf{Z}}$ with $c(1000^i) = z_i$ for $i \geq 0$, and $\operatorname{supp} c$ is the union of $\{1000^i : i \geq 0\}$ and an intricate subset of $\{k \in \mathbf{Z} : k < 0\}$. One of us (Wilson) generalized Weaver's technique to show that for $\mathscr{G}$ as above, there are large families of exotic idempotents in $\mathbf{C}^{\mathscr{G}}$ whose support is not in any proper subgroup $\mathscr{G}'$ of $\mathscr{G}$ having a complementary subgroup $\mathscr{G}''$ for which $\mathscr{G} = \mathscr{G}' + \mathscr{G}''$ is a direct sum. Returning to the group $\mathbf{Z}$, for $N \geq 8$ and $\{r_i\}_{i\geq 0}$ a sequence in $\mathbf{N}$ with r_i in a small neighborhood of N^i, there are exotic idempotents c in $\mathbf{C}^{\mathbf{Z}}$ supported on the union of $\{r_i : i \geq 0\}$ and a subset of the negative integers. By varying the choice of nonvanishing sequences $\{z_i\}_{i\geq 0}$ for which $c(r_i) = z_i$, there are exotic idempotents in $\mathbf{C}^{\mathbf{Z}}$ with rapid decay at $\pm\infty$ and other idempotents with exponential growth at $\pm\infty$. But none of these exotic idempotents are in $\ell^2(\mathbf{Z})$, and, more generally, none are of the form $R^k c(k)$, $k \in \mathbf{Z}$ for some $c \in \ell^2(\mathbf{Z})$. Whether or not there exist idempotents in $\mathbf{C}^{\mathbf{Z}}$ which are neither exotic nor dilates of $\ell^2(\mathbf{Z})$ idempotents is unknown.

The generalization of Weaver's technique for producing large families of exotic idempotents in $\mathbf{C}^{\mathscr{G}}$, $\mathscr{G}$ as above, uses subsets $S \subset \mathscr{G}$ for which both S and $S+S$ are graded in a certain way by finite subsets. Specifically, S is the disjoint union of a subset $R = \{r_i : i \geq 0\}$ of $\mathscr{G}\backslash\{0\}$ and finite subsets I_i, $i \geq 1$, of $\mathscr{G}\backslash\{0\}$, and $S+S$ is the disjoint union of the subsets $J_i = r_i + I_{i+1}$, $i \geq 0$. Moreover, with $I_0 = \emptyset$ and $S_i = \{r_i\} \cup I_i$, $i \geq 0$, each of the sets $\{r_i, 2r_i\}$, I_i, and $I_i + I_i$ are subsets of J_i for $i \geq 0$, and for $0 < j < i$, $S_j + S_i$ is also a subset of J_i. Hence, for $j > i+1$, $(S_j + S) \cap J_i = \emptyset$, and the only members of $(S_{i+1} + S_{i+1}) \cap J_i$ are those of the form $r_{i+1} + l$ with $l \in I_{i+1}$. In particular, S is sparse and, for each choice of $\{z_i\}_{i\geq 0}$, with $z_i \neq 0$ for $i \geq 1$, there is a unique idempotent $c \in \mathbf{C}^{\mathscr{G}}$, with $\operatorname{supp} c \subseteq S$ and $c(r_i) = z_i$ for $i \geq 1$. Indeed, c is obtained by inductive solutions of the equation $(c - c * c)\chi_{J_i} = 0$, $i \geq 0$ with each such equation defining the values of c on I_{i+1} in terms of previously determined values on $S_0 \cup S_1 \cup \cdots \cup S_i$. Curiously, the coefficients of c on members of I_{i+1} are ratios of monomials $i+1$ of complex variables. Hence, the coefficients are a multivariate meromorphic function. For $\mathscr{G} = \mathbf{Z}$ and $r_i = N^i$ with $N \geq 8$, intricate combinatoric relations inductively define the smallest subsets J_i of

$$\left\{k \in \mathbf{Z} : N^{i-1/2} < |k| < N^{i+1/2}\right\}_{i\geq 0}$$

for which, with $I_{i+1} = J_i - r_{i+1}$, $S = \{N^i : i \geq 0\} \cup \bigcup_{i=1}^{\infty} I_i$ satisfies the above grading properties, and, for each nonzero sequence $\{z_i\}_{i\geq 0}$, there is an idempotent c with $c(N^i) = z_i$ and $\operatorname{supp} c = S$.

References

1. Chen, W., Itoh, S.: A sampling theorem for shift-invariant subspace. IEEE Trans. Signal Process. **52**(10), 4643–4648 (2006)
2. Gröchenig, K., Janssen, A.J.E.M., Kaiblinger, N., Pfander, G.: Note on B-splines, wavelet scaling functions, and Gabor frames. IEEE Trans. Inf. Theor. **49**, 3318–332 (2003)
3. Hernández, E., Šikić, H., Weiss, G., Wilson, E.: Cyclic subspaces for unitary representations of LCA groups; generalized Zak transforms. Colloq. Math. **118**(1), 313–332 (2010)
4. Hernández, E., Šikić, H., Weiss, G., Wilson, E.: On the properties of the integer translates of a square integrable function. Harmonic Analysis and Partial Differential Equations, 233–249, Contemporary Mathematics, vol. 505, American Mathematical Society, Providence, RI (2010)
5. Hernández, E., Šikić, H., Weiss, G., Wilson, E.: The Zak transform(s) Fourier analysis and convexity, 151–157. Appl. Numer. Harmon. Anal. BirkhŁuser Boston, Boston (2011)
6. Janssen, A.J.E.M.: The Zak transform: a signal transform for sampled time-continuous signals. Philips J. Res. **43**, 23–49 (1988)
7. Janssen, A.J.E.M.: The Zak transform and sampling theorems for wavelet subspaces. IEEE Trans. Signal Process. **41**, 3360–3364 (1993)
8. Šikić, H., Wilson, E.: Lattice invariant subspaces and sampling. Appl. Comput. Harmon. Anal. **31**(1), 26–43 (2011)

References

Part II
Remote Sensing

Remote sensing is a means of collecting data about objects without coming into contact with them. Typically the setting is the earth's surface and atmosphere, garnering information dealing with geology, meteorology, oceans and glaciers, natural disasters, and the classification and detection problems associated with man-made issues. Remote sensing technology is phenomenally varied. For example, the airborne photography technology, going back to Tournachon's aerial photographs in 1858 from a balloon, has given rise to more recent satellite, radar, and lidar methodologies to obtain data.

Under the general rubric of *remote sensing*, the five chapters of Part II address fundamental modern issues in the field dealing with compressive sensing, hyper-spectral image data, synthetic-aperture radar (SAR) imagery, and the role of radar waveforms for imaging.

Margaret Cheney, Brett Borden, and Ling Wang develop a theory for the imaging of moving targets that encompasses the impact of spatial, temporal, and spectral properties of scattered waves. The authors, with their broad range of expertise, make a convincing argument for such a theory in which an end goal is the formation of high-resolution moving-target images in the setting of different waveforms at different locations received at different places. An in-depth physics-based mathematical model is developed integrating effects due to waveforms, wave propagation, as well as the impact of spatial diversity of transmitters and receivers. The methodology is for the most part based in classical deep analysis with a segue relating the resulting point spread function (PSF) with the narrowband ambiguity function. The PSF characterizes the behavior of the imaging system by relating the phase-space reflectivity distribution to the actual image. The authors' analysis of the PSF is a tour de force ranging from the theoretical to a panorama of essential examples.

Christopher J. Deloye, J. Christopher Flake, David Kittle, Robert S. Rand, and David J. Brady tie together imaging spectrometers to analyze hyperspectral imagery data in the context of compressive sensing. The authors deal with a class of spectrometers, a so-called coded aperture snapshot spectral imagery (CASSI) sensing system, which can be integrated into the basic compressive sensing theory of Donoho, Candès, Romberg, and Tao in a natural way. In fact, the CASSI system can satisfy the restricted isometry property (RIP) with high probability, giving rise to sparse solutions for data reconstruction. A feature of the chapter is the authors' understanding of hyperspectral imagery, allowing them genuinely to apply CASSI reconstructions for pixel classification performance. It is natural to quantify sparsity in addressing the complexity of hyperspectral cubes; it is exciting to see results using such a theoretical foundation.

David B. Gillis and Jeffrey H. Bowles are also experts in the study of hyperspectral imagery with a deep knowledge of the interaction of theory and data, as well as the techniques required to analyze given gigabyte-sized images in efficient ways. Their chapter is a lucid introduction, honed by experience, of the complex methodologies used to fathom hyperspectral data in terms of dimensionality reduction. Here we see the reliable, and sturdy, but often limited, principal component analysis (PCA), along with linear mixing models (LMM), methods for end-member determination and the more general blind source separation (BSS) problem. The authors also describe in context more recent ideas dealing with manifold learning and nonlinear kernel methods for dimension reduction and natural but not yet highly developed spatial-spectral methods. The perspective is wonderful and the examples are luminous by virtue of the authors' understanding.

JOHN GREER rounds out our three chapters dealing with modern techniques used to extract salient information from hyperspectral imagery data in efficient, useful, time-sensitive ways. His specific topic is hyperspectral demixing, and he brings to bear first-class theoretical expertise and invaluable experimental experience in sparse reconstruction methodologies. Orthogonal matching pursuit (OMP) and basis pursuit (BP), which can be so valuable in applications such as transform-based encoding, e.g., with JPEG 2000, take a backseat to the author's ingenious sparse demixing (SD) algorithm.

LING WANG, CAN EVREN YARMEN, and BIRSEN YAZICI present a unified theory of passive SAR imagery data analysis. The physical backdrop is the emerging area of opportunistic sensing, which is a natural component of modern remote sensing given sensing interactions and the fragility of primary sensors in active environments. The mathematical background is dazzling, using microlocal analysis to approximate measurement models formulated in terms of Fourier integral operators (FIOs). The authors' technology for passive SAR can be viewed as a major application of their imaging theory given in terms of both ultra-narrowband continuous-wave waveforms and wideband pulsed waveforms. The mathematical theory in which images are formed by means of the aforementioned FIOs, called filtered-backprojection operators, gives way to compelling numerical simulations illuminating (sic) the power of the theory.

Multistatic Radar Waveforms for Imaging of Moving Targets

Margaret Cheney, Brett Borden, and Ling Wang

Abstract We develop a linearized imaging theory that combines the spatial, temporal, and spectral aspects of scattered waves. We consider the case of fixed sensors and a general distribution of objects, each undergoing linear motion; thus the theory deals with imaging distributions in phase space. We derive a model for the data that is appropriate for narrowband waveforms in the case when the targets are moving slowly relative to the speed of light. From this model, we develop a phase-space imaging formula that can be interpreted in terms of filtered back projection or matched filtering. For this imaging approach, we derive the corresponding phase-space point-spread function. We show plots of the phase-space point-spread function for various geometries. We also show that in special cases, the theory reduces to (a) range-Doppler imaging, (b) inverse synthetic aperture radar (ISAR), (c) synthetic aperture radar (SAR), (d) Doppler SAR, and (e) tomography of moving targets.

Keywords Multistatic radar • Active imaging • Moving target • Point-spread function (PSF) • Ambiguity function • SAR • ISAR

M. Cheney (✉)
Department of Mathematics, Colorado State University, 80523 Fort Collins, CO, USA
e-mail: Cheney@math.colostate.edu

B. Borden
Physics Department, Naval Postgraduate School, Monterey, CA 93943, USA
e-mail: Bhborden@nps.edu

L. Wang
College of Information Science and Technology, Nanjing University of Aeronautics and Astronautics, Nanjing 210016, China
e-mail: wanglrpi@gmail.com

T.D. Andrews et al. (eds.), *Excursions in Harmonic Analysis, Volume 1*,
Applied and Numerical Harmonic Analysis, DOI 10.1007/978-0-8176-8376-4_7,

1 Introduction

The use of radar for detection and imaging of moving targets is a topic of great interest [1–29]. It is well known that radar signals have two important attributes, namely the time delay, which provides information on the target range, and the Doppler shift, which can be used to infer target downrange velocity. Classical "radar ambiguity" theory [30–32] shows that the transmitted waveform determines the accuracy to which target range and velocity can be obtained from a backscattered radar signal. This theory provides an understanding of how to exploit temporal and spectral attributes of radar data.

Another well-developed theory for radar imaging is that of synthetic-aperture radar (SAR) imaging. SAR combines high-range-resolution measurements from a variety of locations to produce high-resolution radar images [21, 33–35]. SAR is thus a theory for combining temporal and spatial attributes of radar data. SAR, however, cannot handle the case when there is unknown relative motion between the target and radar platform; research addressing this includes [3, 7–22]. This research generally does not consider the possibility of using waveforms other than high-range-resolution ones.

When high-Doppler-resolution waveforms are transmitted from a moving platform, the Doppler-shifted returns can also be used to form images of a stationary scene [36]. This theory provides a way to exploit spectral and spatial attributes of radar data.

These three theories are depicted on the coordinate planes in Fig. 1. None of these theories address the possibility of forming high-resolution moving-target images in the case when different waveforms are transmitted from different spatial locations and received at spatially distributed receivers. Such a theory is needed to address questions such as:

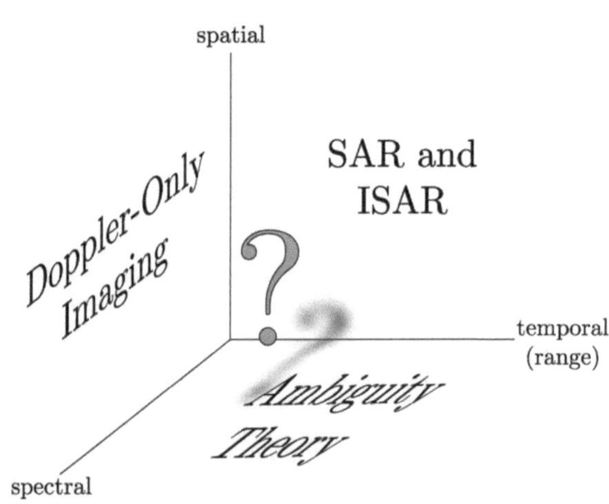

Fig. 1 A notional diagram showing how our work (depicted by the *question mark*) relates to standard radar imaging methods

Fig. 2 This shows an example of the geometry we consider. The antennas are fixed and illuminate a scene in which there are multiple moving targets

- In a system with multiple transmitters and receivers positioned at different locations, which transmitters should transmit which waveforms?
- How many transmitters are needed, and where should they be positioned?
- How can data from such a system be used to form an image of unknown moving targets?
- What is the resolution (in position and in velocity) of such a system?

Some work has been done to develop such a theory:

- Ambiguity theory for bistatic systems has been developed in [34,35,37,38]; such systems allow for estimation of only one component of the velocity vector.
- For multistatic systems, the work [23–29] developed methods for moving-target *detection*.
- Theory for use of a multistatic system for imaging of a *stationary* scene is well known; see, for example, [39–42].
- Multistatic *imaging* of *moving* targets (phase-space imaging) was developed in [43] for the case of fixed transmitters and receivers (see Fig. 2). This theory combines spatial, temporal, and spectral attributes of radar data; in particular the theory exploits the actual Doppler shift from moving targets. In appropriate special cases, this theory reduces to the standard imaging methods shown in Fig. 1.

The work [43] is extended in this chapter, which sets forth the basic ideas in the special cases that are of interest to radar-based imaging, and undertakes an initial numerical exploration of the properties of the imaging system. In particular, we show that under some circumstances, this approach allows for both (spatial) image formation and also estimation of the full vector velocities of multiple targets.

Table 1 Table of notation

Symbol	Designation
$\boldsymbol{x}$	Position (three-dimensional vector)
t	Time (scalar)
$\boldsymbol{y}$	Transmitter position (three-dimensional vector)
$s_{\boldsymbol{y}}(t)$	Waveform transmitted by a transmitter located at $\boldsymbol{y}$
$-T_{\boldsymbol{y}}$	Starting time of the waveform $s_{\boldsymbol{y}}$
$\boldsymbol{v}$	Velocity (three-dimensional vector) of a moving scatterer
$q_{\boldsymbol{v}}(\boldsymbol{x})$	Distribution of scatters with position $\boldsymbol{x}$ and velocity $\boldsymbol{v}$ at time $t=0$
$\alpha_{\boldsymbol{x},\boldsymbol{v}}$	Doppler scale factor for a scatter at position $\boldsymbol{x}$ moving with velocity $\boldsymbol{v}$
$\psi(t,\boldsymbol{x})$	Wavefield at time t and position $\boldsymbol{x}$
$\psi^{\text{in}}(t,\boldsymbol{x},\boldsymbol{y})$	Free-space field at $\boldsymbol{x}$ generated by a source at $\boldsymbol{y}$
$\boldsymbol{z}$	Receiver position (three-dimensional vector)
$\boldsymbol{R}_{\boldsymbol{x},\boldsymbol{z}}$	$\boldsymbol{x}-\boldsymbol{z}$, three-dimensional vector from $\boldsymbol{z}$ to $\boldsymbol{x}$
$\mu_{\boldsymbol{z},\boldsymbol{v}}$	$1+\widehat{\boldsymbol{R}}_{\boldsymbol{x},\boldsymbol{z}}\cdot\boldsymbol{v}/c$
$\psi^{\text{sc}}(t,\boldsymbol{z},\boldsymbol{y})$	Scattered field at the receiver location $\boldsymbol{z}$ due to the field transmitted from position $\boldsymbol{y}$.
$\alpha_{\boldsymbol{x},\boldsymbol{v}}$	Doppler scale factor
$\boldsymbol{p}$	Position (three-dimensional vector) in the reconstructed scene
$\boldsymbol{u}$	Velocity (three-dimensional vector) of object in reconstructed scene
$I_{\boldsymbol{u}}$	Reconstruction of reflectivity of objects moving with velocity $\boldsymbol{u}$
J	Geometry-dependent weighting function for improving the image quality
K	Point spread function (PSF)
$A_{\boldsymbol{y}}$	Radar ambiguity function for the waveform used by the transmitter located at $\boldsymbol{y}$
θ	Angle of reconstructed position vector $\boldsymbol{p}$
ϕ	Angle of reconstructed velocity $\boldsymbol{u}$

In Sect. 2, we develop a physics-based mathematical model that incorporates not only the waveform and wave propagation effects due to moving scatterers but also the effects of spatial diversity of transmitters and receivers. In Sect. 3, we show how a matched-filtering technique produces images that depend on target velocity, and we relate the resulting point-spread function (PSF) to the classical radar ambiguity function. We show four-dimensional plots of the PSF for three different geometries. Finally, in Sect. 4, we list some conclusions.

2 Model for Data

We model wave propagation and scattering by the scalar wave equation for the wavefield $\psi(t,\boldsymbol{x})$ due to a source waveform $\tilde{s}(t,\boldsymbol{x})$:

$$[\nabla^2 - c^{-2}(t,\boldsymbol{x})\partial_t^2]\psi(t,\boldsymbol{x}) = \tilde{s}(t,\boldsymbol{x})\,. \tag{1}$$

For example, a single isotropic source located at $\boldsymbol{y}$ transmitting the waveform $s(t)$ starting at time $t=-T_{\boldsymbol{y}}$ could be modeled as $\tilde{s}(t,\boldsymbol{x}) = \delta(\boldsymbol{x}-\boldsymbol{y})s_{\boldsymbol{y}}(t+T_{\boldsymbol{y}})$, where the subscript $\boldsymbol{y}$ reminds us that different transmitters could transmit different waveforms. For simplicity, in this discussion we consider only localized isotropic sources; the work can easily be extended to more realistic antenna models [46].

A single scatterer moving at velocity $\boldsymbol{v}$ corresponds to an index-of-refraction distribution $n^2(\boldsymbol{x}-\boldsymbol{v}t)$:

$$c^{-2}(t,\boldsymbol{x}) = c_0^{-2}[1+n^2(\boldsymbol{x}-\boldsymbol{v}t)], \tag{2}$$

where c_0 denotes the speed of light in vacuum. We write $q_{\boldsymbol{v}}(\boldsymbol{x}-\boldsymbol{v}t) = c_0^{-2}n^2(\boldsymbol{x}-\boldsymbol{v}t)$. To model multiple moving scatterers, we let $q_{\boldsymbol{v}}(\boldsymbol{x}-\boldsymbol{v}t)\,\mathrm{d}^3x\mathrm{d}^3v$ be the corresponding quantity for the scatterers in the volume $\mathrm{d}^3x\mathrm{d}^3v$ centered at $(\boldsymbol{x},\boldsymbol{v})$. Thus $q_{\boldsymbol{v}}$ is the distribution in *phase space*, at time $t=0$, of scatterers moving with velocity $\boldsymbol{v}$. Consequently, the scatterers at $\boldsymbol{x}$ give rise to

$$c^{-2}(t,\boldsymbol{x}) = c_0^{-2} + \int q_{\boldsymbol{v}}(\boldsymbol{x}-\boldsymbol{v}t)\,\mathrm{d}^3v. \tag{3}$$

We note that the physical interpretation of $q_{\boldsymbol{v}}$ involves a choice of time origin. The choice that is particularly appropriate, in view of our assumption about linear target velocities, is a time during which the wave is interacting with targets of interest. This implies that the activation of the antenna at $\boldsymbol{y}$ takes place at a negative time which we denote by $-T_{\boldsymbol{y}}$; we write $\tilde{s}(t,\boldsymbol{x}) = s_{\boldsymbol{y}}(t+T_{\boldsymbol{y}})\delta(\boldsymbol{x}-\boldsymbol{y})$. The wave equation corresponding to (3) is then

$$\left[\nabla^2 - c_0^{-2}\partial_t^2 - \int q_{\boldsymbol{v}}(\boldsymbol{x}-\boldsymbol{v}t)\,\mathrm{d}^3v\,\partial_t^2\right]\psi(t,\boldsymbol{x}) = s_{\boldsymbol{y}}(t+T_{\boldsymbol{y}})\delta(\boldsymbol{x}-\boldsymbol{y}). \tag{4}$$

From this point on, we drop the subscript on c_0 so that c denotes the speed of light in vacuum.

In the *absence* of scatterers, the field from the antenna is

$$\psi^{\mathrm{in}}(t,\boldsymbol{x},\boldsymbol{y}) = -\frac{s_{\boldsymbol{y}}(t+T_{\boldsymbol{y}}-|\boldsymbol{x}-\boldsymbol{y}|/c)}{4\pi|\boldsymbol{x}-\boldsymbol{y}|}. \tag{5}$$

(Details are given in the Appendix.)

We write $\psi = \psi^{\mathrm{in}} + \psi^{\mathrm{sc}}$ and neglect multiple scattering (i.e., use the weak-scatterer model) to obtain the expression for the scattered field ψ^{sc}:

$$\psi^{\mathrm{sc}}(t,\boldsymbol{z},\boldsymbol{y}) = \int \frac{\delta(t-t'-|\boldsymbol{z}-\boldsymbol{x}|/c)}{4\pi|\boldsymbol{z}-\boldsymbol{x}|}\int q_{\boldsymbol{v}}(\boldsymbol{x}-\boldsymbol{v}t')\,\mathrm{d}^3v \times \frac{\ddot{s}_{\boldsymbol{y}}(t'+T_{\boldsymbol{y}}-|\boldsymbol{x}-\boldsymbol{y}|/c)}{4\pi|\boldsymbol{x}-\boldsymbol{y}|}\,\mathrm{d}^3x\,\mathrm{d}t', \tag{6}$$

where the dots denote derivatives with respect to time. (For details of the derivation, see the Appendix.)

In Eq. (6), we make the change of variables $\boldsymbol{x} \mapsto \boldsymbol{x}' = \boldsymbol{x} - \boldsymbol{v}t'$, (i.e., change the frame of reference to one in which the scatterer $q_{\boldsymbol{v}}$ is fixed) to obtain

$$\psi^{\mathrm{sc}}(t,\boldsymbol{z},\boldsymbol{y}) = -\int \frac{\delta(t-t'-|\boldsymbol{x}'+\boldsymbol{v}t'-\boldsymbol{z}|/c)}{4\pi|\boldsymbol{x}'+\boldsymbol{v}t'-\boldsymbol{z}|} \int \frac{q_{\boldsymbol{v}}(\boldsymbol{x}')}{4\pi|\boldsymbol{x}'+\boldsymbol{v}t'-\boldsymbol{y}|} \times \ddot{s}_{\boldsymbol{y}}(t'+T_{\boldsymbol{y}}-|\boldsymbol{x}'+\boldsymbol{v}t'-\boldsymbol{y}|/c)\,\mathrm{d}^3x'\,\mathrm{d}^3v\,\mathrm{d}t'. \tag{7}$$

The physical interpretation of (7) is as follows: The wave that emanates from $\boldsymbol{y}$ at time $-T_{\boldsymbol{y}}$ encounters a target at time t'; this target, during the time interval $[0,t']$, has moved from $\boldsymbol{x}'$ to $\boldsymbol{x}'+\boldsymbol{v}t'$; the wave scatters with strength $q_{\boldsymbol{v}}(\boldsymbol{x}')$ and then propagates from position $\boldsymbol{x}'+\boldsymbol{v}t'$ to $\boldsymbol{z}$, arriving at time t. Henceforth we will drop the primes on $\boldsymbol{x}$.

Next we assume that the target is slowly moving. More specifically, we assume that $|\boldsymbol{v}|t$ and $k|\boldsymbol{v}|^2t^2$ are much less than $|\boldsymbol{x}-\boldsymbol{z}|$ and $|\boldsymbol{x}-\boldsymbol{y}|$, where $k=\omega_{\max}/c$, with $\omega_{\max}$ the (effective) maximum angular frequency of all the signals $s_{\boldsymbol{y}}$. In this case, expanding $|\boldsymbol{z}-(\boldsymbol{x}+\boldsymbol{v}t')|$ around $t'=0$, yields

$$|\boldsymbol{z}-(\boldsymbol{x}+\boldsymbol{v}t')| \approx R_{\boldsymbol{x},\boldsymbol{z}} + \widehat{\boldsymbol{R}}_{\boldsymbol{x},\boldsymbol{z}}\cdot\boldsymbol{v}t', \tag{8}$$

where $\boldsymbol{R}_{\boldsymbol{x},\boldsymbol{z}} = \boldsymbol{x}-\boldsymbol{z}$, $R_{\boldsymbol{x},\boldsymbol{z}} = |\boldsymbol{R}_{\boldsymbol{x},\boldsymbol{z}}|$, and $\widehat{\boldsymbol{R}}_{\boldsymbol{x},\boldsymbol{z}} = \boldsymbol{R}_{\boldsymbol{x},\boldsymbol{z}}/R_{\boldsymbol{x},\boldsymbol{z}}$.

We can carry out the t' integration in (7) as follows: we make the change of variables

$$t'' = t - t' - \left(R_{\boldsymbol{x},\boldsymbol{z}} + \widehat{\boldsymbol{R}}_{\boldsymbol{x},\boldsymbol{z}}\cdot\boldsymbol{v}t'\right)/c \tag{9}$$

which has the corresponding Jacobian

$$\left|\frac{\partial t'}{\partial t''}\right| = \frac{1}{|\partial t''/\partial t'|} = \frac{1}{1+\widehat{\boldsymbol{R}}_{\boldsymbol{x},\boldsymbol{z}}\cdot\boldsymbol{v}/c}.$$

We denote the denominator of this Jacobian by $\mu_{\boldsymbol{z},\boldsymbol{v}}$. The argument of the delta function in (7) contributes only when $t''=0$; Eq. (9) with $t''=0$ can be solved for t' to yield

$$t' = \frac{t - R_{\boldsymbol{x},\boldsymbol{z}}/c}{1+\widehat{\boldsymbol{R}}_{\boldsymbol{x},\boldsymbol{z}}\cdot\boldsymbol{v}/c}. \tag{10}$$

The argument of s in Eq. (7) is then

$$\begin{aligned} t'+T_{\boldsymbol{y}} - \frac{|\boldsymbol{x}+\boldsymbol{v}t'-\boldsymbol{y}|}{c} &\approx t'+T_{\boldsymbol{y}} - \left(\frac{R_{\boldsymbol{x},\boldsymbol{y}}}{c} + \frac{\widehat{\boldsymbol{R}}_{\boldsymbol{x},\boldsymbol{y}}\cdot\boldsymbol{v}}{c}t'\right) \\ &= \frac{1-\widehat{\boldsymbol{R}}_{\boldsymbol{x},\boldsymbol{y}}\cdot\boldsymbol{v}/c}{1+\widehat{\boldsymbol{R}}_{\boldsymbol{x},\boldsymbol{z}}\cdot\boldsymbol{v}/c}\left(t - \frac{R_{\boldsymbol{x},\boldsymbol{z}}}{c}\right) - \frac{R_{\boldsymbol{x},\boldsymbol{y}}}{c} + T_{\boldsymbol{y}}, \end{aligned} \tag{11}$$

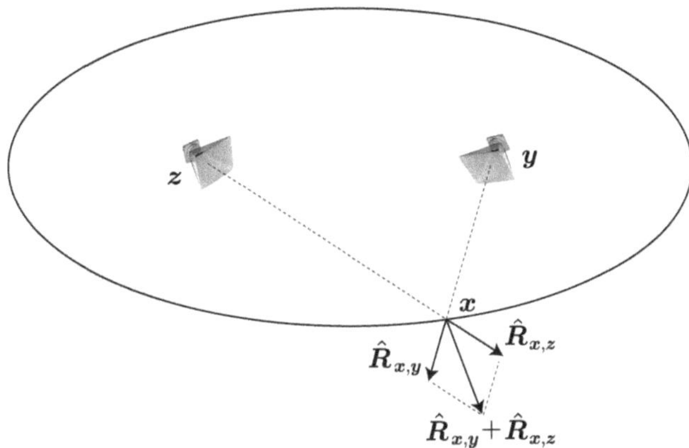

Fig. 3 The bistatic range is the sum of the distances $R_{x,z}$ and $R_{x,y}$, and the bistatic bisector is the vector $\widehat{\boldsymbol{R}}_{x,y}+\widehat{\boldsymbol{R}}_{x,z}$

where we have used (8) and (10). The quantity

$$\alpha_{x,v} = \frac{1-\widehat{\boldsymbol{R}}_{x,y}\cdot \boldsymbol{v}/c}{1+\widehat{\boldsymbol{R}}_{x,z}\cdot \boldsymbol{v}/c} \approx 1-(\widehat{\boldsymbol{R}}_{x,y}+\widehat{\boldsymbol{R}}_{x,z})\cdot \boldsymbol{v}/c \tag{12}$$

is the *Doppler scale factor*. Note that for $|\boldsymbol{v}|/c \ll 1$, the Doppler scale factor (12) can be written approximately as $\alpha_{x,v} \approx 1+\beta_{x,v}$, where

$$\beta_{x,v} = -(\widehat{\boldsymbol{R}}_{x,y}+\widehat{\boldsymbol{R}}_{x,z})\cdot \boldsymbol{v}/c. \tag{13}$$

For a narrowband signal with carrier frequency ω_y, the quantity $\omega_y\beta_{x,v}$ is the (angular) *Doppler shift*. We note that the Doppler scale factor depends on the bistatic bisector vector $\widehat{\boldsymbol{R}}_{x,y}+\widehat{\boldsymbol{R}}_{x,z}$ (see Fig. 3).

With (11) and the notation of Eqs. (12) and (7) becomes

$$\psi^{\mathrm{sc}}(t,\boldsymbol{z},\boldsymbol{y}) = \int \frac{\ddot{s}_y\left[\alpha_{x,v}\left(t-R_{x,z}/c\right)-R_{x,y}/c+T_y\right]}{(4\pi)^2 R_{x,z}R_{x,y}\mu_{z,v}} q_v(\boldsymbol{x})\,\mathrm{d}^3x\,\mathrm{d}^3v. \tag{14}$$

We recognize (14) as the superposition of time-delayed and Doppler-scaled copies of the (second derivative of the) transmitted waveform. The key feature of this model is that it correctly incorporates the positions of the transmitters and receivers, the transmitted waveform, and the target position and velocity. In particular, the time delay and Doppler scale depend correctly on the target and sensor positions and on the target velocity.

If we denote the transmitters by $\boldsymbol{y}_1, \boldsymbol{y}_2, \ldots, \boldsymbol{y}_M$, then at receiver $\boldsymbol{z}$, the data we measure is the sum $\sum_m \psi^{\mathrm{sc}}(t, \boldsymbol{z}, \boldsymbol{y}_m)$.

We denote the receivers by $\boldsymbol{z}_1, \boldsymbol{z}_2, \ldots, \boldsymbol{z}_N$; for ease in notation, we replace $\boldsymbol{y}_m$ by m and $\boldsymbol{z}_n$ by n. Below we assume that we can measure the data $\psi^{\mathrm{sc}}(t,n,m) = \psi^{\mathrm{sc}}(t, \boldsymbol{z}_n, \boldsymbol{y}_m), n = 1,2,\ldots N$, and $m = 1,2,\ldots M$. Here we are implicitly assuming that, at the receiver located at $\boldsymbol{z}_n$, we can identify the part of the signal $\sum_m \psi^{\mathrm{sc}}(t, \boldsymbol{z}_n, \boldsymbol{y}_m)$ that is due to the transmitter at $\boldsymbol{y}_m$. In the case of multiple transmitters, this identification can be accomplished, for example, by having different transmitters operate at different frequencies or possibly by quasi-orthogonal pulse-coding schemes.

3 Imaging

We now address the question of extracting information from the scattered waveform described by Eq. (14).

We will form a (coherent) image as a filtered adjoint or weighted matched filter [44]. Note that in general the desired phase-space image is six-dimensional; if all targets are assumed to move within a two-dimensional plane, then the desired phase-space image is four-dimensional.

3.1 Imaging Formula

We form an image $I_{\boldsymbol{u}}(\boldsymbol{p})$ of the objects with velocity $\boldsymbol{u}$ that, at time $t = 0$, were located at position $\boldsymbol{p}$. Thus $I_{\boldsymbol{u}}(\boldsymbol{p})$ is constructed to be an approximation to $q_{\boldsymbol{u}}(\boldsymbol{p})$.

We form an image by matched filtering and summing over all transmitters and receivers. Matched filtering corresponds to taking the inner product with the expected signal due to a delta-like target. From (14) with $q_{\boldsymbol{v}}(\boldsymbol{x}) = \delta(\boldsymbol{x} - \boldsymbol{p})\delta(\boldsymbol{v} - \boldsymbol{u})$, we see that the predicted signal from a single point target at the phase-space position $(\boldsymbol{p}, \boldsymbol{u})$ (i.e., a target at position $\boldsymbol{p}$ traveling with velocity $\boldsymbol{u}$) is proportional to (the second time derivative of)

$$s_m \left[\alpha_{\boldsymbol{p},\boldsymbol{u}} (t - R_{\boldsymbol{p},n}/c) - R_{\boldsymbol{p},m}/c + T_y\right]. \tag{15}$$

To obtain a phase-space image $I_{\boldsymbol{u}}(\boldsymbol{p})$, we take the inner product of the data ψ^{sc} with (15) and sum over all transmitters and receivers:

$$I_{\boldsymbol{u}}(\boldsymbol{p}) = (4\pi)^2 \sum_{n=1}^{N} \sum_{m=1}^{M} R_{\boldsymbol{p},n} R_{\boldsymbol{p},m} \mu_{n,\boldsymbol{u}} \alpha_{\boldsymbol{p},\boldsymbol{u}} J_{n,m}$$
$$\times \int s_m^* \left[\alpha_{\boldsymbol{p},\boldsymbol{u}} (t - R_{\boldsymbol{p},n}/c) - R_{\boldsymbol{p},m}/c + T_m\right] \psi^{\mathrm{sc}}(t,n,m)\, \mathrm{d}t, \tag{16}$$

where $J_{n,m}$ is a geometry-dependent weighting function that can be inserted to improve the image [44]. The weights $R_{\boldsymbol{p},n}R_{\boldsymbol{p},m}\mu_{n,\boldsymbol{u}}$ are included so that they will cancel the corresponding factors in the data when $\boldsymbol{p} = \boldsymbol{x}$ and $\boldsymbol{u} = \boldsymbol{v}$.

The operation (16) amounts to geometry-corrected and phase-corrected matched filtering with a time-delayed, Doppler-scaled version of the transmitted waveform. Note that we are assuming a coherent system, i.e., that a common time clock is available to all sensors, so that we are able to form the image (16) with the correct phase relationships. If the system is not coherent, then magnitudes can be taken in (16) before the summation is performed; this will result in a degraded image.

3.2 Image Analysis

In order to characterize the behavior of this imaging system, we substitute Eq. (14) into Eq. (16), obtaining

$$I_{\boldsymbol{u}}(\boldsymbol{p}) = \int K(\boldsymbol{p},\boldsymbol{u};\boldsymbol{x},\boldsymbol{v})\, q_{\boldsymbol{v}}(\boldsymbol{x})\,\mathrm{d}^3x\,\mathrm{d}^3v \tag{17}$$

where K, the PSF, is

$$K(\boldsymbol{p},\boldsymbol{u};\boldsymbol{x},\boldsymbol{v}) = -\sum_{n=1}^{N}\sum_{m=1}^{M} J_{n,m}\int s_m^*\left[\alpha_{\boldsymbol{p},\boldsymbol{u}}(t - R_{\boldsymbol{p},n}/c) - R_{\boldsymbol{p},m}/c + T_m\right]$$
$$\ddot{s}_m\left[\alpha_{\boldsymbol{x},\boldsymbol{v}}(t - R_{\boldsymbol{x},n}/c) - R_{\boldsymbol{x},m}/c + T_m\right]\mathrm{d}t\,\frac{R_{\boldsymbol{p},n}R_{\boldsymbol{p},m}\mu_{n,\boldsymbol{u}}}{R_{\boldsymbol{x},n}R_{\boldsymbol{x},m}\mu_{n,\boldsymbol{v}}}\alpha_{\boldsymbol{p},\boldsymbol{u}}. \tag{18}$$

The PSF characterizes the behavior of the imaging system in the sense that it contains all the information about how the true phase-space reflectivity distribution $q_{\boldsymbol{v}}(\boldsymbol{x})$ is related to the image $I_{\boldsymbol{u}}(\boldsymbol{p})$. If $q_{\boldsymbol{v}}(\boldsymbol{x})$ is a delta-like point target at $(\boldsymbol{x},\boldsymbol{v})$, then the PSF $K(\boldsymbol{p},\boldsymbol{u};\boldsymbol{x},\boldsymbol{v})$ is the phase-space image of that target.

With a change of variables, (18) can be written as

$$\boxed{K(\boldsymbol{p},\boldsymbol{u};\boldsymbol{x},\boldsymbol{v}) = \sum_{n=1}^{N}\sum_{m=1}^{M}\mathscr{A}_n\left(\frac{\alpha_{\boldsymbol{p},\boldsymbol{u}}}{\alpha_{\boldsymbol{x},\boldsymbol{v}}}, \Delta\tau_{m,n}(\boldsymbol{x},\boldsymbol{p},\boldsymbol{u},\boldsymbol{v})\right)\mathscr{R}_{m,n}(\boldsymbol{x},\boldsymbol{p},\boldsymbol{u},\boldsymbol{v}),} \tag{19}$$

where the wideband radar ambiguity function [45] is

$$\mathscr{A}_m(\sigma,\tau) = \int \dot{s}_m^*(\sigma t - \tau)\dot{s}_m(t)\,\mathrm{d}t, \tag{20}$$

where

$$\mathscr{R}_{m,n}(\boldsymbol{x},\boldsymbol{p},\boldsymbol{u},\boldsymbol{v}) = \frac{J_{n,m} R_{\boldsymbol{p},n} R_{\boldsymbol{p},m} \mu_{n,\boldsymbol{u}} \alpha_{\boldsymbol{p},\boldsymbol{u}}}{R_{\boldsymbol{x},n} R_{\boldsymbol{x},m} \mu_{n,\boldsymbol{v}} \alpha_{\boldsymbol{x},\boldsymbol{v}}} \tag{21}$$

and where the delay (second) argument of the ambiguity function can be written as

$$\begin{aligned} -\Delta\tau_{m,n} &= \alpha_{\boldsymbol{p},\boldsymbol{u}} \frac{R_{\boldsymbol{x},n} - R_{\boldsymbol{p},n}}{c} + \left[\frac{\alpha_{\boldsymbol{p},\boldsymbol{u}}}{\alpha_{\boldsymbol{x},\boldsymbol{v}}} \left(\frac{R_{\boldsymbol{x},m}}{c} - T_m \right) - \frac{R_{\boldsymbol{p},m}}{c} + T_m \right] \\ &= \alpha_{\boldsymbol{p},\boldsymbol{u}} \frac{R_{\boldsymbol{x},n} - R_{\boldsymbol{p},n}}{c} + \frac{R_{\boldsymbol{x},m} - R_{\boldsymbol{p},m}}{c} + \left(\frac{\alpha_{\boldsymbol{p},\boldsymbol{u}}}{\alpha_{\boldsymbol{x},\boldsymbol{v}}} - 1 \right) \left(\frac{R_{\boldsymbol{x},m}}{c} - T_m \right). \end{aligned} \tag{22}$$

The last term of (22) can be made negligible by activating the transmitters at times $-T_m$ chosen so that $R_{\boldsymbol{x},m}/c - T_m$ is negligible.

> The multistatic phase-space point-spread function is a weighted coherent sum of radar ambiguity functions evaluated at appropriate arguments.

3.3 Examples of the Point-Spread Function

The PSF contains all the information about the performance of the imaging system. Unfortunately it is difficult to visualize this PSF because it depends on so many variables. In the case when the positions and velocities are restricted to a known plane, the PSF is a function of four variables.

We would like to know whether we can find both the position and velocity of moving targets. Ideally, the PSF is delta-like, and so we can obtain both position and velocity. If, however, the PSF is ridge-like, then there will be uncertainty in some directions or in some combination of positions and velocities.

In order to look for possible ridge-like behavior, we write the PSF as

$$K(\boldsymbol{p},\boldsymbol{u};\boldsymbol{x},\boldsymbol{v}) = K\big(|\boldsymbol{p}|(\cos\theta,\sin\theta), |\boldsymbol{u}|(\cos\phi,\sin\phi), \boldsymbol{x}, \boldsymbol{v}\big). \tag{23}$$

We plot the PSF for a fixed target position $\boldsymbol{x}$ and target velocity $\boldsymbol{v}$. We then sample θ and ϕ at intervals of $\pi/4$, and for each choice of θ and ϕ, we plot $|\boldsymbol{p}|$ versus $|\boldsymbol{u}|$. This process results in $9 \times 9 = 81$ plots of $|\boldsymbol{p}|$ versus $|\boldsymbol{u}|$. Finally, to show the entire four-dimensional space at a glance, we display all the 81 plots simultaneously on a grid, arranged as shown in Fig. 4.

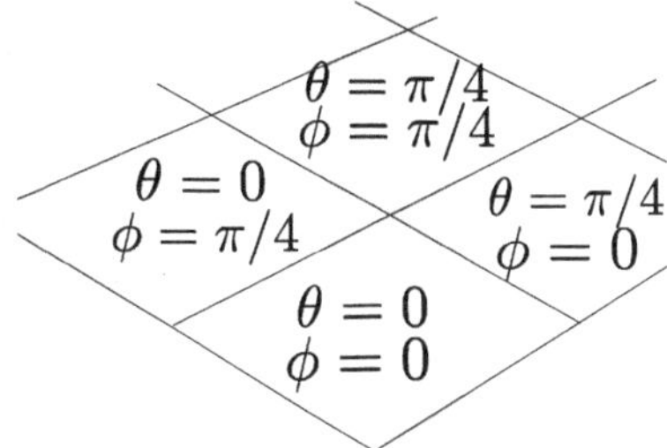

Fig. 4 This shows how our figures display the four-dimensional point-spread function (23)

3.3.1 Simulation Parameters

Our strategy in the simulations is to use a delta-like ambiguity function, and investigate the effect of geometry on the overall PSF. In all cases, we use a transmit time of $T_y = 0$:

- *Waveforms:* Two waveforms of unit amplitude are used.
  Waveform 1 is a high-range-resolution chirp of duration 9.2×10^{-6} s and bandwidth 200 MHz. It is sampled at 250 MHz (2,300 sample points).
  Waveform 4 is a single long CW pulse, of duration 0.05 s, sampled at 5 kHz (250 sample points). It has high Doppler resolution.
- *Target location and velocity:* The target location is (225 m, 45°), and its velocity is (20 m/s, 0°).

3.4 Examples

3.4.1 Two Transmitters, One Receiver

For the simulation of two transmitters and one receiver, the two transmitters are located at (10,000 m, 0) and (−10,000 m, 0), respectively, and the receiver is located at (0, 10,000 m) Ê (see Fig. 5).

The PSFs for a single transmitter and two receivers are shown in Figs. 6 through 8. Figure 6 shows the PSF when both transmitters transmit waveform 1, Fig. 7 shows the PSF when both transmitters transmit waveform 4, and Fig. 8 shows the PSF when one transmitter transmits waveform 1 and the other transmits waveform 4.

We see that for this geometry, the PSF is ridge-like. Whereas a high-range-resolution waveform provides only range information, it appears that the use of high-Doppler-resolution waveforms may be able to provide not only velocity information but also some range information. The fact that the location ridges (Figs. 6 and 8) are higher than the velocity ridges (Figs. 7 and 8) may be explained

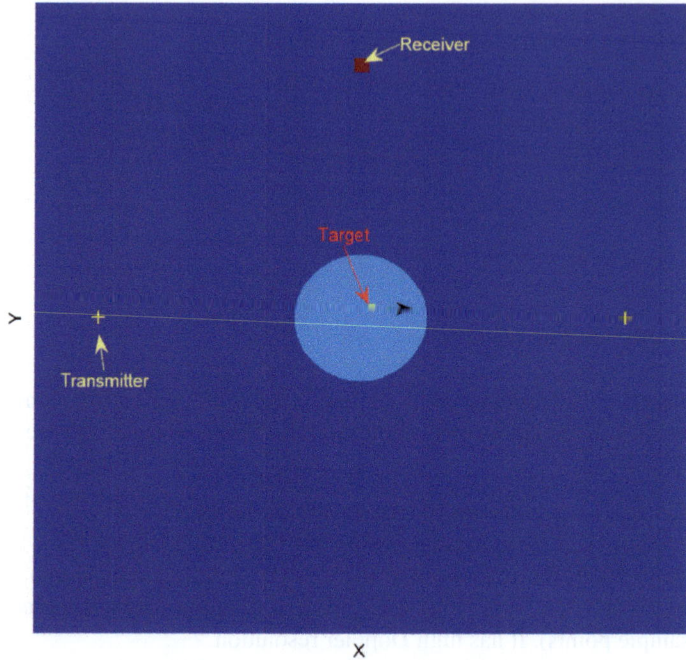

Fig. 5 This shows the geometry for the two-transmitter, single-receiver case, together with the target and region of interest (not to scale)

by the fact that the total power of (the discretized version of) waveform 1 is greater than that of (the discretized version of) waveform 4. This suggests that balancing the power of the various transmitters may be important.

The case in which the power is balanced is shown in Fig. 9. Note that since we assume that the signal due to each transmitter is known, we can do the power balancing in the reconstruction process rather than modifying the actual transmitted power.

3.4.2 Two Transmitters, Two Receivers

For the two-transmitter, two-receiver case, the two transmitters are located at (10,000 m, 0) and (0, 10,000 m), and the two receivers are located at (10,000 m, 0) and (−10,000 m, 0), respectively. See Fig. 10.

Figure 11 shows the combined PSF when the transmitter on the x-axis transmits waveform 1 and the transmitter on the y-axis transmits a multiple of waveform 4.

Comparing Figs. 8 and 11, we see that adding a receiver weakens the ambiguities.

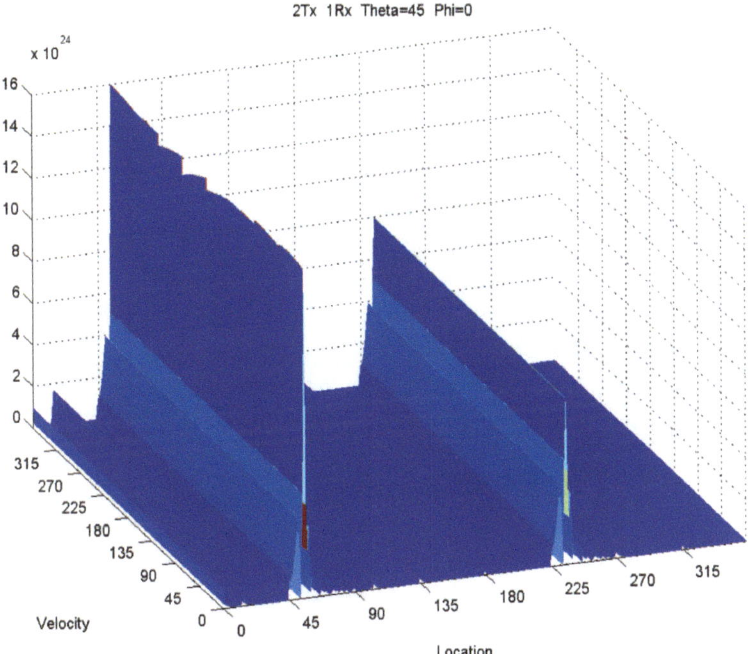

Fig. 6 This shows the combined point-spread function for the two-transmitter, single-receiver case, when both transmitters transmit waveform 1

3.4.3 Circular Geometry

We also considered a circular arrangement of eight transmitters and ten receivers. The transmitters are equally spaced around a circle of radius 10,000 m; the receivers are equally spaced around a circle of radius 9,000 m. (See Fig. 12.) The scene of interest has radius 1,000 m.

We see from Fig. 13 that the velocity information cannot be obtained from the high-range-resolution waveform.

We note that both the position and the velocity can be resolved well if the same high-Doppler-resolution waveform is used for each transmitter (Figs. 14 and 15).

4 Conclusions and Future Work

We have developed a linearized imaging theory that combines the spatial, temporal, and spectral aspects of scattered waves.

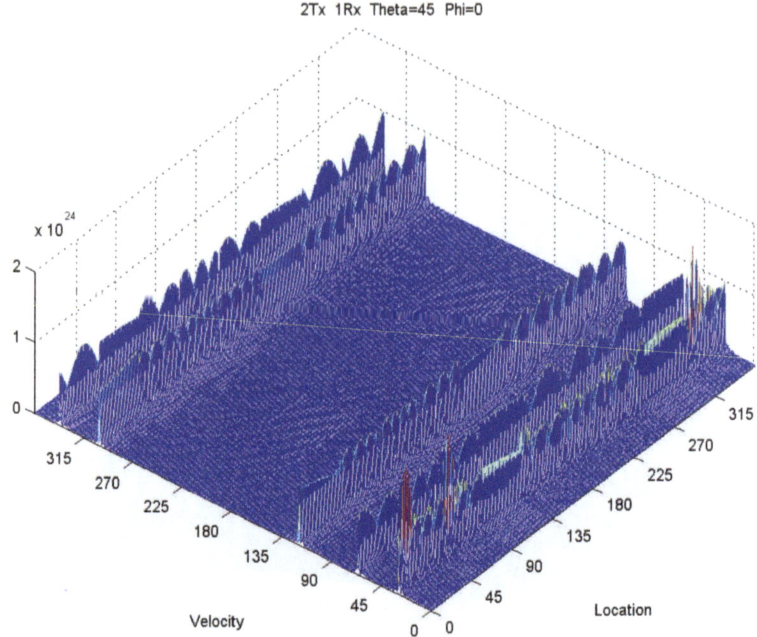

Fig. 7 This shows the combined point-spread function for the two-transmitter, single-receiver case, when both transmitters transmit waveform 4

This imaging theory is based on the general (linearized) expression we derived for waves scattered from moving objects, which we model in terms of a distribution in phase space. The expression for the scattered waves is of the form of a Fourier integral operator; consequently we form a phase-space image as a filtered adjoint of this operator or weighted matched filter.

The theory allows for activation of multiple transmitters at different times, but the theory is simpler when they are all activated so that the waves arrive at the target at roughly the same time.

We conclude that a single kind of high-range-resolution waveform should be avoided if both the position and the velocity are to be reconstructed. Furthermore, we see that a single kind of high-Doppler-resolution waveform can reconstruct not only the velocity but also the position. This may be related to the theory of Doppler SAR imaging [36].

We leave for the future an investigation of the effect of relative waveform power on the imaging results.

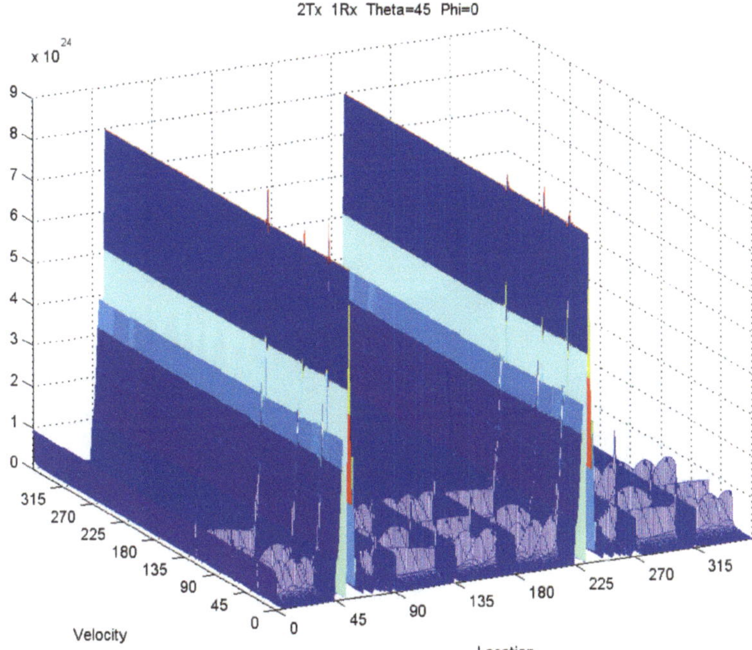

Fig. 8 This shows the combined point-spread function for the two-transmitter, single-receiver case, when the transmitter on the x-axis transmits waveform 1 and the transmitter on the y-axis transmits waveform 4

Appendix Details of the Derivation of (5) and (6)

A.1 The Transmitted Field

We consider the transmitted field ψ^{in} in the absence of any scatterers; ψ^{in} satisfies the version of Eq. (1) in which c is simply the constant background speed c_0, namely

$$[\nabla^2 - c_0^{-2}\partial_t^2]\psi^{\text{in}}(t,\boldsymbol{x},\boldsymbol{y}) = s_{\boldsymbol{y}}(t+T_{\boldsymbol{y}})\delta(\boldsymbol{x}-\boldsymbol{y}). \tag{A.1}$$

Henceforth we drop the subscript on c.

We use the Green's function g, which satisfies

$$[\nabla^2 - c^{-2}\partial_t^2]g(t,\boldsymbol{x}) = -\delta(t)\delta(\boldsymbol{x})$$

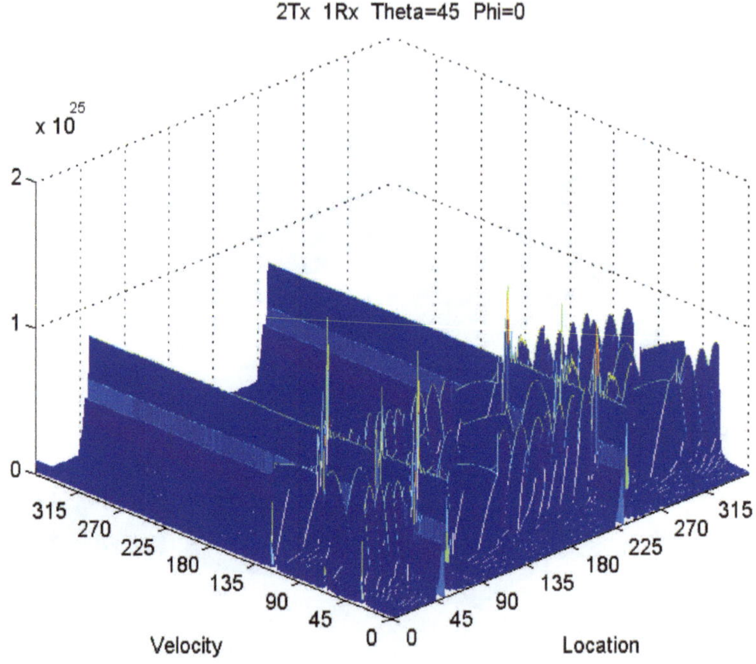

Fig. 9 This shows the combined point-spread function for the two-transmitter, single-receiver case, when the transmitter on the x-axis transmits waveform 1 and the transmitter on the y-axis transmits a power-balanced multiple of waveform 4

and is given by

$$g(t,\boldsymbol{x}) = \frac{\delta(t - |\boldsymbol{x}|/c)}{4\pi|\boldsymbol{x}|}. \tag{A.2}$$

We use (A.2) to solve (A.1), obtaining

$$\begin{aligned}\psi^{\text{in}}(t,\boldsymbol{x},\boldsymbol{y}) &= -\int g(t-t',|\boldsymbol{x}-\boldsymbol{x}'|)s_{\boldsymbol{y}}(t'+T_{\boldsymbol{y}})\delta(\boldsymbol{x}'-\boldsymbol{y})\,\mathrm{d}t'\,\mathrm{d}\boldsymbol{x}' \\ &= -\frac{s_{\boldsymbol{y}}(t+T_{\boldsymbol{y}}-|\boldsymbol{x}-\boldsymbol{y}|/c)}{4\pi|\boldsymbol{x}-\boldsymbol{y}|}.\end{aligned} \tag{A.3}$$

A.2 The Scattered Field

We write $\psi = \psi^{\text{in}} + \tilde{\psi}^{\text{sc}}$, which converts (1) into

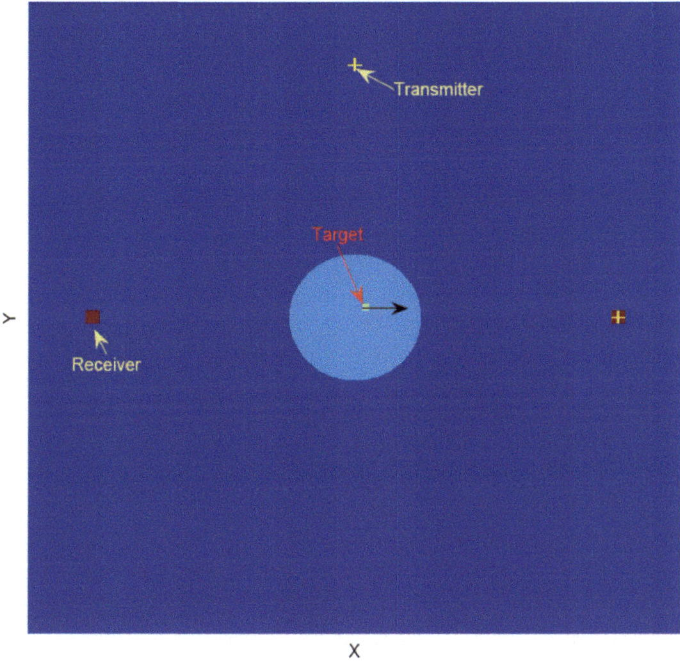

Fig. 10 This shows the geometry for the two-transmitter, two-receiver case, together with the target and region of interest (not to scale)

$$[\nabla^2 - c^{-2}\partial_t^2]\tilde{\psi}^{\mathrm{sc}} = \int q_{\boldsymbol{v}}(\boldsymbol{x} - \boldsymbol{v}t)\,\mathrm{d}^3 v\,\partial_t^2 \psi. \tag{A.4}$$

The map $q \mapsto \tilde{\psi}^{\mathrm{sc}}$ can be linearized by replacing the full field ψ on the right side of (A.4) by ψ^{in}. (This is the Born or single-scattering approximation.) The resulting differential equation we solve with the help of the Green's function; the result is

$$\psi^{\mathrm{sc}}(t,\boldsymbol{z}) = -\int g(t-t', |\boldsymbol{z}-\boldsymbol{x}|) \int q_{\boldsymbol{v}}(\boldsymbol{x} - \boldsymbol{v}t')\,\mathrm{d}^3 v\,\partial_{t'}^2 \psi^{\mathrm{in}}(t', \boldsymbol{x})\,\mathrm{d}t'\,\mathrm{d}^3 x, \tag{A.5}$$

which is (6).

Acknowledgements We are grateful to the Air Force Office of Scientific Research[1] for supporting this work under agreement FA9550-09-1-0013 and to the China Scholarship Council for supporting L.W.'s stay at Rensselaer Polytechnic Institute.

[1]Consequently the US government is authorized to reproduce and distribute reprints for governmental purposes notwithstanding any copyright notation thereon. The views and conclusions contained herein are those of the authors and should not be interpreted as necessarily representing

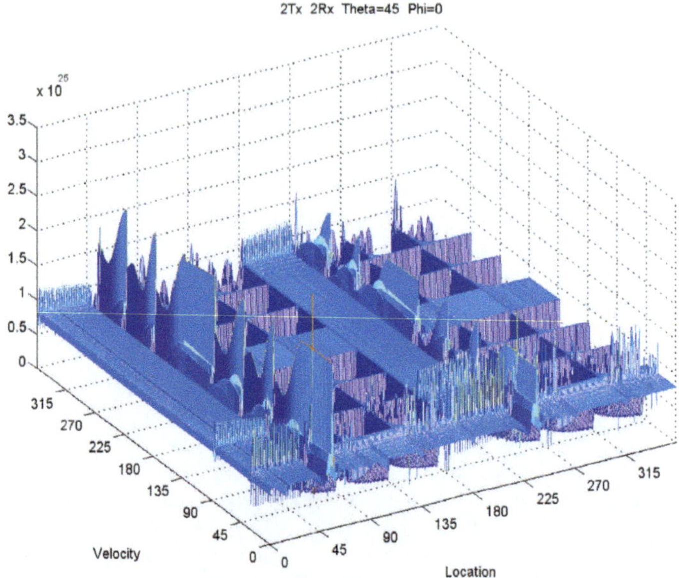

Fig. 11 This shows the combined point-spread function for the two-transmitter, two-receiver case, when the transmitter on the x-axis transmits waveform 1 and the transmitter on the y-axis transmits a power-balanced multiple of waveform 4

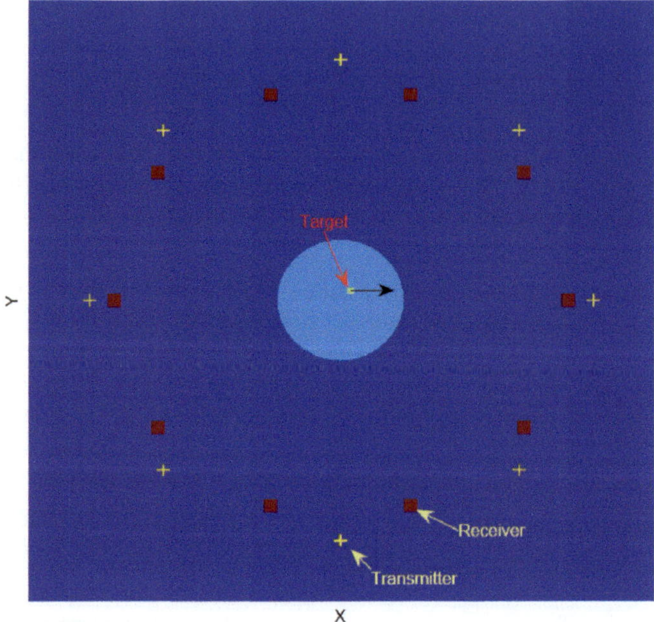

Fig. 12 This shows the circular geometry (not to scale)

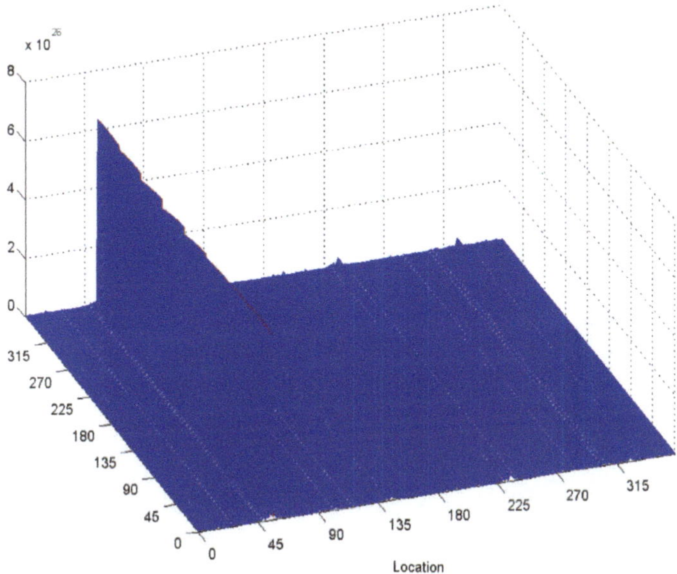

Fig. 13 This shows the combined point-spread function for the circular geometry when all transmitters are transmitting waveform 1

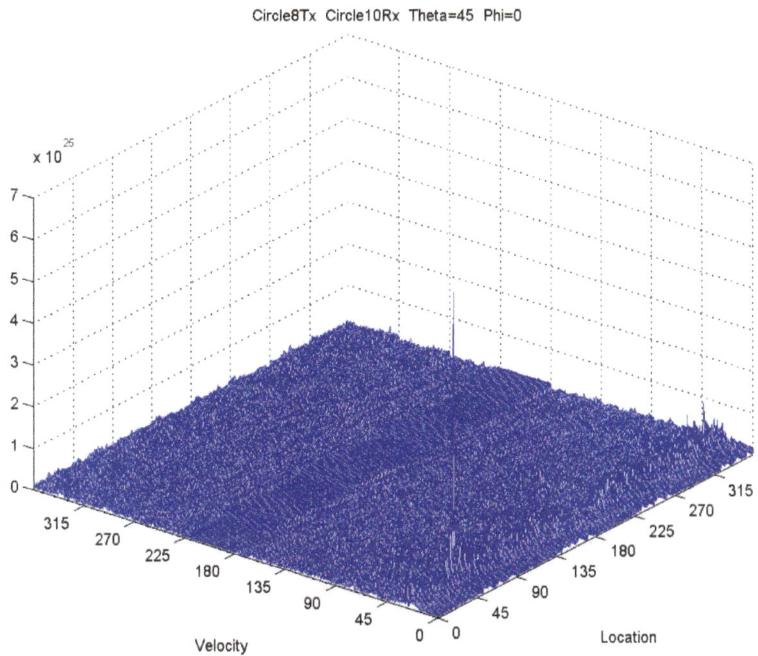

Fig. 14 This shows the combined point-spread function for the circular geometry when all transmitters are transmitting waveform 4

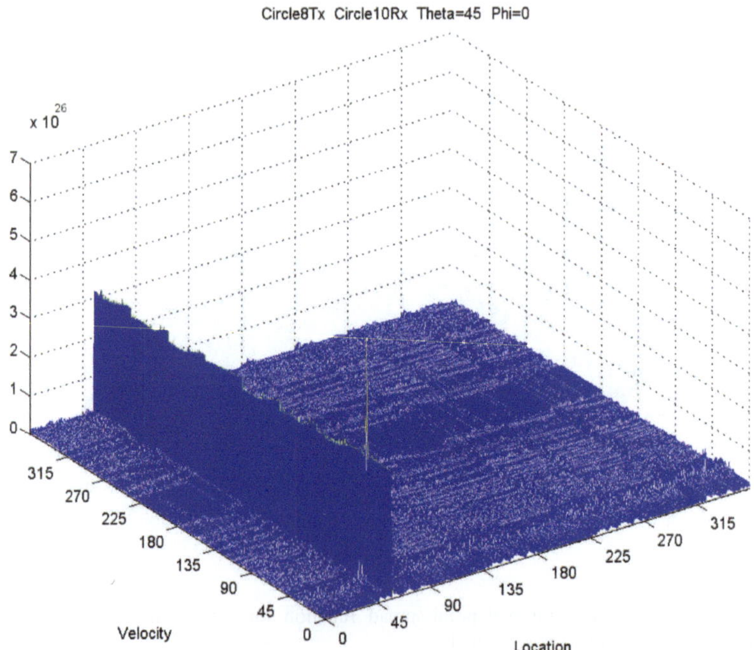

Fig. 15 This shows the combined point-spread function for the circular geometry when every other transmitter transmits waveform 1 and the others transmit a power-balanced multiple of waveform 4

References

1. Skolnik, M.I.: Radar Handbook, 3rd Edition. McGraw-Hill Professional, New York (2008)
2. Brennan, L.E., Reed, I.S.: IEEE Trans. Aerosp. Electronic Syst. **9**, 237 (1973)
3. Raney, R.K.: IEEE Trans. Aerosp. Electronic Syst. **7**(3), 499 (1971)
4. Ward, J.: Space-time Adaptive processing for airborne radar. Technique Rep. 1015, MIT Lincoln Lab., Lexington, MA (1994)
5. Klemm, R.: Principles of Space-Time Adaptive Processing. Institution of Electrical Engineers, London (2002)
6. Cooper, J.: IEEE Trans. Antennas and Propag. **AP-28**(6), 791 (1980)
7. Werness, S., Carrara, W., Joyce, L., Franczak, D.: IEEE Trans. Aerosp. Electronic Syst. **26**(1), 57 (1990)
8. Yang, H., Soumekh, M.: IEEE Trans. Image Process. **2**(1), 80 (1993)
9. Barbarossa, S.: Proc. Inst. Elect. Eng. F **139**, 79 (1992)
10. Barbarossa, S.: Proc. Inst. Elect. Eng. F **139**, 89 (1992)
11. Friedlander, B., Porat, B.: IEE Proc. Radar, Sonar Navigation **144**, 205 (1997)
12. Perry, R.P., Dipietro, R.C., Fante, R.L.: IEEE Trans. Aerosp. Electronic Syst. **35**(1), 188 (1999)
13. Jao, J.K.: IEEE Trans. Geosci. Remote Sens. **39**(9), 1984 (2001)

---

the official policies or endorsements, either expressed or implied, of the Air Force Research Laboratory or the US government.

14. Fienup, J.R.: IEEE Trans. Aerosp. Electronic Syst. **37**(3), 794 (2001)
15. Dias, J.M.B., Marques, P.A.C.: IEEE Trans. Aerosp. Electronic Syst. **39**(2), 604 (2003)
16. Kircht, M.: IEE Proc. Radar, Sonar Navigation **150**(1), 7 (2003)
17. Pettersson, M.: IEEE Trans. Aerosp. Electronic Syst. **40**, 780 (2004)
18. Minardi, M.J., Gorham, L.A., Zelnio, E.G. In: Proceedings of SPIE, vol. 5808, pp. 156–165 (2005)
19. Barbarossa, S., Farina, A.: IEEE Trans. Aerosp. Electronic Syst. **30**(2), 341 (1998)
20. Soumekh, M.: Fourier Array Imaging. Prentice-Hall, Englewood Cliffs (1994)
21. Soumekh, M.: Synthetic Aperture Radar Signal Processing with MATLAB Algorithms. Wiley-Interscience, New York (1999)
22. Stuff, M., Biancalana, M., Arnold, G., Garbarino, J. In: Proceedings of IEEE Radar Conference, pp. 94–98 (2004)
23. Himed, B., Bascom, H., Clancy, J., Wicks, M.C.: Sensors, Systems, and Next-Generation Satellites V. In: Proceedings of SPIE, vol. 4540, pp. 608–619 (2001)
24. Bradaric, I., Capraro, G.T., Weiner, D.D., Wicks, M.C. In: Proceedings of IEEE Radar Conference 2006, pp. 106–113 (2006)
25. Bradaric, I., Capraro, G.T., Wicks, M.C. In: Proceedings of Asilomar Conference (2007)
26. Adve, R.S., Schneible, R.A., Wicks, M.C., McMillan, R. In: Proceedings of 1st Annual IEE Waveform Diversity Conference, Edinburgh (2004)
27. Adve, R.S., Schneible, R.A., Genello, G., Antonik, P. In: 2005 IEEE International Radar Conference Record, pp. 93–97 (2005)
28. Landi, L., Adve, R.S. In: 2007 International Waveform Diversity and Design Conference, pp. 13–17 (2007)
29. Capraro, G.T., Bradaric, I., Weiner, D.D., Day, J.P.R., Wicks, M.C. In: International Waveform Diversity and Design Conference, Lihue, HI (2006)
30. Cook, C.E., Bernfeld, M.: Radar Signals. Academic, New York (1965)
31. Woodward, P.M.: Probability and Information Theory, with Applications to Radar. McGraw-Hill, New York (1953)
32. Levanon, N.: Radar Principles. Wiley, New York (1998)
33. Franceschetti, G., Lanari, R.: Synthetic Aperture Radar Processing. CRC Press, New York (1999)
34. Willis, N.J.: Bistatic Radar. Artech House, Norwood (1991)
35. Willis, N.J.: Bistatic Radar in Radar Handbook, 2nd edn. In: Skolnik, M. I. (ed.) McGraw-Hill, New York (1990)
36. Borden, B., Cheney, M.: Inverse Probl. **21**, 1 (2005)
37. Willis, N.J., Griffiths, H.D.: Advances in Bistatic Radar. SciTech Publishing, Raleigh (2007)
38. Tsao, T., Slamani, M., Varshney, P., Weiner, D., Schwarzlander, H., Borek, S.: IEEE Trans. Aerosp. Electronic Syst. **33**, 1041 (1997)
39. Colton, D., Kress, R.: Inverse Acoustic and Electromagnetic Scattering Theory. Springer (1992)
40. Devaney, A.J.: Opt. Lett. **7**, 111 (1982)
41. Devaney, A.J.: Ultrason. Imaging **4**, 336 (1982)
42. Varsolt, T., Yazici, B., Cheney, M.: Inverse Probl. **24**(4), 045013 (28 pp.) (2008)
43. Cheney, M., Borden, B.: Inverse Probl. **24**, 035005(1 (2008)
44. Nolan, C.J., Cheney, M.: Inverse Probl. **18**, 221 (2002)
45. Swick, D.: A Review of Wideband Ambiguity Functions. (Naval Research Laboratory Rep. 6994, 1969)
46. Webster, T., Liwei, Xu., Cheney, M.: IEEE Radar Conference (RADAR), Radar Div., Naval Res. Lab., Conference Publications. Washington, DC, USA 332–337 (2012)

Exploitation Performance and Characterization of a Prototype Compressive Sensing Imaging Spectrometer

Christopher J. Deloye, J. Christopher Flake, David Kittle, Edward H. Bosch, Robert S. Rand, and David J. Brady

Abstract The coded aperture snapshot spectral imager (CASSI) systems are a class of imaging spectrometers that provide a first-generation implementation of compressive sensing themes to the domain of hyperspectral imaging. Via multiplexing of information from different spectral bands originating from different spatial locations, a CASSI system undersamples the three-dimensional spatial/spectral data cube of a scene. Reconstruction methods are then used to recover an estimate of the full data cube. Here we report on our characterization of a CASSI system's performance in terms of post-reconstruction image quality and the suitability of using the resulting data cubes for typical hyperspectral data exploitation tasks (e.g., material detection, pixel classification). The data acquisition and reconstruction process does indeed introduce trade-offs in terms of achieved image quality and the introduction of spurious spectral correlations versus data acquisition speedup and the potential for reduced data volume. The reconstructed data cubes are of sufficient quality to perform reasonably accurate pixel classification. Potential avenues to improve upon the usefulness of CASSI systems for hyperspectral data acquisition and exploitation are suggested.

C.J. Deloye
The MITRE Corporation, 7515 Colshire Drive, McLean, VA 22102, USA
e-mail: cdeloye@mitre.org

J.C. Flake • E.H. Bosch (✉) • R.S. Rand
National Geospatial-Intelligence Agency, Springfield, VA, USA
e-mail: Edward.H.Bosch@nga.mil

D. Kittle • D.J. Brady
Fitzpatrick Institute for Photonics and Department of Electrical and Computer Engineering, Duke University, 129 Hudson Hall, Durham, NC 27708, USA
e-mail: dbrady@duke.edu

T.D. Andrews et al. (eds.), *Excursions in Harmonic Analysis, Volume 1*,
Applied and Numerical Harmonic Analysis, DOI 10.1007/978-0-8176-8376-4_8,

Keywords Imaging spectrometer • Hyperspectral imaging • Compressive sensing

1 Introduction

Candès et al. [4] and Donoho's [7] seminal work on compressive sensing have led to significant developments involving sub-Nyquist sampling rates which can potentially impact sensing modalities that require a large number of costly and time-intensive measurements. These consequential results are paving the way for sensing devices with faster acquisition rates, lower power consumption, decreased need for large data storage, or faster transmission/downlink rates.

Such developments have the potential to positively impact the data acquisition rates, data transmission latencies, and storage requirements for so-called hyperspectral imaging systems. These are imaging spectrometers that typically collect hundreds of individual spectral bands for each spatial pixel in an acquired scene. The spectral information obtained allows one to detect and identify target materials in a scene or to classify spatial locales based on spectral similarity. This technology has found applications in the identification of ground surface mineral deposits, monitoring vegetation health, estimating crop yield, monitoring chemical and oil spills, and disaster relief efforts, among others. The data generation rate from these systems is significant: each pixel's spectral information takes ~ 0.1–1 kB depending on number of bands and data precision; a full three-dimensional (two spatial/one spectral) "data cube" can thus be many hundreds of megabytes. However, most scenes are typically highly correlated both spatially and spectrally, so much of this information is redundant.

Applying compressive sensing ideas to realize the above-mentioned benefits in the realm of hyperspectral imaging relies on taking advantage of this intrinsic redundancy. Initial attempts in this direction are already under way. For example, Castrodad et al. [6] and Krishnamurthy et al. [12] both have considered different modes of simulating compressively sensed hyperspectral data. Both studies demonstrated the ability to exploit data cubes reconstructed from a sparse sampling. Castrodad et al. demonstrated accurate classification results even when only 20 % of the full data cube was used in reconstructions. While showing initial promise, actually moving from these simulations to a functioning real-world compressive sensing imaging spectrometer is a tall order. Specifically, the physical limitations of actual optical systems make it difficult to design a system that satisfies the requirements of the pure compressive sensing theory (e.g., the measurement matrix satisfying the restricted isometry property (RIP) or having a small coherence between the measurement matrix and the data representation basis).

With that said, several optical imaging systems that exploit scene redundancy to reduce the amount of information collected have been built. The Rice single pixel camera is an instrument that reduces information by randomly multiplexing spatial information together, with a design benefit of keeping the cost down for expensive sensing elements [21]. It is also possible to enhance existing sensing systems by adding compressive sensing structure. For example, the coded strobing photography system increases temporal resolution of video by taking advantage of periodic motion and utilizing a coded strobing light into the imaging system [18]. There is also a class of systems, the coded aperture snapshot spectral imagers (CASSI), explicitly designed to collect compressive measurements of a hyperspectral data cube. The full data cube can then be reconstructed from the reduced set of measurements. The design of the camera gives it advantages when imaging dynamic scenes, imaging in low-light situations, and in obtaining low-resolution persistent observations.

In this contribution we report on a study to characterize the performance of the single-disperser CASSI design [20]. In Sect. 2 we describe the mathematical underpinnings of this system's design and data reconstruction methods and draw out the connections these have with the compressive sensing theory. Then we describe in Sect. 3 the design goals and implementation of our experiment. In Sects. 4 and 5 we discuss the CASSI system's performance in terms of reconstructed image spatial quality and the ability to exploit the spectral information to classify pixels based on materials present in the scene. We summarize and conclude in Sect. 6.

2 Coded Aperture Snapshot Spectral Imager

The single-disperser CASSI sensor can be modeled as a sequence of consecutive mathematical operations. A scene is imaged with a standard lens onto a coded aperture. The transmitted light is then spectrally filtered to a light wavelength bandpass between 450–650 nm, relayed to a prism which disperses the light by wavelength. The dispersion also causes each pixel on the imaging CCD to receive a multiplexed signal; that is, each pixel receives light from a range of spatial locations with each location contributing a different wavelength of light into the pixel.

2.1 Forward Model

Consider the scene represented as a spectral volume $I(x',y',\lambda)$ that describes the reflective intensity from the location in the scene that is projected onto the (x',y') location of the coded aperture mask at some wavelength $\lambda \in [\Lambda_1,\Lambda_2]$. The following calculations will adhere to the schematic diagram in Fig. 1. The aperture code can be described spatially as a binary-valued function $A(x',y') \in \{0,1\}$. For a fixed wavelength we can write the reimaged and spatially modulated scene projected forward from the aperture mask location as

$$S(x',y',\lambda) = I(x',y',\lambda)A(x',y'). \tag{1}$$

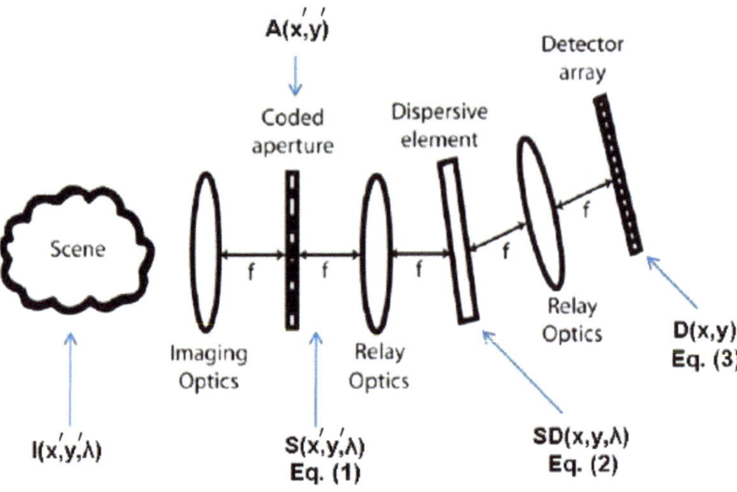

Fig. 1 Schematic of the CASSI forward model from [20]

The prism dispersion can be described by the function $p(\lambda)$, and thus we can write the idealized spectral density at a point (x,y) on the detector for the wavelength λ as

$$SD(x,y,\lambda) = I(x'+p(\lambda),y',\lambda)A(x'+p(\lambda),y') = S(x'+p(\lambda),y',\lambda), \quad (2)$$

where we have modeled the optical propagation through the imaging optics and disperser as

$$\begin{aligned} SD(x,y,\lambda) &= \int_R \int_R S(x',y',\lambda)\delta(x'-(x+p(\lambda)))\delta(y'-y)\,\mathrm{d}y'\mathrm{d}x' \\ &= \int_R \int_R I(x',y',\lambda)A(x',y')\delta(x'-(x+p(\lambda)))\delta(y'-y)\mathrm{d}x'\mathrm{d}y' \\ &= I(x+p(\lambda),y,\lambda)A(x+p(\lambda),y). \end{aligned} \quad (3)$$

The imaging system multiplexes neighboring spectra due to the shear from the dispersive element and this integration can be written down as

$$D(x,y) = \int_{\Lambda_1}^{\Lambda_2} SD(x,y,\lambda)\,\mathrm{d}\lambda, \quad (4)$$

where we have not accounted for any wavelength or spatially dependent response of the detector array.

If we consider the volume to be a voxelized hypercube, we can discretize the forward model by taking into account the physical expanse of the scene, coded aperture, and detector array. This is illustrated in Fig. 2. If the size of a pixel on the

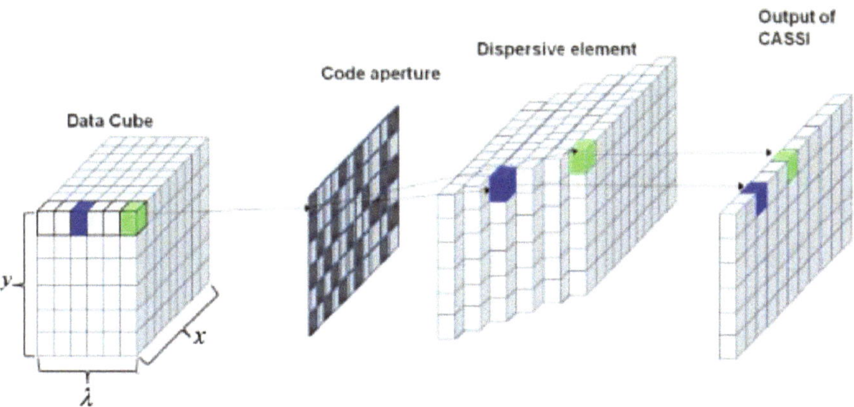

Fig. 2 Discretized schematic of CASSI forward model from [1]

detector array is ρ, then we can integrate over the multiplexed scene across array elements as

$$d[i,j] = \int_R \int_R D(x,y) \chi_{[\rho(i-\frac{1}{2}),\rho(i+\frac{1}{2})] \times [\rho(j-\frac{1}{2}),\rho(j+\frac{1}{2})]} \mathrm{d}x\mathrm{d}y, \tag{5}$$

where χ is the characteristic function

$$\chi_A(x) = \begin{cases} 1 & x \in A \\ 0 & x \notin A \end{cases}.$$

Here the pixels are perfectly square and have no spacing between consecutive elements, and $d[i,j]$ are the intensity measurements made by the detector array at position (i,j).

We can also discretize the aperture code as a mask of equally sized square holes. If we ensure that the size of the aperture code features is suitably large compared to the wavelengths passing through them, then we can simplify the model by not representing diffraction in the system. We set the feature size of the mask elements to be η. Then

$$A(x',y') = \sum_{i,j} a_{i,j} \chi_{[\eta(i-\frac{1}{2}),\eta(i+\frac{1}{2})] \times [\eta(j-\frac{1}{2}),\eta(j+\frac{1}{2})]}, \tag{6}$$

where the binary aperture code elements are $a_{i,j} \in \{0,1\}$. The final detector measurements are then

$$\begin{aligned} d[i,j] &= \int_R \int_R D(x,y) \chi_{[\rho(i-\frac{1}{2}),\rho(i+\frac{1}{2})] \times [\rho(j-\frac{1}{2}),\rho(j+\frac{1}{2})]} \mathrm{d}x\mathrm{d}y \\ &= \int_R \int_R \int_{\Lambda_1}^{\Lambda_2} SD(x,y,\lambda) \chi_{[\rho(i-\frac{1}{2}),\rho(i+\frac{1}{2})] \times [\rho(j-\frac{1}{2}),\rho(j+\frac{1}{2})]} \mathrm{d}\lambda \mathrm{d}x\mathrm{d}y; \end{aligned}$$

substituting with Eq. (2) yields

$$= \int_R \int_R \int_{\Lambda_1}^{\Lambda_2} I(x+p(\lambda),y,\lambda) A(x+p(\lambda),y) \chi_{[\rho(i-\frac{1}{2}),\rho(i+\frac{1}{2})i]\times[\rho(j-\frac{1}{2}),\rho(j+\frac{1}{2})]} \, \mathrm{d}\lambda \mathrm{d}x\mathrm{d}y;$$

finally substituting with Eq. (6) gives

$$= \sum_{i,j} a_{i,j} \int_R \int_R \int_{\Lambda_1}^{\Lambda_2} I(x+p(\lambda),y,\lambda) a_{i,j} \chi_{[\eta(i-\frac{1}{2})-p(\lambda),\eta(i+\frac{1}{2})-p(\lambda)]\times[\eta(j-\frac{1}{2}),\eta(j+\frac{1}{2})]}$$
$$\cdot \chi_{[\rho(i-\frac{1}{2}),\rho(i+\frac{1}{2})]\times[\rho(j-\frac{1}{2}),\rho(j+\frac{1}{2})]} \, \mathrm{d}\lambda \mathrm{d}x\mathrm{d}y.$$

2.2 Inversion of the CASSI Model

The forward model of the CASSI sensing system takes a data cube I and maps it onto the detector response d. We represent this process with the operator H and write $HI = d$. This operator is underdetermined and, in general, solving the inverse problem is challenging. Recent results in the compressive sensing and sparse representation literature have focused attention on methods and situations in which this inverse problem can be made well posed and solved with techniques from optimization theory [4,5,7].

A simple form of the problem can be written down in the following way. Suppose f is a signal

$$f : Z_N \to R,$$

and $\Omega \subset \widehat{Z_N}$ is a small subset of the dual-group

$$\widehat{Z_N} = \{\psi : Z_N \to C \,|\, |\psi(n)| = 1 \text{ and } \psi \text{ is a continuous homomorphism}\}.$$

We wish to understand when it is possible to discern all of f from only the small collection of samples $\hat{f}$ restricted to the set Ω. In general this cannot be done as restricting the values of the Fourier transform essentially filters the data and when the content in f is large, the inverse transform cannot recreate the signal. Donoho, Candès, Romberg, and Tao discovered that this problem can be solved exactly with a high probability if the underlying signal is sparse and the set Ω is sufficiently sized. Moreover, they showed that this solution can be retrieved by an application of convex minimization.

If we write the discrete Fourier transform (DFT) as F and the subselection of the elements in Ω as the operator Φ, then we can write the minimization problem as

$$\min \|g\|_{\ell_0} \text{ subject to } \Phi F g = y, \tag{7}$$

where $y = \hat{f}\mid_{\Omega}$. The solution to this problem is not computationally tractable. Part of the innovation in the papers listed above is that the solution to (7) can be realized as the more easily implemented problem

$$\min \|g\|_{\ell_1} \text{ subject to } \Phi F g = y. \tag{8}$$

The solution to (7) is equivalent to (8) with high probability when Φ satisfies the RIP and the number of elements in Ω satisfies

$$|\Omega| \geq C\mu^2(\Phi, F) S \log N, \tag{9}$$

where C is a constant that depends on the probability of successful reconstruction, $\mu(\Phi, F)$ is the coherence term, and S is the underlying sparsity of the signal.

Definition 1. The coherence between two representations Φ and F is

$$\mu(\Phi, F) = \max_{i \neq j} \left| \langle \Phi_{i*}, F_{*j} \rangle \right|,$$

where Φ_{i*} is the ith row of Φ and F_{*j} is the jth column of F.

Definition 2. A matrix $\Phi \in R^{m\times n}$ satisfies the RIP for an integer $s < n$ if $\exists \delta_s > 0$ such that for every $m \times s$ submatrix of Φ, Φ_s and for every g,

$$(1 - \delta_s)\|g\|_{\ell_2}^2 \leq \|\Phi_s g\|_{\ell_2}^2 \leq (1 + \delta_s)\|g\|_{\ell_2}^2.$$

It should be noted that the related problem

$$\min \|g\|_{TV} \text{ subject to } \Phi F g = y \tag{10}$$

can be solved by applying problem (8) to the derivative of g given sufficient smoothness conditions. This is important for our application of the theory since the total variation norm regularizes tend to produce excellent visual fidelity in image reconstruction problems [17].

The CASSI instrument can be realized as an application of compressive sensing by observing that many of the natural scenes the system may wish to image are spatially compressible in any given wavelength. The number of measurements from the CCD are smaller than would be needed to completely describe the volume I and so we have a system that incoherently samples a small collection of measurements from a spatially compressible set of images. Given how much redundancy exists between various layers of the hypercube, it makes sense that the cube is compressible in some wavelet basis. Therefore, if we write this transform as W, we can decompose the volume into $WI = \Gamma$ and then write the CASSI system as

$$HI = d \Longrightarrow \left(HW^{-1}\right)\Gamma = d;$$

here H acts like the subselection matrix Φ and W^{-1} as the basis F from the simple model in Eq. (8). If the coded aperture is chosen as a binary random pattern from an appropriate distribution, H will satisfy the RIP with high probability [2]. Similarly a basis can be found that both sparsifies the volume I and has minimal coherence with the sensing matrix H, the discrete cosine transform being an example [19]. The reconstruction inverse problem can then be written in the form of Eq. (8):

$$W^{-1}\left(\arg\min_{\Gamma} \| \left(HW^{-1}\right)\Gamma - y\|_{\ell_2} + \tau\|\Gamma\|_{\ell_1}\right), \tag{11}$$

where τ is appropriately chosen so that the unique solution from this unconstrained convex problem coincides with the solution obtained from the similar constrained problem. Equation (11) is how the designers of the CASSI system originally cast the reconstruction problem [10, 20], using the gradient projection for sparse reconstruction method [9] to solve this inverse problem. The ℓ_1 regularizer tends to produce wavelet coefficients that are sparse, which is a desirable property given our assumptions on the underlying scene. However, if we instead wish to place higher priority on obtaining band-wise reconstructions that are sparse in the spatial gradient, we can cast the inverse problem in the form of Eq. (10):

$$\arg\min_{I} \|HI - y\|_{\ell_2} + \tau\|I\|_{TV}, \tag{12}$$

where the total variation regularizer encourages exactly this sparsity of the spatial gradient in the reconstructions [3]. Because of this property, we prefer the use of Eq. (12), and it is this convex optimization problem that we use for the reconstructions in this chapter. However, the authors are not aware of any theorems guaranteeing the equivalence of (12) and (11).

3 Experiment Design Summary

With the theoretical discussion of the CASSI imaging spectrometer data acquisition and reconstruction in place, we now turn to our evaluation of a prototype CASSI system. In this evaluation, we considered both spatial image quality measures and the usefulness of the reconstructed data cubes to perform material detection and scene classification. There were two primary challenges to making our experimental tests realistic. First, we were restricted to using a laboratory space, so direct imaging within typical remote-sensing environments with all their inherent spatial and spectral complexity was not possible. Second, the CASSI prototype used was only sensitive to the visible band, a region of the spectrum where many materials do not have identifying spectral features.

The goal then was to develop targets that would provide some degree of the spatial and spectral complexity found in an actual remote-sensing environment. Capturing both types of complexity at the same time was not achievable.

Fig. 3 The two targets used in this experiment. *Left panel*: the 6-ink printout of the overhead suburban scene. *Right panel*: the petri dish target containing real materials

As a compromise, we developed two targets that independently captured these different forms of complexity. The first target—intended to provide a spatially complex scene—was a large-format ($\approx 813 \times 914$ mm) six-color print of a typical suburban landscape obtained from high-resolution overhead imagery available from U.S. Geological Survey websites. The other target provided more spectral complexity and consisted of several materials in various combinations sealed within petri dishes that were then mounted on plywood. Images of the two targets are shown in Fig. 3.

The materials used in the petri dish target included construction sand, dried moss, green and brown paint chips, red fabric, a rare-earth compound neodymium oxide (Nd_2O_3), and polystyrene beads coated with a UV fluorescing paint. The sand and plywood served as our background materials. The moss provided a natural foliage spectrum; the paint chips, a set of man-made materials with colors similar to the foliage; the red fabric, a man-made material with a spectra distinct from the foliage; and the Nd_2O_3 provided a substance with numerous sharp absorption features in the visible window. The fluorescent paint was intended to provide a sharp emission feature; however, the paint's fluorescence efficiency was not large enough to allow the emission feature to be visible relative to the intensity of the illuminating lights. The set of spectra provided by these materials include both unique elements (the Nd_2O_3 and red fabric) and several sets of mutual "confusers," that is, materials with very similar spectra (sand, plywood, and paint chips). In total, the real materials "petri dish" target consisted of 85 petri dishes, the contents of which were chosen to provide tests of material detection and classification against variations in object size, total dish coverage fraction, and relative material abundances.

The CASSI instrument used in our tests had a spectral bandpass between ≈ 460 and 650 nm divided into 47 spectral bands of nonuniform width. The individual bandwidths varied from ≈2 nm at the blue end of this range to 8 nm at the red end. The system's native instantaneous field of view (IFOV), that is, the half-angle extent of the cone visible to an individual pixel based solely on the detector's pixel pitch and the system's effective focal length, was 0.0004 radians. However, the coded aperture mask had a minimum feature size of 3 pixels when projected onto the detector array. This impacts the resolution achieved in the reconstructions, increasing it to a nominal effective IFOV of ≈0.001 radians. As discussed below, the effective IFOV actually achieved does vary from this nominal value.

For our data collection, the targets were wall mounted and illuminated with 16 Solux 50 W 4,700 K flood lamps. The target-to-objective distance was 5.2 m. With the nominal CASSI effective IFOV, this gives a ground sample distance (GSD, the physical length of the target that is imaged onto the height of a single pixel) of ≈5.2 mm. We acquired multiple snapshots of each target with the CASSI system. Between each snapshot, the coded mask's position in the focal-plane of the objective lens was altered. We reconstructed the three-dimensional spatial/spectral data cubes using between 1 and 64 of these snapshots by finding the minimum of Eq. (12) on a band-by-band basis. We also acquired multiband imagery of both targets with a conventional digital imaging system to serve as a reference point against which the CASSI reconstructions could be compared.

4 Image Spatial Quality Evaluation

Figure 4 compares a 64-snapshot CASSI reconstruction of the overhead suburban scene with the reference image data (scaled and registered to the CASSI reconstructions) in the blue, green, and red channels. The wavelengths of the corresponding CASSI band centers are indicated on the figure. One immediate comment is that the CASSI reconstructions achieve a higher spatial resolution in the longer wavelength bands. This is not too surprising given that the CASSI bands are wider at longer wavelengths and the overall transmission efficiency of the optics and detector drop off below 500 nm. In other words, the shorter wavelength bands suffer from a lower signal-to-noise ratio and this impacts the reconstruction quality. However, even in the longest-wave band displayed in Fig. 4, the CASSI reconstructions show a resolution degradation relative to the scaled reference data.

4.1 Relative Resolution Analysis

One question of interest then is what resolution can a CASSI reconstruction obtain relative to the system's native resolution. To estimate the answer to this question, we employed a method developed by Nunez et al. [14] for calculating the relative resolution between a test image and a higher-resolution reference image of the same scene. This method employs an "á trous" discrete wavelet transform of the higher-

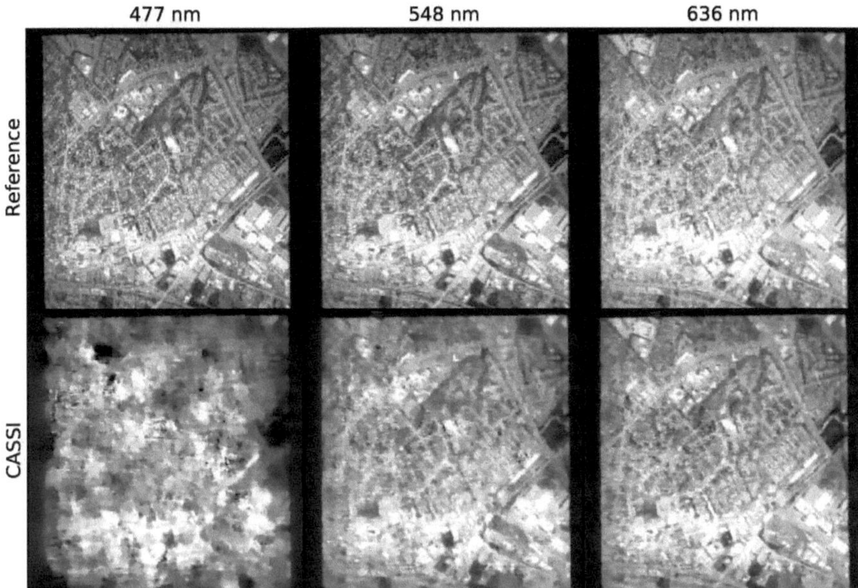

Fig. 4 A comparison between the reference and CASSI images of the overhead suburban scene in three bands corresponding to *blue*, *green*, and *red* channels, respectively. The CASSI images were reconstructed from 64 snapshots

resolution reference image to produce a series of approximating images with increasingly lower resolution. The wavelet transform employed is dyadic so that the relative resolution between successive image approximations is a factor of 1/2. That is, at the wavelet function scale ι, where ι is simply the number of successive applications of the transform required to produce a given approximating image, the resolution of the resulting approximation is a factor of $2^{-\iota}$ relative to the original image. Details of the application of this "á trous" transform algorithm to produce the lower-resolution image stack can be found in [15].

We estimate the test images' relative resolution by first calculating the correlation coefficient between the test image and each of the approximations in the lower-resolution wavelet image stack. The resulting correlation coefficients are interpolated as a function of ι. The relative resolution is then determined from the ι value, $\iota_{\max}$, at which the correlation coefficient interpolation achieves its maximum.

Using the conventional camera images (again, rescaled and registered to the CASSI reconstructions) as our reference, we compared the relative resolution of the CASSI reconstructions of both targets at the three CASSI spectral bands centered on wavelengths $\lambda = 477{,}548$, and 636 nm as a function of number of snapshots used in the reconstructions. Figure 5 summarizes our results in the form of the resolution of the reference images relative to the CASSI reconstructions (i.e., plotted there is $2^{\iota_{\max}}$, not $2^{-\iota_{\max}}$) as a function of snapshot number in each spectral band.

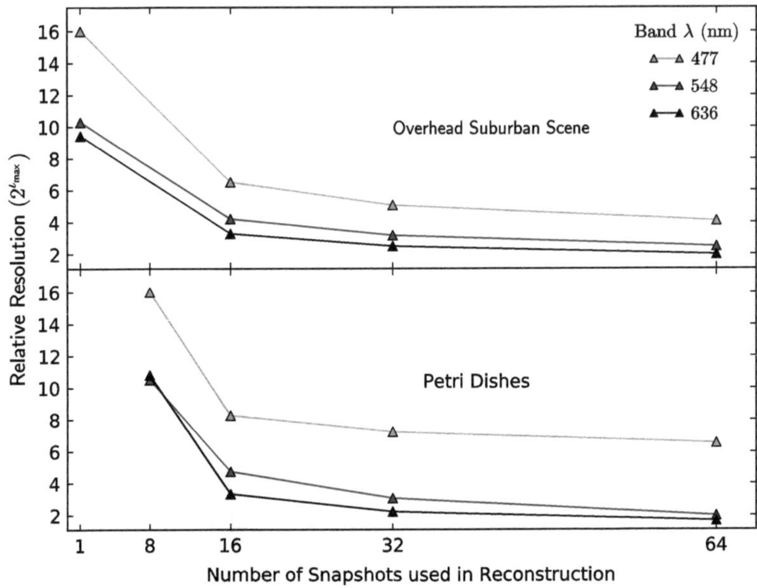

Fig. 5 The resolution of the reference images relative to the CASSI reconstructions (i.e., $2^{l_{max}}$) in the three designated spectral bands as a function of snapshot number. The *upper panel* shows the results for the overhead suburban scene, the *lower panel* those for the petri dish target

The poorer resolution achieved by the reconstructions in the blue follow our expectations from Fig. 4. The blue band resolution is between 2/3 and 1/4 that achieved in the red with a trend to poorer relative performance with decreasing number of snapshots used in the reconstruction. Also apparent is the general trend of improving resolution with snapshot number: between 16 and 64 snapshots, the suburban scene reconstruction resolution improved $\approx$1.7 times, while the petri dish reconstructions improved by a factor of $\approx$2.2.

The original GSD of the reference images was a factor of $\approx$1.16 larger than that of the CASSI system's native resolution. Accounting for this additional factor, the above results indicate that for 64-snapshot reconstructions, the CASSI system used in this experiment can deliver a red-band image resolution that is $\approx$1.9–2.2 times poorer than the system's native resolution. For the 32-snapshot reconstructions, the image resolution is $\approx$2.6–2.9 times reduced from the system's native resolution.

4.2 *Image Fourier Power Spectra*

A more complete description of the spatial characteristics of the CASSI reconstructions can be obtained via analysis of the individual band images' spatial Fourier transforms. Here, specifically, we focus on comparisons between the Fourier power spectral density of a set of reconstructions and reference images. To avoid potential confusion of terminology, we detail how we calculate the power spectral density below.

The image in a single band can be represented by a function, $\mathscr{F}(x,y)$, giving the intensity at each point on the discrete set of the image's pixel coordinates (x,y). The two-dimensional DFT of the image, $\tilde{\mathscr{F}}(k_x,k_y)$, is then a complex-valued function of the discrete set of horizontal and vertical wavenumbers k_x, k_y. The power present at (k_x,k_y) is defined as

$$P(k_x,k_y) = \tilde{\mathscr{F}}^*(k_x,k_y)\tilde{\mathscr{F}}(k_x,k_y), \tag{13}$$

where $\tilde{\mathscr{F}}^*$ indicates the complex conjugate of $\tilde{\mathscr{F}}$. We now estimate the imaged scene's azimuthally averaged Fourier power spectral density, $\mathrm{d}P/\mathrm{d}k$, by taking concentric radial annuli in (k_x,k_y)-space, summing the power present in each annulus, and dividing this total power by the area of the annulus. That is, we bin our discrete set of (k_x,k_y) by radial wavenumber, $k = \sqrt{k_x^2+k_y^2}$. The power in the ℓth such radial bin is

$$P_\ell \equiv \sum_{(k_x,k_y)\in k_\ell} \tilde{\mathscr{F}}^*(k_x,k_y)\tilde{\mathscr{F}}(k_x,k_y)\,, \tag{14}$$

where k_ℓ designates the set of all (k_x, k_y) coordinate pairs that fall within ℓth bin. We then divide P_ℓ by the area of the ℓth bin's cylindrical annulus. In this process, we apply a normalization so that the total power has value unity

$$P_{\mathrm{tot}} \equiv \sum_\ell P_\ell = 1\,. \tag{15}$$

The corresponding definition of $\mathrm{d}P/\mathrm{d}k$ in the continuum limit would be given by

$$2\pi \int_0^{k_{\max}} \frac{\mathrm{d}P}{\mathrm{d}k} k\mathrm{d}k = P_{\mathrm{tot}} = 1\,, \tag{16}$$

where $k_{\max} = 1/\sqrt{2}$ is the maximum value attained by k.

Figure 6 shows several comparisons between the $\mathrm{d}P/\mathrm{d}k$ of red-band images for both targets. Figures 6a, b highlight how the CASSI reconstructions' $\mathrm{d}P/\mathrm{d}k$ vary with snapshot number (with the $\mathrm{d}P/\mathrm{d}k$ of the reference photo's red channel shown for comparison). These panels demonstrate that there is an ordering of power at $k \gtrsim 0.05$ with snapshot number, that is, the relative amount of mid-to-high-frequency power increases with number of snapshots. This is obviously consistent with the improved resolution with snapshot number seen above. Additionally, the distinguishing feature between reconstructions with differing snapshot number is the relative rate of fall off in $\mathrm{d}P/\mathrm{d}k$ between $0.05 \lesssim k \lesssim 0.2$. For $k \gtrsim 0.2$, the CASSI $\mathrm{d}P/\mathrm{d}k$ all tend to have the same slope, at least for reconstructions based on 8 or more snapshots. That is, the relative scaling high-frequency power with k appears to be independent of the number of snapshots once a sufficient number of snapshots are used.

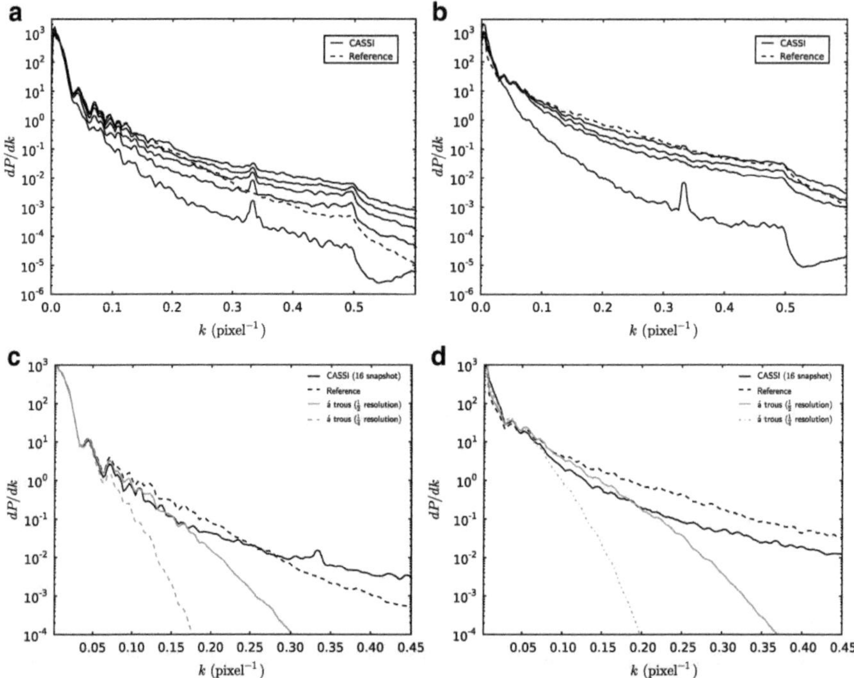

Fig. 6 Panel (**a**): the normalized, discrete Fourier power spectral densities, d$P/$dk of reconstructions using 1, 8, 16, 32, and 64 snapshots for the petri dish target's red-band (*solid lines*). The *dashed line* shows the d$P/$dk of this target's reference photograph red channel resized and registered to the CASSI data. Panel (**b**): same as (a), but for the overhead suburban scene. In this case, there is no 8-snapshot reconstruction. Panels (**c**) and (**d**): comparisons showing the impact of the reference image "átrous" wavelet resolution degradation relative to CASSI reconstructions for the petri dish and overhead suburban targets, respectively

As a consequence of this, it seems that the information present in CASSI reconstructions at $k \gtrsim 0.1$–0.2 is essentially uncorrelated with the scene being imaged. This point is highlighted in Fig. 7a which compares the d$P/$dk of the reference photos and 64-snapshot CASSI reconstructions for both targets. From this figure, the differences in the intrinsic d$P/$dk between the scenes is clear with the highly cluttered overhead suburban scene containing significantly more high-frequency structure. For $k \lesssim 0.15$, the CASSI reconstructions' d$P/$dk agree well with those of the reference image, but diverge for larger k. Despite the intrinsic difference in the two scenes' structures, the two CASSI reconstruction d$P/$dk appear to share the same high-frequency scaling.

This fact is confirmed in Fig. 7b where the CASSI reconstructions' d$P/$dk for both targets and all snapshot numbers are plotted. A different multiplicative factor has been applied to each reconstruction's d$P/$dk in order to scale the d$P/$dk to approximately the same values at $k > 0.2$. Apart from the one-snapshot reconstructions, *all* the CASSI reconstructions share a general high-frequency

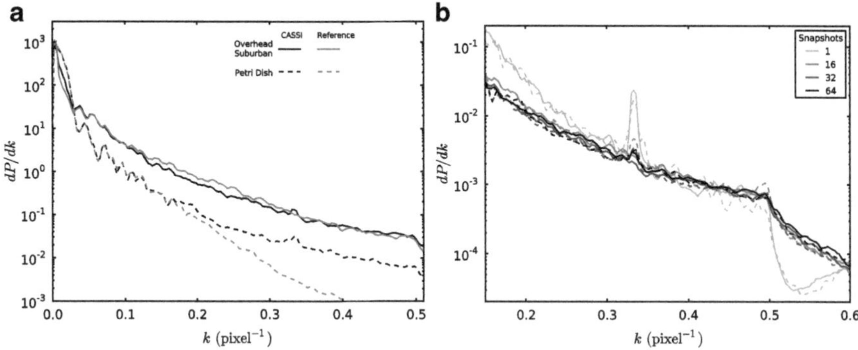

Fig. 7 Panel (**a**): the red-band dP/dk for both targets highlighting the differences in their intrinsic high spatial-frequency content. Panel (**b**): the red-band dP/dk of all CASSI reconstructions of both targets norm scaled so that their high-frequency tails have the same amplitude

behavior scaling. We do not have a definitive explanation for the underlying mechanism producing this behavior but surmise that the main driver has more to do with the reconstruction step as opposed to the data acquisition methodology. As such, we note that the actual value of the high-frequency dP/dk slope is likely sensitive to the reconstruction parameter τ (see Eq. 12).

5 Spectral Exploitation of CASSI Data Cubes

Our evaluation of the suitability of CASSI reconstructions for spectral exploitation applications considered pixel classification performance, pixel demixing/abundance estimates, and quantifying detectability limits. Here we report only on our classification studies.

There is currently no protocol in place for producing radiometrically calibrated reconstructions from the CASSI instrument. This precludes use of library reflectance spectra as exemplars/endmembers in the classification process. The requisite exemplar spectra for each material class were therefore obtained in-scene via the following method. From the 64-snapshot reconstruction (i.e., the reconstruction with the highest spatial resolution), for each of the eight material classes in the scene (brown paint, green paint, red fabric, polystyrene, Nd_2O_3, moss, sand, and plywood), we identified $\approx$20–200 "pure" pixels that contained only the specified material class. These pixels were selected primarily on the basis of spatial location: interior pixels of the larger paint chips and fabric swatches, pixels from the sand-only dishes, etc. For each of the CASSI reconstructions used for spectral exploitation, this collection of pixel sets were used to calculate scene-specific class exemplars by taking the mean of each class' set of pixels.

To assess the accuracy of our classification results, we developed a set of "ground truth" pixels in the following manner. During our data collection, we also obtained a higher spatial resolution image of the petri dish target with our reference camera system. From these images, we identified ≈600–1200 pure pixels for each material class by a combination of visual inspection and knowledge of each dish's contents. From these pixel sets we created a ground truth map in the frame of the high-resolution image. We registered the high-spatial-resolution image to the CASSI reconstructions and applied the resulting transform to project our ground truth map onto the CASSI reconstructions. By the end of this process, the resulting ground truth map in the CASSI data frame contained ≈200–600 pixels in each material class.

We ran four different supervised classification methods on the 32-snapshot reconstruction, using this scene's set of exemplar spectra calculated as described above. We focus on the 32-snapshot reconstruction since it represents a truly "compressive" measurement in terms of total data volume collected. That is, with 47 bands, conventional collection methods must obtain $47 \times n_{\text{pixels}}$ measurements (where n_{pixels} is the total number of pixels occupied by the scene), so the 32-snapshot reconstruction represents a savings factor of $32/47 = 0.681$ in terms of data volume; since the aperture mask is only 50 % transmissive, the actual information content collected for the 32-snapshot is only 0.34 that collected by conventional methods. By the same token, a 64-snapshot reconstruction is definitively not compressive as it requires 1.36 times as much data volume as conventional acquisition methods.

The classification methods used were the Euclidean minimum distance method, the spectral angle mapper (SAM) method [8], an adaptive coherence estimator (ACE) method [11, 13, 16], and a matched filter (MF) method [16]. The minimum distance method uses the Euclidean distance metric between target and exemplar spectra in the vector space formed by taking each of the N spectral bands as a component of an N-dimensional vector. The SAM method's metric is the angle between the target and exemplar spectra in this N-dimensional vector space. The ACE and MF, while also relying on essentially angle-based metrics, represent more sophisticated approaches in that both take into account the spectral statistics of the scene; details of these methods can be found in the references.

For each classifier method/material class pairing, we constructed receiver operating characteristic (ROC) curves using our ground truth map to characterize the classifier's performance. From the ROC curves, we determined the optimal value of the classifier's detection thresholds by identifying the point on each ROC curve that maximizes the difference between the true- and false-positive rates (P_{TP} and P_{FP}, respectively). We took the detection threshold producing this point as the optimal value for each material class. Figure 8 summarizes the results of this process in terms of the maximum $P_{\text{TP}} - P_{\text{FP}}$ value for the non-background material classes (i.e., excluding sand and plywood) for the four classifiers.

Some insight into the relative performance of the classification methods in this figure can be obtained by comparing the material classes' mean spectra. Foliage, brown paint, sand, polystyrene, and plywood all have similar shapes in the visible from a spectral angle perspective, but differ in relative overall reflectivity.

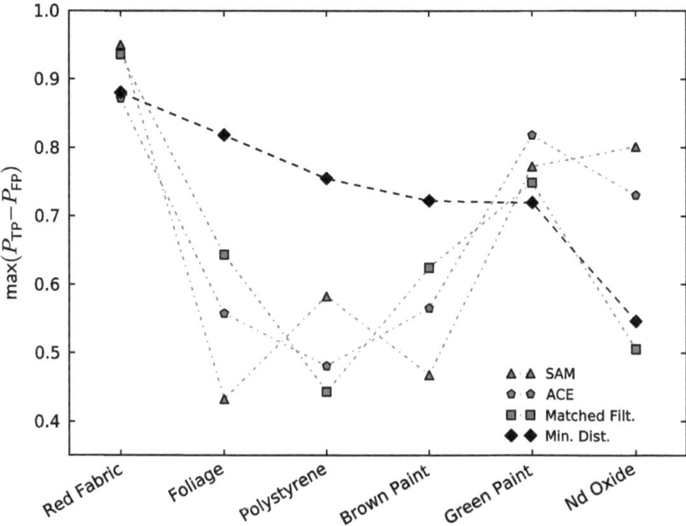

Fig. 8 A summary of the four supervised classification methods' performance when applied to the 32-snapshot CASSI reconstruction. Shown is the maximal value of the difference between true detection and false alarm rates ($P_{TP} - P_{FP}$) for each classifier on each non-background material class. The minimum distance classifier produced the best results on average across these material classes

Plywood and polystyrene cluster together at the high-end of scene brightness; sand and brown paint are nearly indistinguishable in the lower-middle of the brightness range; foliage, the darkest material, is well separated from the others at the bottom of the brightness range. The green paint's brightness is similar to that of sand and brown paint, but its spectrum has a change in slope at the red end of the visible that provides for some separation in spectral angle. The Nd_2O_3 is well separated by spectral angle, but overlaps the polystyrene and plywood at high mixing fractions; sand becomes an increasingly important confuser as the Nd_2O_3 mixing fraction decreases. The red fabric, however, has a rather unique spectral shape and large enough dynamic range in brightness across the visible that it is not easily confused with other materials in either a Euclidean distance or angle based metric.

Based on this discussion, it is not surprising that all four methods classify the red fabric rather accurately. The SAM, ACE, and MF methods perform relatively poorly on the foliage, polystyrene, and brown paint classes as these classes are separated primarily by overall brightness, not spectral shape. Correspondingly, the minimum distance classifier provides the best performance on these classes. On the green paint and Nd_2O_3, this relative ordering switches and the angle-based methods perform better, generally speaking. As an illustration of the results that we obtained with these methods, we show in Fig. 9 two portions of the classification map produced with the minimum distance algorithm. The corresponding confusion matrix is provided in Table 1. Together these show that there is a larger degree

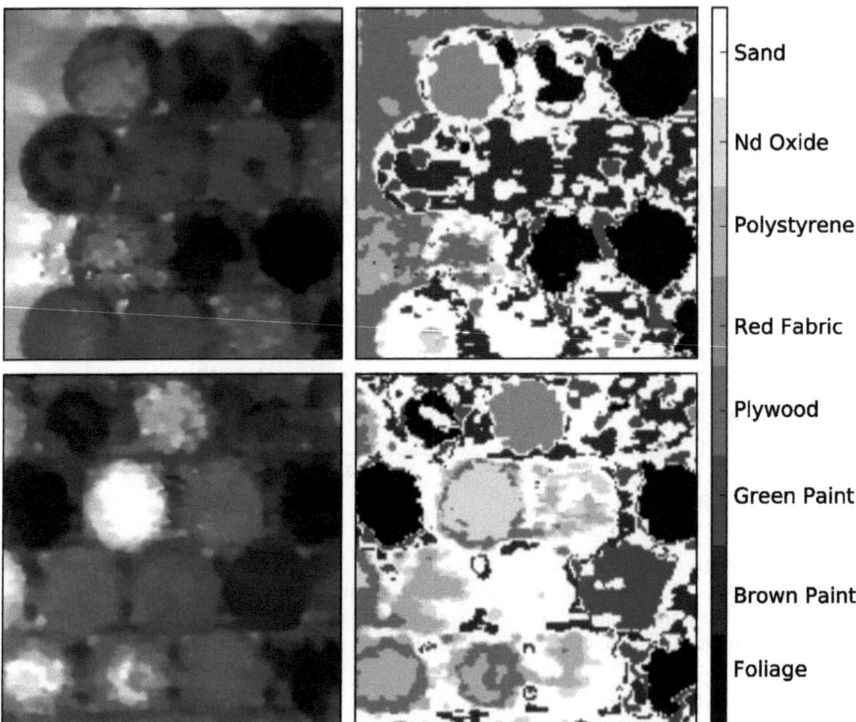

Fig. 9 Selected segments of the petri dish target 32-snapshot reconstructed image (*left panels*) alongside the same segments of the classification map (*right panels*) determined by the minimum distance algorithm using the optimal thresholds for each class

Table 1 The confusion matrix for the minimum distance classifier on the 32-snapshot reconstruction of the petri dish target at the optimal classifier thresholds for each material

	Ground truth								
Classification	Green paint	Brown paint	Red fabric	Poly-styrene	Moss	Nd_2O_3	Sand	Ply-wood	Totals
Green paint	0.778	0.005	–	–	0.025	–	0.025	0.011	0.093
Brown paint	0.049	0.775	0.026	–	0.007	0.004	0.172	0.030	0.105
Red fabric	–	–	0.698	–	–	–	0.011	–	0.077
Polystyrene	–	–	0.004	0.748	–	0.060	0.035	0.126	0.117
Moss	0.078	0.005	–	–	0.946	–	0.005	–	0.115
Nd_2O_3	–	–	0.011	0.032	–	0.365	0.022	0.028	0.056
Sand	0.093	0.216	0.261	–	0.021	0.302	0.684	0.267	0.255
Plywood	–	–	–	0.221	–	0.271	0.046	0.538	0.183

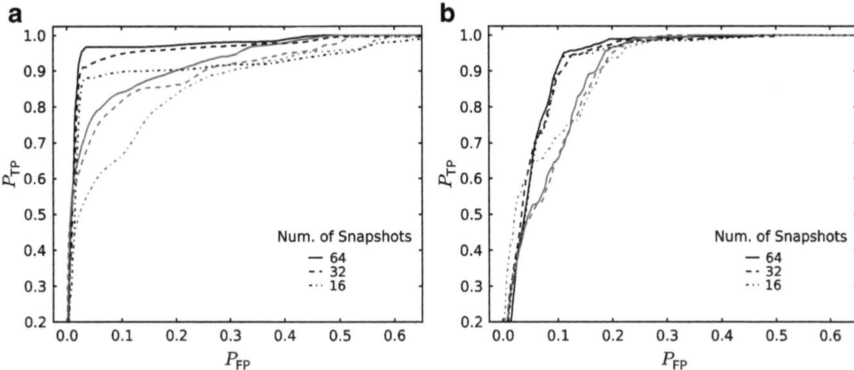

Fig. 10 Minimum distance classifier ROC curves for several of the material classes as a function of number of snapshots used in the reconstruction. Panel (**a**): red fabric (*black lines*) and green paint (*gray lines*). Panel (**b**): foliage (*black lines*) and polystyrene (*gray lines*)

of misclassification with the background classes not captured in the ROC curve analysis. This is due primarily to the ground truth set containing only $\approx 2\,\%$ of the total pixels in the image and is not an issue unique to CASSI reconstructions.

All of this demonstrates that the CASSI data reconstructions can be successfully exploited to classify pixels based on their spectra. Our focus has been on the 32-snapshot reconstruction as this provides a mild savings in data collection relative to that required for a 47-band standard imaging spectrometer to image the scene. There is the question of how does classification performance vary with number of snapshots. Figure 10 provides some initial data on this question by showing the minimum distance classifier results on red fabric, green paint, foliage, and polystyrene for reconstructions based on 16, 32, and 64 snapshots. The results in this figure are mixed: red fabric and green paint (Fig. 10a) both show a drop-off in performance with decreasing snapshot number, while the performance on foliage and polystyrene (Fig. 10b) show no appreciable change. The cause of these different behaviors is unclear but could include differences in proximity of confuser classes, differences in size distributions of the materials (e.g., sub-GSD paint chips being more rapidly blurred into the background with decreasing spatial resolution), or even the specific choice of ground truth pixels.

6 Discussion and Summary

In our analysis of the prototype CASSI multiplexed, coded-mask imaging spectrometer system, we considered two basic performance metrics: spatial quality of the reconstructed images and the accuracy with which the reconstructed data cubes could be exploited to perform pixel classification.

The spatial resolution of the CASSI reconstructions generally is lower than the intrinsic resolution of the underlying optical system and is a function of both wavelength and number of snapshots used. Due to a combination of the λ-dependent sensitivity of the detector and nonuniform bandwidths, the delivered resolution is markedly better in the red portions of the spectrum; for a 32-snapshot reconstruction, the resolution is a factor ≈ 2–7 poorer than the intrinsic resolution moving from the red to the blue end of the spectrum. We also found that the CASSI reconstructions we analyzed all shared a common scaling in their Fourier power spectral density at high spatial frequencies, irrespective of the imaged scene's spatial-frequency distribution. We hypothesize that this imposed high spatial-frequency scaling is an artifact of the reconstruction process and likely has a strong dependence on τ.

We showed that the spectral information in the reconstructed data cubes can be successfully exploited to perform pixel classification. We found, somewhat surprisingly, that the simplest classification method that we applied, the minimum distance method, produced the best classification accuracy on average. The fact the typically more accurate spectral-angle-based methods did not generally perform well here may in part be due to the reconstruction methodology we use. This method introduces band-by-band correlations that can reduce the inter class differences in a spectral-angle-based metric, leading to a greater degree of class confusion and poorer classification performance.

We have demonstrated the potential for a CASSI system to deliver the functionality demanded of imaging spectrometers—albeit with the trade-offs of somewhat reduced spatial resolution, the additional processing-time costs involved in performing the reconstructions, and the likely introduction of reconstruction errors. In spite of these trade-offs, a CASSI system could deliver advantages over conventional imaging spectrometers in certain circumstances, in particular under low-light conditions or situations where very rapid data acquisition is required (e.g., to capture information on rapidly moving objects). Future development of CASSI technology will thus involve several connected fronts: first, identifying application environments where the CASSI data acquisition mode provides an inherent advantage; second, implementing engineering optimizations to mitigate some of the trade-offs discussed here (at least as far as such trade-offs impact detector performance in a given application); and third, exploring whether alternate reconstruction algorithms might deliver improved results relative to that of the currently implemented method.

References

1. Arguello, H., Gonzalo, A.: Code aperture design for compressive spectral imaging. In: 18th European Signal Processing Conference, Aalbork, Denmark (2010)
2. Baraniuk, R.: Compressive sensing [Lecture Notes]. IEEE Signal Process. Mag. **24**(4), 118–121 (2007)
3. Bioucas-Dias, J., Figueiredo, M.: A new TwIST: two-step iterative shrinkage/thresholding algorithms for image restoration. IEEE Trans. Image Process. **16**(12), 2992–3004 (2007)

4. Candes, E., Romberg, J., Tao, T.: Robust uncertainty principles: exact signal reconstruction from highly incomplete frequency information. IEEE Trans. Inf. Theor. **52**(2), 489–509 (2006)
5. Candès, E.J., Romberg, J.K., Tao, T.: Stable signal recovery from incomplete and inaccurate measurements. Comm. Pure Appl. Math. **59**(8), 1207–1223 (2006)
6. Castrodad, A., Xing, Z., Greer, J., Bosch, E., Carin, L., Sapiro, G.: Discriminative sparse representations in hyperspectral imagery. In: 2010 IEEE International Conference on Image Processing, pp. 1313–1316. IEEE (2010)
7. Donoho, D.: Compressed sensing. IEEE Trans. Inf. Theor. **52**(4), 1289–1306 (2006)
8. Kruse, F., Lefkoff, A., Boardman, J., Heidebrecht, K., Shapiro, A., Barloon, P., Goetz, A.: The spectral image processing system (SIPS)-interactive visualization and analysis of imaging spectrometer data. Remote Sens. Environ. **44**(2–3), 145–163 (1993)
9. Figueiredo, M.A.T., Nowak, R.D., Wright, S.J.: Gradient projection for sparse reconstruction: application to compressed sensing and other inverse problems. IEEE J. Sel. Top. Signal Process. **1**(4), 586–597 (2007)
10. Gehm, M.E., John, R., Brady, D.J., Willett, R.M., Schulz, T.J.: Single-shot compressive spectral imaging with a dual-disperser architecture. Opt. Express **15**(21), 14,013 (2007)
11. Kraut, S., Scharf, L., Butler, R.: The adaptive coherence estimator: a uniformly most-powerful-invariant adaptive detection statistic. IEEE Trans. Signal Process. **53**(2), 427–438 (2005)
12. Krishnamurthy, K., Raginsky, M., Willett, R.: Hyperspectral target detection from incoherent projections. In: 2010 IEEE International Conference on Acoustics, Speech and Signal Processing, pp. 3550–3553. IEEE (2010)
13. Manolakis, D., Marden, D., Shaw, G.A.: Hyperspectral image processing for automatic target detection applications. Lincoln Lab. J. **14**(1), 79–116 (2003)
14. Nunez, J., Fors, O., Otazu, X., Pala, V., Arbiol, R., Merino, M.: A wavelet-based method for the determination of the relative resolution between remotely sensed images. IEEE Trans. Geosci. Remote Sens. **44**(9), 2539–2548 (2006)
15. Nunez, J., Otazu, X., Fors, O., Prades, A., Pala, V., Arbiol, R.: Multiresolution-based image fusion with additive wavelet decomposition. IEEE Trans. Geosci. Remote Sens. **37**(3), 1204–1211 (1999)
16. Rand, R.S.: A neural network approach for improved detector performance of spectral matched filters in hyperspectral imagery. In: Proceedings of SPIE **7457**, 74,570T–74,570T–11 (2009)
17. Rudin, L., Osher, S., Fatemi, E.: Nonlinear total variation based noise removal algorithms. Physica D: Nonlinear Phenom. **60**(1–4), 259–268 (1992)
18. Veeraraghavan, A., Reddy, D., Raskar, R.: Coded strobing photography: compressive sensing of high speed periodic videos. IEEE Trans. Pattern Anal. Mach. Intell. **33**(4), 671–86 (2011)
19. Wagadarikar, A.: Compressive spectral and coherence imaging. Ph.D. thesis, Duke University (2009)
20. Wagadarikar, A., John, R., Willett, R., Brady, D.: Single disperser design for coded aperture snapshot spectral imaging. Appl. Opt. **47**(10), B44 (2008)
21. Wakin, M., Laska, J., Duarte, M., Baron, D., Sarvotham, S., Takhar, D., Kelly, K., Baraniuk, R.: An Architecture for Compressive Imaging. In: 2006 International Conference on Image Processing, pp. 1273–1276. IEEE (2006)

An Introduction to Hyperspectral Image Data Modeling

David B. Gillis and Jeffrey H. Bowles

Abstract Hyperspectral data sets collect light from an object over a large number of narrowly spaced, contiguous wavelengths. The resulting data set associates with each spatial pixel (x,y) an n-dimensional vector (or spectrum) $(z_1,\dots,z_n)$, where z_i is the intensity at wavelength λ_i and n is the number of spectral channels (or bands). The large increase in spectral content can be used to develop improved image processing algorithms; however, to fully exploit this information, mathematical models and algorithms are needed that can handle the large amount of data associated with each image. Models should also be able to fully use the highly structured, spatial/spectral nature of the data. In this chapter, we present a general overview of hyperspectral imaging and review a number of existing data models that have been presented to analyze such data.

Keywords Hyperspectral imaging • Linear and nonlinear modeling • Spatial/spectral data analysis

1 Introduction

Hyperspectral imagers (HSI) are a relatively new type of imaging sensor that captures light over a wide range of wavelengths in many (typically hundreds) of contiguous bands. By way of analogy, a black-and-white (or gray scale) image can be considered a way of associating with each pixel in a given image a scalar value (e.g., the intensity or brightness of the pixel), while a color image typically associates a 3-vector (namely, intensities in the red, blue, and green wavelength

---

D.B. Gillis (✉) • J.H. Bowles
Naval Research Laboratory, Washington, DC 20375, USA
e-mail: David.Gillis@nrl.navy.mil; Jeffrey.Bowles@nrl.navy.mil

T.D. Andrews et al. (eds.), *Excursions in Harmonic Analysis, Volume 1*,
Applied and Numerical Harmonic Analysis, DOI 10.1007/978-0-8176-8376-4_9,

areas) to each pixel. Hyperspectral imaging extends this by associating to each pixel an n-vector, with components equal to the wavelength of the captured light.

The increased information available in a hyperspectral image holds promise for a wide number of image processing applications, including, among many others, classification and segmentation, change detection, and target identification.

The increased information comes at a price, however; typical HSI images are on the order of several hundred megabytes, and gigabyte-sized images are not unusual. To fully analyze data sets of this size and complexity in a reasonable amount of time require mathematical models and algorithms that can efficiently exploit both the spatial and spectral information inherent in the data.

In this chapter, we present a general background of hyperspectral imaging and review of a number of the mathematical models that have been introduced to aid in the analysis of HSI data.

2 Hyperspectral Imaging

Hyperspectral image data consists of a set of vectors (or spectra) that are associated with a given spatial position or pixel. Intuitively, the data may be thought of as a three-dimensional "cube," with x and y representing the spatial coordinates and z the associated spectrum. For a fixed wavelength z, the associated (x,y) pairs form a (gray scale) image, commonly called a band image; conversely, for a fixed (x,y), the pixel $P(x,y) = (\lambda_1,\ldots,\lambda_n) \in \mathbb{R}^n$ defines the spectrum for the given pixel. We note that the spectra are a function of the object that occupies a given pixel, and thus different pixels will generally have different spectra. Similarly, the band images associated with different wavelengths will vary, as different objects reflect light differently (Fig. 1). The physical units of the data are generally either in radiance or reflectance; for a water-based scene, the data may also be in remote sensing reflectance.

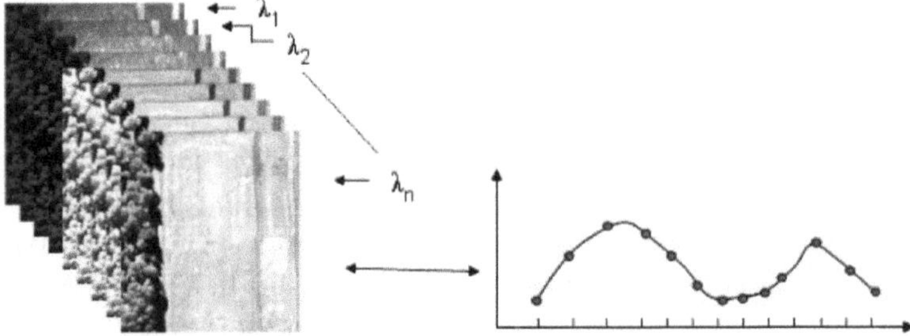

Fig. 1 Two views of hyperspectral imagery. On the *left*, HSI as a series of gray scale images. Each image corresponds to a given wavelength λ. On the *right*, each pixel in the image defines a vector (spectrum) over the wavelength range

In this section, we present an overview of how the data is generated and discuss the difference between the various types of units. We note that, in this chapter, we are focused mainly on reflective, remote-sensing HSI; that is, we generally assume the data has been collected from an airborne sensor (located on a plane or satellite) that is pointed at the surface of the Earth and collects sunlight reflected from the Earth somewhere over the wavelength range 300–2,500 nm. There are other types of HSI data; for example, there exist HSI sensors that collect data in the 5–10μm range (known as longwave) that is primarily a function of the thermal properties (used, e.g., in detecting chemical properties of particles in the atmosphere); similarly, in HSI microscopy, the incoming illumination may be something other than sunlight.

Recall that our Sun outputs a steady stream of electromagnetic radiation in the form of photons of varying wavelengths. The solar spectrum, or distribution of the photons as a function of wavelength, can be estimated as a black-body with a temperature of around 5800 K. The Sun's radiation that reaches the top of the Earth's atmosphere ranges from a low of around 100 nm (in the ultraviolet portion of the EM spectrum) to as high as 1 mm, with the majority being in 300–2,500 nm range.

As photons enter the atmosphere, they may interact (i.e., be scattered or absorbed) with particles in the atmosphere or, more likely, continue down to the Earth's surface. At this point, they will be reflected or absorbed, governed by the optical properties of whatever object they hit on the ground. Reflected light will travel back through the atmosphere and eventually captured by the sensor. Each pixel in the image corresponds to a particular area on the ground; the exact size of that area (known as the ground sample distance or GSD) is determined by the optics of the imager, as well as the height it is flying at. Modern-day sensors typically have GSDs on the order of a few meters to as large as a kilometer.

The "raw" signal collected at the sensor is comprised of photons reflected back to it from the ground, as well as additional photons that have been scattered from the atmosphere into the sensor's field of view (generally known as path radiance). The sensor converts the incoming photons into "counts" or digital numbers that are stored as binary data. The raw numbers are then converted into radiance units via machine calibration procedures.

In general, hyperspectral data is distributed in this form (e.g., radiance units). In many cases, however, it is preferable to work in reflectance units. To be more precise, different materials will reflect various wavelengths of light differently, depending on their chemical properties. This variation of reflectivity vs. wavelength defines a unique spectral reflectance signature that can be used to identify the object. For example, to our eyes grass is green due to the fact that it reflects green wavelengths (around 500 nm) much more strongly than the other visible wavelengths. However, a well-camouflaged tank also appears green, making it difficult to identify if it is in a large grass of field. As we look at more and more wavelengths, however, it becomes easier to identify differences between the two.

With this in mind, we would like to convert the radiance (i.e., the amount of collected light) from the sensor into a measure of the reflectivity of the object on the ground that was imaged. The process of converting radiance to reflectivity is known as *atmospheric correction*. There are a number of algorithms available to do this, and we will not pursue them in this chapter. We do note that atmospheric correction is non-trivial; it involves knowing (or estimating) several atmospheric parameters. Some of the algorithms include the empirical line method (ELM) [27,42], fast line-of-sight atmospheric analysis of spectral hypercubes (FLAASH) [2], the atmosphere removal algorithm (ATREM) [16], Tafkaa [17], and quick atmospheric correction (QUACK) [5].

3 Notation

We begin with some notation that we will use throughout the rest of this chapter. Recall that a hyperspectral data cube consists of a series of gray scale images, one for each wavelength of light that is collected by the sensor. Following tradition, we call the rows of each image the lines and the columns the samples; the spectral coordinates are the bands or wavelengths. The image thus consists of a total of $N_s \times N_l \times N_b$ elements, where N_s, N_l, and N_b are the number of samples, lines, and bands, respectively. We will generally reserve m and $n = N_s \times N_l$ to be number of bands and spatial pixels, respectively. Image spectra will be written as column vectors $\mathbf{x} = (x_1, \ldots, x_m)^t \in \mathbb{R}^m$. We will generally use linear indexing so that the image spectra are $\mathbf{x}_1, \ldots, \mathbf{x}_n$. When necessary, we will be slightly sloppy and use $\mathbf{x}_{i,j}$ to denote the spectrum at sample i and column j. (Note that this is the reverse of standard mathematical notation.) Finally, we will let X denote the m-by-n matrix that contains the image spectra as its columns

$$X = \left[\mathbf{x}_1 \cdots \mathbf{x}_n\right].$$

4 Linear Models

As we have noted, the high dimensionality of HSI data is both a curse and a blessing; the data contains a great deal of information, but the high dimensionality greatly increases the computational complexity and storage cost, as well as subtler effects in modeling (the so-called curse of dimensionality). For this reason, one of the basic data models in HSI is to try to find a lower-dimensional representation of the data that retains most of the important information in the scene. Dimensionality reduction techniques are important in their own right, as they often offer a better view of the data, as well as reducing random noise from the sensor. They are also commonly used as first step in a number of different exploitation algorithms, as well as being useful for compression, which can be used to reduce the storage costs of the data.

Given their usefulness, it is not surprising that there have been a large number of dimensionality reduction algorithms presented in the literature (though not always explicitly listed as such). Of these, the vast majority are linear models, which is the focus of this section. In this section we discuss newer, nonlinear models which expand on these.

In very general terms, linear models treat image spectra $\mathbf{x} \in \mathbb{R}^m$ as a collection of points in m-dimensional space; the goal is to find some k-dimensional (possibly affine) subspace that "best" represents the data, for some definition of best. It is important to note that, by moving to spectral space, all spatial information is lost. In a certain sense, point models throw away half of the information in a scene; however, as we will see, we can sometimes recover this information by choosing the right basis for the given subspace.

4.1 Principal Components and Singular Value Decomposition

One of first linear models used in HSI processing, and probably still the most popular, is known as principal components analysis (PCA) [24]. PCA uses the eigenvectors of the data covariance matrix to define an orthogonal projection onto an (affine) subspace. Formally, let

$$\bar{\mathbf{x}} = \sum_{i}^{n} \mathbf{x}_i$$

be the mean of the image spectra, and let

$$\Sigma = \frac{1}{n-1} \sum_{i}^{n} (\mathbf{x}_i - \bar{\mathbf{x}})(\mathbf{x}_i - \bar{\mathbf{x}})^t$$

be the m-by-m covariance matrix. Note that Σ is symmetric and positive semi-definite; it follows that it has a full set of orthonormal eigenvectors and real, nonnegative eigenvalues $\lambda_1 \geq \lambda_2 \geq \cdots \lambda_m \geq 0$. Let $W, D = \mathrm{diag}(\lambda_1, \ldots, \lambda_n)$ be the corresponding eigen decomposition $\Sigma W = WD$. Then the PCA transformation P is defined as

$$\mathbf{y} = P(\mathbf{x}) = W^{-1}(\mathbf{x} - \bar{\mathbf{x}}).$$

Alternatively, let

$$\bar{M} = [\mathbf{x}_1 - \bar{\mathbf{x}} \cdots \mathbf{x}_n - \bar{\mathbf{x}}]$$

be the mean-centered matrix of the image spectra and let $\bar{M} = WSV^t$ be the singular value decomposition. Then

$$\Sigma = \bar{M}\bar{M}^t = WSW^t.$$

In hyperspectral data, it is very often the case that the first few eigenvalues dominate the rest, that is, only the first few (typically between 10 and 30) eigenvalues are significantly greater than 0. In this sense, we can say that the first k eigenvectors contain most of the information in the data, and the remaining dimensions can be safely ignored. If we let W_k be the m-by-k matrix containing only the first k eigenvectors, then the reduced PCA projection is given by

$$\mathbf{y} = P_k(\mathbf{x}) = W_k(\mathbf{x} - \bar{\mathbf{x}}).$$

Note that P_k projects the m-dimensional spectrum x to the k-dimensional vector y.

PCA has a number of useful properties; for example, it can be shown that the transformed data is uncorrelated; also, the first principal component has the maximum variance, the second has the maximal variance in any direction perpendicular to the first, etc. However, PCA generally does not respect the spatial component of the data. For example, usually only the first few components contain any meaningful spatial information; the remaining components are basically noise. PCA also tends to get confused by multimodal distributions, such as scenes containing varying backgrounds. Also, by using only the dominant eigenvectors, small but significant differences between various elements can be suppressed. In particular, statistical outliers are ignored; in many cases (such as anomaly detection), these are the most interesting pixels in a given scene.

For these reasons, a number of variations and extensions to PCA have been proposed. Perhaps the best known is the maximum (or minimum) noise transform (MNF) [19]. This transform separates image spectra into signal and noise components and then maximizes the signal-to-noise ratios of the projected band images. The resulting transform generally produces a set of band images that are highly spatially correlated.

4.2 *The Linear Mixing Model*

4.2.1 Linear Mixing

The linear mixing model [26] is based on the notion of *mixed pixels*. Intuitively, mixed pixels arise when two or more objects occupy a single pixel of an image. For example, consider a scene that contains a large grassy field with a small river running through it. Away from the river, the image pixels will contain only a single spectrum, namely, grass (we ignore for the moment any variation in the grass signature that may occur). Similarly, pixels that are in the middle of the river will contain only water spectra. Now consider pixels on the riverbank; it is easy to imagine that a single pixel will contain both grass and water. In this case, the measured spectrum $\mathbf{x}_m$ will be a weighted sum of the grass spectrum $\mathbf{x}_g$ and the water spectrum $\mathbf{x}_w$. The weight associated to each component is equal to the proportion p of the area of the pixel that the component occupies; thus we can write the measured spectrum as $\mathbf{x}_m = p\mathbf{x}_g + (1-p)\mathbf{x}_w$.

More generally, a spectrum $\mathbf{x}$ containing k components can be written as

$$\mathbf{x} = \sum_{i=1}^{k} \alpha_i \mathbf{x}_i,$$

where, by the proportionality constraint, $\alpha_i \geq 0$ and $\sum \alpha_i = 1$. We note that this is a *local* condition, since we are simply assuming that the measured spectrum is a linear combination of the various elements in the pixel.

The main assumption of the LMM is that there exists a *global* set of component spectra, known as endmembers, such that every pixel may be written as a sum of these spectra. Formally, if we let $\mathbf{e}_1, \ldots, \mathbf{e}_k$ represent the endmembers, then the LMM says that each image pixel $\mathbf{x}_j$ may be written as the sum

$$\mathbf{x}_j = \sum_{i=1}^{k} \alpha_{i,j} \mathbf{e}_i, \tag{1}$$

$$\alpha_i \geq 0, \sum \alpha_{i,j} = 1, \tag{2}$$

Intuitively, the endmembers represent the major materials (grass, water, asphalt, etc.) within a scene, and the scalars α_i, usually referred to as the abundance coefficients, represent how much of a given material lies within a given pixel. One nice feature of the LMM is that it defines a set of gray scale images, one for each endmember, that maps out the spatial concentrations of the various materials (Fig. 3). These *abundance maps* are often useful in various postprocessing tasks.

In order to implement the LMM, the two tasks that need to be done are to determine the endmembers and then to estimate the abundance coefficients (generally known as *unmixing*). We examine each of these steps below.

4.2.2 Endmember Determination

There are many endmember algorithms that have been presented in the literature. In this section, we present a few of the more common ones; we do not mean to suggest that this list is anywhere near complete.

A central theme in the majority of endmember selection schemes is that, in the fully constrained LMM, the endmembers form the vertices of a k-dimensional simplex within m-dimensional band space. To find the endmembers of a given scene, then, one attempts to find a simplex which best encapsulates the data; the vertices of this simplex will be the endmembers (Fig. 2).

One of the first hyperspectral endmember detections schemes is the pixel purity index (PPI) [6]. To run PPI, the data are first projected to a low-dimensional subspace (the dimensionality of which is determined by the user) via PCA or MNF. Next, a random set of k unit vectors (known as "skewers") are generated, and the data is projected onto each of the skewers. Define the set of all extreme points for each projection as S; note that S may have as many as $2k$ members but will generally have fewer since the same data point can be extremal in different projections.

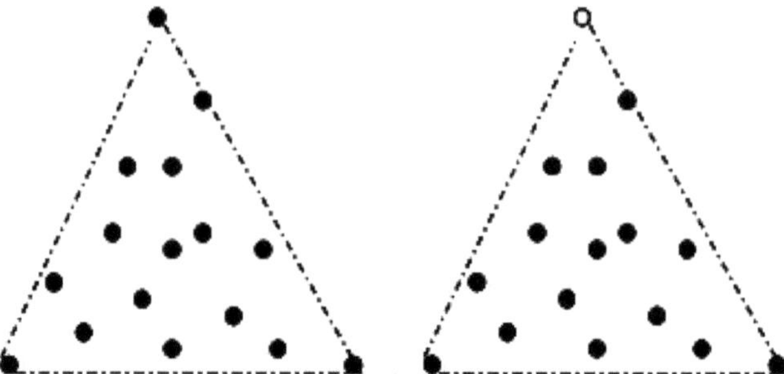

Fig. 2 The linear mixing model assumes that the data points (*black dots*) can be encapsulated within a k-simplex (*dotted line*). The vertices of the simplex are known as endmembers and represent the pure materials in the scene. All other points can be represented as a convex combination of the endmembers. It is often assumed that the endmembers are actual data points (*left*); however, in certain situations, only mixtures are present in scene (*right*), and the missing endmember (*open dot*) must be estimated

The PPI score for each member in S is simply the number of skewers for which the point is extremal. Finally, choose a suitable threshold for PPI and define the endmembers to be the set of points in S that are greater than or equal to the threshold. Note that the threshold is often chosen to be one, in which case every member of the set S becomes an endmember. Since the original publication of the PPI algorithm, a number of extensions have been added.

Another well-known algorithm is the N-FINDR routine [48]. The aim of this algorithm is to choose a set of k points from the data such that the simplex formed from these points is of maximal volume. An exact solution of this problem would require n choose k calculations, which is clearly not feasible for even small images. To get around this, a greedy approach is used. In particular, an initial set of k spectra are chosen at random from the data. All but the first member is held fixed, and each of the remaining image pixels are substituted (one by one) for the first element, and the volume of the resulting simplex is calculated. The image pixel that produces the largest volume simplex is then chosen to replace the first endmember. Next, the second endmember is allowed to vary, and the process repeated. This continues until the last endmember has been varied; then we move back to the first endmember. The process is continued until no new changes occur, which often takes only one or two passes through the endmember set.

In both PPI and N-FINDR (and, in fact, in most endmember selection schemes), it is implicitly assumed that every major constituent in the scene fully fills at least pixel. This assumption is due to the fact that these models only choose endmembers from the pixels in the scene. However, this may not be the case, especially when the GSD of the sensor is large. As an example, consider a cornfield; assuming the rows of corn are tightly spaced but separated from each other, then each pixel will likely contain both corn plants and the underlying dirt; no pixel will contain exclusively corn or dirt. In this case, each pixel will be a mixture of the corn and dirt spectra, but these spectra will not be in the data. In this case, we say the corn spectrum is a *virtual* endmember (Fig. 2).

Finding virtual endmembers can be challenging. One approach that explicitly looks for these types of endmembers is the ORASIS algorithm [7]. The ORASIS algorithm begins by using a vector quantization approach to identify unique pixels in the scene; in particular, the algorithm begins by running through the data and removing redundant spectra (those that are spectrally similar to previously seen pixels). A low-dimensional representation of the remaining spectra is constructed using a bottom-up, modified Gramm-Schmidt algorithm. Finally, a simplex is constructed that encloses the data. The vertices of the simplex are defined using convex combinations of the data points; in particular, the ORASIS algorithm is able to find "missing" endmembers and construct virtual endmembers without assuming they are in data.

Finally, we note that in all of the models presented in this section (and, again, in most endmember selection schemes in the literature) the data are modeled solely as points in band space; in particular, all spatial information is lost. Incorporating the spatial information is not easy; however, a few attempts have been made, including the morphological approach of [39].

4.2.3 Unmixing

Once the endmembers for a given scene have been determined, the next step is to estimate the abundance coefficients for each image pixel. Since the endmembers do not in general span the entire space of the data, no exact solution is available, and the abundances must be chosen to optimize some error criterion.

The most common estimate seeks to minimize the least squares or L^2 norm. In particular, the coefficients α_i are chosen to minimize the residual error

$$r = \|\mathbf{x} - \sum \alpha_i \mathbf{e}_i\|_2,$$

where $\|\cdot\|_2$ is the standard 2-norm.

If neither of the constraints (Eq. 2) are used, then the solution is easily found using the pseudo-inverse. In particular, if we let

$$\mathbf{e} = [\mathbf{e}_1, \ldots, \mathbf{e}_k]$$

be the m-by-k matrix whose columns are given by the endmembers and let $\alpha = (\alpha_1, \ldots, \alpha_k)$ the k-vector containing the abundances, then it is easy to show [29] that

$$\alpha = \left(\mathbf{e}^t \mathbf{e}\right)^{-1} \mathbf{e}^t \mathbf{x}$$

is the optimal least-squares solution. Similarly, if only the sum-to-one constraint is needed, then the abundances can be found by affine projection. If the nonnegativity constraint is desired, then no closed form solution exists, and numerical optimization routines, such as nonnegative least squares, must be used [21].

We note that L^2-based solutions tend to be dense, that is, most of the abundances tend to be nonzero, even when enforcing the nonnegativity constraint. Intuitively, this does not make a lot of sense, since this implies that each pixel contains at least a small amount of every endmember material. In reality, we would expect a sparse solution, that is, most abundances are zero (implying that the associated endmember is absent), and only a small number of coefficients are nonzero.

There have been a few attempts made to deal with this issue. Some of the earlier methods used statistical tests to estimate which (and/or how many) endmembers were present in a given pixel and then unmixed using only those endmembers. Other approaches looked at pairs (or even larger groupings) to unmix.

In recent years, the field of compressed sensing has brought new attention to L^1 (and L^0) minimization. Various methods have been used to apply L^1 minimization to unmix the data. We note, however, L^1 minimization cannot be used with the sum-to-one constraint, which does have physical meaning. We note that L^1 minimization has been taken further and used to find the endmembers (ref).

4.2.4 Endmember Variation

One of the main drawbacks of the linear mixing model is that it implicitly assumes that each constituent in a given scene can be modeled by a single endmember vector. In reality, different pixels belonging to the same general class will show some variation. For example, if a scene contains, say, a large field of grass, the individual pixels will be highly similar but not quite exactly the same. Similarly, a forest scene will have pixels that have the same general shape but will vary (more than the grass). The only way the traditional LMM can deal with this is to find multiple endmembers, all representing the same general class (Fig. 3).

There have been a few different ways of dealing with this variation presented [46]. One of the more common approaches is to model the endmembers statistically [4, 15]; usually, it is assumed that the pixels corresponding to a given class can be modeled as a multivariate Gaussian model, and the scene as a whole can be represented using a Gaussian mixing model. Once the endmembers have been found, the abundances can be derived via various optimization techniques, usually based on expectation-maximization (EM).

An alternative approach is to model the endmember variation geometrically Given the high (spectral) correlation of the pixels corresponding to the same general class, the distribution tends to be highly singular and can be well approximated by a low-dimensional subspace. The generalized mixing model [18] uses this approach to model each endmember as a low-dimensional subspace. One complication arising from this generalization is unmixing; note that in the LMM, each endmember has a natural "scale" (namely, the norm of the vector). Moving to subspaces removes this scale. To deal with this, one can either use affine subspaces or, alternatively, use oblique projections to decompose a given pixel.

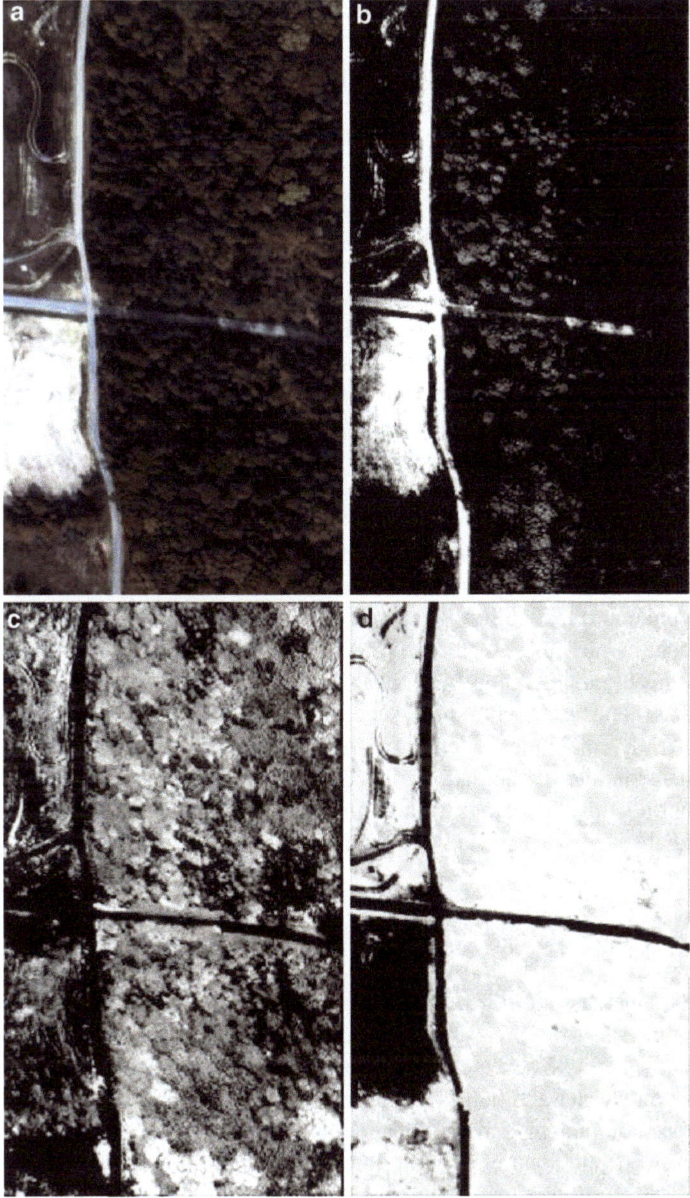

Fig. 3 Examples of linear mixing. (**a**) The original HSI image. (**b**) and (**c**) are examples of unmixing using the (traditional) linear mixing model. The endmember in (b) corresponds to dirt/asphalt and is able to identify most pixels in this class. The endmember in (c) corresponds to a forest spectrum; note that it does a relatively poor job of finding all members; this is due to the variability in the class. (**d**) Shows an example of a grouped vegetation endmember from the generalized linear mixing model

4.2.5 Blind Source Separation and ICA

The determination of endmembers in hyperspectral imagery can be considered as a special case of the blind source (or signal) separation (BSS) problem. In signal processing, the BSS problem is to separate a set of (unknown) signals from a mixed data source. The usual example is the "cocktail party problem," in which a number of individuals are all talking at once; a microphone placed in the room would capture a mix of the individual conversations, and the goal is to separate this source into its individual components. Note that humans can do this with very little effort, but automating this procedure is not easy. A wide number of approaches have been tried with some success (including PCA), but BSS is still an active research problem.

One of the more successful approaches to the problem is independent component analysis (ICA) [14]. The underlying assumption of ICA is that the signals are statistically independent. The standard (linear) ICA model is

$$\mathbf{x} = A\mathbf{s},$$

where $\mathbf{x} \in \mathbb{R}^m$ is the known source (in HSI, $\mathbf{x}$ is simply an image pixel), A is the (unknown) m-by-k mixing matrix, whose columns are the unmixed columns, and $\mathbf{s} \in \mathbb{R}^k$ are the corresponding signal weights). The aim is to find A such that $\mathbf{s}$ is as independent as possible. A number of different methods have been introduced to solve this problem, including JADE [13] and FastICA [22].

ICA has been used with some success in HSI [37]. The main problems are that, generally, the independence of the endmembers is questionable and the high dimensionality of the data (which greatly increases the computational cost). To account for this, the data is usually first whitened and also projected to a lower-dimensional space, usually via PCA or MNF.

4.2.6 Nonlinear Mixing

The fundamental assumption of the linear mixing model is that in a pixel containing multiple materials, the measured spectrum is in fact a linear combination of the component spectra. When the components are separated spatially (e.g., in a pixel that is say half water and half grass), then the incoming light will generally reflect off only one component, and the linear hypothesis is valid.

However, if the incoming light interacts with multiple materials, then the resulting spectrum is generally not linear, and the LMM is not applicable. Such mixtures are usually called intimate mixtures [26]; examples include particulate mineral mixtures (e.g., different grains of sand on a beach) and tree canopies (where the incoming light can be bounced around multiple times before being reflected back to the sensor).

4.3 Other Linear Methods

Principal components and the linear mixture model are the two most popular linear models being used in HSI data analysis, but a number of other models have been proposed.

One common technique is band selection [25], in which one tries to find an optimal subset of the original bands of the image; note that band selection can be thought as a projection into the subspace spanned by the chosen bands. Band selection can be useful in its own right as a dimensionality reduction step. It can also be used to identify which particular wavelengths are most useful for a given problem. Once these wavelengths have been identified, they can be used for example to build a specialized multispectral sensor, which is generally easier to build, and can have higher spatial resolution.

Note that the units in hyperspectral data (whether radiance or reflectance) are physical measurements and thus nonnegative. Under the assumptions of the (constrained) linear mixing model (ref), the m-by-n data matrix M containing the image spectra can be factored as an m-by-k matrix E (the endmembers) and a k-by-n matrix A (the abundance coefficients)

$$M = EA.$$

Since the endmembers represent actual objects in the scene, E must be nonnegative; similarly, since the abundances represent fractional amounts, A is also nonnegative. Thus the linear mixing model can be considered as a special case of nonnegative matrix factorization [32]. A number of different algorithms, not specifically designed for HSI, have been introduced. A number of authors have used these techniques to factor HSI [43].

Concluding this section is projection pursuit (PP), in which the subspace containing the data is built up in single steps, by finding "interesting" directions, removing the data component in that direction, and repeating. Different definitions of interesting are possible and will generally lead to different data sets. We note that PCA can be thought as a special case of PP, in which the variance is the object of interest. Examples of PP in HSI data analysis include [23].

5 Nonlinear Models

In the previous section, we assumed that hyperspectral data can be modeled linearly; in particular, we assumed that the high-dimensional HSI data can be accurately modeled as lying within a relatively low-dimensional subspace. In many situations, this approach, while perhaps not exact, is a reasonable approximation. However, there are other situations where this model breaks down. For example, as we mentioned previously, the presence of multiple objects in a single pixel can lead to nonlinear mixing. Other nonlinearities can arise from multiple scattering and adjacency affects. To better account for these situations, attempts to model HSI using nonlinear structures have been developed.

5.1 Manifold Learning

The basic assumption in manifold learning is that the data have a low intrinsic dimensionality, that is, it is assumed that the data lie on some low-dimensional surface (another way of saying this is that the data has only a few degrees of freedom). In contrast to linear models, manifold learning allows the surface to be curved, that is, the data is assumed to lie on some low-dimensional manifold. The goal of manifold learning is to be able to find the coordinate structure of the underlying manifold and to project the data into these new coordinates. In this sense, manifold learning can be thought of as a nonlinear dimensionality reduction technique.

Over the last several years, there has been a tremendous interest in manifold learning techniques. Many of these methods can be considered as generalizations of metric multidimensional scaling (MMS) [33]. In MMS, we begin with high-dimensional data points $\mathbf{x}_i \in \mathbb{R}^m$ (as usual, in HSI, the $\mathbf{x}$ are simply the spectra), and the pair-wise distances $d_{i,j} = d(x_i, x_j)$ (under some metric d) are calculated and stored in an n-by-n matrix D (here, n is the number of points). The data are centered, and the eigenvectors corresponding to the k largest eigenvalues are found. If we let E denote the n-by-k matrix whose columns are the eigenvectors, then the rows of E define a k-dimensional representation of the original data.

In classical MMS, the distance metric is simply the Euclidean norm, and it can be shown that the reduced data are optimal in the sense of preserving interpoint distances. (In this particular case, the reduced data is basically just PCA in disguise.)

In recent years, this approach has been generalized to other distance functions (which are not, strictly speaking, metrics). One of the best known algorithms is Isomap [47], in which the interpoint distances are estimates of the geodesic distance on the underlying manifold. In particular, to calculate the distances between points, a local neighborhood graph is first calculated, in which points are connected only to their closest spectral neighbors. Since by definition a manifold is locally flat (e.g., isomorphic to $\mathbb{R}^k$), the geodesic distance locally can be estimated by the standard Euclidean distance. To estimate the distance between points outside the local neighborhood, standard graph shortest path algorithms are used. Once all the interpoint distances are calculated, the matrix D is formed, and the data is projected exactly as in MMS.

One of the main drawbacks of Isomap, especially with regard to HSI processing, is that the data scales quadratically (or worse) in both storage and complexity. For example, images often contain 500,000 or more pixels; the resulting distance matrix has 250,000,000,000 or more elements. To handle this, data is often subsampled [45] to reduce the number of pixels (known as landmarks) that need to be processed. Once the projection has been defined, the remaining pixels can be reduced using weighted sums of the landmark pixels. Our colleague Charles Bachmann has been using this approach to develop extremely efficient algorithms that can handle very large HSI images [3].

5.2 *Kernel Methods*

Kernel methods [9] are a family of techniques that map (nonlinearly) the original data into a much higher dimensional space; the aim is to find a suitable representation of the data such that linear methods can be used to analyze the data.

To be more precise, let $\mathbf{x}_1,\ldots,\mathbf{x}_n \in \mathbb{R}^m$ be our original spectra; we wish to find an inner product space H of dimensionality $N \gg n$ and a mapping

$$\Phi : \mathbb{R}^n \Rightarrow H.$$

Suppose now that we have an algorithm that is linear in nature, that is, the only computations that need to be done are inner products (and, by extension, matrix-vector products; note that this includes in particular all of the models of the previous section). If we wanted to apply this to the transformed data, we would need to compute inner products of the form

$$\langle \Phi(\mathbf{x}), \Phi(\mathbf{y}) \rangle.$$

Note this computation is expensive since we are working in the very-high-dimensional space H.

The so-called "kernel trick" allows us to get around this difficulty while still benefiting from the increased dimensionality. In particular, it can be shown that, under certain conditions, the inner product can be calculated without even knowing the mapping Φ; in particular, Mercer's theorem says that the inner products in H can be replaced by a scalar function on $\mathbb{R}^n x \mathbb{R}^n$, that is, there is a function $K : \mathbb{R}^n x \mathbb{R}^n \Rightarrow \mathbb{R}$ such that

$$K(\mathbf{x}_i, \mathbf{x}_j) = \langle \Phi(\mathbf{x}_i), \Phi(\mathbf{x}_j) \rangle.$$

To apply the kernel trick, one simply needs to choose a suitable kernel K that satisfies the appropriate conditions. Some of the more popular kernels are the polynomial kernels $K(\mathbf{x},\mathbf{y}) = (\mathbf{x}^t\mathbf{y} + c)^d$ and the Gaussian or radial basis function

$$K(\mathbf{x},\mathbf{y}) = \mathrm{e}^{\frac{\|\mathbf{x}-\mathbf{y}\|}{c}}.$$

Kernel-based methods have applied to a wide variety of HSI applications, including "kernelized" versions of PCA [44], endmember selection [8, 38], RX anomaly detection [28], and classification via support vector machines (SVM) [10, 34].

6 Spatial-Spectral Models

As we have mentioned, the vast majority of current HSI data models begin by assuming the spectra $\mathbf{x}_i \in \mathbb{R}^m$ are points in some m-dimensional vector space. The problem with this approach is that all of the spatial information is lost; essentially, these models throw away half of the available information in the data.

Developing mathematical models that can fully incorporate both the spatial and spectral models would thus seem to offer the best hope for extracting the most amount of meaningful information from a given scene. Unfortunately, developing models that can do this is not easy. In this section, we give a brief review of the current state of the art in this direction.

One of the first attempts to use spatial-spectral models was to restrict attention to a spatially local neighborhood of a given pixel and then to use spectral modeling. The stereotypical example of this approach is the RX anomaly detection algorithm [40]. In anomaly detection, the aim is to look for image pixels that are significantly different than the remaining background. The RX algorithm solves this problem by examining each pixel in a given image; for each candidate pixel, the statistics of spatially neighboring pixels (as points in spectral space) are estimated, and the Mahalanobis distance between the background mean and target pixel is calculated. Pixels which are anomalous should have relative high scores (i.e., be significantly different than the background). This approach is easily extended to other statistically based algorithms that rely on estimating the background variation.

Although neighborhood-based approaches are often called spatial-spectral techniques, in reality, the actual models are still point structures in spectral space. In the last few years, research has turned to finding true spatial/spectral models for HSI data. For example, morphologically based methods have been used for both segmentation and endmember extraction. Other techniques that combine linear mixing with spatial information include Markov models for classification and sharpening. Several compression schemes have been introduced, using wavelets, discrete cosine transform, and others.

Two relatively new spatial/spectral models that deserve attention are based on graphs and tensors. In [11, 12] the authors model a given image as a weighted graph, with pixels as nodes and weights defined as a (kernelized, see section above) function of spatial and spectral similarity. Spectral graph methods (see section above) are then used to segment the image. In the tensor approach [41], one models the entire image as a 3-tensor

$$M = \left(a_{i,j,k} \in \mathbb{R}^{N_s \times N_l \times N_b}\right).$$

A 3-tensor can be considered as a generalization of vectors (1-tensor) and matrices (2-tensor). Under this formulation, linear algebraic techniques (such as PCA and nonnegative matrix factorization) can be extended to multilinear algebra (multiway PCA, Tucker decompositions).

7 Models for Water-Based Scenes

Although we have not been explicit about it, we have generally assumed that the data we are analyzing is land based; in particular, we assume that the incoming light passes through only one medium (the atmosphere). When looking at water-based scenes, however, the situation is much more complicated. We still have the usual issues with atmospheric effects; now we also need to incorporate the effects of the water body on the incoming and reflected light.

Oceanographers generally divide the world's water into two classes: Case 1 waters are the open ocean, far from shore, and unaffected by coastal processes. Case 2 waters are everything else, including near-shore oceanic water, rivers, and estuaries [35].

The optical properties of Case 1 waters tend to be relatively simple to analyze; the only real variable that needs to be modeled is the amount of phytoplankton and associated chlorophyll. Multispectral sensors such as MODIS and SeaWiFS have been used for years to model Case 1 waters, with much success.

Case 2 waters, however, are much more complicated. Instead of just chlorophyll, the incoming light field can be scattered and absorbed by a variety of other materials, including suspended sediments (such as sand particles) and (color) dissolved organic materials (cdom). In addition, the light may reach the bottom and be reflected off the sea floor and eventually reach the sensor.

Remote sensing of the ocean (often called "ocean color") attempts to use measured spectra from a scene to derive the physical parameters associated with that scene; this includes measurements such as the amount of chlorophyll, tss, and cdom, as well as depth (or bathymetry) and the nature of the bottom. Given all of these complications, it is not surprising that multispectral data does not have enough information to fully understand the scene, and hyperspectral data is needed. In HSI ocean sensing, the signature of interest is what is known as the remote sensing reflectance R_{rs}, which is the water leaving radiance, measured directly above the surface of the water, divided by the downwelling radiance. This normalization is done to eliminate any illumination effects, and allows us to concentrate on the scattering and absorption properties of the water itself, which are known as the inherent optical properties (IOP) of the water (Fig. 4).

Assuming the IOPs, depth, and bottom type are known, the R_{rs} spectrum can be predicted using the radiative transfer equation (RTE) [35]. In practice, we seek to invert the RTE, that is, we begin with a measured spectrum and seek the IOPs and bottom parameters. Unfortunately, direct inversion RTE (an integrodifferential equation) is not possible, and other methods are needed.

The most common method is to use parameterized equations to estimate the IOPs in terms of chlorophyll, tss, and cdom concentrations and then use forward-modeling techniques to create spectra with known parameters. To recover the parameters for a given target spectrum, one seeks to find parameters that, when plugged into the forward model, produce a spectrum that "matches" the target spectrum, according to some metric.

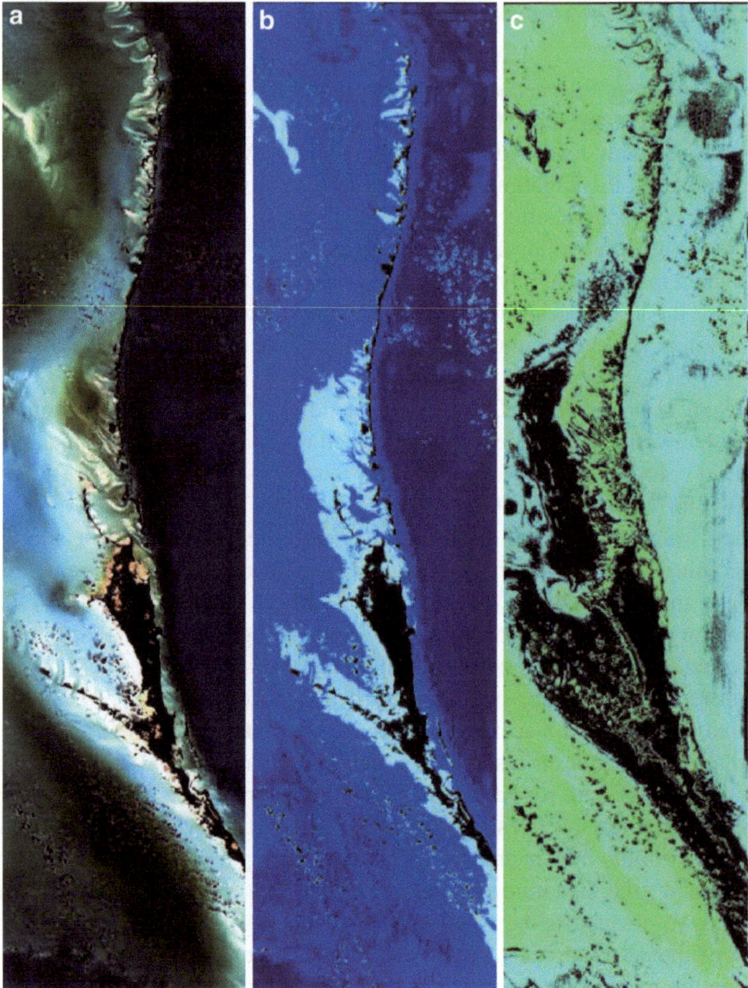

Fig. 4 Examples of IOP characterization via large lookup tables. (**a**) Original HSI image of Lee Stocking Island, Bermuda, captured by the HICO sensor. (**b**) and (**c**) Derived bathymetry and chlorophyll concentration maps, respectively

In order to run the forward modeling, there are a couple of options. The first is to use a simplified approximation of the RTE (known as semi-analytical (SA) models) that can quickly produce spectra for a given input [30, 31]. If we let $\alpha \in \mathbb{R}^k$ be the set of input parameters to the model and let $F : \mathbb{R}^k \Rightarrow \mathbb{R}^m$ denote the forward model (so that $F(\alpha)$ is the output spectrum), then the goal is to choose the α that minimizes

$$\|F(\alpha) - T\|,$$

where $T \in \mathbb{R}^n$ is the given target spectrum. This minimization can be done in reasonable time using standard convex optimization routines, such as Levenberg-Marquardt.

An alternative approach is to forward model the spectrum by numerically solving the RTE equation, using software such as HydroLight [1]. The drawback of this approach is that calculating even a single spectrum for a given set of parameters can take a relatively large amount of time, making numerical optimization techniques implausible. However, once a spectrum has been produced, it can always be stored and used in the future. This idea has lead to the use of very large lookup table approaches to IOP inversion of HSI data [36]. Using HydroLight, we can precompute large numbers of forward modeled with known parameters. With this table in place, we can then take an unknown target spectrum and simply search the table to see if it "matches" any entry; if so, we can assume that the parameters that went into the modeled spectrum are the same as the target spectrum. The main drawbacks of this approach are that we need a very large number (millions) of precomputed spectra to account for all of the variability in nature, which in turn leads to searching issues (basically, we are performing nearest-neighbor search in high dimensions, a notoriously difficult problem). Also, there is no guarantee that the target spectrum has been precomputed; if not, then no "good" match will be found, and no information about the IOPs can be derived.

8 Resources

In our final section we present additional resources, including software, data, and some general references on HSI.

The most common software package for hyperspectral analysis is the commercial ENVI package from ITT, which is unfortunately fairly expensive. Some free alternatives include MultiSpec from Purdue, Tetracorder from the USGS, and the open source Opticks. There are also a couple of MATLAB packages available, including the (free) MATLAB Hyperspectral Toolbox. Most (but not all) data is distributed in binary formats that are relatively well documented; many of them can be read directly from MATLAB.

Acquiring data can still be difficult. JPL has a few free AVIRIS data sets available on their website; more data is available free for graduate students doing research. NASA has extensive data from the hyperion satellite available (as well as tons of multispectral data). Both the hypercube and multicube softwares come with (limited) sample data. The SpecTIR company also has a number of sample cubes available for download from their website. Both USGS and JPL have libraries of reflectance spectrum for many common materials.

For the interested reader, more complete discussions of many of the topics in this chapter can be found in [20, 26].

References

1. http://www.sequoiasci.com/products/hl-radiative.cmsx
2. Adler-Golden, S., Matthew, M., Bernstein, L., Levine, R., Berk, A., Richtsmeier, S.: Atmospheric correction for short-wave spectral imagery based on modtran4. JPL Publ. **99**(17), 21–29 (1999)
3. Bachmann, C., Ainsworth, T., Fusina, R.: Exploiting manifold geometry in hyperspectral imagery. IEEE Trans. Geosci. Remote Sens. **43**(3), 441–454 (2005). DOI 10.1109/TGRS.2004.842292
4. Beaven, S., Stein, D., Hoff, L.: Comparison of gaussian mixture and linear mixture models for classification of hyperspectral data. In: Geoscience and Remote Sensing Symposium, 2000. Proceedings. IGARSS 2000. IEEE 2000 International, vol. 4, pp. 1597 –1599 (2000). DOI 10.1109/IGARSS.2000.857283
5. Bernstein, L., Adler-Golden, S., Sundberg, R., Levine, R., Perkins, T., Berk, A., Ratkowski, A., Felde, G., Hoke, M.: A new method for atmospheric correction and aerosol optical property retrieval for vis-swir multi- and hyperspectral imaging sensors: Quac (quick atmospheric correction). pp. 3549–3552 (2005)
6. Boardman, J.: Geometric mixture analysis of imaging spectrometry data. In: Geoscience and Remote Sensing Symposium, 1994. IGARSS '94. Surface and Atmospheric Remote Sensing: Technologies, Data Analysis and Interpretation., International, vol. 4, pp. 2369 –2371 (1994). DOI 10.1109/IGARSS.1994.399740
7. Bowles, J.H., Gillis, D.B.: An optical real-time adaptive spectral identification system (orasis). In: Chang, C.I. (ed.) Hyperspectral Data Exploitation, pp. 77–106. Wiley, Hoboken (2005)
8. Broadwater, J., Chellappa, R., Banerjee, A., Burlina, P.: Kernel fully constrained least squares abundance estimates. In: Igarss: 2007 IEEE International Geoscience and Remote Sensing Symposium, vol. 1–12: Sensing and Understanding Our Planet, IEEE International Symposium on Geoscience and Remote Sensing (IGARSS), pp. 4041–4044. IEEE (2007)
9. Burges, C.: A tutorial on support vector machines for pattern recognition. Data Min. Knowl. Discov. **2**(2), 121–167 (1998). DOI 10.1023/A:1009715923555
10. Camps-Valls, G., Bruzzone, L.: Kernel-based methods for hyperspectral image classification. IEEE Trans. Geosci. Remote Sens. **43**(6), 1351–1362 (2005). DOI 10.1109/TGRS.2005.846154
11. Camps-Valls, G., Marsheva, T.V.B., Zhou, D.: Semi-supervised graph-based hyperspectral image classification. IEEE Trans. Geosci. Remote Sens. **45**(10), 3044–3054 (2007). DOI 10.1109/TGRS.2007.895416
12. Camps-Valls, G., Shervashidze, N., Borgwardt, K.M.: Spatio-spectral remote sensing image classification with graph kernels. IEEE Geosci. Remote Sens. Lett. **7**(4), 741–745 (2010). DOI {10.1109/LGRS.2010.2046618}
13. Cardoso, J.F.: Infomax and maximum likelihood for blind source separation. IEEE Signal Process. Lett. **4**(4), 112 –114 (1997). DOI 10.1109/97.566704
14. Comon, P.: Independent component analysis, a new concept? (1992)
15. Eismann, M.T., Stein, D.W.J.: Stochastic mixture modeling. In: Chang, C.I. (ed.) Hyperspectral Data Exploitation, pp. 77–106. Wiley, Hoboken (2005)
16. Gao, B.C., Heidebrecht, K., Goetz, A.: Derivation of scaled surface reflectances from aviris data. Remote Sens. Environ. **44**(2–3), 165–178 (1993)
17. Gao, B.C., Montes, M., Davis, C., Goetz, A.: Atmospheric correction algorithms for hyperspectral remote sensing data of land and ocean. Remote Sens. Environ. **113**(1), S17–S24 (2009)
18. Gillis, D., Bowles, J., Ientilucci, E.J., Messinger, D.W.: A generalized linear mixing model for hyperspectral imagery **6966**(1), 69,661B (2008). DOI 10.1117/12.782113. URL http://dx.doi.org/doi/10.1117/12.782113
19. Green, A., Berman, M., Switzer, P., Craig, M.: A transformation for ordering multispectral data in terms of image quality with implications for noise removal. IEEE Trans. Geosci. Remote Sens. **26**(1), 65–74 (1988). DOI 10.1109/36.3001

20. Harsanyi, J.C., Chang, C.I.: Hyperspectral image classification and dimensionality reduction: an orthogonal subspace projection approach. IEEE Trans. Geosci. Remote Sens. **32**, 779–785 (1994). DOI 10.1109/36.298007
21. Heinz, D., Chang, C.I.: Fully constrained least squares linear spectral mixture analysis method for material quantification in hyperspectral imagery. IEEE Trans. Geosci. Remote Sens. **39**(3), 529–545 (2001)
22. Hyvï¿½rinen, A., Oja, E.: Independent component analysis: algorithms and applications. Neural Netw. **13**(4–5), 411–430 (2000). DOI 10.1016/S0893-6080(00)00026-5. URL http://www.sciencedirect.com/science/article/pii/S0893608000000265
23. Jimenez, L.: Hyperspectral data analysis and supervised feature reduction via projection pursuit. IEEE Trans. Geosci. Remote Sens. **37**(6), 2653–2667 (1999)
24. Jolliffe, I.T.: Principal Component Analysis. Springer, New York (2002)
25. Keshava, N.: Distance metrics and band selection in hyperspectral processing with applications to material identification and spectral libraries. IEEE Trans. Geosci. Remote Sens. **42**(7), 1552–1565 (2004). DOI 10.1109/TGRS.2004.830549
26. Keshava, N., Mustard, J.: Spectral unmixing. IEEE Signal Process. Mag. **19**(1), 44–57 (2002)
27. Kruse, F., Kierein-Young, K., Boardman, J.: Mineral mapping at cuprite, nevada with a 63-channel imaging spectrometer. Photogramm. Eng. Remote Sens. **56**(1), 83–92 (1990)
28. Kwon, H., Nasrabadi, N.: Kernel RX-algorithm: a nonlinear anomaly detector for hyperspectral imagery. IEEE Trans. Geosci. Remote Sens. **43**(2), 388–397 (2005). DOI 10.1109/TGRS.2004.841487
29. Lawson, C.L., Hanson, R.J.: Solving least squares problems. Classics in Applied Mathematics. SIAM, Philadelphia (1995)
30. Lee, Z., Carder, K., Mobley, C., Steward, R., Patch, J.: Hyperspectral remote sensing for shallow waters. i. a semianalytical model. Appl. Opt. **37**(27), 6329–6338 (1998)
31. Lee, Z., Carder, K., Mobley, C., Steward, R., Patch, J.: Hyperspectral remote sensing for shallow waters: 2. deriving bottom depths and water properties by optimization. Appl. Opt. **38**(18), 3831–3843 (1999)
32. Lee, D., Seung, H.: Learning the parts of objects by non-negative matrix factorization. Nature **401**(6755), 788–791 (1999)
33. Mardia, K., Kent, J., Bibby, J.: Multivariate Analysis. Academic, London (1979)
34. Melgani, F., Bruzzone, L.: Classification of hyperspectral remote sensing images with support vector machines. IEEE Trans. Geosci. Remote Sens. **42**(8), 1778–1790 (2004). DOI 10.1109/TGRS.2004.831865
35. Mobley, C.D.: Light and Water. Academic, San Diego (1994)
36. Mobley, C., Sundman, L., Davis, C., Bowles, J., Downes, T., Leathers, R., Montes, M., Bissett, W., Kohler, D., Reid, R., Louchard, E., Gleason, A.: Interpretation of hyperspectral remote-sensing imagery by spectrum matching and look-up tables. Appl. Opt. **44**(17), 3576–3592 (2005). DOI 10.1364/AO.44.003576
37. Nascimento, J., Dias, J.: Does independent component analysis play a role in unmixing hyperspectral data? IEEE Trans. Geosci. Remote Sens. **43**(1), 175–187 (2005). DOI 10.1109/TGRS.2004.839806
38. Plaza, A., Benediktsson, J.A., Boardman, J.W., Brazile, J., Bruzzone, L., Camps-Valls, G., Chanussot, J., Fauvel, M., Gamba, P., Gualtieri, A., Marconcini, M., Tilton, J.C., Trianni, G.: Recent advances in techniques for hyperspectral image processing. Remote Sens. Environ. **113**, S110–S122 (2009). DOI 10.1016/j.rse.2007.07.028
39. Plaza, A., Martiï¿½nez, P., Peï¿½rez, R., Plaza, J.: Spatial/spectral endmember extraction by multidimensional morphological operations. IEEE Trans. Geosci. Remote Sens. **40**(9), 2025–2041 (2002)
40. Reed, I., Yu, X.: Adaptive multiple-band cfar detection of an optical pattern with unknown spectral distribution. IEEE Trans. Acoustics, Speech Signal Process. **38**(10), 1760 –1770 (1990). DOI 10.1109/29.60107

41. Renard, N., Bourennane, S., Blanc-Talon, J.: Denoising and dimensionality reduction using multilinear tools for hyperspectral images. IEEE Geosci. Remote Sens. Lett. **5**(2), 138–142 (2008). DOI 10.1109/LGRS.2008.915736
42. Roberts, D., Yamaguchi, Y., Lyon, R.: Calibration of airborne imaging spectrometer data to percent reflectance using field spectral measurements. In: Proceedings of the 19th International Symposium on Remote Sensing of Environment, pp. 679–688 (1985)
43. Sajda, P., Du, S., Parra, L.: Recovery of constituent spectra using non-negative matrix factorization. In: Unser, M.A., Aldroubi, A., Laine, A.F. (eds.) Wavelets: Applications in Signal and Image Processing X, Pts 1 and 2, Proceedings of the Society of Photo-Optical Instrumentation Engineers (SPIE), vol. 5207, pp. 321–331. SPIE (2003). DOI 10.1117/12.504676
44. Scholkopf, B., Smola, A., Muller, K.: Nonlinear component analysis as a kernel eigenvalue problem. Neural Comput. **10**(5), 1299–1319 (1998). DOI 10.1162/089976698300017467
45. Silva, V.D., Tenenbaum, J.B.: Global versus local methods in nonlinear dimensionality reduction. In: Advances in Neural Information Processing Systems 15, pp. 705–712. MIT Press, Cambridge (2003)
46. Somers, B., Asner, G., Tits, L., Coppin, P.: Endmember variability in spectral mixture analysis: A review. Remote Sens. Environ. **115**(7), 1603–1616 (2011)
47. Tenenbaum, J.B., de Silva, V., Langford, J.C.: A Global Geometric Framework for Nonlinear Dimensionality Reduction. Science **290**, 2319–2323 (2000)
48. Winter, M.E.: N-findr: an algorithm for fast autonomous spectral end-member determination in hyperspectral data, pp. 266–275 (1999)

Hyperspectral Demixing: Sparse Recovery of Highly Correlated Endmembers

John B. Greer

Abstract We apply three different sparse reconstruction techniques to spectral demixing. Endmembers for these signatures are typically highly correlated, with angles near zero between the high-dimensional vectors. As a result, theoretical guarantees on the performance of standard pursuit algorithms like orthogonal matching pursuit (OMP) and basis pursuit (BP) do not apply. We evaluate the performance of OMP, BP, and a third algorithm, sparse demixing (SD), by demixing random sparse mixtures of materials selected from the USGS spectral library (Clark et al., USGS digital spectral library splib06a. U.S. Geological Survey, Digital Data Series 231, 2007). Examining reconstruction sparsity versus accuracy shows clear success of SD and clear failure of BP. We also show that the relative geometry between endmembers creates a bias in BP reconstructions.

Keywords Hyperspectral demixing • Sparse demixing (SD) • Correlated endmembers • Basis pursuit (BP) • Orthogonal matching pursuit (OMP)

1 Introduction

Sparsity arises in many important problems in mathematics and engineering. Recent algorithms for finding sparse representations of signals have achieved success in applications including image processing [15], compression [22], and classification [19]. These algorithms, including basis pursuit (BP) [7] and various matching pursuit (MP) methods [16], are guaranteed to converge to correct solutions for problems that meet criteria established, for example, in [6,23]. In practice, however,

J.B. Greer (✉)
National Geospatial-Intelligence Agency, 7500 Geoint Drive, Springfield, VA, USA
e-mail: John.B.Greer@nga.mil

T.D. Andrews et al. (eds.), *Excursions in Harmonic Analysis, Volume 1*, Applied and Numerical Harmonic Analysis, DOI 10.1007/978-0-8176-8376-4_10,

they are often applied to problems that do not meet those criteria. One such example is the source separation problem of spectral demixing of hyperspectral images (HSI).

Spectral demixing is the identification of the materials, called endmembers, comprising a hyperspectral pixel, and their fractional abundances. Images captured by hyperspectral sensors such as airborne visible/infrared imaging spectrometer (AVIRIS) [14], hyperspectral mapper (HyMap) [9], and hyperspectral digital imagery collection experiment (HYDICE) [1] have pixels containing spectral measurements for hundreds of narrowly spaced wavelengths. Ideally these measurements could be used to identify materials by comparing directly with a spectral library, trading a difficult computer vision problem for relatively straightforward spectral analysis. In reality, measured signatures rarely correspond to spectra of pure materials. HSI cameras take images with high spectral resolution at the expense of low spatial resolution; for example, AVIRIS has a 20-m ground resolution when flown at high altitude (20 km) [14]. As a result, measured spectra often correspond to mixtures of many materials.

1.1 The Linear Mixture Model

Even though HSI sensors generally measure nonlinear combinations of the constituent materials' spectra, HSI analysts often assume linear mixing of pure signals [13, 25]. This assumption holds if the materials occur in spatially separated regions with negligible light scattering.

Let E denote an n-by-k matrix of endmembers $E = [\mathbf{e}_i]_1^k$. The linear mixture model (LMM) assumes that every pixel signature $\mathbf{x} \in \mathbb{R}^n$ has an abundance vector $\alpha \in \mathbb{R}^k$ satisfying

$$x = E\alpha + \eta, \tag{1}$$

where η is a small error term. Ideally, endmembers correspond to pure materials, but they more likely represent common mixtures of materials. The abundance vector α gives the relative quantities of the materials making up the mixture.

We consider three common versions of the LMM, each with its own set of constraints on α and E :

(LMM 1) $\alpha_i \geq 0$, $\sum \alpha_i = 1$, and $\operatorname{rank}(E) \geq k-1$.
(LMM 2) $\alpha_i \geq 0$, $\sum \alpha_i \leq 1$, and E has full rank.
(LMM 3) $\alpha_i \geq 0$ and E has full rank.

The rank constraints ensure uniqueness of each signature's abundance vector. Each of these models has different assumptions on the physical properties of the endmembers in E. For LMM 1, we assume either full illumination of every pixel or that at least one endmember represents shade. LMM 2 assumes that E contains spectral signatures corresponding to materials lit as brightly as the brightest pixels in the image. Darker pixels have abundance vectors with $\sum \alpha_i < 1$. LMM 3 allows pixels brighter than any of the $\mathbf{e}_i$. The rank restrictions typically pose no problem, since $k \ll n$ in practice.

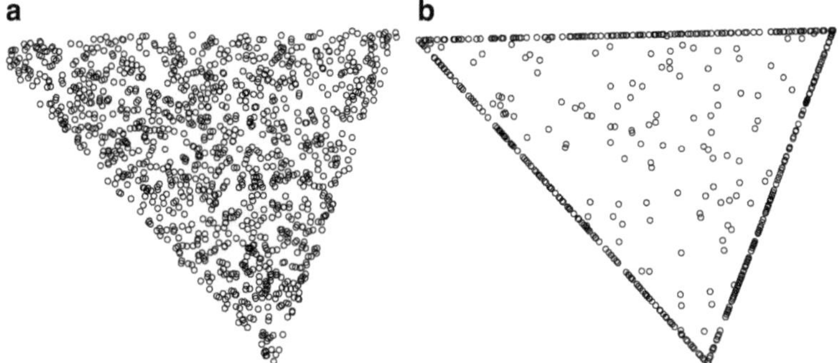

Fig. 1 The *left* shows a standard depiction of the simplex model for HSI. We argue that the depiction on the *right* is more realistic. Since mixtures on the simplex interior contain all of the endmembers, they should rarely occur in natural images

Each LMM has its own constraint set $\mathscr{A}$ for the abundance vectors. Define

$$S = \{\mathbf{y} \quad | \quad \mathbf{y} = E\alpha, \alpha \in \mathscr{A}\}. \tag{2}$$

In LMM2 1 and 2, S describes a simplex. In LMM 1, S is a $(k-1)$-dimensional simplex with corners given by the k columns of E. For LMM 2, S is a k-dimensional simplex determined by E and the origin. The two problems are mathematically equivalent: we can write LMM 1 as LMM 2 by translating the origin to an arbitrary endmember in E, then removing that column from E. Similarly, we can write LMM 2 as LMM 1 by adding a column of zeros to E. In LMM 3, S is the wedge determined by the columns of E. Notice that for a given image, we can rescale the columns of E so that $\sum \alpha_i \leq 1$ for all pixels within the image. We therefore focus on LMMs 1 and 2. In particular, we assume E has full rank.

Demixing often requires learning the endmembers as well as the abundances (blind source separation), but throughout this chapter, we assume known endmembers. See [2, 11, 17, 18, 21, 26] for more information on learning endmembers.

1.2 Sparse Mixtures

Figure 1a shows a standard idealized scatter plot of LMM 1 [3, 4, 13, 26]. In such illustrations, authors typically distribute most pixels throughout the simplex interior. We argue that Fig. 1b illustrates hyperspectral data more accurately. Any interior point of a simplex is a combination of all the endmembers. Physically, this means that the region captured by the pixel contains samples of each endmember within the scene. We suspect that such pixels rarely occur. Instead, most pixels contain a strict subset of the scene's endmembers and thus lie on the simplex boundary. The simplex interior should be nearly empty.

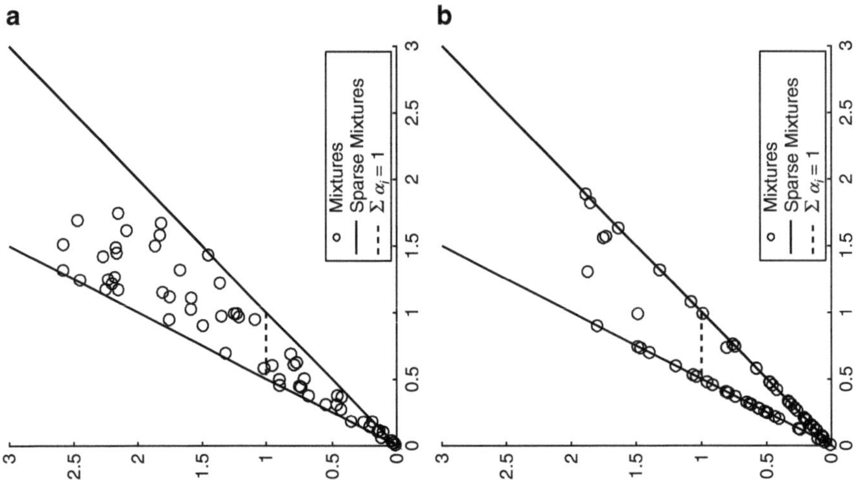

Fig. 2 The *left* depicts LMM 3. We show the line $\sum \alpha_i = 1$, which forms a boundary for LMM 2. In that case all mixtures must lie below this line. The *right* shows sparse mixtures for LMM 3

LMM 2 determines another simplex, but one face of this simplex corresponds to admissible abundance vectors satisfying $\sum \alpha_i = 1$. These vectors are not necessarily sparse. For LMM 3, sparse signals again occur on the boundary of S. Figure 2 illustrates sparse and non-sparse mixtures for LMM 3 with the extra boundary of LMM 2 included for reference.

Traditional pixel demixing algorithms minimize error, using nonnegative least squares (NLS). This might not achieve the most realistic results, however, since we expect some error due to model uncertainty and noise. Assuming that most pixels are made up of only a few endmembers, we may instead seek a balance between error and sparsity. We allow some error in the reconstruction if it comes with a sparser mixture.The ideal mixture is the sparsest mixture with small, but acceptable, error.

1.3 Outline

We evaluate three different algorithms for calculating sparse abundance vectors: a basis pursuit (BP) algorithm that uses the L^1 of the abundance vectors [12], a greedy algorithm called sparse demixing [11], and a natural extension of orthogonal matching pursuit (OMP) to the endmember problem. Theorems guaranteeing the convergence of OMP and BP to accurate sparse mixtures all require low mutual information of the set of vectors being searched over. In practice, they are often used with highly correlated vectors. One such example is hyperspectral demixing.

Section 2 briefly describes the three algorithms and summarizes results from [11] that show that BP preferentially selects endmembers based on the relative geometry between endmembers. Section 4 demonstrates the relative performance of OMP, BP, and SD by demixing spectra with known abundances. We choose a set of endmembers from the USGS Library [8], then randomly select a sparse matrix A. We look at this problem both with and without Gaussian noise. Since these algorithms are intended to find sparse mixtures that accurately approximate the signature, we judge success by examining reconstruction sparsity versus accuracy.

2 Spectral Demixing

Given a matrix of endmembers E and a spectral signature $\mathbf{x}$, HSI analysts typically demix pixels with NLS [13]. The NLS approximation $\mathbf{y}$ of $\mathbf{x}$ solves $\mathbf{y} = E\alpha$ for

$$\alpha = \arg \min_{\alpha' \in \mathscr{A}} \| E\alpha' - \mathbf{x} \|_2 . \tag{3}$$

Defining S by (2), we rewrite (3) as

$$\mathbf{y} = \arg\min_{\mathbf{y}' \in S} \| \mathbf{y}' - \mathbf{x} \|_2 . \tag{4}$$

This quadratic programming problem has a simple geometric solution. Let $\hat{\mathbf{x}}$ denote the orthogonal projection of $\mathbf{x}$ on the column space of E. For problems LMMs (2) and (3),

$$\hat{\mathbf{x}} = E\left(E^\top E\right)^{-1} E^\top \mathbf{x}.$$

Note that since E has full rank, $E^\top E$ is invertible. Since $\hat{\mathbf{x}}$ is the orthogonal projection of $\mathbf{x}$, $\mathbf{y}$ solves (4) if and only if

$$\mathbf{y} = \arg\min_{\mathbf{y}' \in S} \| \mathbf{y}' - \hat{\mathbf{x}} \|_2 . \tag{5}$$

If $\hat{\mathbf{x}}$ lies in S, then $\mathbf{y} = \hat{\mathbf{x}}$. If $\hat{\mathbf{x}}$ lies outside S, then (5) gives the closest point $\mathbf{y}$ to $\hat{\mathbf{x}}$ on the boundary of S.

3 Applying Sparse Coding to the Demixing Problem

We assume that the abundance vector α satisfies (1) for some η with

$$\| \eta \|_2 = \varepsilon > 0.$$

Unlike standard demixing, we assume that the correct abundance vector is the sparsest one that gives an approximation within ε of the measured spectrum. Minimizing the L^2 (Euclidean) norm generally does not give sparse mixtures. The NLS constraint $y \in S$, however, automatically enforces sparsity for some mixtures. If a pixel lies exactly on the boundary of S, then NLS correctly recognizes it as a sparse mixture. However, sensor noise, measurement errors, and model inaccuracy likely prevent such cases. Even when a pixel contains only a few endmembers, these errors push the pixel off the boundary of S. If they push the pixel outside S, then NLS gives the correct sparse solution. If they push the pixel inside S, NLS gives a mixture of all the endmembers.

The sparsest abundance vector giving an approximate mixture with error ε is

$$\alpha = \arg\min_{\beta \in \mathscr{A}} \{\| \beta \|_0 \quad | \quad \| E\beta - \mathbf{x} \|_2 < \varepsilon\} \tag{6}$$

for $\varepsilon > 0$. The L^0 semi-norm of α, $\| \alpha \|_0$, is the number of nonzero components of α. Minimizing the nonconvex L^0 semi-norm is NP-hard, so matching pursuit algorithms only find approximate solutions to the problem. One of these algorithms, OMP has been shown to solve (6) for some matrices E [23]. Minimizing the L^1 norm [defined by (8)] also gives sparse solutions for some matrices E, and it has the mathematical advantage of convexity [5, 10].

In this section, we describe three algorithms for calculating sparse abundance vectors. The first is a basis pursuit (BP) algorithm that uses the L^1 norm. The others, OMP and sparse demixing (SD), find approximate solutions to (6).

3.1 Basis Pursuit

In [12], Guo et al. calculated sparse abundance vectors by minimizing

$$\alpha = \arg\min_{\beta \in \mathscr{A}} \lambda \| \beta \|_1 + \frac{1}{2} \| E\beta - \mathbf{x} \|_2^2 . \tag{7}$$

In each constraint set $\mathscr{A}$, $\alpha_i \geq 0$, so

$$\| \alpha \|_1 := \sum_i |\alpha_i| = \sum_i \alpha_i. \tag{8}$$

Note that the L^1 term makes no meaningful contribution to (7) for LMM 1, since $\| \alpha \|_1 = 1$ for all admissible α. Whenever discussing BP, we assume LMM 2 or LMM 3.

Unfortunately, for general endmember matrices E, (7) does not necessarily give sparse abundance vectors. In fact, Greer[11] shows that (7) gives sparser solutions than NLS only for certain cases. For many other cases, including the L^1 norm *reduces* sparsity. We briefly describe a rationale for this and refer to [11] for details.

The set $\mathscr{A}$ of admissible abundance vectors determines a subset S of the column space of E [see (2)]. Since E has full rank, every $\mathbf{y} \in S$ has a unique abundance vector α satisfying $E\alpha = \mathbf{y}$. In fact, $\alpha(y) = \Phi y$ for

$$\Phi = \left(E^\top E\right)^{-1} E^\top .$$

We call the abundance vector components, α_i, coefficient functions. Each coefficient function is linear, with $\alpha_i(\mathbf{e}_i) = 1$ and $\alpha_i(\mathbf{y}) = 0$ for all $\mathbf{y}$ on the $(k-1)$-dimension hyperplane determined by $\mathbf{e}_{j \neq i}$ and the origin.

For $\mathbf{y} \in S$ define

$$\phi(\mathbf{y}) := \| \alpha(\mathbf{y}) \|_1 = \sum_i \alpha_i = \sum_{i,j} \Phi_{ij} y_j. \tag{9}$$

In particular, ϕ is a linear function of $\mathbf{y}$ in S with $\nabla\phi = \Phi^\top \mathbf{1}$, where $\mathbf{1}$ denotes a column vector of ones. Since $\phi(\mathbf{e}_i) = 1$ for every column $\mathbf{e}_i$ of E, and $\phi\,(0) = 0$, $\nabla\phi$ is normal to the $(k-1)$-dimensional hyperplane determined by the columns of E. Thus the L^1 term's effect depends entirely on the geometry of the endmembers in E.

Define

$$F(\mathbf{s}) = \lambda\phi(\mathbf{s}) + \frac{1}{2} \| \mathbf{s} - \mathbf{x} \|_2^2 .$$

Solving (7) is equivalent to solving

$$\mathbf{y} = \arg\min_{\mathbf{s} \in S} F(\mathbf{s}). \tag{10}$$

The convex function F has a global minimum in $\mathbb{R}^k$ at

$$\mathbf{y} = \mathbf{x} - \lambda\nabla\phi. \tag{11}$$

Compare (11) with the minimum of the NLS optimization function,

$$G(\mathbf{s}) = \frac{1}{2} \| \mathbf{s} - \mathbf{x} \|_2^2 .$$

The minimum of $G(\mathbf{s})$ occurs inside S only when $\mathbf{x}$ lies inside S. The minimum of F can occur on the interior of S even for cases where $\mathbf{x}$ lies on the exterior—cases where NLS gives a sparse solution. If $\mathbf{y}$ lies inside S, then the L^1 reconstruction does not give a sparser representation than NLS. This happens, for example, for any point $\mathbf{x}$ on the interior of S that is further than $\lambda\,|\nabla\phi|$ from the boundary of S. On the other hand, the L^1 reconstruction of $\mathbf{x}$ is less accurate than the NLS solution, which gives $\mathbf{x}$ the exact answer. For these points, NLS is clearly the better method.

The L^1 norm's ability to increase sparsity depends on the sign of $\nabla\alpha_i \cdot \nabla\phi$ for each i. Figure 3 shows the effects of the L^1 norm for a problem with two endmembers giving coefficient functions, α_1 and α_2, with $\nabla\alpha_1 \cdot \nabla\phi < 0$ and

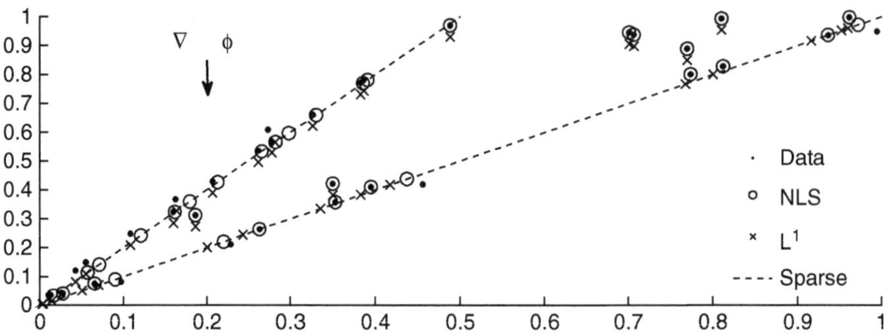

Fig. 3 Failure of L^1 demixing. For this pair of endmembers, NLS produces sparser mixtures for points near the left-hand boundary

$\nabla\alpha_2 \cdot \nabla\phi > 0$. Let $\mathbf{y}_\lambda(\mathbf{x})$ solve (11) for given λ and $\mathbf{x}$. Suppose $\nabla\alpha_i \cdot \nabla\phi > 0$. Then for any given $\mathbf{x}$, $\alpha_i(\mathbf{y}_\lambda(\mathbf{x}))$ monotonically decreases to 0 as λ increases. The same does not hold if $\nabla\alpha_i \cdot \nabla\phi < 0$. In this case, x close to the boundary $\alpha_i = 0$, either inside or outside S, can yield minima $\mathbf{y}_\lambda(\mathbf{x})$ that lie on the interior of S. Section 4.2 will demonstrate how this affects demixing.

3.2 *Orthogonal Matching Pursuit*

OMP is a greedy algorithm that iteratively increases the number of nonzero components in α while minimizing the approximation's residual error at each step. The linear mixing model constraints require modifying the OMP algorithm (introduced in [27]). This section describes one natural modification.

Suppose we have a spectral signature x, a set of endmembers $\{e_i\}$ and an error bound ε. For a subset of endmembers, Λ_k, let x_k be the NLS reconstruction of x over Λ_k. Define

$$r_k = x - x_k,$$

and set

$$\Lambda_0 = \left\{ \arg\min_i \| x - e_i \|_2 \right\}.$$

Until $\| r_k \|_2 < \varepsilon$, OMP sequentially improves the approximation x_k by setting

$$\Lambda_{k+1} = \Lambda_k \cup \left\{ \arg\max_i \frac{r_i^\top e_i}{\| e_i \|_2} \right\}. \tag{12}$$

The standard OMP algorithm uses the absolute value of the inner product, but the LMMs nonnegativity constraint leads to better results without the absolute value.

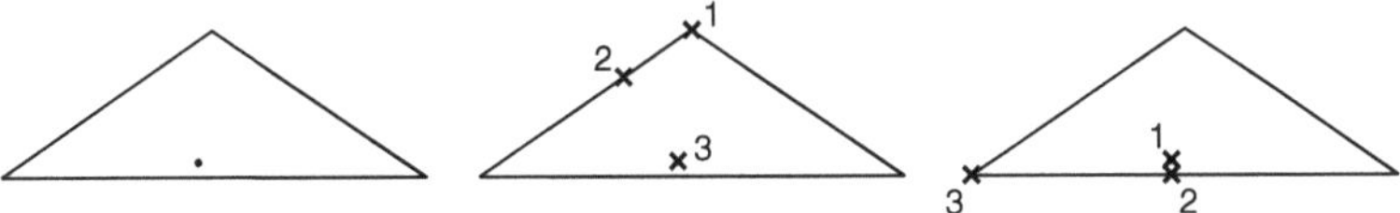

Fig. 4 Consider the simplex and data point shown in the *left* figure. This point is nearly a mixture of the two materials forming the left side of the simplex. The *middle* figure shows the approximations produced in three iterations of OMP. The *right* shows the approximations produced in three iterations of SD. Notice that the second iteration of SD gives the closest sparse mixture, which OMP never produces

3.3 Sparse Demixing

Sparse demixing (SD) uses the greedy approach of OMP, but works in the opposite direction: it begins with a representation over all endmembers, then removes endmembers one by one until reaching the sparsest representation within a specified accuracy. SD first performs NLS over the full set of endmembers, then removes the endmember corresponding to the smallest component of α. Next, it performs NLS on the simplex determined by this smaller set of endmembers. This process repeats, in each iteration removing the endmember corresponding to the smallest abundance value, until the approximation leaves the accuracy range specified by ε in (6). See [11] for details on the application of SD to all three LMMs.

SD has some advantage over OMP. Its initial step gives the widely used NLS solution. For examples like the one in Fig. 4, OMP produces mixtures containing all the endmembers, even though sparsity assumptions for (1) place $\mathbf{x}$ on the simplex's left edge. SD gives the desired mixture in two iterations.

SD takes advantage of key differences between pixel demixing and standard sparse reconstruction problems. In its intended applications, OMP searches large overcomplete sets of vectors for spanning subsets. SD is impractical for such applications, but HSI demixing usually involves far fewer endmembers than the number of spectral bands. Thus, starting with all the endmembers and sequentially eliminating them is feasible.

4 Numerical Experiments

To evaluate the three algorithms describe in Sect. 2, we demix random mixtures satisfying LMM 1 and LMM 2. Each mixture consists of a sparse subset of 14 endmembers chosen from the USGS spectral libraries [8]: nine minerals and five vegetation endmembers. Figure 5 shows the spectral signatures of these endmembers. The spectra from these libraries each have 450 bands ranging from 0.395 to 2.56 μm. These materials differ enough to distinguish between them in HSI (see, for example, [20]).

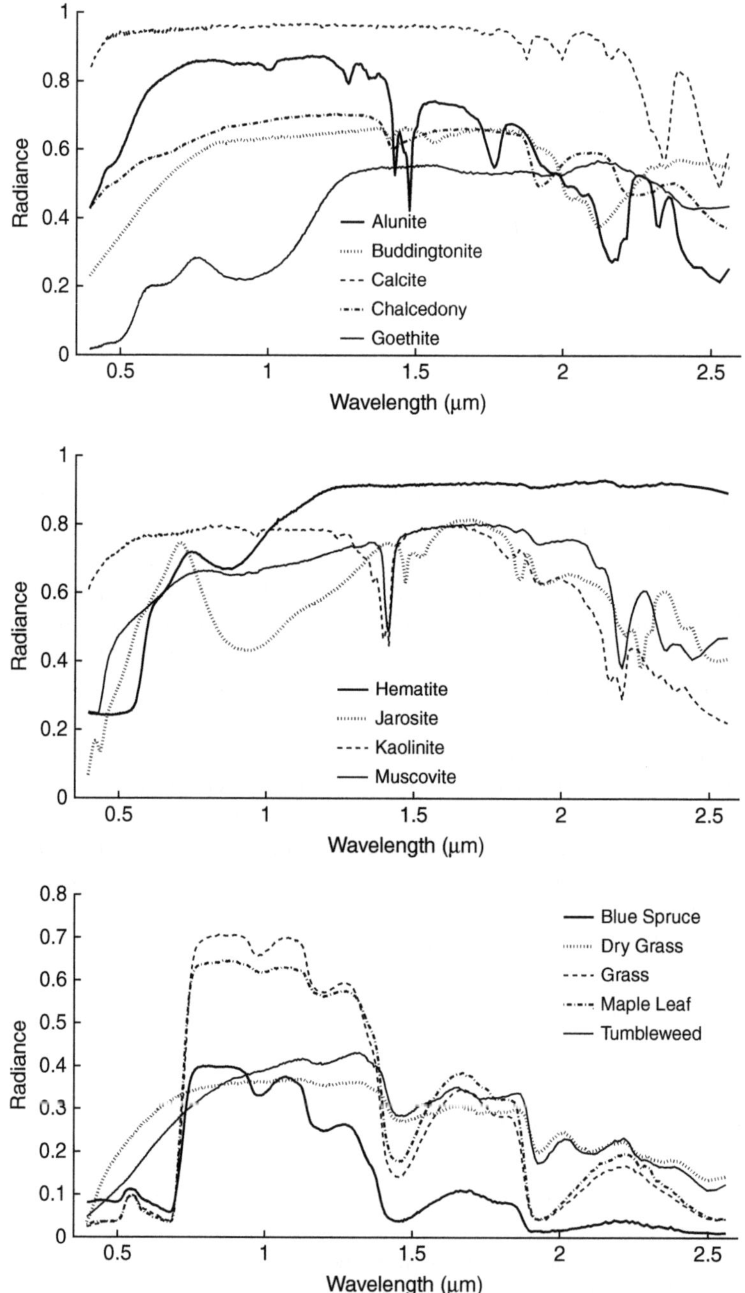

Fig. 5 Endmember plots. The spectra used as endmemebrs for the numerical tests. Spectra consist of minerals and vegetation from the USGS spectral library [8]

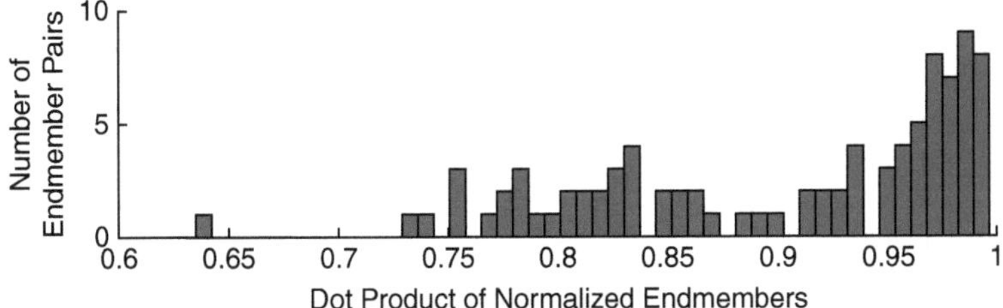

Fig. 6 The linearly independent spectra are highly coherent. Most of the endmembers nearly align with each other, as shown by this histogram of dot products of unique endmember pairs

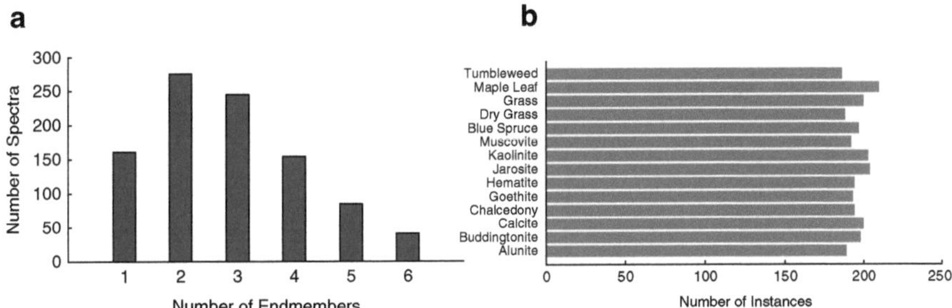

Fig. 7 Distribution of endmembers in the LMM 1 example

Although linearly independent, the spectra are highly correlated, with angles between many pairs of the spectra near zero. See Fig. 6 for a histogram of the dot products between the 14 (normalized) spectra. Notice that those dot products are all larger than $\frac{1}{2}$. Due to this high correlation, support recovery theory for BP and OMP does not apply [23, 24]. The examples demonstrate the failure of BP.

4.1 LMM 1 ($\sum \alpha_i = 1$)

The first example uses mixtures satisfying LMM 1. We determined the set of spectra, X, by randomly selecting a 14-by-1,000 matrix of abundances. Each entry had a 20 % chance of being nonzero, with nonzero entries distributed uniformly between 0 and 1. Columns with all zeros were eliminated. This example has 961 nonzero spectra. Finally, we scaled each column so that its abundances added to 1. Figure 7 shows the distribution of endmembers in X. It also shows the distribution of the number of endmembers comprising each spectral signature in X : each is a mixture of between 1 and 6 endmembers.

Figure 8 shows results of demixing X with no added noise. In this case, NLS gives the exact abundances of X. As discussed in Sect. 3.1, for all λ, BP gives the same solution as NLS for LMM 1. Each curve in Fig. 11 is parametrized by ε,

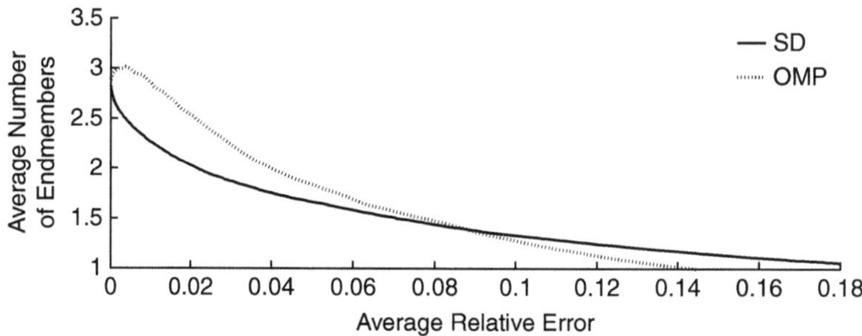

Fig. 8 Performance of OMP and SD on random sparse mixtures satisfying LMM 1 with no additional noise. In this case, NLS gives the exact solution, which corresponds to the intersection of the SD and OMP curves at $\varepsilon = 0$

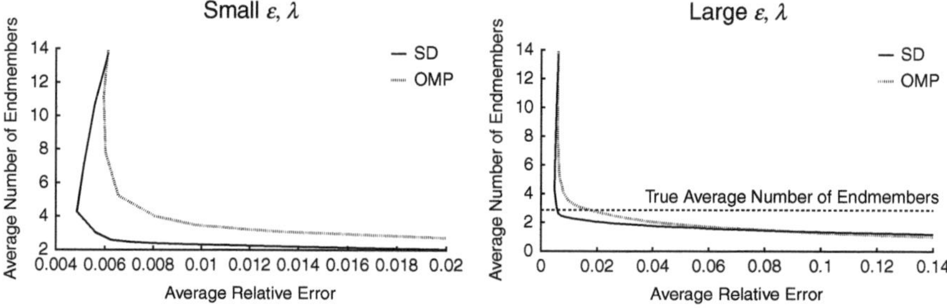

Fig. 9 Performance of SD and OMP on sparse LMM 1 mixtures plus Gaussian noise for small (*left*) and large (*right*) ε and λ. Errors are calculated with respect to the non-noisy spectra. The SD and OMP curves intersect at $\varepsilon = 0$, which is the NLS solution

the amount of error allowed for the sparse approximations [see (6)]. Increasing ε sacrifices some of that accuracy for sparsity. This increase of ε corresponds to more iterations in SD and fewer iterations in OMP. The curves intersect at $\varepsilon = 0$, which is the NLS solution. Curves lying closer to the origin correspond to methods that find sparser solutions with greater accuracy. In this case SD performs better than OMP until both reach the large-error regime—about 9 % relative error.

We next add Gaussian noise with a standard deviation of 0.03 to X. NLS does not give the correct solution for this more realistic scenario. Noise has pushed some of the spectra in X to the interior of the endmember-determined simplex, making NLS choose a mixture that is non-sparse and incorrect. We use the non-noisy signatures in X to calculate errors. Figure 9 shows that SD improves the accuracy of the abundances while simultaneously decreasing the number of endmembers. OMP does not perform as well, but it still shows an initial drop in the number of endmembers with very little increase in error.

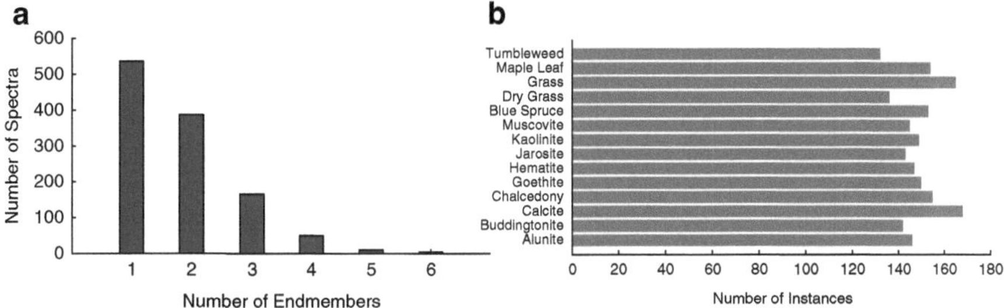

Fig. 10 Distribution of endmembers for the LMM 2 example

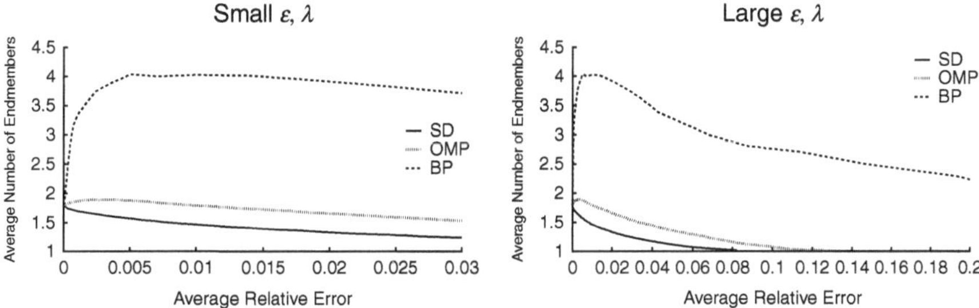

Fig. 11 Comparison of BP, SD, and OMP for random sparse spectra satisfying LMM 2 with no added noise for small (*left*) and large (*right*) ε and λ. Notice that BP does not offer any improved sparsity over NLS, which corresponds to the intersection of all three curves

4.2 LMM 2 ($\sum \alpha_i \leq 1$)

For this example, we used the process in Sect. 4.1 to randomly select a 15-by-1,500 sparse abundance matrix, with the extra row corresponding to an endmember of all zeros. After scaling and eliminating all-zero columns, we removed the 15th row. The resulting set, X, contains 1,152 spectral signatures satisfying LMM 2. Figure 10 shows the distribution of endmembers across X.

Figure 11 shows how BP, OMP, and SD all perform on X without added noise. NLS again gives the exact solution. Both OMP and SD depend on the parameter ε, with $\varepsilon = 0$ giving the NLS solution [see (6)] and increased ε giving sparser solutions. The curve for BP depends on λ, with $\lambda = 0$ corresponding to NLS [see (7)]. BP performs poorly. In fact, increasing λ increases both errors and the number of endmembers. The set X contains only exact sparse mixtures that lie on the simplex boundary. As discussed in Sect. 3.1, as λ increases, BP drives many of these spectra to the simplex interior.

BP preferentially selects the ith endmember if $\nabla\alpha_i \cdot \nabla\phi < 0$, for abundance α_i and ϕ defined by (9). Figure 12 demonstrates this phenomenon. The curve plots the number of BP approximations that contain each given endmember as λ increases. For very large λ, nearly all the BP approximations contain the same endmember.

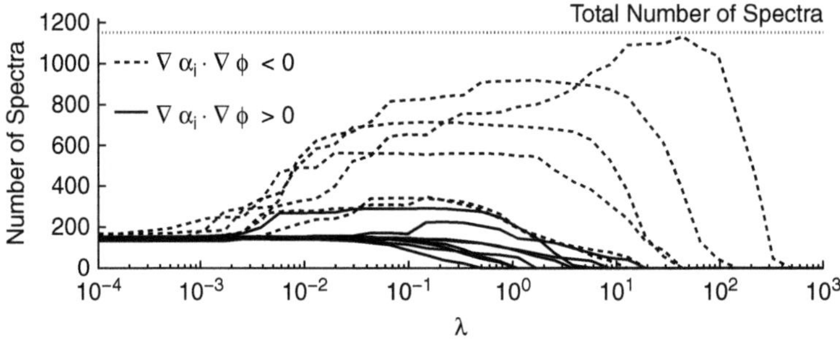

Fig. 12 BP's selection of endmembers for LMM 2 spectra without noise. Increasing λ should increase sparsity, meaning abundances of endmembers should shift to zero. However, the effect on each endmember e_i depends on $\nabla\alpha_i \cdot \nabla\phi$ for abundance α_i and ϕ defined by (9). BP prefers some endmembers over others. In this example, the actual endmembers are distributed nearly uniformly, but BP loses that statistical pattern as λ increases

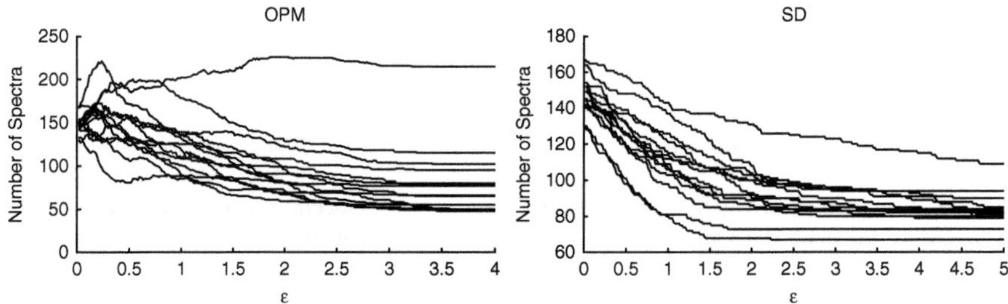

Fig. 13 The selection of endmembers by OMP and SD. As ε increases, SD decreases the number triggered for each endmember. OMP performs somewhere between SD and BP. Some endmembers show brief increases, and one never drops below its original value

SD and OMP perform very differently. Figure 13 tracks the selection of endmembers by SD and OMP for each value of ε. SD consistently decreases each endmember's number of substantiations as ε increases. It shows no obvious bias. On the other hand, OMP treats some endmembers differently, with substantiations of one of the endmembers increasing with ε. Nevertheless, OMP shows far less bias than BP.

We next added Gaussian noise (standard deviation 0.03) to the signatures in X. In this more realistic case, NLS gives incorrect mixtures. All three curves intersect at the NLS solution, $\lambda = \varepsilon = 0$. Errors are calculated with respect to the exact, non-noisy mixtures. Again SD shows the best performance. Both SD and OMP provide more accurate solutions than NLS. BP does not perform nearly as well. It does, however, improve the sparsity as λ increases. This example shows a greater difference in performance between BP and SD than shown in [11]. This is likely because that paper measures error as the distance between the reconstruction and the pixel, which in this example corresponds to the difference between the reconstruction and the noisy signature (Fig. 14).

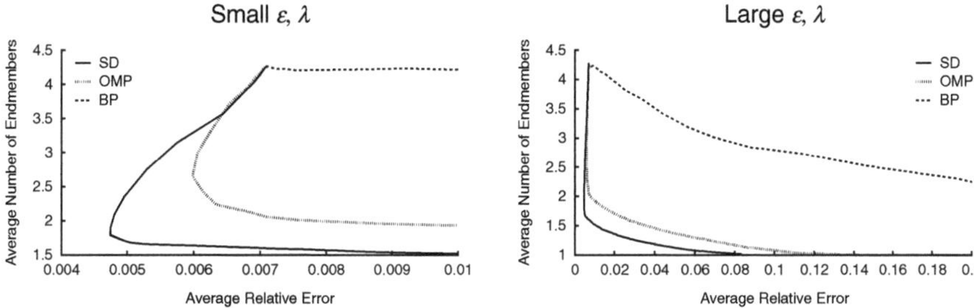

Fig. 14 Performance of BP, OMP, and SD on LMM 2 with added Gaussian noise. In all cases, increasing λ or ε improves sparsity. Note, however, that for small, positive ε, both SD and OMP improve the error by approximating with sparser mixtures

5 Conclusions

This chapter evaluates the ability of sparse reconstruction algorithms to find sparse mixtures of endmembers, which are typically highly correlated. Although restricted to hyperspectral demixing, the work may give some insight into the more general problem of sparse reconstruction over coherent sets. In this case, which certainly is not unique to the HSI problem, we have no theory guaranteeing that standard pursuit algorithms will provide sparse and accurate reconstructions. This chapter's examples show the failure of BP, and some success with OMP, for the endmember problem. It's natural to wonder about their relative performance for other problems. There may also be other application-specific pursuit algorithms that, like SD, offer superior performance searching for sparse support over sets of correlated vectors.

References

1. Basedow, R.W., Carmer, D.C., Anderson, M.E.: HYDICE system: implementation and performance. In: Descour, M.R., Mooney, J.M., Perry, D.L., Illing, L.R. (eds.) Proceedings of SPIE, vol. 2480 (1995)
2. Berman, M., Kiiveri, H., Lagerstrom, R., Ernst, A., Dunne, R., Huntington, J.: Ice: a statistical approach to identifying endmembers. IEEE Trans. Geosci. Remote Sens. **42**, 2085–2095 (2004)
3. Boardman, J.W.: Analysis, understanding and visualization of hyperspectral data as convex sets in n-space. In: Descour, M., Mooney, J., Perry, D., Illing L. (eds.) Proceedings of SPIE, vol. 2480 (1995)
4. Bowles, J.H., Gillis, D.B.: An optical real-time adaptive spectral identification system (ORASIS). In: Change, C.I. (ed.) Hyperspectral Data Exploitation: Theory and Applications. Wiley, Hoboken (2007)
5. Candes, E.J., Romberg, J., Tao, T.: Robust uncertainty principles: exact signal reconstruction from highly incomplete frequency information. IEEE Trans. Inform. Theor. **52**(2), 489–509 (2006)

6. Candes, E.J., Romberg, J., Tao, T.: Stable signal recovery from incomplete and inaccurate measurements. Commun. Pure. Appl. Math. **59**(8), 1208–1223 (2006)
7. Chen, S.S., Donoho, D.L., Saunders, M.A.: Atomic decomposition by basis pursuit. SIAM J. Scientific Comput. **20**, 33–61 (1998)
8. Clark, R.N., Swayze, G.A., Wise, R., Livo, E., Hoefen, T., Kokaly, R., Sutley, S.J.: USGS digital spectral library splib06a. U.S. Geological Survey, Digital Data Series 231 (2007)
9. Cocks, T., Jenssen, R., Stewart, A., Wilson, I., Shields, T.: The HyMap airborne hyperspectral sensor: The system, calibration and performance. In: First EARSEL Workshop on Imaging Spectroscopy. SPIE, Zurich (1998)
10. Donoho, D.L., Tanner, J.: Sparse non-negative solutions of underdetermined linear equations by linear programming. Proc. Natl. Acad. Sci. **102**(27), 9446–9451 (2005)
11. Greer, J.B.: Sparse demixing of hyperspectral images. IEEE Trans. Image Process. **21**(1), 219–228 (2012)
12. Guo, Z., Wittman, T., Osher, S.: L^1 unmixing and its application to hyperspectral image enhancement. In: Shen, S., Lewis, P. (eds.) Proceedings of SPIE, vol. 7334 (2009)
13. Keshava, N., Mustard, J.F.: Spectral unmixing. IEEE Signal Process. Mag. **19**(1), 44–57 (2002)
14. Laboratory, J.P.: AVIRIS homepage: http://aviris.jpl.nasa.gov
15. Mairal, J., Elad, M., Sapiro, G.: Sparse representation for color image restoration. IEEE Trans. Image Process. **17**(1), 53–69 (2008)
16. Mallat, S., Zhang, Z.: Matching pursuits with time-frequency dictionaries. IEEE Trans. Signal Process. **41**(12), 3397–3415 (1993)
17. Moussaoui, S., et al.: On the decomposition of Mars hyperspectral data by ICA and Bayesian positive source separation. Neurocomputing **71**, 2194–2208 (2008)
18. Nascimento, J., Bioucas-Dias, J.: Vertex component analysis: a fast algorithm to unmix hyperspectral data. IEEE Trans. Geosci. Remote Sens. **43**(4), 898–910 (2005)
19. Ramirez, I., Sprechmann, S., Sapiro, G.: Classification and clustering via dictionary learning with structured incoherence and shared features. In: IEEE Conference on Computer Vision and Pattern Recognition (CVPR) (2010)
20. Resmini, R.G., Kappus, M.E., Aldrich, W.S., Harsanyi, J.C., Anderson, M.: Mineral mapping with hyperspectral digital imagery collection experiment (HYDICE) sensor data at Cuprite, Nevada, U.S.A. Int. J. Remote Sens. **18**(7), 1553–1570 (1997)
21. Robila, S.A., Maciak, L.G.: Considerations on parallelizing nonnegative matrix factorization for hyperspectral data unmixing. IEEE Geosci. Remote Sens. Lett. **6**(1), 57–61 (2009)
22. Taubman, D., Marcellin, M.: JPEG2000: Image Compression Fundamentals, Standards and Practice. Kluwer Academic Publishers, Boston (2001)
23. Tropp, J.: Greed is good: algorithmic results for sparse approximation. IEEE Trans. Info. Theor. **50**(10), 2231–2242 (2004)
24. Tropp, J.: Just relax: Convex programming methods for identifying sparse signals in noise. IEEE Trans. Info. Theor. **52**(3), 1030–1051 (2004)
25. Verrelst, J., Clevers, J., Schaepman, M.: Merging the Minnaert-k parameter with spectral unmixing to map forest heterogeneity with CHRIS/PROBA data. IEEE Trans. Geosci. Remote Sens. **48**(11), 4014–4022 (2010)
26. Winter, M.E.: N-FINDR: An algorithm for fast autonomous spectral end-member determination in hyperspectral data. In: Descour, M., Shen, S. (eds.) Proceedings of SPIE, vol. 3753 (1999)
27. Zhang, Z.: Matching pursuit. Ph.D. thesis, Courant Institute, New York University (1993)

Theory of Passive Synthetic Aperture Imaging

Ling Wang, Can Evren Yarman, and Birsen Yazıcı

Abstract We present a unified theory for passive synthetic aperture imaging based on inverse scattering, estimation-detection theory, and microlocal analysis. Passive synthetic aperture imaging uses sources of opportunity for illumination and moving receivers to measure scattered field. We consider passive airborne receivers that fly along arbitrary, but known, flight trajectories and static or mobile sources of opportunity transmitting two types of waveforms: Single-frequency or ultra-narrowband continuous-wave (CW) waveforms and wideband pulsed waveforms. Our theory results in two new and novel synthetic aperture imaging modalities: *Doppler synthetic aperture hitchhiker* (DSAH) that uses single-frequency or ultra-narrowband CW waveforms, and *synthetic aperture hitchhiker* (SAH) that uses wideband pulsed waveforms. We use inverse scattering and estimation-detection theory to develop measurement models in the form of Fourier integral operators (FIOs) for DSAH and SAH. These models are based on windowed, scaled, and translated correlations of the measurements from two different receiver locations. This processing removes the transmitter-related terms from the phase of the resulting FIOs that map the radiance of the scene to correlated measurements. We use microlocal analysis to develop approximate inversion formulas for these FIOs. The inversion formulas involve backprojection of the correlated measurements onto

L. Wang
College of Electronic and Information Engineering, Nanjing University of Aeronautics and Astronautics, Nanjing 210016, People's Republic of China
e-mail: wanglrpi@gmail.com

C.E. Yarman
WesternGeco-Schlumberger, Houston, TX 77042, USA
e-mail: cyarman@slb.com

B. Yazıcı (✉)
Department of Electrical, Computer and System Engineering,
Rensselaer Polytechnic Institute, Troy, NY 12180, USA
e-mail: yazici@ecse.rpi.edu

T.D. Andrews et al. (eds.), *Excursions in Harmonic Analysis, Volume 1*,
Applied and Numerical Harmonic Analysis, DOI 10.1007/978-0-8176-8376-4_11,

certain manifolds where the passive range and passive Doppler are constant for SAH and DSAH imaging, respectively. We present resolution analysis and numerical simulations to demonstrate our theoretical results. While we focus primarily on the passive synthetic aperture radar, the theory we present is also applicable to other wave-based passive synthetic aperture imaging problems such as those in acoustics and geophysics.

Keywords Passive imaging • Passive radar • Synthetic aperture imaging • Microlocal analysis • Fourier integral operator (FIO) • Doppler synthetic aperture hitchhiker (DSAH) • Synthetic aperture hitchhiker (SAH) • Passive iso-range contour • Passive iso-Doppler contour • Filtered-backprojection (FBP) • Scene radiance

1 Introduction

With the rapid growth of illumination sources of opportunity, such as broadcasting stations, mobile phone base stations, as well as relatively low cost and rapid deployment of receivers, there has been a growing interest in passive detection and imaging applications in recent years [1, 2, 5–7, 9–14, 17–20, 23, 27–29, 32, 36].

Most of the existing passive imaging methods are focused on the detection of scatterers with stationary receivers [2, 5–7, 9–14, 17–20, 27–29, 32, 36]. Recently, a number of methods for passive synthetic aperture were introduced [1, 23, 38, 40].

In this chapter, we presented a unified theory of passive synthetic aperture imaging based on inverse scattering theory, estimation-detection theory and microlocal analysis. Our theory facilitates resolution analysis and relates backprojection-based image reconstruction to statistical beamforming methods as well as to ambiguity theory [21, 22, 31, 34, 35, 37]. It is applicable to passive imaging with both cooperative and noncooperative sources of illumination where the location of the sources and transmitted waveforms are unknown. The theory can be also viewed as a limiting case of the passive imaging and detection methods that we developed for sparsely distributed receivers [36]. It results in new and novel passive synthetic aperture imaging modalities [38, 40] with several advantages over the existing passive radar detection methods. (See [36, 38, 40] for a comparative review of related work.)

We consider multiple receivers moving along arbitrary, but known, trajectories over a non-flat topography and two types of illumination sources of opportunity: Single-frequency or ultra-narrowband continuous-wave (CW), and wideband pulsed waveforms of opportunity. Due to the high Doppler resolution nature of the single-frequency or ultra-narrowband CW waveforms, we refer to the modality that uses these waveforms as the *Doppler synthetic aperture hitchhiker* (DSAH) [40]. Due to the high-range resolution nature of the wideband pulsed waveforms, we refer to the modality that uses wideband pulsed waveforms as the *Range synthetic aperture hitchhiker* or simply the *synthetic aperture hitchhiker* (SAH) [38].

For each pair of receivers, we correlate the windowed signal obtained from one of the receivers with the *windowed, scaled, and translated* version of the received signal from another receiver. We express the relationship between the scene radiance and the correlated measurements in the form of Fourier integral operators (FIOs). The correlation of received signals removes the transmitter related terms from the phase component of the resulting FIOs. As a result these FIOs can be inverted approximately by using microlocal techniques without the knowledge of the location of the transmitters. The DSAH measurement model does not rely on the start–stop approximation and is based on the fast-time Doppler, while the SAH measurement model relies on the start–stop approximation. In this context the SAH measurement model can be derived from the DSAH measurement model by setting the fast-time Doppler variable to unity.

The high-frequency analysis of the DSAH and SAH FIOs shows that the correlated measurements are the projections of the scene radiance onto the passive iso-Doppler and passive iso-range curves in DSAH and SAH imaging, respectively.

We use microlocal techniques to develop filtered-backprojection type approximate inversions of DSAH and SAH FIOs. The reconstructed images preserve the location and orientation of the visible edges of the scene radiance. Additionally, the reconstruction formulas can be implemented efficiently using the fast-backprojection algorithms [8]. Our unified approach to passive imaging readily facilitates resolution analysis that is consistent with the ambiguity theory [22, 31, 37].

While we focused primarily on passive synthetic aperture radar, the theory of DSAH and SAH imaging and the resulting methods and algorithms are also applicable to other wave-based passive imaging problems, such as those that arise in geophysics or acoustics.

The organization of the chapter is as follows: In Sect. 2, we derive and analyze the leading order contributors of the measurement models for DSAH and SAH. In Sect. 3, we develop filtered-backprojection type image formation methods for DSAH and SAH, respectively. In Sect. 4, we analyze the resolution of DSAH and SAH imaging. In Sect. 5, we present numerical simulations to demonstrate the performance of the DSAH and SAH imaging methods. Finally, in Sect. 6, we conclude our discussion.

2 Measurement Model

We use the following notational conventions throughout the paper. The bold Roman, bold italic, and Roman lowercase letters are used to denote variables in $\mathbb{R}^3$, $\mathbb{R}^2$, and $\mathbb{R}$, respectively, i.e., $\mathbf{z} = (\boldsymbol{z}, z) \in \mathbb{R}^3$, with $\boldsymbol{z} \in \mathbb{R}^2$ and $z \in \mathbb{R}$. The calligraphic letters ($\mathscr{F}, \mathscr{K}$, etc.) are used to denote operators.

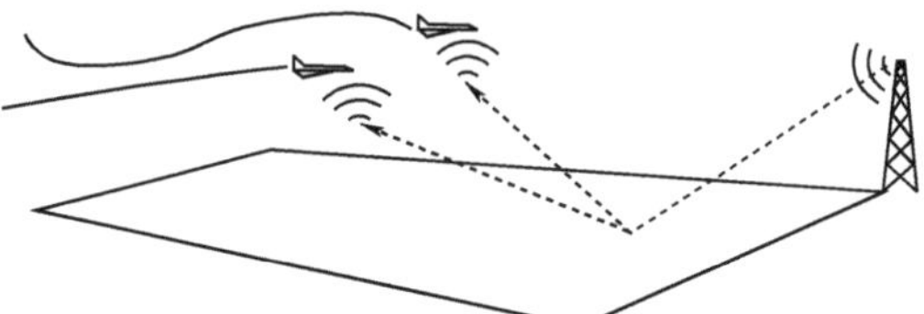

Fig. 1 An illustration of the imaging geometry

Given a pair of transmitter and receiver antennas located at $\mathbf{T}$ and $\mathbf{R}$, respectively, we model the received signal by [26]

$$f(t,\mathbf{R},\mathbf{T}) \approx \int \frac{\mathrm{e}^{\mathrm{i}\omega(t-(|\mathbf{R}-\mathbf{z}|+|\mathbf{z}-\mathbf{T}|)/c_0)}}{(4\pi)^2|\mathbf{R}-\mathbf{z}|\,|\mathbf{z}-\mathbf{T}|}\,\omega^2\hat{p}(\omega) \times J_{\mathrm{tr}}\left(\omega,\widehat{\mathbf{z}-\mathbf{T}},\mathbf{T}\right) J_{\mathrm{rc}}\left(\omega,\widehat{\mathbf{z}-\mathbf{R}},\mathbf{R}\right) V(\mathbf{z})\,\mathrm{d}\omega\,\mathrm{d}\mathbf{z}, \qquad (1)$$

or in time-domain by

$$f(t,\mathbf{R},\mathbf{T}) \approx \int \frac{\ddot{p}(t-(|\mathbf{R}-\mathbf{z}|+|\mathbf{z}-\mathbf{T}|)/c_0)}{(4\pi)^2|\mathbf{R}-\mathbf{z}|\,|\mathbf{z}-\mathbf{T}|} \times J_{\mathrm{tr}}\left(\widehat{\mathbf{z}-\mathbf{T}},\mathbf{T}\right) J_{\mathrm{rc}}\left(\widehat{\mathbf{z}-\mathbf{R}},\mathbf{R}\right) V(\mathbf{z})\,\mathrm{d}\mathbf{z}, \qquad (2)$$

where t denotes time, c_0 denotes the speed of light in free space, $V(\mathbf{z})$ is the reflectivity function, $\hat{p}$ denotes the Fourier transform the transmitted waveform, $p(t)$, J_{tr}, and J_{rc} are the transmitter and receiver antenna beam pattern related terms, respectively.

We denote the earth's surface by $\mathbf{z} = (\boldsymbol{z}, \psi(\boldsymbol{z})) \in \mathbb{R}^3$, where $\boldsymbol{z} \in \mathbb{R}^2$ and $\psi : \mathbb{R}^2 \to \mathbb{R}$ is a known function for the ground topography. Furthermore, we assume that the scattering takes place in a thin region near the surface. Thus, the reflectivity function is in the form

$$V(\mathbf{z}) = \rho(\boldsymbol{z})\delta(z - \psi(\boldsymbol{z})). \qquad (3)$$

Our passive imaging theory is applicable to both mobile and stationary sources of opportunity. However, for the rest of our discussion, we assume that there is a single, stationary transmitter of opportunity illuminating the scene. This allows us to simplify the analysis and distill the important aspects that can readily be generalized using the ideas similar to the ones presented in our work [38].

Let $\mathbf{T} \in \mathbb{R}^3$ denote the location of the transmitter of opportunity and let there be N airborne receivers, each traversing a smooth trajectory $\boldsymbol{\gamma}_i(t')$, $i = 1,\ldots,N$ as shown in Fig. 1. Then, we denote the received signal at the ith receiver starting at time $t' = s$ by

$$
\begin{aligned}
e_i(t+s) &= f(t+s, \boldsymbol{\gamma}_i(t+s), \mathbf{T}) \\
&= \int \frac{\ddot{p}(t+s-(|\boldsymbol{\gamma}_i(t+s)-\mathbf{z}|+|\mathbf{z}-\mathbf{T}|)/c_0)}{(4\pi)^2|\boldsymbol{\gamma}_i(t+s)-\mathbf{z}|\,|\mathbf{z}-\mathbf{T}|} \\
&\quad \times J_{\text{tr}}\left(\widehat{\mathbf{z}-\mathbf{T}}, \mathbf{T}\right) J_{\text{rc}}\left(\widehat{\mathbf{z}-\boldsymbol{\gamma}_i(t+s)}, \boldsymbol{\gamma}_i(t+s)\right) \rho(\mathbf{z})\, \mathrm{d}\mathbf{z}.
\end{aligned} \tag{4}
$$

Note that the time variable t' represents the absolute time, while t represents the relative time within the time interval staring at $t' = s$.

For a finite and relatively short interval, we use the Taylor series expansion around $t = 0$,

$$
\boldsymbol{\gamma}_i(t+s) = \boldsymbol{\gamma}_i(s) + \dot{\boldsymbol{\gamma}}_i(s)t + \cdots \tag{5}
$$

to approximate

$$
|\boldsymbol{\gamma}_i(t+s)-\mathbf{z}| \approx |\boldsymbol{\gamma}_i(s)-\mathbf{z}| + \widehat{\boldsymbol{\gamma}_i(s)-\mathbf{z}} \cdot \dot{\boldsymbol{\gamma}}_i(s)t\,. \tag{6}
$$

Substituting the approximation (6) into (4), we have

$$
\begin{aligned}
e_i(t+s) &\approx \int \frac{\ddot{p}(\alpha_i(s,\mathbf{z})t+s-(|\boldsymbol{\gamma}_i(s)-\mathbf{z}|+|\mathbf{z}-\mathbf{T}|)/c_0)}{(4\pi)^2|\boldsymbol{\gamma}_i(s)-\mathbf{z}|\,|\mathbf{z}-\mathbf{T}|} \\
&\quad \times J_{\text{tr}}\left(\widehat{\mathbf{z}-\mathbf{T}}, \mathbf{T}\right) J_{\text{rc}}\left(\widehat{\mathbf{z}-\boldsymbol{\gamma}_i(t+s)}, \boldsymbol{\gamma}_i(t+s)\right) \rho(\mathbf{z})\, \mathrm{d}\mathbf{z},
\end{aligned} \tag{7}
$$

where the time dilation

$$
\alpha_i(s,\mathbf{z}) = 1 - \frac{\widehat{\boldsymbol{\gamma}_i(s)-\mathbf{z}} \cdot \dot{\boldsymbol{\gamma}}_i(s)}{c_0} \tag{8}
$$

is the Doppler-scale-factor induced by the movement of the ith receiver.

We define the *windowed, scaled, and translated correlation* of the received signals e_i and e_j by

$$
c_{ij}(s', s, \mu) = \int e_i(t+s')\, e_j^*(\mu t+s)\, \phi(t)\, \mathrm{d}t, \tag{9}
$$

for some $s, s' \in \mathbb{R}$ and $\mu \in \mathbb{R}^+$, $i, j = 1, \ldots, N$, where $\phi(t)$ is a smooth compactly supported temporal windowing function centered at $t = 0$.

In the following sections, we develop mappings that relate the expected value of the correlated measurements c_{ij}, denoted by $E[c_{ij}]$, to the scene to be imaged. We assume that the sources of opportunity are noncooperative, where the location of the transmitter, $\mathbf{T}$, and transmitter antenna beam pattern related term J_{tr} are unknown.

We use a stochastic model for the transmitter antenna beam pattern related term, J_{tr}, and the scene reflectivity, ρ, and assume ρ and J_{tr} are statistically independent, to express $E[c_{ij}]$ as

$$E[c_{ij}](s',s,\mu) \approx \int \ddot{p}(\alpha_i(s,\mathbf{z})t + s' - (|\boldsymbol{\gamma}_i(s') - \mathbf{z}| + |\mathbf{z} - \mathbf{T}|)/c_0)$$
$$\times \ddot{p}^*(\mu\alpha_j(s,\mathbf{z}')t + s - (|\boldsymbol{\gamma}_j(s) - \mathbf{z}'| + |\mathbf{z}' - \mathbf{T}|)/c_0)$$
$$\times \frac{C_\rho(z,z')C_{J_{\mathrm{tr}}}(z,z',\mathbf{T})A_{R_{ij}}(z,z',t,s',s,\mu)}{(4\pi)^4 G_{ij}(z,z',s,s',\mu)} \mathrm{d}z\,\mathrm{d}z'\phi(t)\,\mathrm{d}t\,. \quad (10)$$

C_ρ and $C_{J_{\mathrm{tr}}}$ denote the correlation functions of ρ and J_{tr}, respectively, i.e.,

$$C_\rho(z,z') = E[\rho(z)\rho^*(z')], \quad (11)$$
$$C_{J_{\mathrm{tr}}}(z,z',\mathbf{T}) = E\left[J_{\mathrm{tr}}\left(\widehat{\mathbf{z}-\mathbf{T}},\mathbf{T}\right)J^*_{\mathrm{tr}}\left(\widehat{\mathbf{z}'-\mathbf{T}},\mathbf{T}\right)\right]. \quad (12)$$

$A_{R_{ij}}$ denotes the product of the receiver antenna beam patterns,

$$A_{R_{ij}}(z,z',t,s',s,\mu) = J_{\mathrm{rc}}\left(\mathbf{z} - \widehat{\boldsymbol{\gamma}_i(t+s')}, \boldsymbol{\gamma}_i(t+s')\right)$$
$$\times J^*_{\mathrm{rc}}\left(\mathbf{z}' - \widehat{\boldsymbol{\gamma}_j(\mu t+s)}, \boldsymbol{\gamma}_j(\mu t+s)\right), \quad (13)$$

and G_{ij} is the product of the geometric spreading factors,

$$G_{ij}(\mathbf{z},\mathbf{z}',s',s) = |\mathbf{T}-\mathbf{z}|\,|\mathbf{T}-\mathbf{z}'||\boldsymbol{\gamma}_i(s')-\mathbf{z}|\,|\boldsymbol{\gamma}_j(s)-\mathbf{z}'|. \quad (14)$$

Note that for noncooperative sources of opportunity, $\mathbf{T}$, and thus $|\mathbf{T}-\mathbf{z}|\,|\mathbf{T}-\mathbf{z}'|$, are unknown. For the case of cooperative sources of opportunity where these quantities along with the transmitted antenna beam pattern are assumed to be known, (12) can be modified to include the known quantities.

Next, we make the incoherent-field approximation [3] by assuming that ρ and J_{tr} satisfy the following equalities:

$$C_\rho(z,z') = R_\rho(z)\delta(z-z'), \quad (15)$$
$$C_{J_{\mathrm{tr}}}(z,z',\mathbf{T}) = R_T(z)\delta(z-z'). \quad (16)$$

R_ρ is the average power of the electromagnetic radiation emitted by the scene at location z, and R_T is the average power of the electromagnetic radiation emitted by the transmitter at location $\mathbf{T}$ that is incident on the target surface at z. In this regard, R_ρ is referred to as the *scene radiance* and R_T is referred to as the transmitter irradiance [3].

Substituting (15) and (16) into (10), we obtain

$$E[c_{ij}](s',s,\mu) = \int \ddot{p}(\alpha_i(s',\mathbf{z})t + s' - (|\boldsymbol{\gamma}_i(s') - \mathbf{z}| + |\mathbf{z} - \mathbf{T}|)/c_0)$$
$$\times \ddot{p}^*(\mu\alpha_j(s,\mathbf{z})t + s - (|\boldsymbol{\gamma}_j(s) - \mathbf{z}| + |\mathbf{z} - \mathbf{T}|)/c_0)$$
$$\times \frac{R_\rho(z)R_T(z)A_{R_{ij}}(z,z,t,s',s,\mu)}{(4\pi)^4 G_{ij}(z,z,s,s',\mu)} \mathrm{d}z\,\phi(t)\,\mathrm{d}t \quad (17)$$

for some $s,s' \in \mathbb{R}$, $\mu \in \mathbb{R}^+$ and $i,j = 1,\ldots,N$. We refer to (17) as the correlated measurements.

Our objective is to determine the scene radiance R_ρ given $E[c_{ij}](s',s,\mu)$ for a range of s', s, and μ. In the following two sections, we study two special cases of the measurement model to derive the measurement models for the DSAH and SAH.

2.1 Model for Doppler Synthetic Aperture Hitchhiker Imaging

In DSAH, narrowband or ultra-narrowband CW waveforms of opportunity are used for imaging. Thus,

$$p(t) = \mathrm{e}^{\mathrm{i}\omega_0 t}\tilde{p}(t), \tag{18}$$

where ω_0 denotes the carrier frequency and $\tilde{p}(t)$ is the complex envelope of p, which is a slow varying function of t as compared to $\mathrm{e}^{\mathrm{i}\omega_0 t}$.

Substituting (18) into (17), we express $E[c_{ij}]$ as

$$\begin{aligned} E[c_{ij}(s',s,\mu)] = \frac{\omega_0^4}{(4\pi)^4}&\int \mathrm{e}^{\mathrm{i}\omega_0(\alpha_i(s',\mathbf{z})t+s'-(|\boldsymbol{\gamma}_i(s')-\mathbf{z}|+|\mathbf{T}-\mathbf{z}|)/c_0)} \\ &\times \mathrm{e}^{-\mathrm{i}\omega_0(\mu\alpha_j(s,\mathbf{z})t+s-(|\boldsymbol{\gamma}_j(s)-\mathbf{z}|+|\mathbf{T}-\mathbf{z}|)/c_0)} \\ &\times \frac{R_T(\boldsymbol{z})A_{\tilde{p}}(\boldsymbol{z},\boldsymbol{z},t,s',s,\mu)A_{R_{ij}}(\boldsymbol{z},\boldsymbol{z},t,s',s,\mu)}{G_{ij}(\mathbf{z},\mathbf{z},s',s,\mu)} \\ &\times R_\rho(\boldsymbol{z})\mathrm{d}\boldsymbol{z}\,\phi(t)\,\mathrm{d}t, \end{aligned} \tag{19}$$

where $A_{\tilde{p}}$ is the product of the complex envelope of the transmitted waveform,

$$\begin{aligned} A_{\tilde{p}} = {} & \tilde{p}(\alpha_i(s',\mathbf{z})t+s'-(|\boldsymbol{\gamma}_i(s')-\mathbf{z}|+|\mathbf{T}-\mathbf{z}|)/c_0) \\ & \times\tilde{p}^*(\mu\alpha_j(s,\mathbf{z})t+s-(|\boldsymbol{\gamma}_j(s)-\mathbf{z}|+|\mathbf{T}-\mathbf{z}|)/c_0). \end{aligned} \tag{20}$$

After rearranging the terms in (19), we have

$$\begin{aligned} E[c_{ij}(s',s,\mu)] &\approx \mathscr{F}_{ij}^{\mathrm{DSAH}}[R_\rho](s',s,\mu) \\ &= \int \mathrm{e}^{-\mathrm{i}\varphi_{ij}^{\mathrm{DSAH}}(t,\boldsymbol{z},s',s,\mu)}\, A_{ij}^{\mathrm{DSAH}}(\boldsymbol{z},t,s',s,\mu)R_\rho(\boldsymbol{z})\,\mathrm{d}\boldsymbol{z}\,\mathrm{d}t, \end{aligned} \tag{21}$$

where

$$\varphi_{ij}^{\mathrm{DSAH}}(t,\boldsymbol{z},s',s,\mu) = \omega_0\alpha_j(s,\mathbf{z})t\left[\mu - S_{ij}(s',s,\mathbf{z})\right], \tag{22}$$

with

$$S_{ij}\left(s',s,\mathbf{z}\right) = \frac{\alpha_i\left(s',\mathbf{z}\right)}{\alpha_j\left(s,\mathbf{z}\right)} = \frac{1-\left(\widehat{\boldsymbol{\gamma}_i(s')-\mathbf{z}}\right)\cdot\dot{\boldsymbol{\gamma}}_i(s')/c_0}{1-\left(\widehat{\boldsymbol{\gamma}_j(s)-\mathbf{z}}\right)\cdot\dot{\boldsymbol{\gamma}}_j(s)/c_0}, \tag{23}$$

and

$$A_{ij}^{\mathrm{DSAH}}(z,t,s',s,\mu) = \frac{\omega_0^4 R_T(z) A_{\tilde{p}}(z,z,t,s',s,\mu) A_{R_{ij}}(z,z,t,s',s,\mu)\,\phi(t)}{(4\pi)^4 G_{ij}(\mathbf{z},\mathbf{z},t,s',s,\mu)} \times \mathrm{e}^{\mathrm{i}\omega_0(s'-s-(|\boldsymbol{\gamma}_i(s')-\mathbf{z}|-|\boldsymbol{\gamma}_j(s)-\mathbf{z}|)/c_0)}\,. \tag{24}$$

We refer to $S_{ij}(s',s,\mathbf{z})$ as the *Doppler-hitchhiker-scale-factor.*

For cooperative sources of opportunity, where the transmitter locations and antenna beam patterns are assumed to be known, we treat J_{tr} deterministically and replace $R_T(z)$ with $J_{\mathrm{tr}}(\widehat{\mathbf{z}-\mathbf{T}},\mathbf{T})J_{\mathrm{tr}}^*(\widehat{\mathbf{z}-\mathbf{T}},\mathbf{T})$.

We refer to $\mathscr{F}_{ij}^{\mathrm{DSAH}}$ defined in (21) as the DSAH or *Doppler-hitchhiker* FIO; and $\varphi_{ij}^{\mathrm{DSAH}}$, A_{ij}^{DSAH} as the phase and amplitude terms of the operator $\mathscr{F}_{ij}^{\mathrm{DSAH}}$.

Note that the scaled and translated correlation of the received signal removes all transmitter related terms from the phase of the operator $\mathscr{F}_{ij}^{\mathrm{DSAH}}$.

2.1.1 High-Frequency Analysis of the DSAH FIO and Passive iso-Doppler Contours

We assume that for some m_A^{DSAH}, A_{ij}^{DSAH} satisfy the inequality

$$\sup_{(t,\mu,s',s,z)\in U_{\mathrm{DSAH}}}\left|\partial_t^{\alpha_t}\partial_\mu^{\alpha_\mu}\partial_{s'}^{\beta_1}\partial_s^{\beta_2}\partial_{z_1}^{\varepsilon_1}\partial_{z_2}^{\varepsilon_2}A_{ij}^{\mathrm{DSAH}}(z,t,s',s,\mu)\right| \leq C_A^{\mathrm{DSAH}}(1+t^2)^{(m_A^{\mathrm{DSAH}}-|\alpha_t|)/2}, \tag{25}$$

where U_{DSAH} is any compact subset of $\mathbb{R}\times\mathbb{R}^+\times\mathbb{R}\times\mathbb{R}\times\mathbb{R}^2$ and the constant C_A^{DSAH} depends on $U_{\mathrm{DSAH}}, \alpha_{t,\mu}, \beta_{1,2}$, and $\varepsilon_{1,2}$. In practice, (25) is satisfied for transmitters and receivers are sufficiently far away from the illuminated region.

Under the assumption (25), (21) defines $\mathscr{F}_{ij}^{\mathrm{DSAH}}$ as an FIO whose leading-order contributions come from those points lying in the intersection of the illuminated surface $(z,\psi(z))$ and points that have the same Doppler-hitchhiker-scale-factor, i.e., $\{\mathbf{z}\in\mathbb{R}^3 : S_{ij}(\tau',\tau,\mathbf{z})=\mu\}$. We denote the curves formed by this intersection by

$$F_{ij}^{\mathrm{DSAH}}(s',s,\mu) = \{z : S_{ij}(s',s,\mathbf{z}=(z,\psi(z)))=\mu\}. \tag{26}$$

When the speed of the receivers is much slower than the speed of light c_0, S_{ij} can be approximated as follows:

$$S_{ij}\left(s',s,\mathbf{z}\right) = 1+\frac{\left(\widehat{\boldsymbol{\gamma}_j(s)-\mathbf{z}}\right)\cdot\dot{\boldsymbol{\gamma}}_j(s)/c_0-\left(\widehat{\boldsymbol{\gamma}_i(s')-\mathbf{z}}\right)\cdot\dot{\boldsymbol{\gamma}}_i(s')/c_0}{1-\left(\widehat{\boldsymbol{\gamma}_j(s)-\mathbf{z}}\right)\cdot\dot{\boldsymbol{\gamma}}_j(s)/c_0}$$

$$\approx 1+\left[\widehat{\left(\boldsymbol{\gamma}_j(s)-\mathbf{z}\right)}\cdot\dot{\boldsymbol{\gamma}}_j(s)-(\widehat{\boldsymbol{\gamma}_i(s')-\mathbf{z}})\cdot\dot{\boldsymbol{\gamma}}_i(s')\right]/c_0. \tag{27}$$

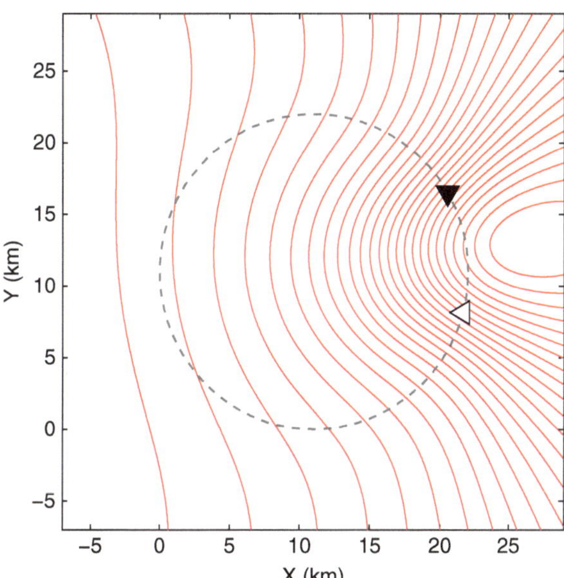

Fig. 2 DSAH iso-Doppler contours $F_{12}(s', s, \mu)$ for the Doppler-hitchhiker-scale-factor $S_{12}(22.0674\,\mathrm{s}, -11.0337\,\mathrm{s}, \mathbf{z})$. Two receivers are traversing a circular flight trajectory (*dashed line*) at the speed of 261 m/s over a flat topography. $\boldsymbol{\gamma}_1(\tilde{s}) = \boldsymbol{\gamma}_C(\tilde{s})$ and $\boldsymbol{\gamma}_2(\tilde{s}) = \boldsymbol{\gamma}_C(\tilde{s} - \pi/4)$, respectively, where *white* and *black triangles* denote the positions of the two receivers at $\tilde{s} = \pi/6$, respectively. (See (23) and (70) for explicit formulae of $S_{12}(22.0674\,\mathrm{s}, -11.0337\,\mathrm{s}, \mathbf{z})$ and $\boldsymbol{\gamma}_C(\tilde{s})$, respectively)

Substituting (27) into

$$S_{ij}(s', s, \mathbf{z}) = \mu\,, \tag{28}$$

multiplying both sides of (28) by ω_0, and rearranging the terms, we have

$$\frac{\omega_0}{c_0}\left[(\widehat{\boldsymbol{\gamma}_i(s') - \mathbf{z}}) \cdot \dot{\boldsymbol{\gamma}}_i(s') - (\widehat{\boldsymbol{\gamma}_j(s) - \mathbf{z}}) \cdot \dot{\boldsymbol{\gamma}}_j(s)\right] = (1-\mu)\omega_0, \tag{29}$$

where the left-hand side of (29) is the hitchhiker Doppler defined in [38] for a fixed frequency. In this regard, we refer to $F_{ij}^{\mathrm{DSAH}}(s', s, \mu)$ as the passive iso-Doppler or *DSAH iso-Doppler* contour. Figure 2 shows the DSAH iso-Doppler contours for two receivers traversing a circular trajectory over a flat topography.

2.2 *Model for Synthetic Aperture Hitchhiker Imaging*

In SAH, wideband pulsed waveforms of opportunity are used for imaging.

Assuming that the velocity, $\dot{\boldsymbol{\gamma}}_j(s)$, of the receivers are much less than the speed of light, we use the "start–stop" approximation, where the receiver is assumed to be stationary within a certain window of time, and approximate the Doppler-scale factor

$$\alpha_i(s, \mathbf{z}) \approx 1. \tag{30}$$

Then the received signal at the ith receiver starting at time $t' = s'$ is approximated by

$$e_i(t+s') \approx \int \frac{\ddot{p}(t+s'-(|\boldsymbol{\gamma}_i(s')-\mathbf{z}|+|\mathbf{z}-\mathbf{T}|)/c_0)}{(4\pi)^2|\boldsymbol{\gamma}_i(s')-\mathbf{z}|\,|\mathbf{z}-\mathbf{T}|} \\ \times J_{\mathrm{tr}}\left(\widehat{\mathbf{z}-\mathbf{T}},\mathbf{T}\right) J_{\mathrm{rc}}\left(\widehat{\mathbf{z}-\boldsymbol{\gamma}_i(s')},\boldsymbol{\gamma}_i(s')\right) \rho(\boldsymbol{z})\,\mathrm{d}\boldsymbol{z}. \tag{31}$$

Similarly, under the start–stop approximation, the Doppler-hitchhiker-scale-factor becomes

$$S_{ij}(s',s,\mathbf{z}) \approx (1-\alpha_i(s',\mathbf{z}))(1+\alpha_j(s,\mathbf{z})) \approx 1. \tag{32}$$

Consequently, by (28), it is sufficient to consider $E[c_{ij}]$ for $\mu = 1$.

In this regard, taking into account the high range resolution of the wideband waveforms, we incorporate the fast-time delay in the forward model of SAH and define

$$d_{ij}(s',s,\tau) = c_{ij}(s',s-\tau,1) = \int e_i(t+s')e_j^*(t+s-\tau)\phi(t)\mathrm{d}t. \tag{33}$$

We refer to (33) as the *spatiotemporal correlation* of e_i and e_j.

Using (31), we approximate the expectation of $d_{ij}(s',s,\tau)$ as

$$E[d_{ij}(s,s',\tau)] = \int \ddot{p}(t+s'-(|\boldsymbol{\gamma}_i(s')-\mathbf{z}|+|\mathbf{z}-\mathbf{T}|)/c_0) \\ \times \ddot{p}^*(t+s-\tau-(|\boldsymbol{\gamma}_j(s)-\mathbf{z}|+|\mathbf{z}-\mathbf{T}|)/c_0) \\ \times \frac{R_\rho(\boldsymbol{z})R_T(\boldsymbol{z})\tilde{A}_{R_{ij}}(\boldsymbol{z},\boldsymbol{z},s',s)}{(4\pi)^4 G_{ij}(\boldsymbol{z},\boldsymbol{z},s',s)}\mathrm{d}\boldsymbol{z}\,\phi(t)\,\mathrm{d}t \\ = \int \mathrm{e}^{\mathrm{i}\omega(t+s'-(|\boldsymbol{\gamma}_i(s')-\mathbf{z}|+|\mathbf{z}-\mathbf{T}|)/c_0)} \\ \times \mathrm{e}^{-\mathrm{i}\omega(t+s-\tau-(|\boldsymbol{\gamma}_j(s)-\mathbf{z}|+|\mathbf{z}-\mathbf{T}|)/c_0)} \\ \times \frac{\omega^2|\hat{p}(\omega)|^2 R_\rho(\boldsymbol{z})R_T(\boldsymbol{z})\tilde{A}_{R_{ij}}(\boldsymbol{z},\boldsymbol{z},s',s)}{(4\pi)^4 G_{ij}(\boldsymbol{z},\boldsymbol{z},s,s')}\mathrm{d}\boldsymbol{z}\,\mathrm{d}\omega\phi(t)\,\mathrm{d}t, \tag{34}$$

where

$$\tilde{A}_{R_{ij}}(\boldsymbol{z},\boldsymbol{z}',s',s) = A_{R_{ij}}(\boldsymbol{z},\boldsymbol{z}',0,s',s,1) \tag{35}$$

and $A_{R_{ij}}$ and G_{ij} are as in (13) and (14). We write

$$E[d_{ij}(s',s,\tau)] \approx \mathscr{F}_{ij}^{\mathrm{SAH}}[R_\rho](s,s',\tau) \\ = \int \mathrm{e}^{-\mathrm{i}\varphi_{ij}^{\mathrm{SAH}}(\omega,\boldsymbol{z},s,s',\tau)} A_{ij}^{\mathrm{SAH}}(\boldsymbol{z},\omega,s,s')R_\rho(\boldsymbol{z})\,\mathrm{d}\boldsymbol{z}\,\mathrm{d}\omega, \tag{36}$$

where

$$\varphi_{ij}^{\mathrm{SAH}}(\omega,z,s',s,\tau)=\omega[r_{ij}(s',s,\mathbf{z})/c_0+s-s'-\tau], \tag{37}$$

with

$$r_{ij}(s',s,\mathbf{z})=|\boldsymbol{\gamma}_i(s')-\mathbf{z}|-|\boldsymbol{\gamma}_j(s)-\mathbf{z}| \tag{38}$$

and

$$A_{ij}^{\mathrm{SAH}}(z,\omega,s',s)=\frac{\omega^4\,|\hat{p}(\omega)|^2 R_T(z)\tilde{A}_{R_{ij}}(z,z,s',s)}{(4\pi)^4 G_{ij}(z,z,s',s)}. \tag{39}$$

We refer to $r_{ij}(s',s,\mathbf{z})$ as the *hitchhiker or passive range*.

We remark that under the start–stop approximation $\varphi_{ij}^{\mathrm{DSAH}}\approx 1$ and $\varphi_{ij}^{\mathrm{SAH}}$ is given by the exponential term in A_{ij}^{DSAH} [see (24)], where s is replaced with $s-\tau$ in SAH to incorporate the fast-time delay information.

Similar to DSAH, for cooperative sources of opportunity, we treat J_{tr} deterministically and replace $\tilde{R}_T(z)$ with $J_{\mathrm{tr}}(\widehat{\mathbf{z}-\mathbf{T}},\mathbf{T})J_{\mathrm{tr}}^*(\widehat{\mathbf{z}-\mathbf{T}},\mathbf{T})$.

We refer to $\mathscr{F}_{ij}^{\mathrm{SAH}}$ defined in (36) as the SAH or *range hitchhiker* or simply hitchhiker FIO; and $\varphi_{ij}^{\mathrm{SAH}}$, A_{ij}^{SAH} as the phase and amplitude terms of the operator $\mathscr{F}_{ij}^{\mathrm{SAH}}$.

Note that the spatiotemporal correlation of the received signal removes all transmitter related terms from the phase of the operator $\mathscr{F}_{ij}^{\mathrm{SAH}}$.

2.2.1 High-Frequency Analysis of the SAH FIO and Passive iso-Range Contours

We assume that for some m_A^{SAH}, A_{ij}^{SAH} satisfy the inequalities

$$\sup_{(\omega,s',s,z)\in U_{\mathrm{SAH}}}\left|\partial_\omega^{\alpha_\omega}\partial_{s'}^{\beta_1}\partial_s^{\beta_2}\partial_{z_1}^{\varepsilon_1}\partial_{z_2}^{\varepsilon_2}A_{ij}^{\mathrm{SAH}}(z,\omega,s',s)\right|\leq C_A^{\mathrm{SAH}}(1+\omega^2)^{(m_A^{\mathrm{SAH}}-|\alpha_\omega|)/2}, \tag{40}$$

where U_{SAH} is any compact subset of $\mathbb{R}\times\mathbb{R}\times\mathbb{R}\times\mathbb{R}^2$; the constant C_A^{SAH} depends on $U_{\mathrm{SAH}},\alpha_\omega,\beta_{1,2},\ \varepsilon_{1,2}$. These assumptions are needed to make various stationary phase calculations hold. In practice, (40) is satisfied for transmitters and receivers sufficiently far away from the illuminated region.

Under the assumption (40), (36) defines $\mathscr{F}_{ij}^{\mathrm{SAH}}$ as an FIO whose leading-order contribution comes from those points lying at the intersection of the illuminated surface and the hyperboloid $\{\mathbf{x}\in\mathbb{R}^3 : r_{ij}(s',s,\mathbf{z})=c_0(\tau+s'-s)\}$. We denote the curves formed by this intersection by

$$H_{ij}(s',s,\tau)=\{z : r_{ij}(s',s,\mathbf{z})=c_0(\tau+s'-s)\} \tag{41}$$

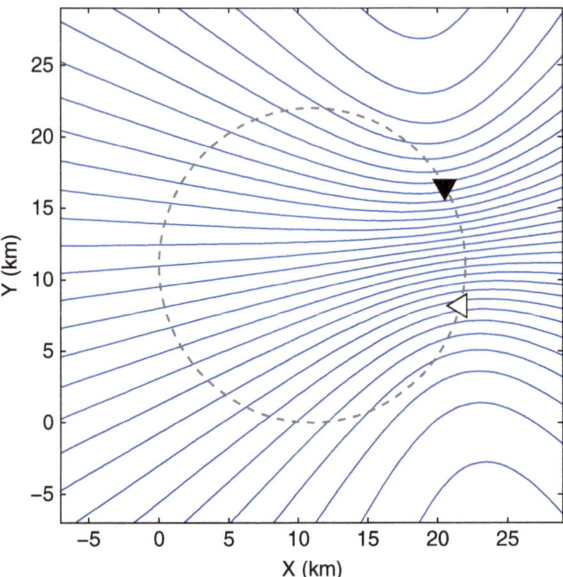

Fig. 3 SAH Iso-range contours $H_{12}(s',s,C)$ for the hitchhiker range $r_{12}(22.0674\,\mathrm{s}, -11.0337\,\mathrm{s}, \mathbf{z})$. Two receivers are traversing a circular flight trajectory (*dashed line*) at the speed of $261\,\mathrm{m/s}$ over a flat topography. $\boldsymbol{\gamma}_1(\tilde{s}) = \boldsymbol{\gamma}_C(\tilde{s})$ and $\boldsymbol{\gamma}_2(\tilde{s}) = \boldsymbol{\gamma}_C(\tilde{s} - \pi/4)$, respectively, where *white* and *black triangles* denote the positions of the two receivers at $\tilde{s} = \pi/6$, respectively. (See (38) and (70) for explicit formulae of $r_{12}(22.0674\,\mathrm{s}, -11.0337\,\mathrm{s}, \mathbf{z})$ and $\boldsymbol{\gamma}_C(\tilde{s})$, respectively)

and refer to $H_{ij}(s',s,\tau)$ as the passive iso-range or *SAH iso-range* contour. For flat topography, $\psi(\boldsymbol{z}) = 0$, the SAH iso-range contours are given by hyperbolas on the plane $z_3 = 0$. We present the iso-range contours for circular receiver flight trajectories over a flat topography in Fig. 3.

Comparing the DSAH and SAH presented in Sects. 2.1 and 2.2, we see that the DSAH imaging does not rely on the start–stop approximation and is based on the fast-time Doppler. In other words, the DSAH imaging method takes into account range variations or Doppler induced due to the movement of the receivers during the reception of a CW waveform. SAH imaging, on the other hand, relies on the start–stop approximation and ignores the range variation due to the movement of the receivers, hence the fast-time Doppler, during the reception of a wideband pulse. Thus, the SAH measurement model can be derived from the DSAH measurement model by setting the Doppler-scale-factor to unity and decoupling the time into fast- and slow-time variables.

3 Image Formation

Our objective is to form an image of the scene radiance $R_\rho(\boldsymbol{z})$ using $E[c_{ij}(s',s,\mu)]$ or $E[d_{ij}(s',s,\tau)]$, $i,j = 1,\ldots,N$ based on the correlated measurement models (21) for DSAH or (36) for SAH, respectively.

Since both $\mathscr{F}_{ij}^{\mathrm{DSAH}}$ and $\mathscr{F}_{ij}^{\mathrm{SAH}}$ are FIOs, we form an image of the scene radiance by other suitably designed FIOs, which we refer to as the *filtered-backprojection* operators. For DSAH, we backproject $E[c_{ij}(s',s,\mu)]$ onto passive iso-Doppler contours defined by $F_{ij}(s',s,\mu)$. For SAH, we backproject $E[d_{ij}(s',s,\tau)]$ onto passive iso-range contours defined by $H_{ij}(s',s,\tau)$ for $i,j=1,\ldots,N$. We form an image of the scene radiance by the superposition of the filtered and backprojected data.

3.1 DSAH Filtered-Backprojection Operator

For DSAH image formation, we invert $E[c_{ij}(s',s,\mu)]$ as follows:

$$\tilde{R}_\rho^{\mathrm{DSAH}}(\boldsymbol{z}) = \sum_{ij} \int \mathscr{K}_{ij}^{\mathrm{DSAH}}[E[c_{ij}]](\boldsymbol{z},s')\mathrm{d}s', \tag{42}$$

where we define

$$\mathscr{K}_{ij}^{\mathrm{DSAH}}[E[c_{ij}]](\boldsymbol{z},s') = \int \mathrm{e}^{\mathrm{i}\varphi_{ij}^{\mathrm{DSAH}}(t,\boldsymbol{z},s',s,\mu)} Q_{ij}^{\mathrm{DSAH}}(\boldsymbol{z},t,s',s)\, E[c_{ij}(s',s,\mu)]\mathrm{d}t\mathrm{d}s\mathrm{d}\mu. \tag{43}$$

We refer to $\mathscr{K}_{ij}^{\mathrm{DSAH}}$ as the DSAH filtered-backprojection operator with respect to the ith and jth receivers with filter Q_{ij}^{DSAH} to be determined below.

We assume that for some m_Q^{DSAH}, Q_{ij}^{DSAH} satisfies the inequality

$$\sup_{(t,s',s,\boldsymbol{z})\in K_{\mathrm{DSAH}}} \left|\partial_t^{\alpha_t}\partial_{s'}^{\beta_1}\partial_s^{\beta_2}\partial_{z_1}^{\varepsilon_1}\partial_{z_2}^{\varepsilon_2} Q_{ij}^{\mathrm{DSAH}}(\boldsymbol{z},t,s',s)\right| \leq C_Q^{\mathrm{DSAH}}(1+t^2)^{(m_Q^{\mathrm{DSAH}}-|\alpha_t|)/2}, \tag{44}$$

where K_{DSAH} is any compact subsets of $\mathbb{R}\times\mathbb{R}\times\mathbb{R}\times\mathbb{R}^2$, and the constant C_Q^{DSAH} depends on $K_{\mathrm{DSAH}}, \alpha_t, \beta_{1,2}, \varepsilon_{1,2}$. The assumption in (44) makes $\mathscr{K}_{ij}^{\mathrm{DSAH}}$ an FIO.

Substituting (21) into (43) and the result back into (42), we obtain

$$\begin{aligned}\tilde{R}_\rho^{\mathrm{DSAH}}(\boldsymbol{z}) &= \sum_{ij} \mathscr{K}_{ij}^{\mathrm{DSAH}}\mathscr{F}_{ij}^{\mathrm{DSAH}}[R_\rho](\boldsymbol{z})\\ &= \int \mathrm{e}^{\mathrm{i}[\varphi_{ij}^{\mathrm{DSAH}}(t,\boldsymbol{z},s',s,\mu)-\varphi_{ij}^{\mathrm{DSAH}}(t',\boldsymbol{z}',s',s,\mu)]}\, Q_{ij}^{\mathrm{DSAH}}(\boldsymbol{z},t,s',s)\\ &\quad \times A_{ij}^{\mathrm{DSAH}}(\boldsymbol{z}',t,s',s,\mu)R_\rho(\boldsymbol{z}')\,\mathrm{d}t'\,\mathrm{d}t\,\mathrm{d}s\,\mathrm{d}s'\,\mathrm{d}\mu\,\mathrm{d}\boldsymbol{z}'.\end{aligned} \tag{45}$$

We use the stationary phase theorem to approximate the t' and μ integrations [4, 15, 16, 24] and obtain

$$\begin{aligned}\tilde{R}_\rho^{\mathrm{DSAH}}(\boldsymbol{z}) \approx \sum_{ij}\int &\mathrm{e}^{\mathrm{i}\omega_0 t\left[1-(\widehat{\boldsymbol{\gamma}_j(s)-\boldsymbol{z}})\cdot\dot{\boldsymbol{\gamma}}_j(s)/c_0\right]\left[S_{ij}(s',s,\boldsymbol{z}')-S_{ij}(s',s,\boldsymbol{z})\right]}\\ &\times Q_{ij}^{\mathrm{DSAH}}(\boldsymbol{z},t,s',s)\,A_{ij}^{\mathrm{DSAH}}(\boldsymbol{z}',t,s',s,S_{ij}(s',s,\boldsymbol{z}'))\,R_\rho(\boldsymbol{z}')\,\mathrm{d}t\,\mathrm{d}s\,\mathrm{d}s'\,\mathrm{d}\boldsymbol{z}'.\end{aligned} \tag{46}$$

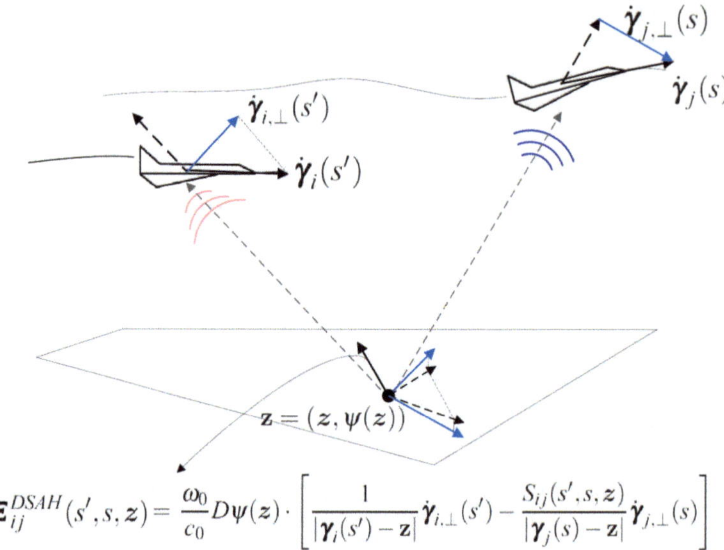

Fig. 4 An illustration of the vector $\boldsymbol{\Xi}_{ij}^{\mathrm{DSAH}}(s', s, \boldsymbol{z})$ in the data collection manifold $\Omega_{ij,\boldsymbol{z}}^{\mathrm{DSAH}}$ for the flat topography, $\psi(\boldsymbol{z}) = 0$. $\boldsymbol{\Xi}_{ij}^{\mathrm{DSAH}}(s', s, \boldsymbol{z})$ is the projection of the difference of the scaled vectors $\dot{\boldsymbol{\gamma}}_{i,\perp}(s')$ and $\dot{\boldsymbol{\gamma}}_{j,\perp}(s)$ onto the tangent plane of the ground topography at $\mathbf{z}$. (See (48) for an explicit form of $\boldsymbol{\Xi}_{ij}^{\mathrm{DSAH}}(s', s, \boldsymbol{z})$)

We linearize $S_{ij}(s', s, \boldsymbol{z}')$ around $\boldsymbol{z}' = \boldsymbol{z}$ and approximate

$$S_{ij}(\tau', \tau, \boldsymbol{z}') - S_{ij}(\tau', \tau, \boldsymbol{z}) \approx (\boldsymbol{z}' - \boldsymbol{z}) \cdot \nabla_{\boldsymbol{z}} S_{ij}(s', s, \boldsymbol{z}). \tag{47}$$

Let

$$\begin{aligned} \boldsymbol{\Xi}_{ij}^{\mathrm{DSAH}}(s', s, \boldsymbol{z}) &= \omega_0 \left[1 - (\widehat{\boldsymbol{\gamma}_j(s) - \mathbf{z}}) \cdot \dot{\boldsymbol{\gamma}}_j(s)/c_0 \right] \nabla_{\boldsymbol{z}} S_{ij}(s', s, \boldsymbol{z}) \\ &= \frac{\omega_0}{c_0} D\psi(\boldsymbol{z}) \cdot \left[\frac{1}{|\boldsymbol{\gamma}_i(s') - \mathbf{z}|} \dot{\boldsymbol{\gamma}}_{i,\perp}(s') - \frac{S_{ij}(s', s, \boldsymbol{z})}{|\boldsymbol{\gamma}_j(s) - \mathbf{z}|} \dot{\boldsymbol{\gamma}}_{j,\perp}(s) \right], \end{aligned} \tag{48}$$

where

$$D\psi(\boldsymbol{z}) = \begin{bmatrix} 1 \ 0 \ \partial\psi(\boldsymbol{z})/\partial z_1 \\ 0 \ 1 \ \partial\psi(\boldsymbol{z})/\partial z_2 \end{bmatrix} \tag{49}$$

and $\dot{\boldsymbol{\gamma}}_{i,\perp}(s')$, $\dot{\boldsymbol{\gamma}}_{j,\perp}(s)$ are the projections of $\dot{\boldsymbol{\gamma}}_i(s')$ and $\dot{\boldsymbol{\gamma}}_j(s)$ onto the planes whose normal vectors are $(\widehat{\boldsymbol{\gamma}_i(s') - \mathbf{z}})$ and $(\widehat{\boldsymbol{\gamma}_j(s) - \mathbf{z}})$, respectively.

We show an illustration of the vector $\boldsymbol{\Xi}_{ij}^{\mathrm{DSAH}}(s', s, \boldsymbol{z})$ in Fig. 4 for two receivers flying over a flat topography.

Substituting (47) and (48) into (46), we obtain

$$\tilde{R}_\rho^{\rm DSAH}(\boldsymbol{z}) \approx \sum_{ij} \int \mathrm{e}^{-\mathrm{i}t(\boldsymbol{z}'-\boldsymbol{z})\cdot \boldsymbol{\Xi}_{ij}^{\rm DSAH}(s',s,\boldsymbol{z})}\, Q_{ij}^{\rm DSAH}(\boldsymbol{z},t,s',s) \times A_{ij}^{\rm DSAH}(\boldsymbol{z},t,s',s) R_\rho(\boldsymbol{z}')\,\mathrm{d}t\,\mathrm{d}s\,\mathrm{d}\boldsymbol{z}'\,\mathrm{d}s' . \tag{50}$$

Note that under the assumptions (44) and (25), (50) shows that $\mathscr{K}_{ij}^{\rm DSAH}\mathscr{F}_{ij}^{\rm DSAH}$ is a pseudodifferential operator [33]. This means that the backprojection operator reconstructs the visible edges of the scene radiance at the correct location and correct orientation.

3.2 SAH Filtered-Backprojection Operator

For SAH imaging, we form an image of the scene radiance as follows:

$$\tilde{R}_\rho^{\rm SAH}(\boldsymbol{z}) = \sum_{ij} \int \mathscr{K}_{ij}^{\rm SAH}[E[d_{ij}]](\boldsymbol{z},s')\,\mathrm{d}s' , \tag{51}$$

where we define

$$\mathscr{K}_{ij}^{\rm SAH}[E[d]](\boldsymbol{z},s') = \sum_{ij} \int \mathrm{e}^{\mathrm{i}\varphi_{ij}^{\rm SAH}(\omega,\boldsymbol{z},s,s',\tau)} Q_{ij}^{\rm SAH}(\boldsymbol{z},\omega,s',s)\,\mathrm{d}(s',s,\tau)\,\mathrm{d}\tau\,\mathrm{d}\omega\,\mathrm{d}s\,\mathrm{d}s'. \tag{52}$$

We refer to $\mathscr{K}_{ij}^{\rm SAH}$ as the SAH filtered-backprojection operator with respect to the ith and jth receivers with filter $Q_{ij}^{\rm SAH}$ to be determined below.

Similarly, we assume that for some $m_Q^{\rm SAH}$, $Q_{ij}^{\rm SAH}$ satisfies the inequality

$$\sup_{(\omega,s',s,\boldsymbol{z})\in K_{\rm SAH}} \left| \partial_\omega^{\alpha_\omega} \partial_{s'}^{\beta_1} \partial_s^{\beta_2} \partial_{z_1}^{\varepsilon_1} \partial_{z_2}^{\varepsilon_2} Q_{ij}^{\rm SAH}(\boldsymbol{z},\omega,s',s) \right| \le C_Q^{\rm SAH}(1+\omega^2)^{(m_Q^{\rm SAH}-|\alpha_\omega|)/2} , \tag{53}$$

where $K_{\rm SAH}$ is any compact subsets of $\mathbb{R}\times\mathbb{R}\times\mathbb{R}\times\mathbb{R}^2$, and the constant $C_Q^{\rm SAH}$ depends on $K_{\rm SAH}, \alpha_\omega, \beta_{1,2}, \varepsilon_{1,2}$. The assumption in (53) makes $\mathscr{K}_{ij}^{\rm SAH}$ an FIO.

Substituting (36) into (52), and using the stationary phase theorem as in DSAH imaging, we approximate

$$\tilde{R}_\rho^{\rm SAH}(\boldsymbol{z}) = \sum_{ij} \mathscr{K}_{ij}^{\rm SAH}\mathscr{F}_{ij}^{\rm SAH}[R_\rho](\boldsymbol{z}) = \int \mathrm{e}^{\mathrm{i}\omega[r_{ij}(s',s,\mathbf{z}')-r_{ij}(s',s,\mathbf{z})]/c_0}\, Q_{ij}^{\rm SAH}(\boldsymbol{z},\omega,s',s) \times A_{ij}^{\rm SAH}(\boldsymbol{z}',\omega,s',s) R_\rho(\boldsymbol{z}')\,\mathrm{d}\omega\,\mathrm{d}s\,\mathrm{d}s'\,\mathrm{d}\boldsymbol{z}' . \tag{54}$$

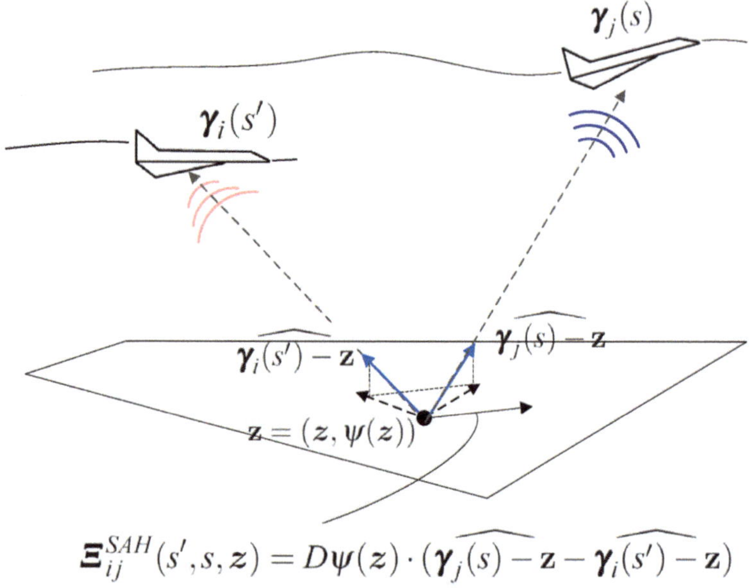

Fig. 5 An illustration of the vector $\boldsymbol{\Xi}_{ij}^{\mathrm{SAH}}(s', s, \boldsymbol{z})$ in the data collection manifold $\Omega_{ij,\boldsymbol{z}}^{\mathrm{SAH}}$ for the flat topography, $\psi(\boldsymbol{z}) = 0$. $\boldsymbol{\Xi}_{ij}^{\mathrm{SAH}}(s', s, \boldsymbol{z})$ is the projection of the difference of the unit vectors $\widehat{\boldsymbol{\gamma}_j(s) - \mathbf{z}}$ and $\widehat{\boldsymbol{\gamma}_i(s') - \mathbf{z}}$ onto the tangent plane of the ground topography at $\mathbf{z}$. (See (56) for an explicit form of $\boldsymbol{\Xi}_{ij}^{\mathrm{SAH}}(s', s, \boldsymbol{z})$)

We linearize $r_{ij}(s', s, \mathbf{z}')$ around $\boldsymbol{z}' = \boldsymbol{z}$ and make the following approximations:

$$r_{ij}(s', s, \mathbf{z}') - r_{ij}(s', s, \mathbf{z}) = (\boldsymbol{z}' - \boldsymbol{z}) \cdot \boldsymbol{\Xi}_{ij}^{\mathrm{SAH}}(s', s, \boldsymbol{z}), \tag{55}$$

where

$$\boldsymbol{\Xi}_{ij}^{\mathrm{SAH}}(s', s, \boldsymbol{z}) = D\psi(\boldsymbol{z}) \cdot \left(\widehat{\boldsymbol{\gamma}_j(s) - \mathbf{z}} - \widehat{\boldsymbol{\gamma}_i(s') - \mathbf{z}} \right) \tag{56}$$

and $D\psi(\boldsymbol{z})$ is given in (49).

For flat topography, we present an illustration of $\boldsymbol{\Xi}_{ij}^{\mathrm{SAH}}(s', s, \boldsymbol{z})$ in Fig. 5. Substituting (55) into (54), we obtain

$$\tilde{R}_{\rho}^{\mathrm{SAH}}(\boldsymbol{z}) = \sum_{ij} \int \mathrm{e}^{\mathrm{i}\omega(\boldsymbol{z}' - \boldsymbol{z}) \cdot \boldsymbol{\Xi}_{ij}^{\mathrm{SAH}}(s', s, \boldsymbol{z})/c_0} Q_{ij}^{\mathrm{SAH}}(\boldsymbol{z}, \omega, s', s) \times A_{ij}^{\mathrm{SAH}}(\boldsymbol{z}, \omega, s', s) R_{\rho}(\boldsymbol{z}) \, \mathrm{d}\omega \, \mathrm{d}s \, \mathrm{d}s' \, \mathrm{d}\boldsymbol{z}' . \tag{57}$$

Under the assumptions (53) and (40), (57) shows that $\mathscr{K}_{ij}^{\mathrm{SAH}} \mathscr{F}_{ij}^{\mathrm{SAH}}$ is a pseudodifferential operator [33]. This means that the backprojection operator reconstructs the visible edges of the scene radiance at the correct location and orientation in SAH imaging.

3.3 Determination of the Filter

$Q_{ij}^{\rm DSAH}$ and $Q_{ij}^{\rm SAH}$ can be determined with respect to various criteria [39]. Ideally, they are chosen such that the leading-order contributions of the point spread functions (PSFs) of $\mathscr{K}_{ij}^{\rm DSAH}$ and $\mathscr{K}_{ij}^{\rm SAH}$ are Dirac-delta functions. This choice of the filters ensures that $\mathscr{K}_{ij}^{\rm DSAH}$ and $\mathscr{K}_{ij}^{\rm SAH}$ reconstruct the visible edges of the scene radiance not only at the correct location and orientation, but also with the correct strength [25, 26, 30, 39].

In DSAH imaging, for each s' and $\boldsymbol{z}$, we make the following change of variables:

$$(t,s) \rightarrow \boldsymbol{\xi}_{ij}^{\rm DSAH} = t\,\boldsymbol{\Xi}_{ij}^{\rm DSAH}(s',s,\boldsymbol{z}) \tag{58}$$

in (50) and obtain

$$\tilde{R}_{\rho}^{\rm DSAH}(\boldsymbol{z}) \approx \sum_{ij}\int_{\Omega_{ij,s',\boldsymbol{z}}^{\rm DSAH}} {\rm e}^{-{\rm i}(\boldsymbol{z}'-\boldsymbol{z})\cdot\boldsymbol{\xi}_{ij}^{\rm DSAH}} Q_{ij}^{\rm DSAH}\left(\boldsymbol{z},t\left(\boldsymbol{\xi}_{ij}^{\rm DSAH}\right),s',s\left(\boldsymbol{\xi}_{ij}^{\rm DSAH}\right)\right)$$
$$\times A_{ij}^{\rm DSAH}\left(\boldsymbol{z},t\left(\boldsymbol{\xi}_{ij}^{\rm DSAH}\right),s',s\left(\boldsymbol{\xi}_{ij}^{\rm DSAH}\right)\right)\left|\frac{\partial(t,s)}{\partial\boldsymbol{\xi}_{ij}^{\rm DSAH}}\right| R_{\rho}(\boldsymbol{z}')\,{\rm d}\boldsymbol{\xi}_{ij}^{\rm DSAH}{\rm d}\boldsymbol{z}'{\rm d}s', \tag{59}$$

where $|\partial(t,s)/\partial\boldsymbol{\xi}_{ij}^{\rm DSAH}|$ is the determinant of the Jacobian that comes from the change of variables given in (58).

The domain of integration in (59) is given as follows:

$$\Omega_{ij,s',\boldsymbol{z}}^{\rm DSAH} = \Big\{\boldsymbol{\xi}_{ij}^{\rm DSAH} = t\,\boldsymbol{\Xi}_{ij}^{\rm DSAH}(s',s,\boldsymbol{z})\,|\,A_{ij}^{\rm DSAH}(\boldsymbol{z},t,s',s) \neq 0,$$
$$(t,s',s) \in (\mathbb{R},\mathbb{R},\mathbb{R})\Big\}. \tag{60}$$

We refer to $\Omega_{ij,s',\boldsymbol{z}}^{\rm DSAH}$ as the DSAH *partial data collection manifold* at $(s',\boldsymbol{z})$ obtained by the ith and jth receivers for a fixed s' and refer to the union $\cup_{ij,s'}\Omega_{ij,s',\boldsymbol{z}}^{\rm DSAH}$ as the DSAH *data collection manifold* at $\boldsymbol{z}$ and denote it by $\Omega_{\boldsymbol{z}}^{\rm DSAH}$. This set determines many of the properties of the reconstructed DSAH image.

To approximate the PSF with the Dirac-delta function, we choose the filter as follows:

$$Q_{ij}^{\rm DSAH}(\boldsymbol{z},t,s',s) = \frac{A_{ij}^{\rm DSAH*}(\boldsymbol{z},t,s',s,\mu)}{|A_{ij}^{\rm DSAH}(\boldsymbol{z},t,s',s,\mu)|^2}\,\frac{\chi_{\Omega_{ij,s',\boldsymbol{z}}}^{\rm DSAH}(\boldsymbol{z},t,s',s)}{|\partial(t,s)/\partial\boldsymbol{\xi}_{ij}^{\rm DSAH}|}, \tag{61}$$

where $\chi_{\Omega_{ij,s',\boldsymbol{z}}}^{\rm DSAH}$ is a smooth cutoff function that is equal to one in the interior of $\Omega_{ij,s',\boldsymbol{z}}^{\rm DSAH}$ and zero in the exterior of $\Omega_{ij,s',\boldsymbol{z}}^{\rm DSAH}$.

Similarly, in SAH Imaging, for each s' and $\boldsymbol{z}$, we make the following change of variables:

$$(\omega, s) \rightarrow \boldsymbol{\xi}_{ij}^{\mathrm{SAH}} = \frac{\omega}{c_0} \boldsymbol{\Xi}_{ij}^{\mathrm{SAH}}(s', s, \boldsymbol{z}) \tag{62}$$

in (57) and obtain

$$\tilde{R}_{\rho}^{\mathrm{SAH}}(\boldsymbol{z}) \approx \sum_{ij} \int_{\Omega_{ij,s',\boldsymbol{z}}^{\mathrm{SAH}}} \mathrm{e}^{\mathrm{i}(\boldsymbol{z}'-\boldsymbol{z})\cdot\boldsymbol{\xi}_{ij}^{\mathrm{SAH}}} Q_{ij}^{\mathrm{SAH}}\left(\boldsymbol{z}, \omega\left(\boldsymbol{\xi}_{ij}^{\mathrm{SAH}}\right), s', s\left(\boldsymbol{\xi}_{ij}^{\mathrm{SAH}}\right)\right)$$
$$\times A_{ij}^{\mathrm{DSAH}}\left(\boldsymbol{z}, \omega\left(\boldsymbol{\xi}_{ij}^{\mathrm{SAH}}\right), s', s\left(\boldsymbol{\xi}_{ij}^{\mathrm{SAH}}\right)\right) \left|\frac{\partial(\omega, s)}{\partial \boldsymbol{\xi}_{ij}^{\mathrm{SAH}}}\right| R_{\rho}(\boldsymbol{z}') \mathrm{d}\boldsymbol{\xi}_{ij}^{\mathrm{SAH}} \mathrm{d}\boldsymbol{z}' \mathrm{d}s', \tag{63}$$

where $\left|\frac{\partial(\omega,s)}{\partial \boldsymbol{\xi}_{ij}^{\mathrm{SAH}}}\right|$ is the determinant of the Jacobian that comes from the change of variables in (62).

In (63), the domain of integration is given as follows:

$$\Omega_{ij,s',\boldsymbol{z}}^{\mathrm{SAH}} = \left\{ \boldsymbol{\xi}_{ij}^{\mathrm{SAH}} = \frac{\omega}{c_0} \boldsymbol{\Xi}_{ij}^{\mathrm{SAH}}(s', s, \boldsymbol{z}) \,|\, A_{ij}^{\mathrm{SAH}}(\boldsymbol{z}, \omega, s', s) \neq 0, \right.$$
$$\left. (\omega, s', s) \in (\mathbb{R}, \mathbb{R}, \mathbb{R}) \right\}. \tag{64}$$

We refer to $\Omega_{ij,s',\boldsymbol{z}}^{\mathrm{SAH}}$ as the SAH *partial data collection manifold* at $(s', \boldsymbol{z})$ obtained by the ith and jth receivers for a fixed s' and refer to the union $\cup_{ij,s'} \Omega_{ij,s',\boldsymbol{z}}^{\mathrm{SAH}}$ as the SAH *data collection manifold* at $\boldsymbol{z}$ and denote it by $\Omega_{\boldsymbol{z}}^{\mathrm{SAH}}$. Again, this set determines many of the properties of the reconstructed SAH image.

Similarly, to approximate the PSF with the Dirac-delta function, we choose the filter as follows:

$$Q_{ij}^{\mathrm{SAH}}(\boldsymbol{z}, \omega, s', s) = \frac{A_{ij}^{\mathrm{SAH}*}(\boldsymbol{z}, \omega, s', s)}{|A_{ij}^{\mathrm{SAH}}(\boldsymbol{z}, \omega, s', s)|^2} \frac{\chi_{\Omega_{ij,s',\boldsymbol{z}}}^{\mathrm{SAH}}(\boldsymbol{z}, \omega, s', s)}{|\partial(\omega, s)/\partial \boldsymbol{\xi}_{ij}^{\mathrm{SAH}}|}, \tag{65}$$

where $\chi_{\Omega_{ij,s',\boldsymbol{z}}}^{\mathrm{SAH}}$ is a smooth cutoff function that is equal to one in the interior of $\Omega_{ij,s',\boldsymbol{z}}^{\mathrm{SAH}}$ and zero in the exterior of $\Omega_{ij,s',\boldsymbol{z}}^{\mathrm{SAH}}$.

Irrespective of the choice of the filters, the filtered-backprojection operators $\mathscr{K}_{ij}^{\mathrm{DSAH}}$ and $\mathscr{K}_{ij}^{\mathrm{SAH}}$ reconstruct the visible edges of the scene radiance at the correct location and correct orientation. With the choice of the filters given in (61) and (65), the resulting image formation method can recover the visible edges not only at the correct location and orientation, but also with the correct strengths.

4 Resolution Analysis

Substituting (61) and (65) into (59) and (63), respectively, we obtain

$$\tilde{R}_\rho^{\mathrm{DSAH}}(\boldsymbol{z}) \approx \sum_{ij} \int_{\Omega^{\mathrm{DSAH}}_{ij,s',\boldsymbol{z}}} \mathrm{e}^{-\mathrm{i}(\boldsymbol{z}'-\boldsymbol{z})\cdot\boldsymbol{\xi}_{ij}^{\mathrm{DSAH}}} R_\rho(\boldsymbol{z}')\,\mathrm{d}\boldsymbol{z}'\,\mathrm{d}\boldsymbol{\xi}_{ij}^{\mathrm{DSAH}}\,\mathrm{d}s', \tag{66}$$

$$\tilde{R}_\rho^{\mathrm{SAH}}(\boldsymbol{z}) \approx \sum_{ij} \int_{\Omega^{\mathrm{SAH}}_{ij,s',\boldsymbol{z}}} \mathrm{e}^{\mathrm{i}(\boldsymbol{z}'-\boldsymbol{z})\cdot\boldsymbol{\xi}_{ij}^{\mathrm{SAH}}} R_\rho(\boldsymbol{z}')\,\mathrm{d}\boldsymbol{z}'\,\mathrm{d}\boldsymbol{\xi}_{ij}^{\mathrm{SAH}}\,\mathrm{d}s'. \tag{67}$$

Equation (66) and (67) show that the DSAH and SAH images, $\tilde{R}_\rho^{\mathrm{DSAH}}$ and $\tilde{R}_\rho^{\mathrm{SAH}}$ are bandlimited versions of R_ρ whose bandwidth are determined by the data collection manifolds $\Omega_{\boldsymbol{z}}^{\mathrm{DSAH}}$ and $\Omega_{\boldsymbol{z}}^{\mathrm{SAH}}$, respectively. The data collection manifolds determine the resolution of the reconstructed images at $\boldsymbol{z}$. The larger the data collection manifold, the better the resolution of the reconstructed image is.

Microlocal analysis of (66) and (67) tell us that an edge at point $\boldsymbol{z}$ is visible in DSAH or SAH image if the direction $\boldsymbol{n_z}$ normal to the edge is contained in $\Omega_{\boldsymbol{z}}^{\mathrm{DSAH}}$ or $\Omega_{\boldsymbol{z}}^{\mathrm{SAH}}$, respectively [25, 26, 30, 39]. Consequently, an edge at point $\boldsymbol{z}$ with $\boldsymbol{n_z}$ normal to edge is visible if there exists i, j, s', s such that $\boldsymbol{\xi}_{ij}^{\mathrm{DSAH}}$ or $\boldsymbol{\xi}_{ij}^{\mathrm{SAH}}$ is parallel to $\boldsymbol{n_z}$.

The bandwidth contribution of $\boldsymbol{\xi}_{ij}^{\mathrm{DSAH}}$ and $\boldsymbol{\xi}_{ij}^{\mathrm{SAH}}$ to a visible edge at $\boldsymbol{z}$ is given by

$$\frac{\omega_0}{c_0} L_\phi \left| D\psi(\boldsymbol{z}) \cdot \left[\frac{1}{|\boldsymbol{\gamma}_i(s') - \mathbf{z}|} \dot{\boldsymbol{\gamma}}_{i,\perp}(s') - \frac{S_{ij}(s', s, \boldsymbol{z})}{|\boldsymbol{\gamma}_j(s) - \mathbf{z}|} \dot{\boldsymbol{\gamma}}_{j,\perp}(s) \right] \right|, \tag{68}$$

$$\frac{B_\omega}{c_0} \left| D\psi(\boldsymbol{z}) \cdot (\widehat{\boldsymbol{\gamma}_j(s) - \mathbf{z}} - \widehat{\boldsymbol{\gamma}_i(s') - \mathbf{z}}) \right|, \tag{69}$$

where L_ϕ denotes the length of the support of $\phi(t)$ and B_ω denotes the bandwidth of the transmitted waveform.

Equation (68) shows that for DSAH imaging, the longer the support of $\phi(t)$ becomes, the larger the magnitude of $\boldsymbol{\xi}_{ij}^{\mathrm{DSAH}}$ is, giving rise to sharper reconstructed edges perpendicular to $\boldsymbol{\xi}_{ij}^{\mathrm{DSAH}}$, $i, j = 1, \ldots, N$. Additionally, the higher the carrier frequency of the transmitted signal ω_0 becomes, larger the magnitude of $\boldsymbol{\xi}_{ij}^{\mathrm{DSAH}}$ is, contributing to higher image resolution.

Equation (69) shows that for SAH imaging, as the bandwidth of the transmitted signal becomes larger, the magnitude of $\boldsymbol{\xi}_{ij}^{\mathrm{SAH}}$ gets larger, which results in higher image resolution. The sharpness of the reconstructed edges is also directly proportional to the bandwidth of the transmitted signal.

Furthermore, we note that in DSAH imaging, the resolution depends on the range via the terms $|\boldsymbol{\gamma}_i(s') - \boldsymbol{z}|$ and $|\boldsymbol{\gamma}_j(s) - \boldsymbol{z}|$ and the velocities of the receivers via the terms $\dot{\boldsymbol{\gamma}}_{i,\perp}$ and $\dot{\boldsymbol{\gamma}}_{j,\perp}$. As the scatterers are further away from the receivers, or the

Table 1 Parameters that affect the DSAH image resolution

Parameter	Increase	Resolution
Carrier frequency: ω_0	↑	↑
Length of the windows L_ϕ	↑	↑
Distance $\lvert\boldsymbol{\gamma}_i(s')-\boldsymbol{z}\rvert$, $\lvert\boldsymbol{\gamma}_j(s)-\boldsymbol{z}\rvert$	↑	↓
Antenna velocity $\dot{\boldsymbol{\gamma}}_i$ or $\dot{\boldsymbol{\gamma}}_j$	↑	↑
Number of s samples	↑	↑
Number of time windows (s')	↑	↑

↑: Increase (higher) ↓: Decrease (lower)

Table 2 Parameters that affect the SAH image resolution

Parameter	Increase	Resolution
Bandwidth of the transmitted waveform: B_ω	↑	↑
Angle between $\widehat{\boldsymbol{\gamma}_j(s)-\boldsymbol{z}}$ and $\widehat{\boldsymbol{\gamma}_i(s')-\boldsymbol{z}}$	↑	↑
Number of s samples	↑	↑
Number of time windows (s')	↑	↑

↑: Increase (higher) ↓: Decrease (lower)

velocities of the receivers decrease, the resolution gets worse due to the decrease in the magnitude of $\boldsymbol{\xi}_{ij}^{\mathrm{DSAH}}$. In SAH imaging, the resolution also depends on the angle between the unit vectors $\widehat{\boldsymbol{\gamma}_j(s)-\mathbf{z}}$ and $\widehat{\boldsymbol{\gamma}_i(s')-\mathbf{z}}$. The larger the angle is, the larger the magnitude of $\boldsymbol{\xi}_{ij}^{\mathrm{SAH}}$ becomes, resulting in better resolution.

Additionally, the increase in the number of s samples and the time windows (indicated by s') used for imaging also leads to a larger data collection manifold in DSAH or SAH imaging, which improves the resolution.

We summarize the parameters that affect the resolution of the reconstructed image in DSAH and SAH imaging in Tables 1 and 2, respectively.

5 Numerical Simulations

We considered a scene of size $[0,22]\times[0,22]\,\mathrm{km}^2$ with flat topography. The scene was discretized into 128×128 pixels, where $[0,0,0]\,\mathrm{km}$ and $[22,22,0]\,\mathrm{km}$ correspond to the pixels $(1,1)$ and $(128,128)$, respectively.

In all the numerical experiments, we used two airborne receivers and a single, stationary transmitter operating either cooperatively or noncooperatively. We assumed that both the receiver and transmitter antennas were isotropic. We assumed that the transmitter was located at $\mathbf{y}_0=(0,0,6.5)\,\mathrm{km}$ and the receivers were traversing the circular trajectory given by

$$\gamma_C(\tilde{s}) = (11+11\cos(\tilde{s}), 11+11\sin(\tilde{s}), 6.5)\,\mathrm{km}. \tag{70}$$

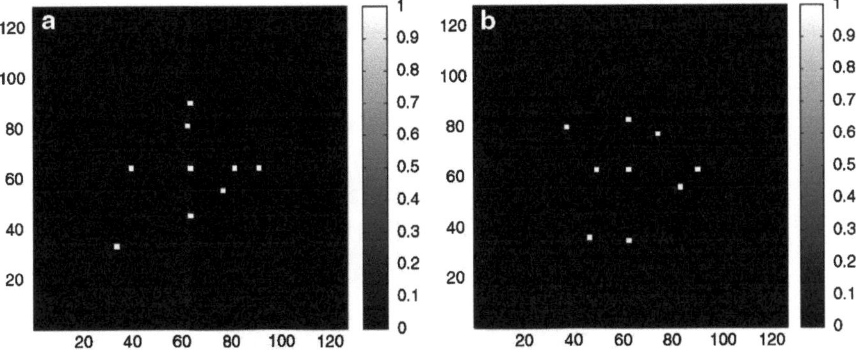

Fig. 6 Discretized scene reflectivity used in the numerical simulations of (**a**) DSAH imaging and (**b**) SAH imaging. $[0,0,0]$ km and $[22,22,0]$ km are located at the *lower left* and *upper right corners*, respectively

Let $\gamma_1(\tilde{s})$ and $\gamma_2(\tilde{s})$ denote the trajectories of the two receivers. We set $\gamma_1(\tilde{s}) = \gamma_C(\tilde{s})$ and $\gamma_2(\tilde{s}) = \gamma_1(\tilde{s} - \frac{\pi}{6})$. Note that the variable $\tilde{s}$ in γ_C is equal to $\frac{V}{R}t$, where V is the speed of the receiver, and R is the radius of the circular trajectory. We set the speed of the two receivers to 261 m/s. We chose the sampling rate of s to be 1.9335 Hz so as to uniformly sample the circular trajectory with 512 points.

In accordance with the incoherent field approximation, we used the following multiple-point-target model for the scene reflectivity,

$$\rho(z) = \sum_{l=1}^{L} g_l \delta(z - z_l), \tag{71}$$

where g_l, $l = 1, \ldots, L$ are independent Gaussian random variables with mean μ_l and variance σ_l^2. The corresponding scene radiance is given by

$$R_\rho(z) = E[\rho(z)\rho^*(z)] = \sum_l (\mu_l^2 + \sigma_l^2)\delta(z - z_l). \tag{72}$$

In our simulations, we considered a deterministic reflectivity and set $\sigma_l^2 = 1$. We used $L = 9$ and approximated the Dirac-delta functions in (72) by square target reflectors of size $344 \times 344\,\text{m}^2$, each having a unit reflectivity, i.e., $\mu_l = 1$, $l = 1, \ldots, 9$.

Figure 6a, b show the scene with targets used in the simulations of the DSAH imaging and SAH imaging, respectively. Figure 7 shows the receiver trajectories and the transmitter antenna location used for DSAH and SAH simulations.

We performed image reconstruction for each s' and coherently superimposed the reconstructed images obtained over a range of s'.

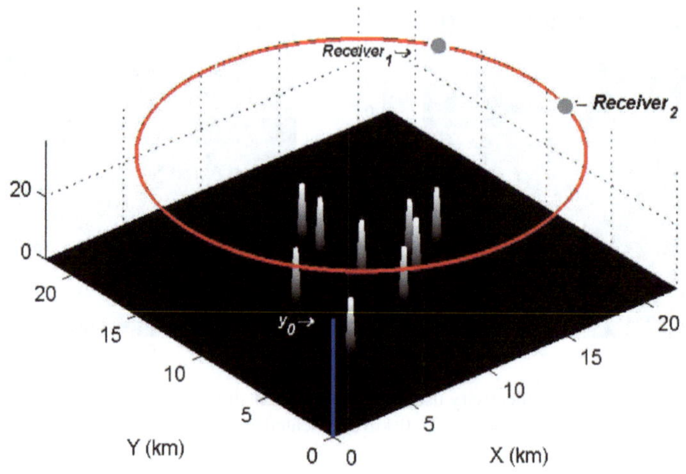

Fig. 7 3-D view of the scene with multiple point targets, illuminated by a single transmitter located at $\mathbf{y}_0 = (0,0,6.5)\,\mathrm{km}$ and the circular receiver trajectory $\gamma_C(\tilde{s}) = (11+11\cos(\tilde{s}), 11+11\sin(\tilde{s}), 6.5)\,\mathrm{km}$, as shown by the *red solid line*. At a certain time instant, two receivers are located at the positions shown in the figure

5.1 Numerical Simulations for DSAH Imaging

We used (9) to generate the data and chose the windowing function ϕ in (9) to be a Hanning function.

The transmitted waveform was assumed to be a single-frequency CW waveform with 800 MHz carrier frequency. The length of the windowing function was set to $L_\phi = 0.0853\,\mathrm{s}$.

For the case of a cooperative transmitter, the reconstructed image is shown in Fig. 8a. It can be seen that the targets are well reconstructed using the DSAH image formation method.

Figure 8b shows the reconstructed image using a noncooperative transmitter. Since the location of the transmitter was assumed to be unknown, the received signal was not compensated for the transmitter related geometric spreading factors. As a result, the targets closer to the transmitter appears brighter in the reconstructed image than those that are further away from the transmitter.

5.2 Numerical Simulations for SAH Imaging

We used (33) to generate the data for performing SAH imaging simulations and chose the windowing function ϕ in (33) to be a Hanning function as in DSAH imaging simulations.

A transmitted pulse at center-frequency 0 Hz with bandwidth equal to 0.873 MHz was used in the simulations.

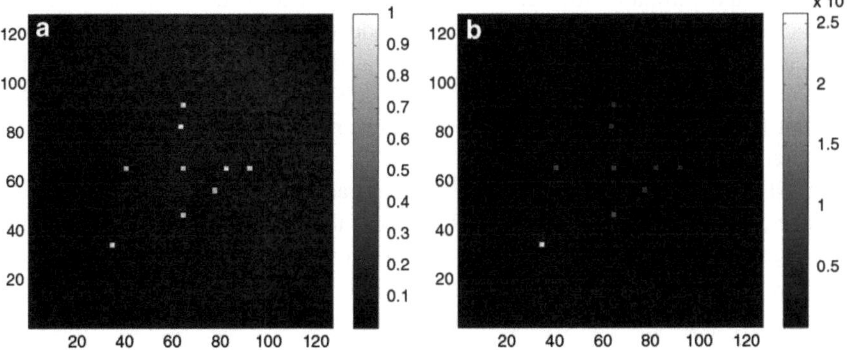

Fig. 8 The reconstructed DSAH images obtained by superposing the images obtained using multiple s' values and two receivers traversing the circular flight trajectories $\gamma_1(\tilde{s})$ and $\gamma_2(\tilde{s})$ as shown in Fig. 7 and a single (**a**) cooperative transmitter, and (**b**) noncooperative transmitter located at $\mathbf{y}_0$

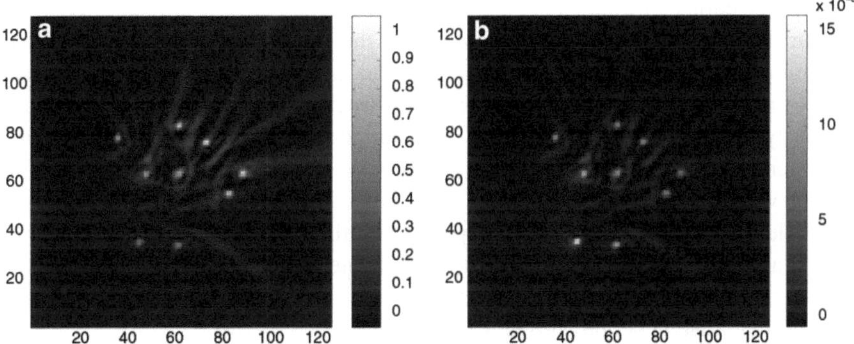

Fig. 9 The reconstructed SAH images obtained by superposing the images obtained using multiple s' values and two receivers traversing the circular flight trajectories $\gamma_1(\tilde{s})$ and $\gamma_2(\tilde{s})$ as shown in Fig. 7 and a single (**a**) cooperative transmitter, and (**b**) noncooperative transmitter located at $\mathbf{y}_0$

The reconstructed images corresponding to the cooperative and noncooperative transmitter cases are shown in Fig. 9a, b, respectively. We see that the targets are reconstructed successfully in both cases and in the noncooperative case, the strength of the targets closer to the transmitter are higher as expected.

6 Conclusion

We presented a unified theory of passive synthetic aperture imaging based on inverse scattering theory, estimation-detection theory and microlocal analysis. Our theory involves development of passive measurement models based on inverse scattering and estimation-detection theory and analytic inversion methods based on microlocal analysis. The measurement models involve windowed, scaled, and translated correlation of the received signals at different receiver locations. This correlation process results in measurement models in the form of FIOs. Taking into account the nature of the waveforms of opportunity, we developed two different measurement models: DSAH FIO and SAH FIO.

DSAH FIO-based model projects the scene radiance onto passive iso-Doppler contours, and SAH FIO-based model projects the scene radiance onto passive iso-range contours. The correlation process removes the transmitter-related terms from the phase of the resulting FIO-based DSAH and SAH measurement models, allowing us to perform backprojection without the knowledge of the transmitter locations.

We used microlocal techniques to backproject the correlated signals onto the passive iso-Doppler contours in DSAH imaging and onto passive iso-range contours in SAH imaging. The filtered-backprojection reconstruction methods for DSAH and SAH have the desirable property of preserving the visible edges of the scene radiance at the correct location and orientation, and at the correct strength with appropriate choice of filters.

Our analysis shows that the resolution of the reconstructed DSAH images is determined primarily by the temporal duration and frequency of the transmitted waveforms, and the resolution of the reconstructed SAH images is determined primarily by the bandwidth of the transmitted waveforms. These results are consistent with the ambiguity theory of the CW or ultra-narrowband waveforms and the wideband waveforms.

While we focused primarily on the passive synthetic aperture radar, the theory of DSAH and SAH imaging introduced in this chapter and the resulting methods and algorithms are also applicable to other wave-based passive imaging problems, such as those that arise in geophysics or acoustics.

References

1. Antoniou, M., Cherniakov, M., Cheng, Hu.: Space-surface bistatic sar image formation algorithms. IEEE Trans. Geosci. Remote Sens. **47**(6), 1827–1843 (2009)
2. Baker, C.J., Griffiths, H.D., Papoutsis, I.: Passive coherent location radar systems. part 2: waveform properties. IEE Proc. Radar Sonar Navig. **152**(3), 160–168 (2005)
3. Barrett, H.H., Myers, K.J.: Foundations of Image Science. Wiley-Interscience, Hoboken (2004)
4. Bleistein, N., Handelsman, R.A.: Asymptotic Expansions of Integrals. Dover, New York (1986)
5. Chetty, K., Woodbridge, K., Guo, H., Smith, G.E.: Passive bistatic WiMAX radar for marine surveillance. In: Proceedings of 2010 IEEE Radar Conference, May (2010)
6. Christiansen, J.M., Olsen, K.E.: Range and Doppler walk in DVB-T based passive bistatic radar. In: Proceedings of 2010 IEEE Radar Conference, May (2010)
7. Coleman, C., Yardley, H.: Passive bistatic radar based on target illumniations by digital audio broadcasting. IET Radar Sonar Navig. **2**(5), 366–375 (2008)
8. Demanet, L., Ferrara, M., Maxwell, N., Poulson, J., Ying, L.: A butterfly algorithm for synthetic aperture radar imaging. SIAM J. Imaging Sci. (2010, Submitted) **5**(1), 203–243 (2012)

9. Falcone, P., Colone, F., Bongioanni, C., Lombardo, P.: Experimental results for OFDM WiFi-based passive bistatic radar. In: Proceedings of 2010 IEEE Radar Conference, May (2010)
10. Garnier, J., Papanicolaou, G.: Passive imaging using cross correlations of ambinent noise signals. In: Proceedings of 2009 3rd IEEE International workshop on Computational Advances in Multi-Sensor Adaptive Processing (CAMSAP), pp. 221–224, December (2009)
11. Garnier, J., Papanicolaou, G.: Passive sensor imaging using cross correlations of noisy signals in a scattering medium. SIAM J. Imaging Sci. **2**(2), 396–437 (2009)
12. Garnier, J., Papanicolaou, G.: Resolution analysis for imaging with noise. Inverse Probl. **26**(074001), 22 (2010)
13. Garnier, J., Solna, K.: Passive imaging and detection in cluttered media. In: Proceedings of 2009 3rd IEEE Int. workshop on Computational Advances in Multi-Sensor Adaptive Processing (CAMSAP), pp. 225–228, December (2009)
14. Griffiths, H.D., Baker, C.J.: Passive coherent location radar systems. part 1: performance prediction. IEE Proc. Radar Sonar Navig. **152**(3), 153–159 (2005)
15. Grigis, A., Sjöstrand, J.: Microlocal Analysis for Differential Operators: An Introduction. London Mathematical Society Lecture Note Series, vol. 196. Cambridge University Press, Cambridge (1994)
16. Guillemin, V., Sternberg, S.: Geometric Asymptotics. American Mathematical Society, Providence (1979)
17. Guo, H., Woodbridge, K., Baker, C.J.: Evaluation of WiFi beacon transmissions for wireless based passive radar. In: Proceedings of 2008 IEEE Radar Conference, May (2008)
18. Harms, H.A., Davis, L.M., Palmer, J.: Understanding the signal structure in DVB-T signals for passive radar detection. In: Proceedings of 2010 IEEE Radar Conference, May (2010)
19. Homer, J., Kubik, K., Mojarrabi, B., Longstaff, I.D., Donskoi, E., Cherniakov, M.: Passive bistatic radar sensing with leos based transmitters. In: Proc. IEEE Int. Geosci. Remote Sens. Symposium **1**, 438–440 (2002)
20. Howland, P.E., Maksimiuk, D., Reitsma, G.: Fm radio based bistatic radar. IEE Proc. Radar Sonar Navig. **152**(3), 107–115 (2005)
21. Kay, S.M.: Fundamentals of Statistical Signal Processing, Vol. I and Vol. II. Prentice Hall, Englewood Cliffs (1998)
22. Lenanon, N., Mozeson, E.: Radar Signals. Wiley-IEEE Press, USA (2004)
23. Li, G., Xu, J., Peng, Y.N., Xia, X.G.: Bistatic linear array SAR for moving target detection, location and imaging with two passive airborne radars. IEEE Trans. Geosci. Remote Sens. **45**(3), 554–565 (2007)
24. Natterer, F., Wübbeling, F.: Mathematical methods in image reconstruction. SIAM, Philadelphia, PA (2001)
25. Nolan, C.J., Cheney, M.: Synthetic aperture inversion. Inverse Probl. **18**, 221–236 (2002)
26. Nolan, C.J., Cheney, M.: Synthetic aperture inversion for arbitrary flight paths and non-flat topography. IEEE Trans. Image Process. **12**, 1035–1043 (2003)
27. O'Hagan, D.W., Baker, C.J.: Passive bistatic radar (PBR) using FM radio illuminators of opportunity. In: Proceedings of 2008 IEEE Radar Conference, May (2008)
28. Pollard, R.: The role of passive radar sensors for air traffic control. In: The Institution of Engineering and Technology Seminar on the Future of Civil Radar (2006)
29. Poullin, D.: Passive detection using digital broadcasters (DAB, DVB) with COFDM modulation. IEE Proc. Radar Sonar Navig. **152**(3), 143–152 (2005)
30. Quinto, E.T.: Singularities of the x-ray transform and limited data tomography in r^2 and r^3. SIAM J. Math. Anal. **24**, 1215–1225 (1993)
31. Skolnik, M.I.: Introduction to Radar Systems, 3rd edn. McGraw-Hill, New York (2002)
32. Tan, D.K.P., Sun, H., Lu, Y., Lesturgie, M., Chan, H.L.: Passive radar using global system for mobile communication signal: theory, implementation and measurements. IEE Proc. Radar Sonar Navig. **152**(3), 116–123 (2005)
33. Treves, F.: Introduction to Pseudodifferential and Fourier Integral Operators, volumes I and II. Plenum Press, New York (1980)

34. Van Veen, B.D., Buckley, K.M.: Beamforming: a versatile approach to spatial filtering. IEEE ASSP Mag. **5**, 4–24 (1988)
35. Voccola, K., Yazici, B., Cheney, M., Ferrara, M.: On the equivalence of the generalized likelihood ratio test and backprojection method in synthetic aperture imaging. In: Proceedings SPIE on Defense and Security Conference, vol. 7335, pp. 1Q1–10, April (2009)
36. Wang, L., Son, I.Y., Yazıcı, B.: Passive imaging using distributed apertures in multiple scattering environments. Inverse Probl. **26**(065002) (2010)
37. Woodward, P.M.: Radar Ambiguity Analysis. Technical Note No.731 (1967)
38. Yarman, C.E., Yazıcı, B.: Synthetic aperture hitchhiker imaging. IEEE Trans. Imaging Process. **17**(11), 2156–2173 (2008)
39. Yazici, B., Cheney, M., Yarman, C.E.: Synthetic-aperture inversion in the presence of noise and clutter. Inverse Probl. **22**, 1705–1729 (2006)
40. Yarman, C.E., Wang, L., Yazıcı, B.: Doppler synthetic aperture hitchhiker imaging. Inverse Probl. **26**(065006), 26 (2010)

Part III
Mathematics of Data Processing

Mathematics of data processing is represented by select topics of harmonic and geometric analysis showcasing the latest developments in this area. Three chapters are devoted to analysis of special classes of polynomials and code design. Two chapters deal with finding the structure of large data sets, and one chapter studies an aspect of a class of stochastic processes.

The first of the six chapters on mathematics of data processing is by GEORGE BENKE. He connects a beam-pattern design problem for thin arrays to the unimodular polynomial problem in harmonic analysis which asks for a trigonometric polynomial $P(X)$ with unit modulus coefficients that has equal values of $|P(X)|$. He extends the construction of Golay-Rudin-Shapiro polynomials to produce trigonometric polynomials of unit coefficients supported on a set of size $N = p^n + O(p^{n/2})$ for a given integer $n \geq 2$ and prime $p \geq 2$.

GUANGLIANG CHEN, ANNA V. LITTLE, and MAURO MAGGIONI analyze large data sets in high dimensions using geometric multi-resolution analysis. Given a point cloud obtained by sampling a lower-dimensional manifold and adding noise, the authors present results on dimension estimation and subsequently on parametric modeling of the underlying manifold. Their approach is constructive, and an algorithm that implements the geometric wavelet transform is also presented. The chapter ends with numerical results on synthetic and real-world data sets, such as the MNIST data set of handwritten digits.

CAROL T. CHRISTOU and GARRY M. JACYNA study the fourth-order structure function of a fractional Brownian motion. Their focus is on the flatness function which is the ratio of the fourth-order structure function to the square of the second-order structure function. The authors show that the flatness function grows unboundedly with the lower band-edge frequency of the high-pass component of the signal.

RONALD COIFMAN and MATAN GAVISH present a framework that integrates methods of harmonic analysis and geometry to problems and concepts from machine learning. The main problem is exemplified by finding structures in point clouds. Using a similarity measure (affinity), the authors construct an approximation of the discretized Laplace-Beltrami operator whose set of eigenfunctions generate the first-level structure of the point cloud.

GREGORY E. COXSON analyzes transformations that preserve good autocorrelation properties of unimodular codes. Specifically he looks for groups of transformations that do not change the magnitude of the peak sidelobe level (PSL). For binary codes there is a single Abelian group of order 8 that preserves the PSL. For quad-phase codes, whose entries belong to $\{1, -1, i, -i\}$, there are four groups (two pairs of isomorphic groups). For the general mth roots of unity codes of odd length there are $4m^2$ groups. The author describes their structure for odd values of m.

DAVID JOYNER studies classes of self-reciprocal polynomials $p(z)$ of degree n that satisfy $z^n p(\frac{1}{z}) = p(z)$. He is particularly interested in self-reciprocal polynomials whose zeros are on the unit circle. The author lists several classes of polynomials such as Littlewood polynomials, zeta polynomials, Duursma zeta polynomials, and Alexander polynomials. He ends his chapter by conjecturing that zeros of self-reciprocal polynomials whose coefficients grow slowly are on the unit circle.

Golay–Rudin–Shapiro Polynomials and Phased Arrays

George Benke

Abstract A single-frequency plane wave propagating at speed c in the direction of the unit vector $\mathbf{N}$ is given by

$$S(\mathbf{X},t) = \exp i\omega\left(t - \frac{\mathbf{N}\cdot\mathbf{X}}{c}\right). \tag{1}$$

Suppose $\{\mathbf{X}_1,\ldots,\mathbf{X}_N\} \subset \mathbf{R}^n$ is a fixed set of locations, called the array, and $w_1,\ldots,w_N \subset \mathbf{C}$ is a set of weights. The linear combination

$$\begin{aligned} G(t) &= \sum_{k=1}^{N} w_k S(\mathbf{X},t) = \sum_{k=1}^{N} w_k \exp(i\omega t)\exp\left(\frac{i\omega}{c}\mathbf{N}\cdot\mathbf{X}_k\right) \\ &= \exp i\omega t \sum_{k=1}^{N} w_k \exp\left(-2\pi i\left(\frac{\mathbf{N}\cdot\mathbf{X}_k}{\lambda}\right)\right) = (\exp i\omega t)H(\mathbf{N}) \end{aligned} \tag{2}$$

is the output of the "phased array". Given unit vector $\mathbf{N}_0$, if we let

$$w_k = \exp\left(2\pi i\left(\frac{\mathbf{N}_0\cdot\mathbf{X}_k}{\lambda}\right)\right) \tag{3}$$

then $H(\mathbf{N}_0) = N$ and $|H(\mathbf{N})| \leq N$ for all $\mathbf{N}$. The "phased array problem" is to find $\{\mathbf{X}_1,\ldots,\mathbf{X}_N\}$ such that $|H(\mathbf{N})$ looks as much like a delta function as possible. We study this problem in the simplified case where $\{\mathbf{X}_1,\ldots,\mathbf{X}_N\}$ lie on a line. In this case, through a mild change of variables, $H(\mathbf{N})$ is replaced by the 1-periodic trigonometric polynomial

$$A(x) = \sum_{k=1}^{N} \mathrm{e}^{2\pi i n_k x}, \tag{4}$$

G. Benke (✉)
Department of Mathematics, Georgetown University, Washington DC 20057, USA
e-mail: benke@georgetown.edu

T.D. Andrews et al. (eds.), *Excursions in Harmonic Analysis, Volume 1*,
Applied and Numerical Harmonic Analysis, DOI 10.1007/978-0-8176-8376-4_12,

where the integers n_k specify the location of the $\mathbf{X}_k$ on the line. Then the problem is to chose the n_k so that $A(x)$ looks like a (periodic) delta function with mass at $x = 0$. In order to achieve a sharp peak at $x = 0$ the n_k must be very spread out. When this is so, the array is referred to as a "thin phased array". However if the n_k are too uniformly spaced there will be large peaks at locations other than $x = 0$. The graph of $|A(x)|$ is fairly insensitive to the distribution of the n_k in the region close to $x = 0$, called the "main lobe region". The complement of the main lobe region is called the "sidelobe region". The problem is to minimize the maximum of $|A(x)|$ in the sidelobe region. Energy considerations show that the maximum of $|A(x)|$ in the sidelobe region cannot be less than $O(\sqrt{N})$. Random choices of the n_k cannot do better than $O(\sqrt{N \log N})$ for sidelobe maxima. In this chapter we give a construction which achieves $O(\sqrt{N})$ sidelobe maxima. Specifically we use the theory of generalized golay–rudin–shapiro polynomials to construct asymptotically optimal thin phased arrays. More precisely, given $n \geq 2$ and prime $p \geq 2$ we construct integer sets $\{n_1, \ldots, n_N\}$ contained in $\{0, 1, \ldots, p^{n+1}\}$ such that

$$\left|\sum_{k=1}^{N} \mathrm{e}^{2\pi i n_k x}\right| \leq \left|\frac{\sin \pi p^{n+1} x}{p \sin \pi x}\right| + O(p^{n/2}),$$

where $N = p^n + O(p^{n/2})$.

Keywords Golay • Rudin • Shapiro • Golay–Rudin–Shapiro • Phased array • Antenna • Unimodular polynomial • Thin array • Array factor • Complementary pair

1 Introduction

In this chapter we discuss the connection between an extremal problem in harmonic analysis and the design of optimal phased arrays. An old question in harmonic analysis, often called the unimodular polynomial problem, is "how close can we come to having a trigonometric polynomial $P(x)$ with unit modulus coefficients also have equal values of $|P(x)|$?". While still not completely solved, the literature on this problem contains may spectacular results [3–7, 10, 14–17]. At the same time engineers have asked "how close can we get to designing phased arrays with flat sidelobes?". A tiny sample of the large literature in this area is [11–13, 19–21, 23–25].

A particularly interesting chapter in the investigation of the unimodular polynomial problem is that of the theory of Golay–Rudin–Shapiro polynomials. In 1949 Golay [6] wrote a paper entitled "Multislit spectroscopy," in which he defined pairs of ± 1 valued sequences such that the sum of the autocorrelations is a discrete δ-function. Such pairs are called complementary sequences and have been studied and extended in various directions [1, 3–5, 7, 8, 14–18, 22].

If the ± 1's from a complementary pair of sequences are the coefficients of a pair of polynomials $P(x)$ and $Q(x)$, then the autocorrelation property is equivalent to $|P(x)|^2 + |Q(x)|^2 \equiv$ constant. Such pairs of polynomials were defined by Shapiro [17] in 1957 and later by Rudin [16] in 1959. These Golay–Rudin–Shapiro polynomials address the unimodular polynomial problem since they satisfy

$$\|P\|_2 \leq \|P\|_\infty \leq \sqrt{2}\|P\|_2. \tag{5}$$

The phase array design problem can be formulated a little more precisely by asking "how close can we come to finding sets of integers $\{n_1, \ldots, n_N\}$ of prescribed density such that

$$\left|\sum_{k=1}^{N} \mathrm{e}^{2\pi i n_k x}\right| \tag{6}$$

looks like a δ-function in the interval $-1/2 \leq x \leq 1/2$?". It turns out that this requires two conflicting properties: (1) the n_k's should be spread evenly over a larger interval of integers, and (2) the n_k's should not have too much arithmetic structure. This problem has been addressed for small N by computationally intensive optimization techniques and searches. However, there is practical as well as theoretical interest in the case where N is large. For such N the computational complexity makes such approaches infeasible. Choosing the sets $\{n_1, \ldots, n_N\}$ randomly has also been considered [2, 9, 13, 19]. However, it can be shown that such methods are necessarily sub-optimal. More precisely, it can be shown that for N element sets, the sidelobes cannot be better than $O(\sqrt{N})$, and random sets can only achieve sidelobes of order $O(\sqrt{N \log N})$.

We will show how generalized Golay–Rudin–Shapiro polynomials can be used to construct sets of integers $\{n_1, \ldots, n_N\}$ such that the sidelobes are $O(\sqrt{N})$ and the density is asymptotically $1/p$, where p is a prime. Our initial construction does not allow us to specify N arbitrarily, instead producing sets of cardinality $N = p^n + O(p^{n/2})$ which are subsets of $\{1, 2, \ldots, p^{n+1}\}$ and produce sidelobes of order $O(p^{n/2})$. However such sets can be concatenated to give sets of arbitrary cardinality N, asymptotic density $1/p$ and sidelobes of order $O(\sqrt{N})$. It is an open question as to how to find sets of cardinality N, sidelobes of order $O(\sqrt{N})$ but arbitrary density.

The remainder of this chapter is structured as follows. Since we wish to make this paper accessible to readers other than antenna engineers, we have included Sect. 2 which gives some background in array theory and which formulates the array design problem as a precise mathematical problem which is solved in Sect. 4. To prepare for and to motivate the work in Sect. 4, we have included Sect. 3, where we briefly review the definition and basic properties of the classical Golay–Rudin–Shapiro polynomials. We then show how the definition and theory of complementary pairs can be extended to complementary triples and by a small example how such polynomials can be used to construct certain subsets of integers. Essentially, this is our construction. However, the claimed properties are not obvious from the example and are proven in the main theoretical Sect. 4. In Sect. 4 we reformulate those parts of generalized Golay–Rudin–Shapiro theory [1] which are relevant to our construction and prove the claimed properties.

2 Some Background on Phased Arrays

A complex time-varying field in an n-dimensional space is a function $S(\mathbf{X},t)$ from $\mathbf{R}^n \times \mathbf{R}$ into the complex numbers $\mathbf{C}$. Given a set $B \in \mathbf{R}^n$, an "antenna supported in B" is an operator which takes $S(\mathbf{X},t)$ restricted to $B \times \mathbf{R}$ and produces some kind of information about the field. If B is a discrete set we have an antenna array. By a phased array we mean an operator which takes the field $S(\mathbf{X},t)$ and produces

$$G(t) = \sum_{k=1}^{N} w_k S(\mathbf{X}_k,t), \tag{7}$$

where $B = \{\mathbf{X}_1,\ldots,\mathbf{X}_N\}$ and $w_1,\ldots,w_N$ are a set of weights. For example, if $\sigma(t)$ is a "signal", then

$$S(\mathbf{X},t) = \sigma\left(\frac{t-\mathbf{N}\cdot\mathbf{X}}{c}\right) \tag{8}$$

corresponds to the signal $\sigma(t)$ propagating with speed c in the direction of the unit vector $\mathbf{N}$. Two useful properties of an antenna are (1) to produce $K\sigma(t)$, an amplified version of the signal, and (2) the vector $\mathbf{N}$ which gives the propagation direction of the signal.

An important special case is the single-frequency complex plane wave

$$S(\mathbf{X},t) = \exp i\omega\left(\frac{t-\mathbf{N}\cdot\mathbf{X}}{c}\right). \tag{9}$$

The output of a phased array sensing this field is

$$G(t) = \sum_{k=1}^{N} w_k S(\mathbf{X}_k,t) = \sum_{k=1}^{N} w_k \exp(i\omega t)\exp\left(\frac{-i\omega}{c}\mathbf{N}\cdot\mathbf{X}_k\right) \tag{10}$$

$$= \exp i\omega t \sum_{k=1}^{N} w_k \exp\left(-2\pi i\left(\frac{\mathbf{N}\cdot\mathbf{X}_k}{\lambda}\right)\right), \tag{11}$$

where $\lambda = 2\pi c/\omega$ is the wavelength of the signal. Let

$$H(\mathbf{N}) = \sum_{k=1}^{N} w_k \exp\left(-2\pi i\left(\frac{\mathbf{N}\cdot\mathbf{X}_k}{\lambda}\right)\right). \tag{12}$$

Then the array output is $H(\mathbf{N})\exp i\omega t$. Thus the array produces an output which is an amplified version of the propagating signal, where the amplification factor $H(\mathbf{N})$ depends on the propagation direction of the signal. The function $H(\mathbf{N})$ defined on the unit sphere in $\mathbf{R}^n$ is called the "antenna pattern" of the array. Note that the antenna pattern depends on the wavelength λ, the array "sensor" locations $\mathbf{X}_1,\ldots,\mathbf{X}_N$ and the weights $w_1,\ldots,w_N$. Note also that $|H(\mathbf{N})| \leq N$ and can equal N if the weights w_k are set such that

$$w_k = \exp\left(2\pi i\left(\frac{\mathbf{N}\cdot\mathbf{X}_k}{\lambda}\right)\right). \tag{13}$$

Given any direction $\mathbf{N}_0$, if weights are set such that

$$w_k = \exp\left(2\pi i\left(\frac{\mathbf{N}_0\cdot\mathbf{X}_k}{\lambda}\right)\right), \tag{14}$$

then the antenna pattern is

$$|H(\mathbf{N})| = \left|\sum_{k=1}^{N} \exp\left(-2\pi i\left(\frac{(\mathbf{N}-\mathbf{N}_0)\cdot\mathbf{X}_k}{\lambda}\right)\right)\right|. \tag{15}$$

In this case we say that the antenna has been "pointed" or "steered" in the direction $\mathbf{N}_0$ since for signals propagating in that direction we obtain the largest possible amplification. Suppose we are interested in the maximum amplification of signals propagating in a certain direction $\mathbf{N}_0$ and the rejection of signals propagating in other directions. In this case we would like $|H(\mathbf{N})|$ to resemble a δ-function on the unit sphere supported at $\mathbf{N}_0$. Of course since $H(\mathbf{N})$ is a trigonometric polynomial this is impossible. So we ask "how close can we come to making $|H(\mathbf{N})|$ resemble a δ-function?"

Since $\mathbf{N}_0$ and λ are given and the w_k are chosen as above to point the antenna in the direction $\mathbf{N}_0$, the array pattern $|H(\mathbf{N})|$ is determined solely by the sensor positions $\{\mathbf{X}_1,\ldots,\mathbf{X}_N\}$. Defining the set $\{\mathbf{X}_1,\ldots,\mathbf{X}_N\}$ so as to produce desired characteristics in the antenna pattern $|H(\mathbf{N})|$ is the "array design problem". In this chapter we consider the "thin array design problem". By this we mean that

1. Sensor positions $\{\mathbf{X}_1,\ldots,\mathbf{X}_N\}$ are a subset of the line $\mathbf{X}(\tau) = \mathbf{X}_0 + \tau\mathbf{v}$ in $\mathbf{R}^n$.
2. The number N of sensors is given.
3. Sensor positions $\{\mathbf{X}_1,\ldots,\mathbf{X}_N\}$ are a subset of the lattice interval $\{\mathbf{X}_0 + n\mathbf{v} | 1 \leq n \leq M\}$ with $\mathbf{X}_1$ corresponding to $n = 0$ and $\mathbf{X}_N$ corresponding to $n = M$. The array is thin if M is substantially larger than N.
4. $|H(\mathbf{N})|$ resembles a δ-function with peak at $\mathbf{N}_0$.

The last point is vague and needs elaboration. This point will be clarified below, but first we make a few simple remarks. Since the sensor positions lie in the lattice interval, each $\mathbf{X}_k$ has the form $\mathbf{X}_k = \mathbf{X}_0 + n_k\mathbf{v}$ where n_k is an integer. Hence the array pattern is

$$|H(\mathbf{N})| = \left|\sum_{k=1}^{N} \exp\left(\frac{-2\pi i}{\lambda}(\mathbf{N}-\mathbf{N}_0)\cdot(\mathbf{X}_0 + n_k\mathbf{v})\right)\right|, \tag{16}$$

$$= \left|\sum_{k=1}^{N} \exp\left(-2\pi i n_k\left(\frac{(\mathbf{N}-\mathbf{N}_0)\cdot\mathbf{v}}{\lambda}\right)\right)\right|. \tag{17}$$

It is customary to choose the lattice length scale to be $\lambda/2$, meaning that $\|\mathbf{v}\| = \lambda/2$. This produces

$$|H(\mathbf{N})| = \left|\sum_{k=1}^{N} \exp\left(-2\pi i n_k\left(\frac{(\mathbf{N}-\mathbf{N}_0)\cos\theta(\mathbf{N})}{2}\right)\right)\right|, \tag{18}$$

where $\theta(\mathbf{N})$ is the angle between $\mathbf{N}-\mathbf{N}_0$ and $\mathbf{v}$. Setting

$$x = \frac{(\mathbf{N}_0-\mathbf{N})\cos\theta(\mathbf{N})}{2} \tag{19}$$

the array pattern in terms of the new variable x is $|A(x)|$ where

$$A(x) = \sum_{k=1}^{N} \mathrm{e}^{2\pi i n_k x}. \tag{20}$$

Since $\mathbf{N}$ and $\mathbf{N}_0$ are unit vectors $-1 \le x \le 1$. The 1-periodic trigonometric polynomial $A(x)$ is called the "array factor" of the array.

So our array design problem amounts to choosing an integer set $\Lambda = \{n_1, \dots, n_N\}$ such that $|A(x)|$ resembles a δ-function. Consider first the example where $n_k = k$. In this case

$$|A(x)| = \left|\sum_{k=1}^{N} \mathrm{e}^{2\pi i k x}\right| = \left|\frac{\sin \pi N x}{\sin \pi x}\right|. \tag{21}$$

In the interval $1/2 \le x \le 1/2$, $A(x)$ is zero for $x = j/N$ for non-zero integers $-N/2 \le j \le N/2$. The part of the graph of $|A(x)|$ between $x = -1/N$ and $x = 1/N$ is called the "main lobe" of the array factor and the part between $x = j/N$ and $x = (j+1)/N$ is called a "sidelobe". Using this terminology, our array design problem has the following objectives:

1. Minimize the width of the main lobe.
2. Maximize the height of the main lobe.
3. Minimize the maximum height of the sidelobes.

Next suppose we consider choosing integers $n_k = pk$ for $k = 1, \dots, N$ and p a fixed positive integer. This produces

$$|A(x)| = \left| \frac{\sin \pi N p x}{\sin \pi p x} \right|, \tag{22}$$

which is a scaled copy of the case just considered. This function has period $1/p$, and therefore in the interval $1/2 < x \leq 1/2$ there are $p-1$ extra copies of the main lobe, which has height N. These copies are called "grating lobes" and work against objective (3) above. On the other hand the width of the main lobe is now $2/Np$ which is a reduction by a factor of p from the original case. This change is an improvement with respect to objective (1) above. The array corresponding to $p=1$ is called a "filled array" while the cases $p>1$ are "sparse arrays" or "thin arrays". The desirable narrow main lobe that occurs when p is large stems from the fact that the N integers in Λ are spread over the large interval $\{p, p+1, \ldots, Np\}$ The length of the interval $(N-1)p$ is called the "aperture" of the array and $1/p$ is called the "density" or "fill factor" of the array. The undesirable grating lobes stem from the regular arithmetic structure of the array. Therefore to meet our design objectives we want to choose integer sets $\Lambda = \{n_1, \ldots, n_N\}$ which are spread uniformly over the integer interval $\{p, p+1, \ldots, Np\}$ but not with arithmetic regularity.

What is the best that we can hope for with a given N and p? The first problem is that there is no good and universally agreed upon definition of main lobe and sidelobe for arbitrary sets Λ. Since values of $A(x)$ near $x=0$ are controlled by the large scale structure of Λ, the main lobe is relatively unaffected by the detailed structure of Λ so long as the integers in Λ are spread fairly uniformly over the aperture. On the other hand the sidelobes not near $x=0$ encode the detailed structural information about Λ, and minimizing the sidelobe maxima amounts to minimizing the arithmetic structure in Λ at all but the largest scales.

To get an idea about the best we can hope for regarding the sidelobe maxima, consider the function

$$b(x) = s\mathbf{1}_{[-1/2,1/Np]}(x) + N\mathbf{1}_{[-1/Np,1/Np]}(x) + s\mathbf{1}_{[1/Np,1/2]}(x) \tag{23}$$

which models an array pattern with a flat-topped main lobe of height N, width $2/Np$ and flat side sidelobe regions of height s. Taking $\|b\|_2^2 = N$ which is the same as $\|A\|_2^2$, we can easily calculate s, and find that

$$s = \left(\frac{1-\frac{2}{p}}{1-\frac{2}{Np}} \right)^{1/2} N^{1/2} = C(N,p)N^{1/2}. \tag{24}$$

Note that

$$C(\mathbf{N},p) \to (1-2/p)^{1/2} \quad \text{as} \quad N \to \infty. \tag{25}$$

Analogous but much more complicated calculations show that the sidelobe maxima for an actual array pattern $|A(x)|$ cannot be less than $C(p)N^{1/2}$ where $C(p)$ is a constant depending on the density $1/p$.

These remarks still do not address the question of where the sidelobes begin. Returning briefly to the case of a filled array where

$$|A(x)| = \left| \frac{\sin \pi N x}{\sin \pi x} \right| \tag{26}$$

and the main lobe is supported in $[-1/N, 1/N]$ we find that $A(3/2N) > 2N/3\pi$, so that the lobe immediately to the right of the main lobe has a maximum which is $O(N)$ rather than $O(\sqrt{N})$. The sidelobe maxima are not $O(\sqrt{N})$ until x is approximately $1/\sqrt{N}$.

Suppose sensor positions are chosen randomly. That is, suppose $N_1, \ldots, N_N$ are independent identically distributed random variables with uniform distribution in the set of integers $\{1, 2, \ldots, Np\}$. Then for each x

$$A(x) = \sum_{k=1}^{N} \mathrm{e}^{2\pi i N_k x} \tag{27}$$

is a random variable with mean

$$E(A(x)) = N \frac{1}{Np} \sum_{n=1}^{Np} \mathrm{e}^{2\pi i n x} = \frac{\sin \pi N p x}{p \sin \pi x}. \tag{28}$$

Note that this expression is similar to the array factor

$$\frac{\sin \pi N p x}{\sin \pi p x} \tag{29}$$

for the uniformly spaced thin array but with the exception that $E(A(x))$ does not have grating lobes. Also $E(A(x))$ is a $1/p$-scaled version of the array factor for a filled array with $\Lambda = \{1, 2, \ldots, Np\}$.

These considerations motivate us to regard

$$\frac{\sin \pi N p x}{p \sin \pi x} \tag{30}$$

as the ideal model array factor that an N-sensor array with density $1/p$ seeks to achieve. We therefore formulate the precise mathematical problem:

> Given integers $N \geq 1$ and $p \geq 2$, find a function $C(p)$ and an algorithm which produces sets of integers $\{n_1, \ldots, n_N\}$ contained in $\{1, 2, \ldots, Np\}$ such that for all $-1/2 \leq x \leq 1/2$

$$\left| \sum_{k=1}^{N} \exp(2\pi i n_k x) \right| \leq \left| \frac{\sin \pi N p x}{p \sin \pi p x} \right| + C(p) N^{1/2}. \tag{31}$$

3 Golay–Rudin–Shapiro Polynomials

The Golay–Rudin–Shapiro polynomials are defined by

$$P_0(x) = 1, \qquad Q_0(x) = 1 \tag{32}$$

and

$$\begin{aligned} P_{n+1}(x) &= P_n(x) + \exp(2\pi i 2^n x)\, Q_n(x) \\ Q_{n+1}(x) &= P_n(x) - \exp(2\pi i 2^n x)\, Q_n(x). \end{aligned} \tag{33}$$

The coefficients of these polynomials are therefore

$$\begin{array}{rl} P_0: & 1 \\ Q_0: & 1 \\ P_1: & 1 \quad 1 \\ Q_1: & 1 \ -1 \\ P_2: & 1 \quad 1 \quad 1 \ -1 \\ Q_2: & 1 \quad 1 \ -1 \quad 1 \\ P_3: & 1 \quad 1 \quad 1 \ -1 \quad 1 \quad 1 \ -1 \quad 1 \\ Q_3: & 1 \quad 1 \quad 1 \ -1 \ -1 \ -1 \quad 1 \ -1 \end{array} \tag{34}$$

Note that

1. The coefficients of P_{n+1} and Q_{n+1} are obtained from the coefficients of P_n and Q_n by a simple "append" rule whereby the coefficients of Q_n are appended to the coefficients of P_n to produce the coefficients of P_{n+1} and the negative of the coefficients of Q_n are appended to the coefficients of P_n to produce the coefficients of Q_{n+1}.
2. The coefficients of P_n and Q_n are +1 or -1
3. P_n and Q_n have exactly 2^n coefficients
4. $\|P_n\|_2 = \|Q_n\|_2 = 2^{n/2}$
5. $|P_n(x)|^2 + |Q_n(x)|^2 = 2\left(|P_{n-1}(x)|^2 + |Q_{n-1}(x)|^2\right)$
6. $|P_n(x)|^2 + |Q_n(x)|^2 = 2^{n+1}$
7. $\|P_n\|_2 \le |P_n(x)| \le 2^{(n+1)/2} = 2^{1/2}\,\|P_n\|_2$
   $\|Q_n\|_2 \le |Q_n(x)| \le 2^{(n+1)/2} = 2^{1/2}\,\|Q_n\|_2$

The inequalities in item 3 above express the remarkable "flatness" property of the Golay–Rudin–Shapiro polynomials. This flatness property stems from the equality in item 3, and that inequality is the parallelogram law for the vectors $P_{n-1}(x)$ and $Q_{n-1}(x)$. Moreover the parallelogram law is an expression of the fact that the matrix $\begin{pmatrix} 1 & 1 \\ 1 & -1 \end{pmatrix}$ is $2^{1/2}$ times the unitary matrix $\begin{pmatrix} 1/\sqrt{2} & 1/\sqrt{2} \\ 1/\sqrt{2} & -1/\sqrt{2} \end{pmatrix}$. These connections are more apparent if basic Golay–Rudin–Shapiro recursion is expressed in matrix from as follows.

$$\begin{pmatrix} P_{n+1}(x) \\ Q_{n+1}(x) \end{pmatrix} = \begin{pmatrix} 1 & 1 \\ 1 & -1 \end{pmatrix} \begin{pmatrix} 1 & 0 \\ 0 & e^{2\pi i 2^n x} \end{pmatrix} \begin{pmatrix} P_n(x) \\ Q_n(x) \end{pmatrix}. \tag{35}$$

These observations suggest a generalization of the Golay–Rudin–Shapiro polynomials whereby the unitary 2×2 matrix $\begin{pmatrix} 1/\sqrt{2} & 1/\sqrt{2} \\ 1/\sqrt{2} & -1/\sqrt{2} \end{pmatrix}$ is replaced by a larger unitary matrix. This idea is more fully developed in [1] and forms the foundation for the developments in array design that follow. As an example, consider the case where the defining matrix is

$$M(r) = \begin{pmatrix} 1 & 1 & 1 \\ 1 & \omega & \omega^2 \\ 1 & \omega^2 & \omega^4 \end{pmatrix}, \tag{36}$$

where $\omega = \exp(2\pi i r/3)$. Note that $M(r)$ is $\sqrt{3}$ times a unitary matrix. Letting P_n, Q_n and R_n be defined by

$$P_0(x) = 1, \quad Q_0(x) = 1, \quad R_0(x) = 1, \tag{37}$$

$$\begin{aligned} P_{n+1}(x) &= P_n(x) + e^{2\pi i 3^n x} Q_n(x) + e^{2\pi i 2\cdot 3^n x} R_n(x) \\ Q_{n+1}(x) &= P_n(x) + \omega e^{2\pi i 3^n x} Q_n(x) + \omega^2 e^{2\pi i 2\cdot 3^n x} R_n(x) \\ Q_{n+1}(x) &= P_n(x) + \omega^2 e^{2\pi i 3^n x} Q_n(x) + \omega^4 e^{2\pi i 2\cdot 3^n x} R_n(x), \end{aligned} \tag{38}$$

it can be shown that

1. The coefficients of P_{n+1}, Q_{n+1} and R_{n+1} are obtained by and append rule similar to the classic Golay–Rudin–Shapiro case.
2. The coefficients of P_n, Q_n and R_n are third roots of unity.
3. P_n, Q_n and R_n have exactly 3^n coefficients.
4. $\|P_n\|_2 = \|Q_n\|_2 = \|R_n\|_2 = 3^{n/2}$
5. $|P_n(x)|^2 + |Q_n(x)|^2 + |R_n(x)|^2 = 3\left(|P_{n-1}(x)|^2 + |Q_{n-1}(x)|^2 + |R_n(x)|^2\right)$
6. $|P_n(x)|^2 + |Q_n(x)|^2 + |R_n(x)|^2 = 3^{n+1}$
7. $\|P_n\|_2 \le |P_n(x)| \le 3^{(n+1)/2} = 3^{1/2}\|P_n\|_2$
   $\|Q_n\|_2 \le |Q_n(x)| \le 3^{(n+1)/2} = 3^{1/2}\|Q_n\|_2$
   $\|R_n\|_2 \le |R_n(x)| \le 3^{(n+1)/2} = 3^{1/2}\|R_n\|_2$

In this example the polynomials are determined by the parameter r. Let $P_{n,r}$ be the polynomial P_n defined above using matrix $M(r)$. The coefficients of $P_{2,0}$, $P_{2,1}$ and $P_{2,2}$ are

$$\begin{aligned} &P_{2,0} : 1\ 1\ 1\ 1\ 1\ \ 1\ 1\ 1\ \ 1 \\ &P_{2,1} : 1\ 1\ 1\ 1\ \zeta\ \zeta^2\ 1\ \zeta^2\ \zeta^4 \\ &P_{2,2} : 1\ 1\ 1\ 1\ \zeta^2\ \zeta^4\ 1\ \zeta^4\ \zeta^8, \end{aligned} \tag{39}$$

where $\zeta = \exp(2\pi i/3)$. Next define

$$\begin{aligned} A_{2,0}(x) &= P_{2,0}(x) + P_{2,1}(x) + P_{2,2}(x) \\ A_{2,1}(x) &= P_{2,0}(x) + \alpha P_{2,1}(x) + \alpha^2 P_{2,2}(x) \\ A_{2,2}(x) &= P_{2,0}(x) + \alpha^2 P_{2,1}(x) + \alpha^4 P_{2,2}(x), \end{aligned} \tag{40}$$

where $\alpha = \exp(-2\pi i/3)$. Computing the $A_{2,r}$ we find the coefficients to be

$$\begin{aligned} A_{2,0} &: 3\ 3\ 3\ 3\ 0\ 0\ 3\ 0\ \ 0 \\ A_{2,1} &: 0\ 0\ 0\ 0\ 0\ 3\ 0\ 3\ \ 0 \\ A_{2,2} &: 0\ 0\ 0\ 0\ 3\ 0\ 0\ 0\ 3. \end{aligned} \tag{41}$$

Note that, remarkably, the coefficients in the $A_{2,r}$ are either 0 or 3. Furthermore, letting

$$S_{2,r} = \left\{ 0 \le k \le 8 \,|\, \hat{A}_{2,r}(k) = 3 \right\}, \tag{42}$$

where $\hat{A}_{2,r}(k)$ is the coefficient of $\exp(2\pi ikx)$ in $A_{2,r}$ we see that $S_{2,0}$, $S_{2,1}$ and $S_{2,2}$ form a partition of $\{0, 1, \ldots, 8\}$.

This example can be extended beyond the case 3×3 case to the $p \times p$ case where p is prime. It turns out that the sets $S_{n,r}$ define thin arrays having asymptotic density $1/p$ with optimal sidelobes. Proving this is the content of the next section.

4 A Thin Phased Array Design Algorithm

In this section we will solve the problem formulated at the end of Sect. 2. In particular we will use the theory of generalized Golay–Rudin–Shapiro polynomials [1] to accomplish this. Our solution is not completely general since we require the p to be a prime number.

Definition 1. For integers $p \ge 1$ and $n \ge 1$ let q and k be integers in the interval $0 \le q \le p^n - 1$ and $0 \le k \le p^n - 1$. Write

$$k = \omega_0 + \omega_1 p + \cdots + \omega_{n-1} p^{n-1} \tag{43}$$

and

$$q = \nu_0 + \nu_1 p + \cdots + \nu_{n-1} p^{n-1} \tag{44}$$

so that the n-tuples $(\omega_0, \ldots, \omega_{n-1})$ and $(\nu_0, \ldots, \nu_{n-1})$ are the base-p digits of k and q, respectively. Define ϕ_n by

$$\phi_n(q,k) = \sum_{j=0}^{n-1} \nu_{n-1-j}\omega_j + \sum_{j=0}^{n-2} \omega_j \omega_{j+1} \tag{45}$$

for $n \ge 2$ and $\phi_1(q,k) = \nu_0 \omega_0$.

Our thin array design algorithm is the following.

Algorithm 1. *Given an integer $n \geq 1$ and a prime $p \geq 2$,*

1. *Choose any $0 \leq q \leq p^n - 1$.*
2. *Choose any $0 \leq l \leq p - 1$.*
3. *Define $E(n,p,l,q)$ by*

$$E(n,p,l,q) = \{0 \leq k \leq p^n - 1 \,|\, \phi_n(q,k) \equiv l \bmod p\}. \tag{46}$$

The set $\Lambda = E(n,p,l,q)$ which determines the set of sensor positions of our array is a subset of the integer interval $0,1,\ldots,p^n-1$. While we cannot say exactly how many elements there are in Λ, we will show that the order is asymptotically p^{n-1} so that the array has asymptotic density $1/p$.

Definition 2. For p, n, k and q as in Definition 1 and $0 \leq r \leq p-1$ define the trigonometric polynomial

$$A_{r,n,q}(x) = \sum_{k=0}^{p^n-1} \exp\left(\frac{2\pi ir}{p}\phi_n(q,k)\right)\exp(2\pi ikx). \tag{47}$$

Theorem 1. *For p, k, q, and r as in Definition 2, and $n \geq 2$*

$$A_{r,n,q}(x) = \sum_{\lambda=0}^{p-1} \exp\left(2\pi i\frac{r\nu_0\lambda}{p}\right)\exp\left(2\pi i\lambda p^{n-1}x\right)A_{r,n-1,S(q,\lambda)}(x) \tag{48}$$

were

$$S(q,\lambda) = ((\nu_1+\lambda)\bmod p) + \nu_2 p + \nu_3 p^2 + \cdots + \nu_{n-1}p^{n-2} \tag{49}$$

is the integer between 0 and $p^{n-1}-1$ with base-p digits

$$((\nu_1+\lambda)\bmod p, \nu_2, \ldots, \nu_{n-1}). \tag{50}$$

Proof.

$$\begin{aligned}\phi_n(q,k) &= \nu_{n-1}\omega_0 + \nu_{n-2}\omega_1 + \cdots + \nu_1\omega_{n-2} + \nu_0\omega_{n-1} \\ &\quad + \omega_0\omega_1 + \cdots + \omega_{n-3}\omega_{n-2} + \omega_{n-2}\omega_{n-1}.\end{aligned} \tag{51}$$

Isolating the term $\nu_0\omega_{n-1}$ and combining the two terms involving ω_{n-2}, we obtain

$$\begin{aligned}\phi_n(q,k) &= \nu_{n-1}\omega_0 + \nu_{n-2}\omega_1 + \cdots + \nu_2\omega_{n-3} + (\nu_1+\omega_{n-1})\,\omega_{n-2} \\ &\quad + \omega_0\omega_1 + \cdots + \omega_{n-4}\omega_{n-3} + \omega_{n-3}\omega_{n-2} \\ &\quad + \nu_0\omega_{n-1}.\end{aligned} \tag{52}$$

Let $\phi_n^*(q,k)$ be defined by

$$\begin{aligned}\phi_n^*(q,k) = & \ \nu_{n-1}\omega_0 + \nu_{n-2}\omega_1 + \cdots + \nu_2\omega_{n-3} + ((\nu_1 + \omega_{n-1}) \bmod p)\,\omega_{n-2} \\ & + \omega_0\omega_1 + \cdots + \omega_{n-4}\omega_{n-3} + \omega_{n-3}\omega_{n-2} \\ & + \nu_0\omega_{n-1}.\end{aligned} \tag{53}$$

Since

$$\exp\left(2\pi i\frac{r}{p}(\nu_1 + \omega_{n-1})\,\omega_{n-2}\right) = \exp\left(2\pi i\frac{r}{p}((\nu_1 + \omega_{n-1}) \bmod p)\,\omega_{n-2}\right) \tag{54}$$

it follows that

$$\exp\left(2\pi i\frac{r}{p}\phi_n(q,k)\right) = \exp\left(2\pi i\frac{r}{p}\phi_n^*(q,k)\right). \tag{55}$$

Note that

$$\phi_n^*(q,k) = \phi_{n-1}(s,m) + \nu_0\omega_{n-1}, \tag{56}$$

where s and m are between 0 and $p^{n-1}-1$ with $n-1$ base-p digits

$$((\nu_1 + \omega_{n-1}) \bmod p, \nu_2, \nu_3, \ldots, \nu_{n-1}) \quad \text{for} \quad s \tag{57}$$

and

$$(\omega_0, \omega_1, \ldots, \omega_{n-2}) \quad \text{for} \quad m. \tag{58}$$

Next, writing $k = m + \lambda p^{n-1}$, where $\lambda = \omega_{n-1}$, we have

$$\begin{aligned}A_{r,n,q}(x) &= \sum_{k=0}^{p^n-1} \exp\left(2\pi i\frac{r}{q}\phi_n(q,k)\right)\exp(2\pi ikx) \\ &= \sum_{\lambda=0}^{p-1}\sum_{m=0}^{p^{n-1}-1} \exp\left(2\pi i\frac{r}{p}\phi_n\left(q, m+\lambda p^{n-1}\right)\right) \\ &\quad \times \exp(2\pi imx)\exp\left(2\pi i\lambda p^{n-1}x\right).\end{aligned} \tag{59}$$

However

$$\begin{aligned}\exp\left(2\pi i\frac{r}{p}\phi_n\left(q, m+\lambda p^{n-1}\right)\right) &= \exp\left(2\pi i\frac{r}{p}\phi_n(q,k)\right) \\ &= \exp\left(2\pi i\frac{r}{p}\phi_n^*(q,k)\right)\end{aligned}$$

$$= \exp\left(2\pi i \frac{r}{p}\left(\phi_{n-1}(s,m) + \nu_0\lambda\right)\right)$$
$$= \exp\left(2\pi i \frac{r}{p}\left(\phi_{n-1}(s,m)\right)\right)$$
$$\times \exp\left(2\pi i \frac{r}{p}\nu_0\lambda\right). \tag{60}$$

Hence

$$A_{r,n,q}(x) = \sum_{\lambda=0}^{p-1} \exp\left(2\pi i \frac{r}{p}\nu_0\lambda\right)\exp\left(2\pi i\lambda p^{n-1}x\right)$$
$$\times \sum_{m=0}^{p^{n-1}-1} \exp\left(2\pi i \frac{r}{p}\phi_{n-1}(s,m)\right)\exp\left(2\pi imx\right)$$
$$= \sum_{\lambda=0}^{p-1} \exp\left(2\pi i \frac{r}{p}\nu_0\lambda\right)\exp\left(2\pi i\lambda p^{n-1}x\right)A_{r,n-1,s}(x). \tag{61}$$

□

Theorem 2. *For p,n, and k as above, let $1 \le r \le p-1$ and let p be prime. Then*

$$\sum_{\nu_0=0}^{p-1} \left|A_{r,n,q}(x)\right|^2 = p\sum_{\nu_1=0}^{p-1} \left|A_{r,n-1,q'}(x)\right|^2, \tag{62}$$

where $q = \nu_0 + \nu_1 p + \cdots + \nu_{n-1}p^{n-1}$ and $q' = \nu_1 + \nu_2 p + \cdots + \nu_{n-1}p^{n-2}$.

Proof. By Theorem 1

$$\sum_{\nu_0=0}^{p-1} \left|A_{r,n,q}(x)\right|^2$$
$$= \sum_{\nu_0=0}^{p-1}\sum_{\lambda=0}^{p-1}\sum_{\lambda'=0}^{p-1} \exp\left(2\pi i \frac{r}{p}\nu_0\lambda\right)\exp\left(2\pi i\lambda p^{n-1}x\right)A_{r,n-1,S(q,\lambda)}(x)$$
$$\times \exp\left(2\pi i \frac{r}{p}\nu_0\lambda'\right)\exp\left(2\pi i\lambda' p^{n-1}x\right)A_{r,n-1,S(q,\lambda')}(x)$$
$$= \sum_{\lambda=0}^{p-1}\sum_{\lambda'=0}^{p-1}\left(\sum_{\nu_0=0}^{p-1} \exp\left(2\pi i \frac{r}{p}\nu_0\left(\lambda-\lambda'\right)\right)\right)\exp\left(2\pi i\left(\lambda-\lambda'\right)p^{n-1}x\right)$$
$$\times A_{r,n-1,S(q,\lambda)}(x)\overline{A_{r,n-1,S(q,\lambda')}(x)}. \tag{63}$$

Consider the p-th roots of unity

$$\exp\left(2\pi i \frac{r(\lambda-\lambda')\,\nu_0}{p}\right) \tag{64}$$

as ν_0 runs from 0 to $p-1$. Since $1 \le r \le p-1$ and $-(p-1) \le \lambda - \lambda' \le (p-1)$ and p is prime, $r(\lambda-\lambda')$ is not a multiple of p unless $\lambda - \lambda' = 0$. Hence if $\lambda \ne \lambda'$ the numbers

$$\exp\left(2\pi i \frac{r(\lambda-\lambda')\,\nu_0}{p}\right) \tag{65}$$

constitute a full set of the p-th roots of unity as ν_0 runs from 0 to $p-1$. Therefore

$$\sum_{\nu_0=0}^{p-1} \exp\left(2\pi i \frac{r(\lambda-\lambda')\,\nu_0}{p}\right) = \begin{cases} p & if \quad \lambda = \lambda' \\ 0 & if \quad \lambda \ne \lambda'. \end{cases} \tag{66}$$

Consequently the right side of Eq. (63) becomes

$$p \sum_{\lambda=0}^{p-1} \left|A_{r,n-1,S(q,\lambda)}(x)\right|^2. \tag{67}$$

However, as λ runs from 0 to $p-1$, $(\nu_1+\lambda) \bmod p$ also takes on all values from 0 to $p-1$. Therefore the previous sum can also be written as

$$p \sum_{\nu_1=0}^{p-1} \left|A_{r,n-1,q'}(x)\right|^2, \tag{68}$$

where $q' = \nu_1 + \nu_2 p + \cdots + \nu_{n-1} p^{n-2}$. □

Corollary 1. *Let p,n, and k be as above, let $1 \le r \le p-1$ and let p be prime. Fix digits $\nu_1, \nu_2, \ldots, \nu_{n-1}$, let ν_0 run from 0 to $p-1$ and set $q = \nu_0 + \nu_1 p + \cdots + \nu_{n-1} p^{n-1}$. Then*

$$\sum_{\nu_0=0}^{p-1} \left|A_{r,n,q}(x)\right|^2 = p^{n+1}. \tag{69}$$

Proof. Repeated application of Theorem 2 gives

$$\sum_{\nu_0=0}^{p-1} \left|A_{r,n,q}(x)\right|^2 = p^{n-1} \sum_{\lambda=0}^{p-1} \left|A_{r,1,\lambda}(x)\right|^2. \tag{70}$$

Since

$$A_{r,1,\lambda}(x) = \sum_{k=0}^{p-1} \exp\left(\frac{2\pi i r \lambda k}{p}\right) \exp(2\pi i k x) \tag{71}$$

we have

$$\sum_{\lambda=0}^{p-1} \left|A_{r,1,\lambda}(x)\right|^2 = \sum_{\lambda=0}^{p-1}\sum_{k=0}^{p-1}\sum_{k'=0}^{p-1} \exp\left(\frac{2\pi i r\lambda\,(k-k')}{p}\right)\exp\left(2\pi i\,(k-k')\,x\right)$$
$$= \sum_{k=0}^{p-1}\sum_{k'=0}^{p-1}\left(\sum_{\lambda=0}^{p-1}\exp\left(\frac{2\pi i r\lambda\,(k-k')}{p}\right)\right)$$
$$\times \exp\left(2\pi i\,(k-k')\,x\right). \tag{72}$$

As in the proof of Theorem 2, the inner sum vanishes unless $k = k'$ in which case it equals p. Therefore, the right side of Eq. (72) becomes p^2. Using this in Eq. (70) gives Eq. (69). □

Corollary 2. *Let p,n and k be as above, let $1 \le r \le p-1$, and let p be prime. Then for $0 \le q \le p^n - 1$*

$$\left|A_{r,n,q}(x)\right| \le p^{(n+1)/2} = p^{1/2}\left\|A_{r,n,q}\right\|_2. \tag{73}$$

Proof. Choosing a single term in the sum in Eq. (69) yields the first inequality. Since the coefficients in $A_{r,n,q}(x)$ all have modulus 1 and $A_{r,n,q}(x)$ has p^n terms, Parseval's theorem gives

$$\left\|A_{r,n,q}\right\|_2 = p^{n/2}. \tag{74}$$

□

Definition 3. For integers $n \ge 1$, $p \ge 2$, $0 \le q \le p^n - 1$ and $0 \le l \le p-1$, define

$$H_{l,n,q}(x) = \frac{1}{p}\sum_{r=0}^{p-1}\exp\left(\frac{-2\pi i r l}{p}\right)A_{r,n,q}(x). \tag{75}$$

Theorem 3. *For integers $n \ge 1$, prime $p \ge 2$, $0 \le q \le p^n - 1$ and $0 \le l \le p-1$ write*

$$H_{l,n,q}(x) = \sum_{\iota=0}^{p^n-1} h_{l,n,q}(k)\exp\left(2\pi i k x\right). \tag{76}$$

then

$$h_{l,n,q}(k) = \begin{cases} 1 & \text{if} \quad \phi_n(q,k) \equiv l \bmod p, \\ 0 & \text{if} \quad \phi_n(q,k) \not\equiv l \bmod p. \end{cases} \tag{77}$$

Proof. From the definition of $A_{r,n,q}(x)$ we have

$$H_{l,n,q}(x) = \frac{1}{p}\sum_{r=0}^{p-1}\sum_{k=0}^{p^n-1} \exp\left(\frac{-2\pi irl}{p}\right)\exp\left(2\pi i\frac{r}{p}\phi_n(q,k)\right)\exp(2\pi ikx)$$
$$= \sum_{k=0}^{p^n-1}\left(\frac{1}{p}\sum_{r=0}^{p-1}\exp\left(2\pi i\frac{r}{p}\left(\phi_n(q,k)-l\right)\right)\right)\exp(2\pi ikx). \quad (78)$$

The inner sum equals 1 if $\phi_n(q,k)-l$ is a multiple of p and equals 0 otherwise. □

Theorem 4. *For n, p, q and l as in Theorem 3, and p prime*

$$\left|H_{l,n,q}(x)\right| \leq \left|\frac{\sin \pi p^n x}{p \sin \pi x}\right| + (p-1)p^{(n-1)/2}. \quad (79)$$

Proof.

$$\left|H_{l,n,q}(x)\right| = \left|\frac{1}{p}\sum_{r=0}^{p-1}\exp\left(\frac{-2\pi irl}{p}\right)A_{r,n,q}(x)\right|$$
$$= \left|\frac{1}{p}A_{0,n,q}(x) + \frac{1}{p}\sum_{r=1}^{p-1}\exp\left(\frac{-2\pi irl}{p}\right)A_{r,n,q}(x)\right|$$
$$\leq \frac{1}{p}\left|A_{0,n,q}(x)\right| + \frac{1}{p}\sum_{r=1}^{p-1}\left|A_{r,n,q}(x)\right|. \quad (80)$$

Note that

$$A_{0,n,q}(x) = \sum_{k=0}^{p^n-1}\exp(2\pi ikx) \quad (81)$$

and therefore

$$\left|A_{0,n,q}(x)\right| = \left|\frac{\sin \pi p^n x}{p \sin \pi x}\right|. \quad (82)$$

This fact together with the estimate in Corollary 2 give the result. □

Definition 4. Let n, p, q and l be integers as in Theorem 3. Define the set

$$S_{l,n,q} = \left\{0 \leq k \leq p^n - 1 \,|\, h_{l,n,q}(k) = 1\right\}. \quad (83)$$

Theorem 5. *Let n, p, q and l be integers as in Theorem 3, and let* $\left|S_{l,n,q}\right|$ *denote the number of elements in the set* $S_{l,n,q}$*. Then*

$$p^{n-1} - (p-1)^2 p^{(n-1)/2} \leq \left|S_{l,n,q}\right| \leq p^{n-1} + (p-1)p^{(n-1)/2}. \quad (84)$$

Proof. Since

$$H_{l,n,q}(x) = \sum_{k \in S_{l,n,q}} \exp(2\pi ikx) \tag{85}$$

it follows that

$$|S_{l,n,q}| = H_{l,n,q}(0). \tag{86}$$

Using the estimate from Theorem 4 gives

$$H_{l,n,q}(0) \le p^{n-1} + (p-1)p^{(n-1)/2} \tag{87}$$

which therefore establishes the upper bound for $|S_{l,n,q}|$. From the definition of $S_{l,n,q}$ and the representation of $h_{l,n,q}(k)$ given in Theorem 3, it follows that the sets $S_{l,n,q}$ for $0 \le l \le p-1$ form a partition of the integer interval $\{0,1,\ldots,p^n-1\}$. Therefore

$$p^n = \sum_{l=0}^{p-1} |S_{l,n,q}| \tag{88}$$

which for any fixed choice of l can be rearranged to be

$$|S_{l,n,q}| = p^n - \sum_{j \neq l} |S_{j,n,q}|. \tag{89}$$

Using the upper bound shown in the earlier part of this proof gives

$$\begin{aligned} |S_{l,n,q}| &\ge p^n - (p-1)\left(p^{n-1} + (p-1)p^{(n-1)/2}\right) \\ &= p^{n-1} - (p-1)^2 p^{(n-1)/2}. \end{aligned} \tag{90}$$

□

References

1. Benke, G.: Generalized Rudin-Shapiro systems. J. Fourier Anal. Appl. **1**(1), 87–101 (1994)
2. Benke, G., Hendricks, W.J.: Estimates for large deviations in random trigonometric polynomials. Siam. J. Math. Anal. **24**, 1067–1085 (1993)
3. Brillhart, J., Lomont, J.S., Morton, P.: Cyclotomic properties of Rudin-Shapiro polynomials. J. Reine Angew. Math. **288**, 37–65 (1976)
4. Byrnes, J.S.: Quadrature mirror filters, low crest factor arrays, functions achieving optimal uncertainty principle bounds, and complete orthonormal sequences—a unified approach. J. Comp. Appl. Harmonic Anal. **1**, 261–266 (1994)
5. Eliahou, S., Kervaire, M., Saffari, B.: A new restriction on the lengths of Golay complementary sequences. J. Combin. Theor. Series A **55**, 49–59 (1990)
6. Golay, M.J.E.: Multislit spectrometry. J. Opt. Soc. Amer. **39**, 437 (1949)
7. Golay, M.J.E.: Complementary series. IRE Trans. Inform. Theor. **IT-7**, 82–87 (1961)

8. Habib, I., Turner, L.: New class of M-ary communication systems using complementary sequences. IEEE Proceedings **3**, 293–300 (1986)
9. Hendricks, W.J.: The totally random versus bin approach to random arrays. IEEE Trans. Ant. Prop. **39**, 1757–1762 (1991)
10. Kahane, J.P.: Sur les polynomes a coefficients unimodulaires. Bull. Lond. Math. Soc. **12**, 321–342 (1980)
11. Kopilovich, L.E., Sodin, L.G.: Linear non-equidistant antenna arrays based on difference sets. Sov. J. Comm. Tech. Electron. **35** (7) 42–49 (1990)
12. Leeper, D.G.: Isophoric arrays – massively thinned phased arrays with well-controlled sidelobes. IEEE Ant. Prop. **47**, 1825–1835 (1999)
13. Lo, Y.T., Agrawal, V.D.: Distribution of sidelobe level in random arrays. Proc. IEEE **57**, 1764–1765 (1969)
14. Mendes-France, M., Tenenbaum, G.: Dimension des courbes planes, papiers plies et suite de Rudin-Shapiro. Bull. Soc. Math. France **109**(2), 207–215 1981
15. Newman, D.J., Byrnes, J.S.: The L^4 norm of a polynomial with coefficients ± 1. The Amer. Math. Monthly **97** (1) 42–45 (1990)
16. Rudin, W.: Some theorems on Fourier coefficients. Proc. Amer. Math. Soc. **10**, 855–859 (1959)
17. Shapiro, H.S.: Extremal problems for polynomials and power series. Thesis, Massachusetts Institute of Technology (1957)
18. Sivaswamy, R.: Multiphase complementary codes. IEEE Trans. Inform. Theor. **IT-24**(5), 546–552 (1978)
19. Steinberg, B.D.: The peak sidelobe of the phased array having randomly located elements. IEEE Trans. Ant. Prop. **AP-20**, 129–136 (1972)
20. Steinberg, B.D.: Principles of Aperture and Array System Design. Wiley, New York (1976)
21. Taki, Y., Miyakawa, H., Hatori, M., Namba, S.: Even-shift orthogonal sequences. IEEE Trans. Inform. Theor. **IT-15**(2), 295–300 (1969)
22. Tseng, C.C., Lui, C.L.: Complementary sets of sequences. IEEE Trans. Inform. Theor. **IT-18**(5), 664–651 (1972)
23. Werner, D.H., Haupt, R.L.: Fractal constructions of linear and planar arrays. Ant. Prop. Int. Symp. 1997. IEEE 1997 Digest' **3**, 1968–1971 (1997)
24. Zhaofei, Cheng, N.: Genetic algorithm in the design of thinned arrays with low sidelobe levels. International Conference on Wireless Communications and Signal Processing, 2009. WCSP 2009, 1–4 (2009)
25. Zhang, L., Jiao, Y.C., Weng, Z.B., Zhang, F.S.: Design of planar thinned arrays using a Boolean differential evolution algorithm. Microwaves, Ant. Prop. **4**(12), 2172–2178 (2010)

Multi-Resolution Geometric Analysis for Data in High Dimensions

Guangliang Chen, Anna V. Little, and Mauro Maggioni

Abstract Large data sets arise in a wide variety of applications and are often modeled as samples from a probability distribution in high-dimensional space. It is sometimes assumed that the support of such probability distribution is well approximated by a set of low intrinsic dimension, perhaps even a low-dimensional smooth manifold. Samples are often corrupted by high-dimensional noise. We are interested in developing tools for studying the geometry of such high-dimensional data sets. In particular, we present here a multiscale transform that maps high-dimensional data as above to a set of multiscale coefficients that are compressible/sparse under suitable assumptions on the data. We think of this as a geometric counterpart to multi-resolution analysis in wavelet theory: whereas wavelets map a signal (typically low dimensional, such as a one-dimensional time series or a two-dimensional image) to a set of multiscale coefficients, the geometric wavelets discussed here map points in a high-dimensional point cloud to a multiscale set of coefficients. The geometric multi-resolution analysis (GMRA) we construct depends on the support of the probability distribution, and in this sense it fits with the paradigm of dictionary learning or data-adaptive representations, albeit the type of representation we construct is in fact mildly nonlinear, as opposed to standard linear representations. Finally, we apply the transform to a set of synthetic and real-world data sets.

Keywords Multiscale analysis • Geometric analysis • High-dimensional data • Covariance matrix estimation

---

G. Chen • A.V. Little • M. Maggioni (✉)
Mathematics and Computer Science Departments, Duke University,
P.O. Box 90320, Durham, NC 27708, USA
e-mail: mauro.maggioni@duke.edu

T.D. Andrews et al. (eds.), *Excursions in Harmonic Analysis, Volume 1,*
Applied and Numerical Harmonic Analysis, DOI 10.1007/978-0-8176-8376-4_13,

1 Introduction

We are interested in developing tools for harmonic analysis and processing of large data set that arise in wide variety of applications, such as sounds, images (RGB or hyperspectral, [16]), gene arrays, EEG signals [9], and manifold-valued data [44], to name a few. These data sets are often modeled as samples from a probability distribution in $\mathbb{R}^D$, but it is sometimes assumed that the support of such probability distribution is in fact a set of low intrinsic dimension, perhaps with some nice geometric properties, for example, those of a smooth manifold.

Approximating and learning functions in high-dimensional spaces is hard because of the curse of high dimensionality, it is natural to try to exploit the intrinsic low dimensionality of the data: this idea has attracted wide interest across different scientific disciplines and various applications. One example of exploitation of low intrinsic dimension is to map the data to low-dimensional space, while preserving salient properties of data [3, 19, 21, 27, 28, 30, 31, 46, 52, 54]. Another example is the construction of dictionaries of functions supported on the data [7, 17, 18, 38–40, 49, 50]. Yet another possibility is modeling the data as a union of low-dimensional subspaces, which is related to the ideas of sparse representations and dictionary learning ([1, 2, 10, 11, 51] and references therein).

When performing dimensionality reduction/manifold learning, the objective is mapping data to a low-dimensional space. The maps used are often nonlinear, and in at least two problems arise: that of extending the map from a training data set to new data points and that of inverting such a map, i.e., going from a low-dimensional representation of a data point back to its higher-dimensional original representation. Both problems seem to be rather hard (depending of course of the map used) and to require some form of high-dimensional interpolation/extrapolation.

We will work directly in the high-dimensional space, but by taking advantage of the assumed low intrinsic dimensionality of the data and its geometry. One advantage of this approach is that while our representations will be low-dimensional, we will not have to produce inverse maps from low dimensions to high dimensions. We construct geometric multi-resolution analysis (GMRA) for analyzing intrinsically low-dimensional data in high-dimensional spaces, modeled as samples from a d-dimensional set $\mathcal{M}$ (in particular, a manifold) embedded in $\mathbb{R}^D$, in the regime $d \ll D$. Data may be sampled from a class of signals of interest; in harmonic analysis, a linear infinite-dimensional function space $\mathcal{F}$ often models the class of signals of interest, and linear representations in the form $f = \sum_i \alpha_i \phi_i$, for $f \in \mathcal{F}$ in terms of a dictionary of atoms $\Phi := \{\phi_i\} \subseteq \mathcal{F}$ are studied. Such dictionaries may be bases or frames and are constructed so that the sequence of coefficients $\{\alpha_i\}_i$ has desirable properties, such as some form of sparsity, or a distribution highly concentrated at zero. Several such dictionaries have been constructed for function classes modeling one- and two-dimensional signals of interest [8, 12, 14, 20, 22, 47] and are proven to provide optimal representations (in a suitably defined sense) for certain function spaces and/or for operators on such spaces. A more recent trend [1, 12, 41–43, 51, 55], motivated by the desire to model classes of signals that are not well modeled by the linear structure of function spaces, has been that of *constructing data-adapted dictionaries*: an algorithm is allowed to see samples from a class of signals $\mathscr{F}$ (not necessarily a linear function space) and constructs a dictionary $\Phi := \{\phi_i\}_i$ that optimizes some functional, such as the sparsity of the coefficients for signals in $\mathscr{F}$.

There are several parameters in this problem: given training data from $\mathscr{F}$, one seeks Φ with I elements, such that every element in the training set may be represented, up to a certain precision ε, by at most m elements of the dictionary. The smaller I and m are, for a given ε, the better the dictionary.

Several current approaches may be summarized as follows [42]: consider a finite training set of signals $X_n = \{x_i\}_{i=1}^n \subset \mathbb{R}^D$, which we may represent by a $\mathbb{R}^{D\times n}$ matrix, and optimize the cost function

$$f_n(\Phi) = \frac{1}{n}\sum_{i=1}^{n} \ell(x_i, \Phi),$$

where $\Phi \in \mathbb{R}^{D\times I}$ is the dictionary, and ℓ a loss function, for example,

$$\ell(x, \Phi) := \min_{\alpha\in\mathbb{R}^I} \frac{1}{2}||x - \Phi\alpha||_{\mathbb{R}^D}^2 + \lambda||\alpha||_1,$$

where λ is a regularization parameter. This is basis pursuit [12] or lasso [53]. One typically adds constraints on the size of the columns of Φ, for example, $||\phi_i||_{\mathbb{R}^D} \leq 1$ for all i, which we can write as $\Phi \in \mathscr{C}$ for some convex set $\mathscr{C}$. The overall problem may then be written as a matrix factorization problem with a sparsity penalty:

$$\min_{\Phi\in\mathscr{C},\alpha\in\mathbb{R}^{I\times n}} \frac{1}{2}||X_n - \Phi\alpha||_F^2 + \lambda||\alpha||_{1,1},$$

where $||\alpha||_{1,1} := \sum_{i_1,i_2} |\alpha_{i_1,i_2}|$. We refer the reader to [42] and references therein for techniques for attacking this optimization problem.

In this chapter we make additional assumptions on the data, specifically that it is well approximated by a smooth low-dimensional manifold, and we exploit this geometric assumption to construct data-dependent dictionaries. We use a multiscale approach that will lead to a GMRA of the data: this is inspired not only by quantitative geometric analysis techniques in geometric measure theory (see, e.g., [25, 32]) but also from multiscale approximation of functions in high dimensions [5, 6]. These dictionaries are structured in a multiscale fashion and, under suitable assumptions on the data, are computed efficiently; the expansion of a data point on the dictionary elements is guaranteed to have a certain degree of sparsity, m, and may also be computed by fast algorithms; the growth of the number of dictionary elements I as a function of ε is controlled depending on geometric properties of the data. This may be thought of as a wavelet analysis for data sets rather than for functions, where the geometry of a set of points is approximated, rather than a single function.

2 Geometric Multi-resolution Analysis

Let μ be a probability measure in $\mathbb{R}^D$ and $\mathscr{M}$ its support. In this chapter we will consider the case in which $\mathscr{M}$ is endowed with the structure of a Riemannian manifold, but the examples will show that the construction is robust enough to extend and be useful when this assumption is severely violated. In this setting we have a Riemannian metric g and a volume measure $d$vol. The geodesic distance on $\mathscr{M}$ associated with g will be denoted by ρ. We shall assume that $d\mu$ is absolutely continuous with respect to $d$vol, with $d\mu/d$vol bounded above and below. We are interested in the case when the "dimension" d of $\mathscr{M}$ is much smaller than the dimension of the ambient space $\mathbb{R}^D$. While d is typically unknown in practice, efficient (multiscale, geometric) algorithms for its estimation are available (see [37], which also contains many references to previous work on this problem), under additional assumptions on the geometry of $\mathscr{M}$.

2.1 *Dyadic Cubes*

We start by constructing *dyadic cubes* on $\mathscr{M}$. This may be thought of as an analogue of dyadic cubes in Euclidean space. It is a collection of (measurable) subsets $\{Q_{j,k}\}_{k\in\mathscr{K}_j, j\geq J_0}$ of $\mathscr{M}$ with the following properties [13, 23, 24]:

- For every $j\in\mathbb{Z}$, $\mu(\mathscr{M}\setminus\cup_{k\in\mathscr{K}_j}Q_{j,k})=0$.
- For $j'\geq j$ and $k'\in\mathscr{K}_{j'}$, either $Q_{j',k'}\subseteq Q_{j,k}$ or $\mu(Q_{j',k'}\cap Q_{j,k})=0$.
- For $j<j'$ and $k'\in\mathscr{K}_{j'}$, there exists a unique $k\in\mathscr{K}_j$ such that $Q_{j',k'}\subseteq Q_{j,k}$.
- Each $Q_{j,k}$ contains a point $c_{j,k}$ such that $B^{\mathscr{M}}_{c_1\cdot 2^{-j}}(c_{j,k})\subseteq Q_{j,k}\subseteq B^{\mathscr{M}}_{2^{-j}}(c_{j,k})$, for a constant c_1 depending on intrinsic geometric properties of $\mathscr{M}$. Here $B^{\mathscr{M}}_r(x)$ is the ρ-ball inside $\mathscr{M}$ of radius $r>0$ centered at $x\in\mathscr{M}$. In particular, we have $\mu(Q_{j,k})\sim 2^{-dj}$.

Let $\mathscr{T}$ be the tree structure associated to the decomposition above: for any $j\in\mathbb{Z}$ and $k\in\mathscr{K}_j$, we let $\mathrm{ch}(j,k)=\{k'\in\mathscr{K}_{j+1}:Q_{j+1,k'}\subseteq Q_{j,k}\}$. We use the notation (j,x) to represent the unique $(j,k(x)), k(x)\in\mathscr{K}_j$ such that $x\in Q_{j,k(x)}$.

2.2 *Multiscale SVD and Intrinsic Dimension Estimation*

An introduction to the use of the ideas we present for the estimation of intrinsic dimension of point clouds is in [37] and references therein (see [35, 36] for previous short accounts). These types of constructions are motivated by ideas in both multiscale geometric measure theory [24, 26, 33] and adaptive approximation of functions in high dimensions[5, 6].

In each dyadic cell $Q_{j,k}$ we consider the mean

$$m_{j,k} := \mathbb{E}_\mu[x|x \in Q_{j,x}] = \frac{1}{\mu(Q_{j,k})}\int_{Q_{j,k}} x\,\mathrm{d}\mu(x) \in \mathbb{R}^D \tag{1}$$

and the local covariance

$$\mathrm{cov}_{j,k} = \mathbb{E}_\mu[(x-m_{j,k})(x-m_{j,k})^*|x \in Q_{j,k}] \in \mathbb{R}^{D\times D}, \tag{2}$$

where vectors in $\mathbb{R}^D$ are considered d-dimensional column vectors. Let the rank-d singular value decomposition (SVD) [29] of $\mathrm{cov}_{j,k}$ be

$$\mathrm{cov}_{j,k} \approx \Phi_{j,k}\Sigma_{j,k}\Phi_{j,k}^*, \tag{3}$$

where $\Phi_{j,k}$ is an orthonormal $D\times d$ matrix and Σ is a diagonal $d\times d$ matrix. Let

$$\mathbb{V}_{j,k} := V_{j,k} + m_{j,k}, \quad V_{j,k} = \langle\Phi_{j,k}\rangle, \tag{4}$$

where $\langle A\rangle$ denotes the span of the columns of A, so that $\mathbb{V}_{j,k}$ is the affine subspace of dimension d parallel to $V_{j,k}$ and passing through $m_{j,k}$. It is an approximate tangent space to $\mathcal{M}$ at location $m_{j,k}$ and scale 2^{-j}; in fact by the properties of the SVD it provides the best $d_{j,k}$-dimensional planar approximation to $\mathcal{M}$ in the least squares sense:

$$\mathbb{V}_{j,k} = \underset{\Pi}{\mathrm{argmin}} \int_{Q_{j,k}} ||x - \mathbb{P}_\Pi(x)||^2\,\mathrm{d}\mu(x), \tag{5}$$

where Π is taken on the set of all affine $d_{j,k}$-planes and $\mathbb{P}_\Pi$ is the orthogonal projection onto the affine plane Π. Let $\mathbb{P}_{j,k}$ be the associated affine projection

$$\mathbb{P}_{j,k}(x) := \mathbb{P}_{\mathbb{V}_{j,k}}(x) = \Phi_{j,k}\Phi_{j,k}^*(x-m_{j,k}) + m_{j,k}, \quad x \in Q_{j,k}. \tag{6}$$

The behavior of the singular values in the matrix $\Sigma_{j,x}$ in Eq. (3) as a function of the scale j, for x fixed, contains a lot of useful information about the geometry of the data around x. In particular they may be used to detect the intrinsic dimension of the data in a neighborhood of x. We need to introduce several definition before stating some results. Because of space constraints, we will consider here the case when $\mathcal{M}$ is a manifold of co-dimension one, leaving the discussion of the general case ($\mathcal{M}$ with arbitrary co-dimension and $\mathcal{M}$ not a manifold) to [2, 37]. Let

$$\lambda = \frac{d}{d+2}, \qquad \kappa = \frac{d}{(d+2)^2(d+4)}\left[\frac{d+1}{2}\sum_{i=1}^{d}\kappa_i^2 - \sum_{i<j}\kappa_i\kappa_j\right],$$

where κ_i's are the sectional curvatures of the manifold. We refer the reader to [37] for an extended discussion of these quantities, which arise naturally in the study of multiscale SVD of manifolds. When $\mathcal{M}$ has co-dimension larger than 1 more complicate functions of the curvatures arise [similar to those in Eq. (18)]; in the non-manifold case a notion of L^2 that generalizes the above may be used [37]. Suppose we sample n points $x_1,\dots,x_n$ i.i.d. from the volume measure on the manifold, and each is perturbed by i.i.d. realizations of white Gaussian noise in $\mathbb{R}^D$ with variance $\sigma^2 I_D$. We denote by $\tilde{X}_{n,\tilde{z},r}$ the set of noisy samples $x_i+\eta_i$ that are in $B_{z+\eta_z}(r)$, where η_z is the noise corresponding to the data point z: this is the data being observed, which is sampled and noisy, at disposal of an algorithm. We denote by $X_{z,r}$ a random variable distributed in $B_z(r)\cap\mathcal{M}$ according to volume measure: this is the ideal data, uncorrupted by noise and sampling. Finally, we let $r_=^2 := r^2 - 2\sigma^2 D$.

Let $r = 2^{-j}$ and $X_{z,r} = \mathcal{M}\cap B_z(r)$. The behavior of the ideal covariance of $X_{z,r}$ (which is comparable to $\mathrm{cov}_{j,k}$) as a function of r reveals interesting properties of the data, for example, it may be used to measure intrinsic dimension and L^2-curvature of $\mathcal{M}$ around a point z, since the d largest singular values will grow quadratically in r, and the remaining ones will measure L^2-curvatures. In particular for r small the largest gap between these singular values will be the dth gap, leading to an estimator of intrinsic dimension. However, since we do not have access to $X_{z,r}$, we are interested in the behavior of the empirical covariance matrix of the noisy samples $\tilde{X}_{n,\tilde{z},r}$ as a function of r. In particular, we ask how close it is to $\mathrm{cov}(X_{z,r})$ and when is the dth gap of $\mathrm{cov}(\tilde{X}_{n,\tilde{z},r})$ the largest, so that we may use it to estimate the intrinsic dimension of $\mathcal{M}$? Observe that while we would like to choose r small, since then the difference in the behavior of the top d singular values and the remaining ones is largest, we are not allowed to do that anymore: having only n samples forces a lower bound on r, since in small balls we will have too small a number of samples to estimate the covariances. Moreover, the presence of noise also puts a lower bound on the interesting range of r: since the expected length of a noise vector is $\sigma\sqrt{D}$, and the covariance of the noise has norm σ, we expect that r should be larger than a function of these quantities in order for $\mathrm{cov}(\tilde{X}_{n,\tilde{z},r})$ to provide meaningful information about the geometric of $\mathcal{M}$.

Here and in what follows C, C_1, and C_2 will denote numerical constants whose value may change with each occurrence.

Theorem 1 ($n\to\infty$). *Fix $z\in\mathcal{M}$; assume $D\geq C$, $\sigma\sqrt{D}\leq\frac{\sqrt{d}}{2\sqrt{2}\kappa}$,*

$$r\in\left(R_{\min}+4\sigma\sqrt{D}+\frac{1}{6\kappa}, R_{\max}-\sigma\sqrt{D}-\frac{1}{6\kappa}\right)\cap\left(3\sigma\left(\sqrt{D}\vee d\right),\frac{\sqrt{d}}{\kappa}\right). \quad (7)$$

Then for n large enough, with probability at least $1-Ce^{-C\sqrt{D}}$, we have

$$||\mathrm{cov}(\tilde{X}_{n,\tilde{z},r})-\mathrm{cov}(X_{z,r_=})||\leq C\left(\frac{\kappa^2 r_=^4}{d}+\sigma^2+\frac{\lambda\kappa r_=^3}{d}\left(\frac{\lambda\kappa r_=}{\lambda^2-C\kappa^2 r_=^2}\wedge 1\right)\right). \quad (8)$$

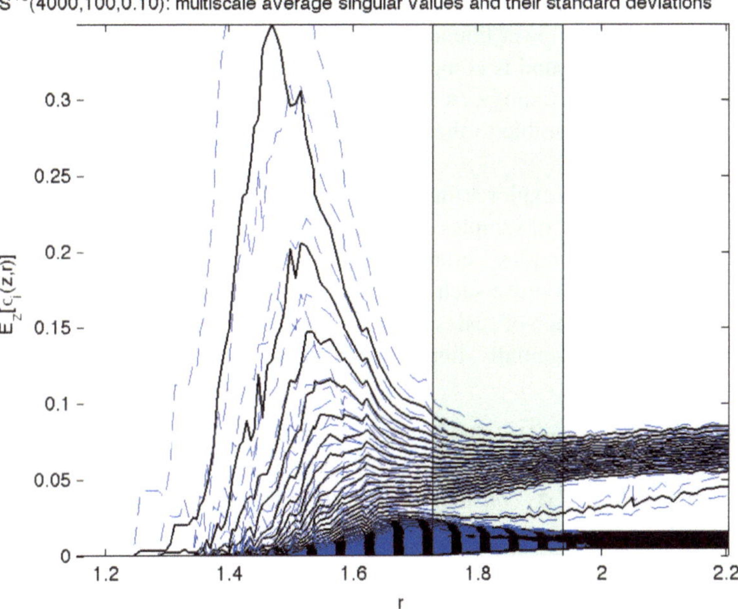

Fig. 1 We consider 4,000 points uniformly sampled on a 16-dimensional unit sphere, embedded in $\mathbb{R}^{100}$, with $\eta \sim 0.1\mathcal{N}(0, I_{100})$ Gaussian noise added to each point. We plot empirical mean (over data points $\tilde{z}$) of the squared singular values of the empirical covariance matrix $\mathrm{cov}(\tilde{X}_{n,\tilde{z},r})$, as a function of r: in the "reasonable" range of scales, above the size of the noise, we see 16 singular values corresponding to the approximate tangent planes, at 17th squared singular value corresponding to curvature, and all the other squared singular values of size comparable to the energy of the noise 10^{-2}. The algorithm detects a range of scales, above the scale of the noise, where the 16th gap between the squared singular values is largest, i.e., noise is small compared to curvature, which is in turn small compared to elongation along the tangent plane. It is remarkable, albeit predicted by our results, that only 4,000 points (typically considered a small number if 16 (even more in 100) dimensions), perturbed by large noise (note that $\mathrm{e}[||\eta||] \sim 1$), are enough to obtain accurate geometric information

Moreover, in the range of scales

$$C_1 \frac{\sigma\sqrt{d}}{\sqrt{\lambda^2 - \delta^2}} \leq r_= \leq C_2 \frac{\lambda^2 - \delta^2}{\lambda\kappa}, \tag{9}$$

$\Delta_k(\tilde{X}_{n,\tilde{z},r})$ is the largest gap, with the probability as above.

Theorem 1 essentially says that if we have $O(d \log d)$ points in $\mu(B_z(r))$, and the noise variance σ is not too large compared to curvature, then the largest gap in the empirical covariance matrix of the data in $B_z(r)$ is the dth gap, with high probability, for r in the range:

$$C_1 \sigma^2 \leq \frac{r^2}{d} \leq C_2 \frac{\lambda^2}{\kappa^2 d}.$$

The upper bound $\frac{\lambda}{\kappa}$ is dictated by curvature, while the lower bound $\sigma\sqrt{D}$ is forced by the noise level: the lower bound is comparable to the size of the covariance of the noise, the upper bound is comparable to the size of the covariance of the data computed at the largest radius λ/κ where the curvature is not too large, and the term in the middle is comparable to the size of the data along the local approximating plane.

Our second theorem explores the regime where the ambient dimension D goes to infinity, but the number of samples n is fixed, dependent on the intrinsic dimension. While of course n samples certainly lie in an n-dimensional affine subspace, because of the ambient noise such subspace is unreliable at small scales, and this regime captures the range of scales where we have independence from the ambient dimension and the essentially linear dependence on d for the minimal needed number of points in $B_z(r_=)$.

Theorem 2 **($D \to \infty$, $\sigma\sqrt{D} = O(1)$).** *Fix $z \in \mathcal{M}$. Let the assumptions of Theorem 1 and the restriction (7) hold. Fix $t \in (C, Cd)$ and assume $\varepsilon := \varepsilon_{r_=,n,t} \leq \frac{1}{2}$. Then for $D \geq C$ and $m \leq D$, and σ_0 constant, for r in the range of scales (7) intersected with*

$$r \in \left(\frac{4\sigma_0 \left(1 \vee \frac{d}{\sqrt{D}} \vee \lambda_{\max}\varepsilon\right)}{\lambda_{\min}^2 - \delta^2 \lambda_{\max}\varepsilon - \frac{\varepsilon^2}{\lambda_{\min}^2}\left(\frac{C\sigma_0 d}{r} \vee \frac{1}{m}\right) - \frac{\sigma_0 \kappa}{t}}, \frac{\frac{\lambda_{\max}}{4} \wedge \sqrt{d}}{\kappa} \right),$$

the following hold, with probability at least $1 - Ce^{-Ct^2}$:

(i) $\Delta_k(\mathrm{cov}(\tilde{X}_{n,\tilde{z},r}))$ *is the largest gap of* $\mathrm{cov}(\tilde{X}_{n,\tilde{z},r})$.

(ii) $\|\mathrm{cov}(\tilde{X}_{n,\tilde{z},r}) - \mathrm{cov}(X_{z,r_=}) - \sigma^2 I_D\| \leq \left(\sigma_0^2\varepsilon + \lambda_{\max}\sigma_0 r + \left(\lambda_{\max} + 2\sigma_0\kappa + \frac{\varepsilon}{m}\right) r^2 + O\left(\frac{r^3}{\varepsilon}\right)\right)\frac{\varepsilon}{d}$.

These bounds, and in fact the finer bounds of [37], may of course be trivially used to obtain perturbation bounds for the empirical noisy tangent spaces estimated by looking at the top d singular vectors of the empirical covariance matrix of the data in $B_{\tilde{z}}(r)$ (a slightly better approach, taken in [37], uses Wieland's lemma before applying the usual sine theorems). It turns out that, since the noise essentially "to first order" adds only a multiple to the identity matrix, the approximate tangent space computed in this fashion is very stable, even in the regime of Theorem 2 [37].

This is the subject of [37], where it is shown that under rather general conditions on the geometry of the data (much more general than the manifold case) and under sampling and ambient noise, one may use these multiscale singular values to estimate the intrinsic dimension of the data. Moreover, under suitable assumptions, the number of samples in a ball around x required in order to do so is linear in the intrinsic dimension and independent of the ambient dimension. We refer the reader to [10, 35–37]. We now proceed by using not only the information in the singular values but also in the singular vectors in the SVD decomposition in Eq. (3).

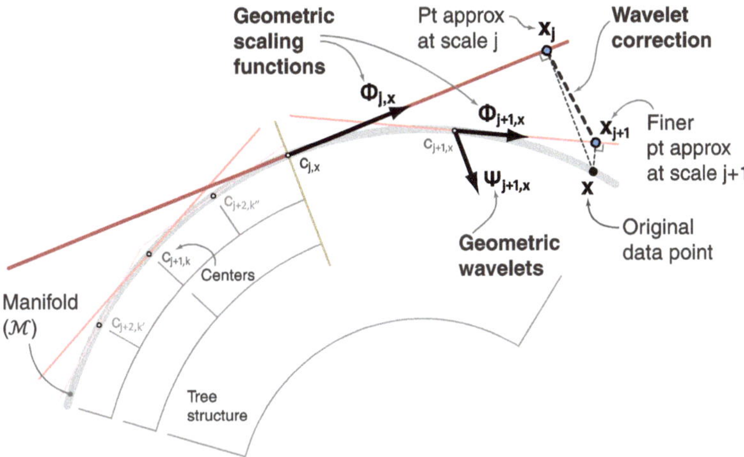

Fig. 2 An illustration of the geometric wavelet decomposition. The centers $m_{j,x}$'s are represented as lying on $\mathcal{M}$ while in fact they are only close to $\mathcal{M}$, and the corresponding planes $\mathbb{V}_{j,x}$ are represented as tangent planes, albeit they are only an approximation to them. Art courtesy of E. Monson

2.3 Geometric Scaling Functions

Then $\mathbb{P}_{j,k}(Q_{j,k})$ is the projection of $Q_{j,k}$ onto the local linear approximation given by the affine subspace in Eq. (4). The fact that this linear subspaces are affine will have various implications in our construction, creating mild nonlinearities and forcing us to construct a different transform and data representation which is not simply in the form of linear combination of certain atoms. On the other hand it seems an extremely natural construction, and not only the nonlinearities involved will not cause conceptual or computational overheads, but in fact we shall obtain algorithms which are faster than those needed to compute sparse linear representations in the standard dictionary learning setting. $\{\Phi_{j,k}\}_{k\in\mathcal{K}_j}$ are the geometric analogue of a family of scaling functions at scale j, and therefore we call them *geometric scaling functions*. They "span" an approximate piecewise linear manifold at scale j

$$\mathcal{M}_j := \{\mathbb{P}_{j,k}(Q_{j,k})\}_{k\in\mathcal{K}_j} \tag{10}$$

Under general conditions, $\mathcal{M}_j \to \mathcal{M}$ in the Hausdorff distance, as $j \to +\infty$. It is natural to define the nonlinear projection of $\mathcal{M}$ onto $\mathcal{M}_j$ by

$$x_j \equiv P_{\mathcal{M}_j}(x) := \mathbb{P}_{j,k}(x), \qquad x \in Q_{j,k}. \tag{11}$$

Note that in general $\mathcal{M}_j$ is not contained in $\mathcal{M}_{j+1}$, due to the nonlinearity of the underlying manifold $\mathcal{M}$. This is important as we move into the next section when we will encode "the difference" between $\mathcal{M}_j$ and $\mathcal{M}_{j+1}$.

2.4 Geometric Wavelets

In wavelet analysis, wavelets span the difference between scaling function spaces and are contained in the finer scale scaling function space. In our setting that would correspond to encoding the difference needed to go from $\mathcal{M}_j$ to $\mathcal{M}_{j+1}$: for a fixed $x \in \mathcal{M}$, $x_{j+1} - x_j \in \mathbb{R}^D$, but in general not contained in $\mathcal{M}_{j+1}$, due to the nonlinearity of $\mathcal{M}_j$ and $\mathcal{M}_{j+1}$. The main observation is that nevertheless the collection of vectors $x_{j+1} - x_j$ for x varying in $Q_{j+1,x}$ is in fact contained in a low-dimensional subspace and may be therefore encoded efficiently in terms of a basis of that subspace. We proceed as follows: for $j \leq J-1$ we let

$$\begin{aligned}
Q_{\mathcal{M}_{j+1}}(x) :&= x_{j+1} - x_j = x_{j+1} - \mathbb{P}_{j,x}(x_{j+1}) + \mathbb{P}_{j,x}(x_{j+1}) - \mathbb{P}_{j,x}(x) \\
&= (I - P_{j,x})(x_{j+1} - c_{j,x}) + P_{j,x}(x_{j+1} - x) \\
&= (I - P_{j,x})(\underbrace{x_{j+1} - c_{j+1,x}}_{\in V_{j+1,x}} + c_{j+1,x} - c_{j,x}) - P_{j,x}(x - x_{j+1}). \qquad (12)
\end{aligned}$$

Let $W_{j+1,x} := (I - P_{j,x}) V_{j+1,x}$, $Q_{j+1,x}$ be the orthogonal projection onto $W_{j+1,x}$, and let $\Psi_{j+1,x}$ be an orthonormal basis for $W_{j+1,x}$, which we will call a *geometric wavelet basis*. Observe $\dim W_{j+1,x} \leq \dim V_{j+1,x} = d_{j+1,x}$. We define several quantities below:

$$\begin{aligned}
t_{j+1,x} :&= c_{j+1,x} - c_{j,x}, w_{j+1,x} := (I - P_{j,x}) t_{j+1,x}; \\
&\mathbb{Q}_{j+1,x}(x) := Q_{j+1,x}(x - c_{j+1,x}) + w_{j+1,x}.
\end{aligned}$$

Then we may rewrite Eq. (12) as

$$\begin{aligned}
Q_{\mathcal{M}_{j+1}}(x) &= \underbrace{Q_{j+1,x}(x_{j+1} - c_{j+1,x})}_{\in W_{j+1,x}} + w_{j+1,x} - P_{j,x}\left(x - x_J + \sum_{l=j+1}^{J-1} (x_{l+1} - x_l)\right) \\
&= \mathbb{Q}_{j+1,x}(x_{j+1}) - P_{j,x} \sum_{l=j+1}^{J-1} (x_{l+1} - x_l) - P_{j,x}(x - x_J) \\
&= \mathbb{Q}_{j+1,x}(x_{j+1}) - P_{j,x} \sum_{l=j+1}^{J-1} Q_{\mathcal{M}_{l+1}}(x) - P_{j,x}(x - x_J), \qquad (13)
\end{aligned}$$

where $J \geq j+1$ is the index of the finest scale (and the last term vanishes as $J \to +\infty$, under general conditions). Note that this multiscale expansion contains terms that involve not only the current scale $j+1$ and the previous scale j but terms from finer scales as well, all the way to the finest scale J. This is once again due to the nonlinearity of $\mathcal{M}$ and of the whole construction: knowing $P_{\mathcal{M}_{j+1}}(x)$ is not enough to construct $P_{\mathcal{M}_j}(x)$, since the whole local nonlinear structure of $\mathcal{M}$ determines the

	J	$J-1$	$J-2$	$\dots$	$j+2$	$j+1$	j	$\dots$
$P_{\mathcal{M}_{J-1}}(x)$	$\mathbb{Q}_{J,x}(x_J)$	$P_{\mathcal{M}_{J-1}}(x)$						
$P_{\mathcal{M}_{J-2}}(x)$	$P_{J-2}(x-x_J)$	$\mathbb{Q}_{J-1,x}(x_{J-1})$	$P_{\mathcal{M}_{J-2}}(x)$					
$\vdots$	$\vdots$	$\vdots$	$\vdots$	$\vdots$	$\vdots$	$\ddots$		
$P_{\mathcal{M}_{j+1}}(x)$	$P_j(x-x_J)$	$P_j Q_{\mathcal{M}_J}(x)$	$P_j Q_{\mathcal{M}_{J+1}}(x)$	$\cdots$	$P_j Q_{\mathcal{M}_{j+2}}(x)$	$\mathbb{Q}_{j+1,x}(x_{j+1})$	$P_{\mathcal{M}_j}(x)$	$\dots$
$\vdots$	$\vdots$	$\vdots$	$\vdots$	$\vdots$	$\vdots$	$\vdots$	$\vdots$	$\ddots$

Fig. 3 We represent in this table the triangular array summarizing the geometric wavelet expansion of a term in the first column in terms of geometric wavelets, according to the multiscale relations (15) and the equalities in Eq. (13)

locally optimal projection $P_{\mathcal{M}_j}(x)$. In [2] we describe a variation of the transform where optimality is relaxed and a "two-scale equation" is obtained.

In terms of the geometric scaling functions and wavelets, the above may be written as

$$x_{j+1} - x_j = \Psi_{j+1,x}\Psi_{j+1,x}^*\left(x_{j+1} - m_{j+1,x}\right) + w_{j+1,x} - \Phi_{j,x}\Phi_{j,x}^* \sum_{l=j+1}^{J-1} Q_{\mathcal{M}_{l+1}}(x) - \Phi_{j,x}\Phi_{j,x}^*\left(x - x_J\right). \tag{14}$$

This shows that the difference $x_{j+1} - x_j$ can be expressed as the sum of a component in $W_{j+1,x}$, a second component that only depends on the cell $(j+1,x)$ (but not on the point x itself) which accounts for the translation of centers and lying in $V_{j,x}^{\perp}$ (but not necessarily in $W_{j+1,x}$), and a sum of projections on $V_{j,x}$ of differences $x_{l+1} - x_l$ at finer scales. By construction we have the two-scale equation

$$P_{\mathcal{M}_{j+1}}(x) = P_{\mathcal{M}_j}(x) + Q_{\mathcal{M}_{j+1}}(x), \quad x \in \mathcal{M} \tag{15}$$

which can be iterated across scales, leading to a multiscale decomposition along low-dimensional subspaces, with efficient encoding and algorithms. We think of $P_{j,k}$ as being attached to the node (j,k) of $\mathcal{T}$ and the $Q_{j+1,k'}$ as being attached to the edge connecting the node $(j+1,k')$ to its parent.

We say that the set of multiscale piecewise affine operators $\{P_{\mathcal{M}_j}\}$ and $\{Q_{\mathcal{M}_{j+1}}\}$ form a **geometric multi-resolution analysis** or GMRA for short.

2.5 Approximation for Manifolds

We analyze the error of approximation to a d-dimensional manifold in $\mathbb{R}^D$ by using geometric wavelets representation. Our analysis gives a full explanation of the examples in Sect. 4.1. We have the following theorem from [2]:

Theorem 3. *Let $(\mathcal{M},\rho,\mu)$ be a compact $\mathcal{C}^{1+\alpha}$ Riemannian manifold of dimension d isometrically embedded in $\mathbb{R}^D$, with $\alpha\in(0,1]$, and μ absolutely continuous with respect to the volume measure on $\mathcal{M}$. Let $\{P_{\mathcal{M}_j},Q_{\mathcal{M}_{j+1}}\}$ be a GMRA for $(\mathcal{M},\rho,\mu)$. For any $x\in\mathcal{M}$, there exists a scale $j_0=j_0(x)$ such that for any $j\geq j_0$ and any $p>0$, if we let $d\mu_{j,x}:=\mu(Q_{j,x})^{-1}d\mu$,*

$$\left|\left|\left|\left|z-P_{\mathcal{M}_j}(z)\right|\right|_{\mathbb{R}^D}\right|\right|_{L^p(Q_{j,x},d\mu_{j,x}(z))}=\left|\left|\left|\left|z-P_{\mathcal{M}_{j_0}}(z)-\sum_{l=j_0}^{j-1}Q_{\mathcal{M}_{l+1}}(z)\right|\right|_{\mathbb{R}^D}\right|\right|_{L^p(Q_{j,x},d\mu_{j,x}(z))}$$
$$\leq||\kappa||_{L^\infty(Q_{j,x})}2^{-(1+\alpha)j}+o(2^{-(1+\alpha)j}). \tag{16}$$

If $\alpha<1$, $\kappa(x)$ depends on the $\mathcal{C}^{1+\alpha}$ norm of a coordinate chart from $T_x(\mathcal{M})$ to $Q_{j,x}\subseteq\mathcal{M}$ and on $\left|\left|\frac{d\mu}{d\mathrm{vol}}\right|\right|_{L^\infty(Q_{j,x})}$.

If $\alpha=1$,

$$\kappa(x)=\left|\left|\frac{d\mu}{d\mathrm{vol}}\right|\right|_{L^\infty(Q_{j,x})}\min(\kappa_1(x),\kappa_2(x)), \tag{17}$$

with

$$\kappa_1(x):=\frac{1}{2}\max_{i\in\{1,\dots,D-d\}}||H_i(x)||;$$

$$\kappa_2^2(x):=\max_{w\in\mathbb{S}^{D-d}}\frac{d(d+1)}{4(d+2)(d+4)}\left[\left|\left|\sum_{l=1}^{D-d}w_lH_l(x)\right|\right|_F^2-\frac{1}{d+2}\left(\sum_{l=1}^{D-d}w_l\mathrm{Tr}(H_l(x))\right)^2\right], \tag{18}$$

and the $D-d$ matrices $H_l(x)$ are the d-dimensional Hessians of $\mathcal{M}$ at x.

Observe that κ_2 can be smaller than κ_1 (by a constant factor) or larger (by factors depending on d^2), depending on the spectral properties and commutativity relations between the Hessians H_l. κ_2^2 may be unexpectedly small, in the sense that it may scale as $d^{-2}r^4$ as a function of d and r, as observed in [37]. For the proof we refer the reader to [2].

3 Algorithms

We present in this section algorithms implementing the construction of the GMRA and the corresponding geometric wavelet transform (GWT).

3.1 Construction of Geometric Multi-resolution Analysis

The first step in the construction of the geometric wavelets is to perform a geometric nested partition of the data set, forming a tree structure. For this end, one may consider various methods listed below:

- Use of METIS [34]: a multiscale variation of iterative spectral partitioning. We construct a weighted graph as done for the construction of diffusion maps [15, 21]: we add an edge between each data point and its k nearest neighbors and assign to any such edge between x_i and x_j the weight $e^{-\|x_i-x_j\|^2/\sigma}$. Here k and σ are parameters whose selection we do not discuss here (but see [45] for a discussion in the context of molecular dynamics data). In practice, we choose k between 10 and 50 and choose σ adaptively at each point x_i as the distance between x_i and its $\lfloor k/2 \rfloor$ nearest neighbor.
- Use of cover trees [4].
- Use of iterated PCA: at scale 1, compute the top d principal components of data and partition the data based on the sign of the $(d+1)$-st singular vector. Repeat on each of the two partitions.
- Iterated k-means: at scale 1 partition the data based on k-means clustering, then iterate on each of the elements of the partition.

Each construction has pros and cons, in terms of performance and guarantees. For (I) we refer the reader to [34], for (II) to [4] (which also discussed several other constructions), and for (III) and (IV) to [48]. Only (II) provides the needed properties for the cells $Q_{j,k}$. However constructed, we denote by $\{Q_{j,k}\}$ the family of resulting dyadic cells and let $\mathscr{T}$ be the associated tree structure, as in Section 2.1.

In Fig. 4 we display pseudo-code for the GMRA of a data set X_n given a precision $\varepsilon > 0$ and a method τ_0 for choosing local dimensions (e.g., using thresholds or a fixed dimension). The code first constructs a family of multiscale dyadic cells (with local centers $c_{j,k}$ and bases $\Phi_{j,k}$) and then computes the geometric wavelets $\Psi_{j,k}$ and translations $w_{j,k}$ at all scales. In practice, we use METIS [34] to construct a dyadic (not 2^d-adic) tree $\mathscr{T}$ and the associated cells $Q_{j,k}$.

3.2 The Fast Geometric Wavelet Transform and Its Inverse

For simplicity of presentation, we shall assume $x = x_J$; otherwise, we may first project x onto the local linear approximation of the cell $Q_{J,x}$ and use x_J instead of x from now on. That is, we will define $x_{j;J} = P_{\mathscr{M}_j}(x_J)$, for all $j < J$, and encode the differences $x_{j+1;J} - x_{j;J}$ using the geometric wavelets. Note also that $\|x_{j;J} - x_j\| \leq \|x - x_J\|$ at all scales.

The geometric scaling and wavelet coefficients $\{p_{j,x}\}, \{q_{j+1,x}\}$, for $j \geq 0$, of a point $x \in \mathscr{M}$ are chosen to satisfy the equations

`GMRA = GeometricMultiResolutionAnalysis` $(X_n, \tau_0, \varepsilon)$

// **Input:**
// X_n: a set of n samples from $\mathcal{M}$
// τ_0: some method for choosing local dimensions
// ε: precision

// **Output:**
// A tree $\mathcal{T}$ of dyadic cells $\{Q_{j,k}\}$, their local means $\{m_{j,k}\}$ and bases $\{\Phi_{j,k}\}$, together with a family of geometric wavelets $\{\Psi_{j,k}\}, \{w_{j,k}\}$

Construct the dyadic cells $Q_{j,k}$ with centers $\{m_{j,k}\}$ and form a tree $\mathcal{T}$.
$J \leftarrow$ finest scale with the ε-approximation property.
Let $\mathrm{cov}_{J,k} = |C_{J,k}|^{-1} \sum_{x \in C_{J,k}} (x - m_{J,k})(x - m_{J,k})^*$, for $k \in \mathcal{K}_J$, and compute $\mathrm{SVD}(\mathrm{cov}_{J,k}) = \Phi_{J,k} \Sigma_{J,k} \Phi_{J,k}$ (where the dimension of $\Phi_{J,k}$ is determined by τ_0).
for $j = J-1$ **down to** 0

  for $k \in \mathcal{K}_j$

    Compute $\mathrm{cov}_{j,k}$ and $\Phi_{j,k}$ as above.
    For each $k' \in \mathrm{ch}(j,k)$, construct the wavelet bases $\Psi_{j+1,k'}$ and translations $w_{j+1,k'}$.

  end

end
For convenience, set $\Psi_{0,k} := \Phi_{0,k}$ and $w_{0,k} := m_{0,k}$ for $k \in \mathcal{K}_0$.

Fig. 4 Pseudo-code for the construction of geometric wavelets

$$P_{\mathcal{M}_j}(x) = \Phi_{j,x} p_{j,x} + m_{j,x}; \tag{19}$$

$$Q_{\mathcal{M}_{j+1}}(x) = \Psi_{j+1,x} q_{j+1,x} + w_{j+1,x} - P_{j,x} \sum_{l=j+1}^{J-1} Q_{\mathcal{M}_{l+1}}(x). \tag{20}$$

The computation of the coefficients, from fine to coarse, is simple and fast: since we assume $x = x_J$, we have

$$\begin{aligned} p_{j,x} &= \Phi_{j,x}^*(x_J - c_{j,x}) = \Phi_{j,x}^*(\Phi_{J,x} p_{J,x} + c_{J,x} - c_{j,x}) \\ &= \left(\Phi_{j,x}^* \Phi_{J,x}\right) p_{J,x} + \Phi_{j,x}^*(c_{J,x} - c_{j,x}). \end{aligned} \tag{21}$$

Moreover the wavelet coefficients $q_{j+1,x}$ [defined in Eq. (20)] are obtained from Eq. (14):

$$q_{j+1,x} = \Psi_{j+1,x}^*(x_{j+1} - c_{j+1,x}) = \left(\Psi_{j+1,x}^* \Phi_{j+1,x}\right) p_{j+1,x}. \tag{22}$$

Note that $\Phi_{j,x}^* \Phi_{J,x}$ and $\Psi_{j+1,x}^* \Phi_{j+1,x}$ are both small matrices (at most $d_{j,x} \times d_{j,x}$) and are the only matrices we need to compute and store (once for all, and only up to a specified precision) in order to compute all the wavelet coefficients $q_{j+1,x}$ and the scaling coefficients $p_{j,x}$, given $p_{J,x}$ at the finest scale.

$\{q_{j,x}\}$ =`FGWT(GMRA,`x`)`

// **Input:** GMRA structure, $x \in \mathcal{M}$
// **Output:** A sequence $\{q_{j,x}\}$ of wavelet coefficients

$p_{J,x} = \Phi_{J,x}^*(x - m_{J,x})$
for $j = J$ **down to** 1

$q_{j,x} = (\Psi_{j,x}^* \Phi_{j,x})\, p_{j,x}$
$p_{j-1,x} = (\Phi_{j-1,x}^* \Phi_{J,x})\, p_{J,x} + \Phi_{j-1,x}^*(m_{J,x} - m_{j-1,x})$

end
$q_{0,x} = p_{0,x}$ (for convenience)

Fig. 5 Pseudo-code for the forward geometric wavelet transform

$\hat{x}$ =`IGWT(GMRA,`$\{q_{j,x}\}$`)`

// **Input:** GMRA structure, wavelet coefficients $\{q_{j,x}\}$
// **Output:** Approximation $\hat{x}$ at scale J

$Q_{J,x} = \Psi_{J,x} q_{J,x} + w_{J,x}$
for $j = J - 1$ **down to** 1

$Q_j(x) = \Psi_{j,x} q_{j,x} + w_{j,x} + \Phi_{j-1,x} \Phi_{j-1,x}^* \sum_{\ell > j} Q_\ell(x)$

end
$\hat{x} = \Psi_{0,x} q_{0,x} + w_{0,x} + \sum_{j>0} Q_j(x)$

Fig. 6 Pseudo-code for the inverse geometric wavelet transform

In Figs. 5 and 6 we display pseudo-codes for the computation of the forward and inverse geometric wavelet transforms (F/IGWT). The input to FGWT is a GMRA object, as returned by `GeometricMultiResolutionAnalysis`, and a point $x \in \mathcal{M}$. Its output is the wavelet coefficients of the point x at all scales, which are then used by IGWT for reconstruction of the point at all scales.

For any $x \in \mathcal{M}_J$, the set of coefficients

$$q_x = (q_{J,x}; q_{J-1,x}; \ldots; q_{1,x}; p_{0,x}) \tag{23}$$

is called the discrete *GWT* of x. Letting $d_{j,x}^w = \text{rank}(\Psi_{j+1,x})$, the length of the transform is $d + \sum_{j>0} d_{j,x}^w$, which is bounded by $(J+1)d$ in the case of samples from a d-dimensional manifold (due to $d_{j,x}^w \leq d$).

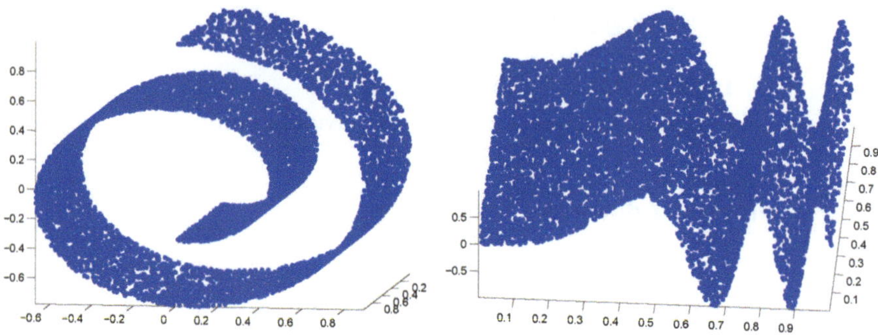

Fig. 7 Toy data sets for the following examples of GMRA

4 Examples

We conduct numerical experiments in this section to demonstrate the performance of the algorithm (i.e., Figs. 4–6).

4.1 Low-Dimensional Smooth Manifolds

To illustrate the construction presented so far, we consider simple synthetic data sets: a *SwissRoll*, an *S-Manifold*, and an *Oscillating2DWave*, all two-dimensional manifolds but embedded in $\mathbb{R}^{50}$ (see Fig. 7). We apply the algorithm to construct the GMRA and obtain the FGWT of the sampled data (10,000 points, without noise) in Fig. 8. We use the manifold dimension $d_{j,k} = d = 2$ at each node of the tree when constructing scaling functions and choose the smallest finest scale for achieving an absolute precision .001 in each case. We compute the average magnitude of the wavelet coefficients at each scale and plot it as a function of scale in Fig. 8. The reconstructed manifolds obtained by the inverse geometric wavelets transform (at selected scales) are shown in Fig. 9, together with a plot of relative approximation errors,

$$\mathscr{E}_{j,2}^{\text{rel}} = \frac{1}{\sqrt{\mathrm{Var}(X_n)}} \sqrt{\frac{1}{n} \sum_{x \in X_n} \left(\frac{||x - P_{j,x}(x)||}{||x||} \right)^2}, \tag{24}$$

where X_n is the training data of n samples. Both the approximation error and the magnitude of the wavelet coefficients decrease quadratically with respect to scale as expected.

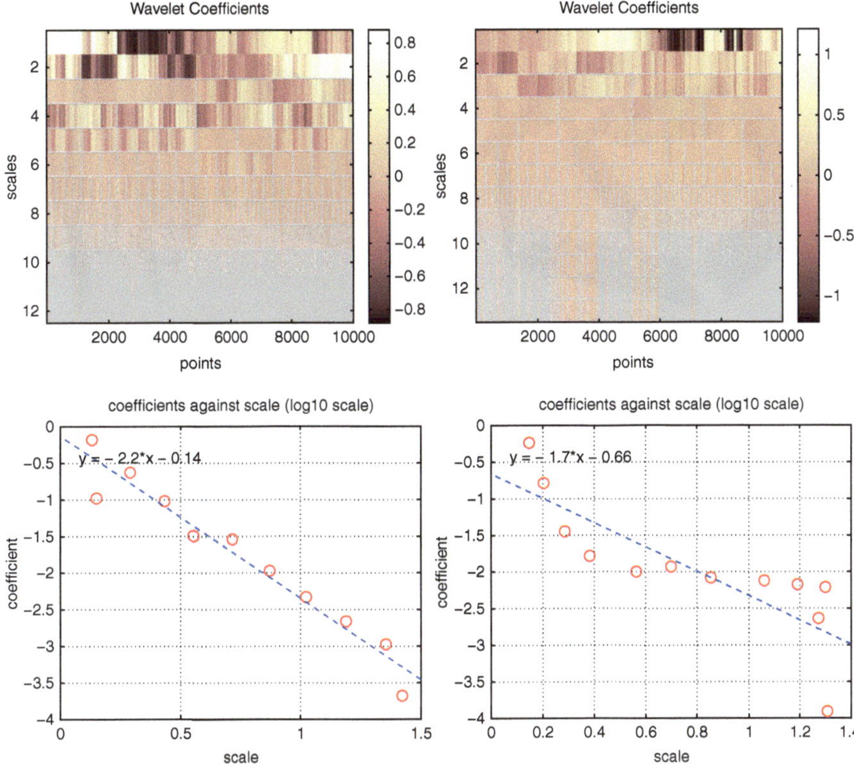

Fig. 8 *Top row*: wavelet coefficients obtained by the algorithm for the three data sets in Fig. 7. The horizontal axis indexes the points (arranged according to the tree), and the vertical axis multi-indexes the wavelet coefficients, from coarse (*top*) to fine (*bottom*) scales: the block of entries $(x,j), x \in Q_{j,k}$ displays $\log_{10}|q_{j,x}|$, where $q_{j,x}$ is the vector of geometric wavelet coefficients of x at scale j (see Sect. 3). In particular, each row indexes multiple wavelet elements, one for each $k \in \mathscr{K}_j$. *Bottom row*: magnitude of wavelet coefficients decreasing quadratically as a function of scale

We threshold the wavelet coefficients to study the compressibility of the wavelet coefficients and the rate of change of the approximation errors (using compressed wavelet coefficients). For this end, we use a smaller precision 10^{-5} so that the algorithm can examine a larger interval of thresholds. We threshold the wavelet coefficients of the *Oscillating2DWave* data at the level .01 and plot in Fig. 10 the reduced matrix of wavelet coefficients and the corresponding best reconstruction of the manifold (i.e., at the finest scale).

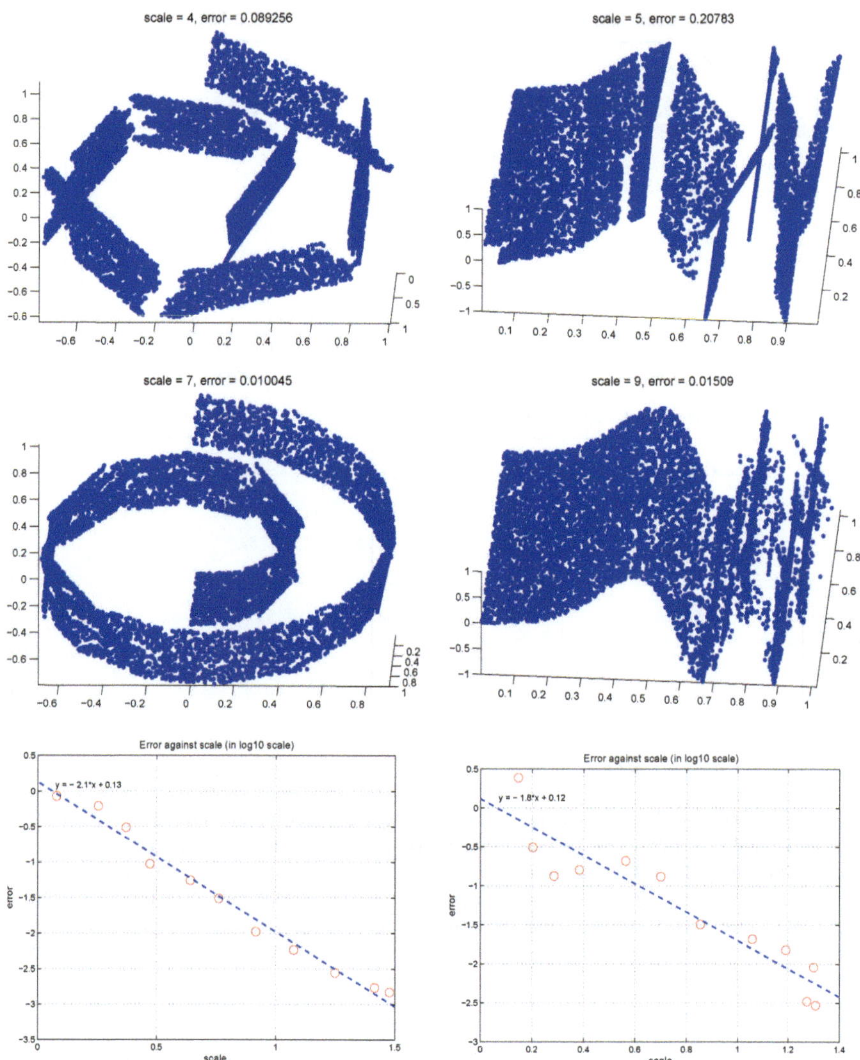

Fig. 9 *Top* and *middle*: reconstructions by the algorithm of the three toy data sets in Fig. 7 at two selected scales. *Bottom*: reconstruction errors as a function of scale

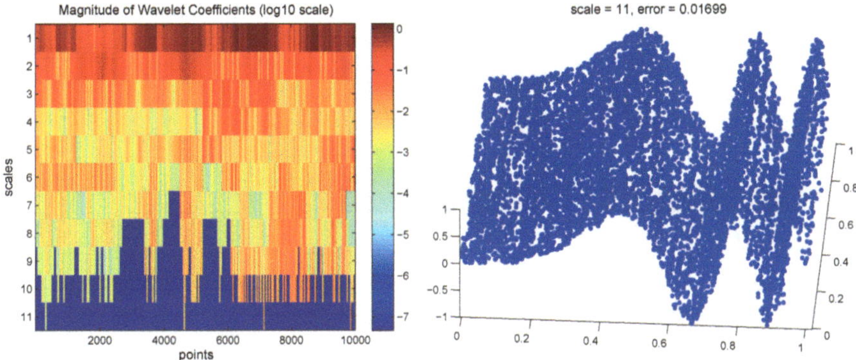

Fig. 10 The wavelet coefficients of the *Oscillating2DWave* data may be thresholded leading to adaptive approximation. *Left*: after sorting the points so that the x-axis orders them as going from *left* to *right* on the manifold, we see that when the manifold oscillates more, larger wavelet coefficients arise at fine scales. By threshold at the level of .01 and prune the dyadic tree accordingly, we reconstruct the manifold at the corresponding precision (*right*)

4.2 Data Sets

4.2.1 MNIST Handwritten Digits

We first consider the MNIST data set of images of handwritten digits,[1] each of size 28×28. We use the digits 0 and 1 and randomly sample for each digit 3,000 images from the database. We apply the algorithm to construct the geometric wavelets and show the wavelet coefficients and the reconstruction errors at all scales in Fig. 11. We select local dimensions for scaling functions by keeping 50 % and 95 % of the variance, respectively, at the nonleaf and leaf nodes. We observe that the magnitudes of the coefficients stop decaying after a certain scale. This indicates that the data is not on a smooth manifold. We expect optimization of the tree and of the wavelet dimensions in future work to lead to a more efficient representation in this case.

We then fix a data point (or equivalently an image), for each digit, and show in Fig. 12 its reconstructed coordinates at all scales and the corresponding dictionary elements (all of which are also images). We see that at every scale we have a handwritten digit, which is an approximation to the fixed image, and those digits are refined successively to approximate the original data point. The elements of the dictionary quickly fix the orientation and the thickness, and then they add other distinguishing features of the image being approximated.

[1] Available at http://yann.lecun.com/exdb/mnist/.

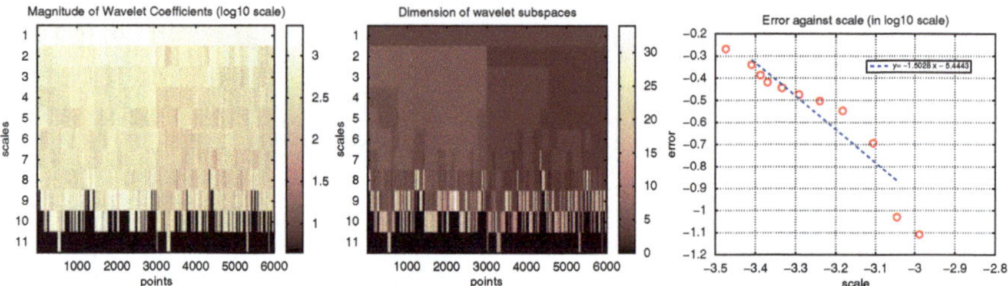

Fig. 11 From *left* to *right*: geometric wavelet representation of the MNIST digits data set for 1 and 0. As usual, the vertical axis multi-indexes the wavelet coefficients, from coarse (*top*) to fine (*bottom*) scales: the block of entries at $(x, j), x \in Q_{j,k}$ is $\log_{10}|q_{j,x}|$, where $q_{j,x}$ is the vector of geometric wavelet coefficients of x at scale j (see Sect. 3). In particular, each row indexes multiple wavelet elements, one for each $k \in \mathcal{K}_j$. *Top right*: dimensions of the wavelet subspaces (with the same convention as in the previous plot): even if the data lies in 784 dimensions, the approximating planes used have mostly dimension 1–6, except for some planes at the leaf nodes. *Rightmost inset*: reconstruction error as functions of scale. The decay is nonlinear and not what we would expect from a manifold structure

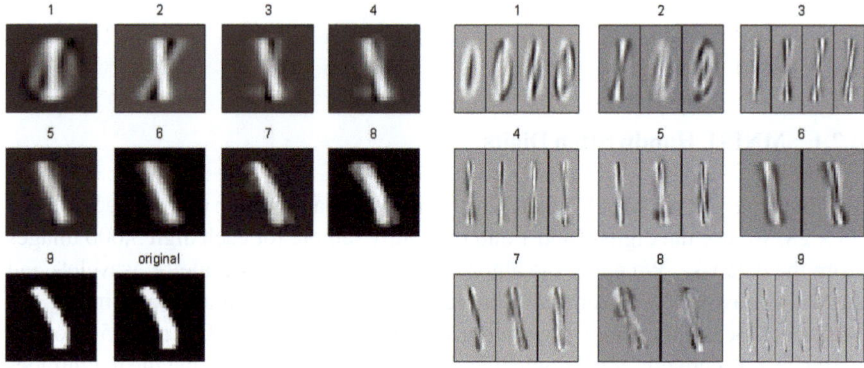

Fig. 12 *Left*: in each figure we plot coarse-to-fine geometric wavelet approximations of the original data point (represented in the last image). *Right*: elements of the wavelet dictionary (ordered from coarsest to finest scales) used in the expansion of the data point on the *left*

Example: A Connection to Fourier Analysis and FFT

Consider band-limited functions of band B:

$$BF_B = \{f : \text{supp.}\hat{f} \subseteq [-B\pi, B\pi]\}.$$

Classes of smooth functions (e.g., $W^{k,2}$) are essentially characterized by their L^2-energy in dyadic spectral bands of the form $[-2^{j+1}\pi, -2^j\pi] \cup [2^j\pi, 2^{j+1}\pi]$, i.e., by the L^2-size of their projection onto $BF_{2^{j+1}} \ominus BF_{2^j}$ (some care is of course needed in that smooth frequency cutoff, but this issue is not relevant for our purposes here). We generate random smooth (band-limited!) functions as follows:

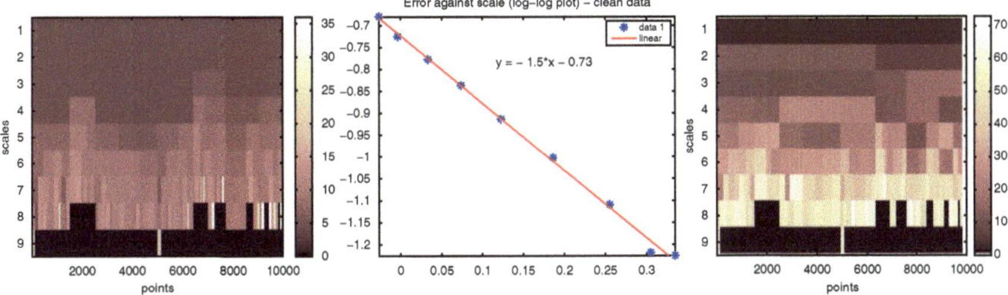

Fig. 13 We construct an orthogonal geometric multi-resolution analysis (see [2]) on a random sample of 10,000 band-limited functions. *Left*: dimension of the GMRA wavelet subspaces. *Center*: approximation error as a function of scale. *Right*: dominant frequency in each GMRA subspace, showing that frequencies are sorted from low (*top*, coarse GMRA scales) to high (*bottom*, fine GMRA scales). This implies that the geometric scaling function subspaces roughly correspond to a Littlewood–Paley decomposition, and the GWT of a function f corresponds to a rough standard wavelet transform

$$f(x) = \sum_{j=0}^{J} a_j(\omega)\cos(jx)$$

with a_j random Gaussian (or bounded) with mean $2^{-\lfloor \frac{j}{J} \rfloor \alpha}$ and standard deviation $2^{-\lfloor \frac{j}{J} \rfloor \alpha} \cdot \frac{1}{5}$. The GMRA associated with a random sample of this family of functions takes advantage of the multiscale nature of the data and organizes this family of functions in a Littlewood–Paley type of decomposition: the scaling function subspace at scale j roughly corresponds to $BF_{2^{j+1}} \ominus BF_{2^j}$, and the GMRA of a point is essentially a block Fourier transform, where coefficients in the same dyadic band are grouped together. Observe that the cost of the GMRA of a point f is comparable to the cost of the fast Fourier transform.

5 Data Representation, Compression, and Computational Considerations

A set of n points in $\mathbb{R}^D$ can trivially be stored in space Dn; if it lies, up to a least squares error ε in a linear subspace of dimension $d_\varepsilon \ll D$, we could encode n points in space $d_\varepsilon(D+n)$ (cost of encoding a basis for the linear subspace, plus encoding of the coefficients of the points on that basis). This is much less than the trivial encoding for $d_\varepsilon \ll D$. It can be shown [2] that the cost of encoding with a GMRA a $\mathscr{C}^2$ manifold $\mathscr{M}$ of dimension d sampled at n points, for a fixed precision $\varepsilon > 0$ and n large, is $O(\varepsilon^{-\frac{d}{2}} dD + nd\log_2 \varepsilon^{-\frac{1}{2}})$.

Also, the cost of the algorithm is

$$O(nD(\log(n) + d^2)) + O_{d,D}(n\log n),$$

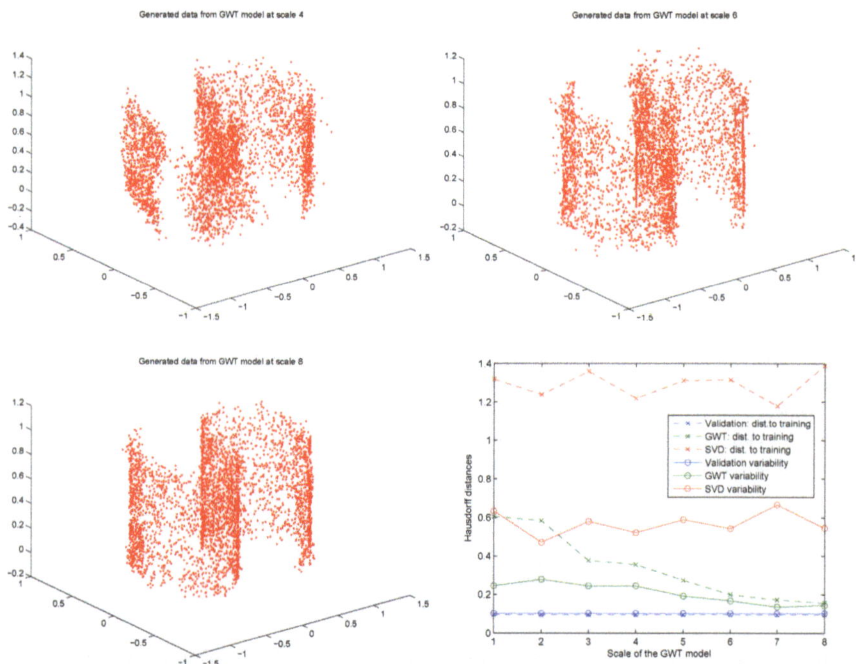

Fig. 14 Approximations of the probability distribution concentrated on a S-shaped 2-dimensional manifold within the GMRA framework. From *left* to *right*, *top* to *bottom*: 4,000 samples drawn from our approximate distribution, constructed at scale 4, 6, and 8, respectively, from 2,000 training samples. *Bottom right*: as a function of scale, the Hausdorff distance between points generated by the SVD model and GWT models and the training data, as well as the Hausdorff distance variability of the generated data and true data. We see that $p_{\mathcal{M}_j}$ has small distance to the training set and decreasingly so for models constructed at finer scales, while p_{SVD_j}, being a model in the ambient space, generates points farther from the distribution. Looking at the plots of the in-model Hausdorff distance variability, we see that such measure increases for $p_{\mathcal{M}_j}$ as a function of j (reflecting the increasing expression power of the model). Samples from the SVD model look like a Gaussian point cloud, as the kernel density estimator did not have enough training samples to adapt to the low-dimensional manifold structure

while the cost of performing the FGWT of a point is

$$O(2^d D \log n + dD + d^2 \log \varepsilon^{-\frac{1}{2}}).$$

The cost of the IGWT is similar but without the first term.

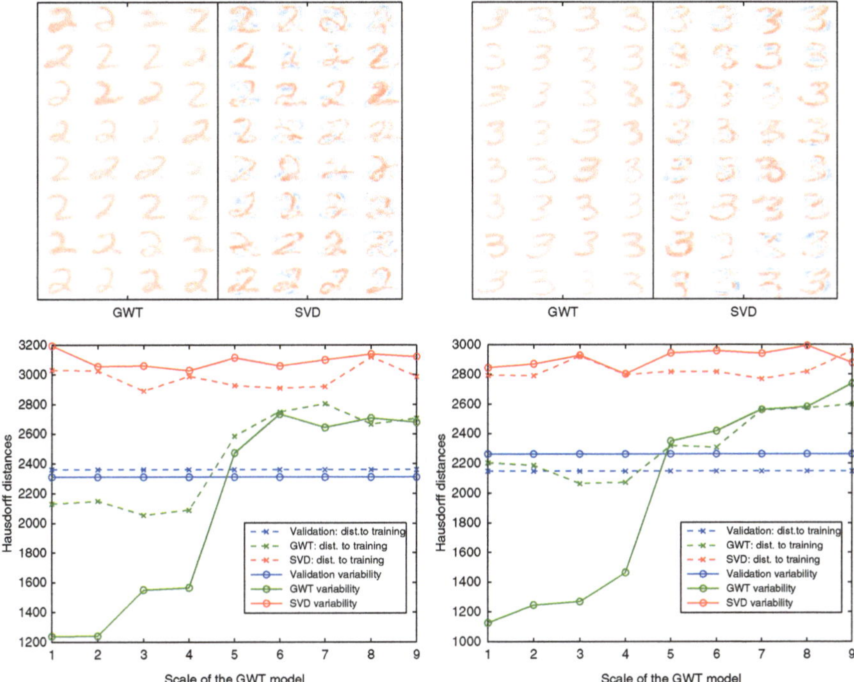

Fig. 15 *Left* and *right columns*: a training set of 2,000 digits 2 (respectively, 3) from the MNIST data set are used to train probability models with GMRA ($p_{\mathcal{M}_j}$, one for each scale j in the GMRA of the training set) and SVD (p_{SVD_j}, one for each GMRA scale, see text). *Left:* 32 digits drawn from $p_{\mathcal{M}_5}$, p_{SVD_5}: the quality of $p_{\mathcal{M}_5}$ is qualitatively better than that of p_{SVD_5}. *Center*: plots of the Hausdorff distance to training set and in-model Hausdorff distance variability. *Right*: a similar experiment with a training set of 2,000 points from a SwissRoll-shaped manifold with no noise: the finest scale GMRA-based models perform best (in terms of both approximation and variability, the SVD-based models are once again unable to take advantage of the low intrinsic dimension)

6 Multiscale Models of Densities

We present a simple example of how our techniques may be used to model measures supported on low-dimensional sets which are well approximated by the multiscale planes we constructed; results from more extensive investigations will be reported in an upcoming publication.

We sample n training points from a point cloud $\mathcal{M}$ and, for a fixed scale j, we consider the coarse approximation $\mathcal{M}_j$ [defined in Eq. (10)], and on each local linear approximating plane $V_{j,k}$ we use the training set to construct a multifactor Gaussian model on $Q_{j,k}$: let $\pi_{j,k}$ be the estimated distribution. We also estimate from the training data the probability $\pi_j(k)$ that a given point in $\mathcal{M}$ belongs to $Q_{j,k}$ (recall that j is fixed, so this is a probability distribution over the $|\mathcal{K}_j|$ labels of the planes at scale j). We may then generate new data points by drawing a $k \in \mathcal{K}_j$ according to π_j

and then drawing a point in $V_{j,k}$ from the distribution $\pi_{j,k}$: this defines a probability distribution supported on $\mathcal{M}_j$ that we denote by $p_{\mathcal{M}_j}$.

In this way we may generate new data points which are consistent with both the geometry of the approximating planes $V_{j,k}$ and with the distribution of the data on each such plane. In Fig. 14 we display the result of such modeling on a simple manifold. In Fig. 15 we construct $p_{\mathcal{M}_j}$ by training on 2,000 handwritten 2s and 3s from the MNIST database, and on the same training set we train two other algorithms: the first one is based on projecting the data on the first a_j principal components, where a_j is chosen so that the cost of encoding the projection and the projected data is the same as the cost of encoding the GMRA up to scale j and the GMRA of the data and then running the same multifactor Gaussian model used above for generating $\pi_{j,k}$. This leads to a probability distribution we denote by p_{SVD_j}. In order to test the quality of these models, we consider the following two measures. The first measure is simply the Hausdorff distance between 2,000 randomly chosen samples according to each model and the training set: this is measuring how close the generated samples are to the training set. The second measure quantifies if the model captures the variability of the true data and is computed by generating multiple point clouds of 2,000 points for a fixed model and looking at the pairwise Hausdorff distances between such point clouds, called the within-model Hausdorff distance variability.

The bias–variance trade-off in the models $p_{\mathcal{M}_j}$ is the following: as j increases the planes better model the geometry of the data (under our usual assumptions), so that the bias of the model (and the approximation error) decreases as j increases; on the other hand the sampling requirements for correctly estimating the density of $Q_{j,k}$ projected on $V_{j,k}$ increases with j as less and less training points fall in $Q_{j,k}$. A pruning greedy algorithm that selects, in each region of the data, the correct scale for obtaining the correct bias–variance trade-off, depending on the samples and the geometry of the data, similar in spirit to the what has been studied in the case of multiscale approximation of functions, will be presented in a forthcoming publication. It should be remarked that such a model would be very complicated in the wavelet domain, as one would need to model very complex dependencies among wavelet coefficients, in both space and scale.

Acknowledgments The authors thank E. Monson for useful discussions. AVL was partially supported by NSF and ONR. GC was partially supported by DARPA, ONR, NSF CCF, and NSF/DHS FODAVA program. MM is grateful for partial support from DARPA, NSF, ONR, and the Sloan Foundation.

References

1. Aharon, M., Elad, M., Bruckstein, A.: K-SVD: Design of dictionaries for sparse representation. In: Proceedings of SPARS 05', pp. 9–12 (2005)
2. Allard, W.K., Chen, G., Maggioni, M.: Multi-scale geometric methods for data sets II: Geometric multi-resolution analysis. Appl. Computat. Harmonic Analysis **32**, 435–462 (2012)

3. Belkin, M., Niyogi, P.: Using manifold structure for partially labelled classification. Advances in NIPS, vol. 15. MIT Press, Cambridge (2003)
4. Beygelzimer, A., Kakade, S., Langford, J.: Cover trees for nearest neighbor. In: ICML, pp. 97–104 (2006)
5. Binev, P., Cohen, A., Dahmen, W., Devore, R., Temlyakov, V.: Universal algorithms for learning theory part i: Piecewise constant functions. J. Mach. Learn. **6**, 1297–1321 (2005)
6. Binev, P., Devore, R.: Fast computation in adaptive tree approximation. Numer. Math. **97**, 193–217 (2004)
7. Bremer, J., Coifman, R., Maggioni, M., Szlam, A.: Diffusion wavelet packets. Appl. Comp. Harm. Anal. **21**, 95–112 (2006) (Tech. Rep. YALE/DCS/TR-1304, 2004)
8. Candès, E., Donoho, D.L.: Curvelets: A surprisingly effective nonadaptive representation of objects with edges. In: Schumaker, L.L., et al. (eds.) Curves and Surfaces. Vanderbilt University Press, Nashville (1999)
9. Causevic, E., Coifman, R., Isenhart, R., Jacquin, A., John, E., Maggioni, M., Prichep, L., Warner, F.: QEEG-based classification with wavelet packets and microstate features for triage applications in the ER, vol. 3. ICASSP Proc., May 2006 10.1109/ICASSP.2006.1660859
10. Chen, G., Little, A., Maggioni, M., Rosasco, L.: Wavelets and Multiscale Analysis: Theory and Applications. Springer (2011) submitted March 12th, 2010
11. Chen, G., Maggioni, M.: Multiscale geometric wavelets for the analysis of point clouds. Information Sciences and Systems (CISS), 2010 44th Annual Conference on. IEEE, 2010.
12. Chen, S.S., Donoho, D.L., Saunders, M.A.: Atomic decomposition by basis pursuit. SIAM J. Sci. Comput. **20**, 33–61 (1998)
13. Christ, M.: A $T(b)$ theorem with remarks on analytic capacity and the Cauchy integral. Colloq. Math. **60–61**, 601–628 (1990)
14. Christensen, O.: An introduction to frames and Riesz bases. Applied and Numerical Harmonic Analysis. Birkhäuser, Boston (2003)
15. Coifman, R., Lafon, S.: Diffusion maps. Appl. Comp. Harm. Anal. **21**, 5–30 (2006)
16. Coifman, R., Lafon, S., Maggioni, M., Keller, Y., Szlam, A., Warner, F., Zucker, S.: Geometries of sensor outputs, inference, and information processing. In: Athale, R.A. (ed.) Proc. SPIE, J. C. Z. E. Intelligent Integrated Microsystems, vol. 6232, p. 623209, May 2006
17. Coifman, R., Maggioni, M.: Diffusion wavelets. Appl. Comp. Harm. Anal. **21**, 53–94 (2006) (Tech. Rep. YALE/DCS/TR-1303, Yale Univ., Sep. 2004).
18. Coifman, R., Maggioni, M.: Multiscale data analysis with diffusion wavelets. In: Proc. SIAM Bioinf. Workshop, Minneapolis (2007)
19. Coifman, R., Maggioni, M.: Geometry analysis and signal processing on digital data, emergent structures, and knowledge building. SIAM News, November 2008
20. Coifman, R., Meyer, Y., Quake, S., Wickerhauser, M.V.: Signal processing and compression with wavelet packets. In: Progress in Wavelet Analysis and Applications (Toulouse, 1992), pp. 77–93. Frontières, Gif (1993)
21. Coifman, R.R., Lafon, S., Lee, A.B., Maggioni, M., Nadler, B., Warner, F., Zucker, S.W.: Geometric diffusions as a tool for harmonic analysis and structure definition of data: Diffusion maps. PNAS **102**, 7426–7431 (2005)
22. Daubechies, I.: Ten lectures on wavelets. Society for Industrial and Applied Mathematics, Philadelphia, PA, USA (1992) ISBN: 0-89871-274-2.
23. David, G.: Wavelets and singular integrals on curves and surfaces. In: Lecture Notes in Mathematics, vol. 1465. Springer, Berlin (1991)
24. David, G.: Wavelets and Singular Integrals on Curves and Surfaces. Springer, Berlin (1991)
25. David, G., Semmes, S.: Analysis of and on uniformly rectifiable sets. Mathematical Surveys and Monographs, vol. 38. American Mathematical Society, Providence (1993)
26. David, G., Semmes, S.: Uniform Rectifiability and Quasiminimizing Sets of Arbitrary Codimension. American Mathematical Society, Providence (2000)
27. Donoho, D.L., Grimes, C.: When does isomap recover natural parameterization of families of articulated images? Tech. Rep. 2002–2027, Department of Statistics, Stanford University, August 2002

28. Donoho, D.L., Grimes, C.: Hessian eigenmaps: new locally linear embedding techniques for high-dimensional data. Proc. Nat. Acad. Sciences **100**, 5591–5596 (2003)
29. Golub, G., Loan, C.V.: Matrix Computations. Johns Hopkins University Press, Baltimore (1989)
30. Jones, P., Maggioni, M., Schul, R.: Manifold parametrizations by eigenfunctions of the Laplacian and heat kernels. Proc. Nat. Acad. Sci. **105**, 1803–1808 (2008)
31. Jones, P., Maggioni, M., Schul, R.: Universal local manifold parametrizations via heat kernels and eigenfunctions of the Laplacian. Ann. Acad. Scient. Fen. **35**, 1–44 (2010) http://arxiv.org/abs/0709.1975.
32. Jones, P.W.: Rectifiable sets and the traveling salesman problem. Invent. Math. **102**, 1–15 (1990)
33. Jones, P.W.: The traveling salesman problem and harmonic analysis. Publ. Mat. **35**, 259–267 (1991) Conference on Mathematical Analysis (El Escorial, 1989)
34. Karypis, G., Kumar, V.: A fast and high quality multilevel scheme for partitioning irregular graphs. SIAM J. Sci. Comput. **20**, 359–392 (1999)
35. Little, A., Jung, Y.-M., Maggioni, M.: Multiscale estimation of intrinsic dimensionality of data sets. In: Proc. A.A.A.I. (2009)
36. Little, A., Lee, J., Jung, Y.-M., Maggioni, M.: Estimation of intrinsic dimensionality of samples from noisy low-dimensional manifolds in high dimensions with multiscale SVD. In: Proc. S.S.P. (2009)
37. Little, A., Maggioni, M., Rosasco, L.: Multiscale geometric methods for data sets I: Estimation of intrinsic dimension, submitted (2010)
38. Maggioni, M., Bremer, J. Jr., Coifman, R., Szlam, A.: Biorthogonal diffusion wavelets for multiscale representations on manifolds and graphs. SPIE, vol. 5914, p. 59141M (2005)
39. Maggioni, M., Mahadevan, S.: Fast direct policy evaluation using multiscale analysis of markov diffusion processes. In: ICML 2006, pp. 601–608 (2006)
40. Mahadevan, S., Maggioni, M.: Proto-value functions: A spectral framework for solving markov decision processes. JMLR **8**, 2169–2231 (2007)
41. Mairal, J., Bach, F., Ponce, J., Sapiro, G.: Online dictionary learning for sparse coding. In: ICML, p. 87 (2009)
42. Mairal, J., Bach, F., Ponce, J., Sapiro, G.: Online learning for matrix factorization and sparse coding. J. Mach. Learn. Res. **11**, 19–60 (2010)
43. Olshausen, B.A., Field, D.J.: Sparse coding with an overcomplete basis set: A strategy employed by V1? Vision Res. **37**, 3311–3325 (1997)
44. Rahman, I.U., Drori, I., Stodden, V.C., Donoho, D.L.: Multiscale representations for manifold-valued data. SIAM J. Multiscale Model. Simul. **4**, 1201–1232 (2005).
45. Rohrdanz, M.A., Zheng, W., Maggioni, M., Clementi, C.: Determination of reaction coordinates via locally scaled diffusion map. J. Chem. Phys. **134**, 124116 (2011)
46. Roweis, S., Saul, L.: Nonlinear dimensionality reduction by locally linear embedding. Science **290**, 2323–2326 (2000)
47. Starck, J.L., Elad, M., Donoho, D.: Image decomposition via the combination of sparse representations and a variational approach. IEEE T. Image Process. **14**, 1570 1582 (2004)
48. Szlam, A.: Asymptotic regularity of subdivisions of euclidean domains by iterated PCA and iterated 2-means. Appl. Comp. Harm. Anal. **27**, 342–350 (2009)
49. Szlam, A., Maggioni, M., Coifman, R., Bremer, J. Jr.: Diffusion-driven multiscale analysis on manifolds and graphs: top-down and bottom-up constructions. SPIE, vol. 5914(1), p. 59141D (2005)
50. Szlam, A., Maggioni, M., Coifman, R.: Regularization on graphs with function-adapted diffusion processes. J. Mach. Learn. Res. **9**, 1711–1739 (2008) (YALE/DCS/TR1365, Yale Univ, July 2006)
51. Szlam, A., Sapiro, G.: Discriminative k-metrics. In: Proceedings of the 26th Annual International Conference on Machine Learning, pp. 1009–1016 (2009)
52. Tenenbaum, J.B., Silva, V.D., Langford, J.C.: A global geometric framework for nonlinear dimensionality reduction. Science **290**, 2319–2323 (2000)

53. Tibshirani, R.: Regression shrinkage and selection via the lasso. J. Royal. Statist. Soc. B **58**, 267–288 (1996)
54. Zhang, Z., Zha, H.: Principal manifolds and nonlinear dimension reduction via local tangent space alignment. SIAM J. Sci. Comput. **26**, 313–338 (2002)
55. Zhou, M., Chen, H., Paisley, J., Ren, L., Sapiro, G., Carin, L.: Non-parametric Bayesian dictionary learning for sparse image representations. In: Neural and Information Processing Systems (NIPS) (2009)

On the Fourth-Order Structure Function of a Fractal Process*

Carol T. Christou and Garry M. Jacyna

Abstract Multifractal processes are key to modeling complex nonlinear systems. MITRE has applied fractal theory to agent-based combat simulations to understand complex behavior on the battlefield. The outstanding features of general fractal processes are long-range correlation and intermittency. If B is the lower band edge frequency of the high-pass signal component, the flatness function $F(B)$, defined as the ratio of the fourth-order moment to the square of the second-order moment of a stationary process, is a measure of the intermittency or burstiness of a random process at small scales. If $F(B)$ increases with no upper bound as B increases, then the process is intertmittent. In this work, we have derived an expression for the fourth-order structure function of the increments of a fractional Brownian motion (fBm) process through the use of integrals over the generalized multispectrum. It was concluded that the flatness function is independent of the lower edge of the high-pass signal component B, as expected of an fBm.

Keywords Complex systems • Agent-based simulation • Multifractal processes • Fractional Brownian motion • Intermittency • Second and fourth-order structure functions • Flatness function • Generalized multispectrum • Long-range correlation • Self-similarity and scale invariance

---

*Approved for Public Release: 12-1203. This work was supported by MITRE internally funded research.

C.T. Christou (✉) • G.M. Jacyna
The MITRE Corporation, 7515 Colshire Drive, McLean, VA 22102, USA
e-mail: christou@mitre.org; gjacyna@mitre.org

T.D. Andrews et al. (eds.), *Excursions in Harmonic Analysis, Volume 1*,
Applied and Numerical Harmonic Analysis, DOI 10.1007/978-0-8176-8376-4_14,

1 Introduction

The use of multifractal processes for modeling complex systems has gained wide acceptance in the mathematics, engineering, and physical sciences communities, as it has grown in leaps and bounds in the past 40 years. Simple fractal processes are divided into deterministic, e.g., the Cantor set, and random fractals, such as Brownian motion and fractional Brownian motion (fBm). Fractal patterns are observed in complex dynamical systems composed of many nonlinearly interacting parts. They often exhibit properties such as long-range correlation and intermittent or bursty behavior. Long-range correlation is the tendency of a system to interact with its parts across extended spatial and temporal scales. Complex patterns emerge that cannot be explained from an analysis of individual parts. Intermittency is an indication of global change in a complex system and is related to the relative sensitivity of the system to small changes in its internal states. MITRE has applied complexity theory to agent-based simulation, where processes unfold in a highly unpredictable manner. One of these modeling areas is combat simulation, where small changes in one part of the battle space produce profound effects in another. Multifractal techniques were necessary in the development of the underlying complexity-based analysis tools.

This short chapter examines the flatness function, a measure of the intermittency of a random process, for an fBm. The focus of the development is the derivation of the fourth-order structure function, on which the flatness depends. The fourth-order structure function is derived by integration in terms of the generalized trispectrum to find the fourth-order correlation function. Section 2 contains the theoretical development, Sect. 3 the results, while the conclusions are contained in Sect. 4.

2 Theory

Important properties of fBm are self-similarity and scale invariance. It obeys the generalized scaling relation:

$$x(\lambda t) \to \text{“law”} \lambda^H x(t), \tag{1}$$

where → “law” means that the relation is a “law” or equivalence for all probability distributions and H is called the Hurst parameter. Brownian motion is a special case of this scaling law when $H = 1/2$.

Scaling properties of the correlation function are related to analogous scaling properties of the power spectrum of a random fractal process. The second-order structure function, formally defined, is the second moment of the increment process, $E(|x(t+\tau)-x(t)|^2)$. For the increments of an fBm process:

$$E[|x(t+\tau)-x(t)|^2] = \sigma^2 V_H |\tau|^{2H}, \tag{2}$$

where V_H is a function of the Hurst parameter and $E[x]$ denotes the expectation value of x. Therefore, even though fBm is not stationary, its increments are. The flatness function $F(B)$ for a zero-mean stationary random process $x(t)$ is defined as the ratio of the fourth-order moment to the square of the second-order moment [1]:

$$F(B) = E\left[x_B^4(t)\right]/E\left[x_B^2(t)\right]^2, \tag{3}$$

where B represents the lower band edge frequency of the high-pass component of the signal and $x_B(t)$ represents the signal components above that frequency. Let $X(f)$ denote the Fourier transform of the waveform $x(t)$. The high-pass component of the waveform is obtained using an ideal high-pass filter with perfect response characteristics:

$$x_B(t) = \int_{|f|\geq B} X(f)\exp(2\pi i f t)\mathrm{d}f. \tag{4}$$

The flatness function is a measure of the *intermittency* or *burstiness* of a random process at small scales. If $F(B)$ increases without bound as B increases, the multifractal process is said to be intermittent. If the process $x(t)$ is not stationary, then the stationary increments $x(t+\tau)-x(t)$ are used. Here we use the stationary increments $(x_B(t+\tau)-x_B(t))$ of the generally nonstationary process $x_B(t)$:

$$F(B) = E\left[(x_B(t+\tau)-x_B(t))^4\right]/E\left[(x_B(t+\tau)-x_B(t))^2\right]^2, \tag{5}$$

where τ is the time increment. As $B\tau \to 0$, we regain the full signal, and $E\left[(x_B(t+\tau)-x_B(t))^4\right] \to E\left[(x(t+\tau)-x(t))^4\right]$. In this chapter, we examine the case of $B\tau \to 0$, for which the multifractal scaling law predicts $E\left[(x(t+\tau)-x(t))^q\right] = c(q)|\tau|^{\tau(q)+1}$, where $c(q)$ is a constant and $\tau(q)$ is the *scaling* or *generating* function [2]. In order to calculate the flatness function, one must derive expressions for both the second- and fourth-order structure functions of a process.

We assume a real fBm model for $x(t)$, which is useful in describing processes with long-term correlations, i.e., $1/\omega^n$ spectral behavior, where ω is the angular frequency. Because fBm is not a stationary process, however, it is difficult to define or interpret a power spectrum. In [3], it was shown that it is possible to construct fBm from a white-noise-type process through a stochastic integral in frequency of a stationary uncorrelated random process, in this case the time increments of fBm. A spectral representation of $x(t)$ was obtained assuming it is driven by a stationary white noise process $W(t)$, not necessarily Gaussian. Specifically, it was shown that

$$x(t) = \frac{1}{2\pi}\int_{-\infty}^{\infty}(\mathrm{e}^{i\omega t}-1)\left(\frac{1}{i\omega}\right)^n \mathrm{d}\beta(\omega), \tag{6}$$

where $n = H+1/2$, H being the fractal scaling parameter or Hurst coefficient, $0 < H < 1$. The quantity

$$\beta(\omega) \equiv \int_0^{\omega} F(\omega')\mathrm{d}\omega', \tag{7}$$

is a complex-valued Wiener process in frequency of orthogonal increments, where

$$F(\omega') = \int_{-\infty}^{\infty} e^{-i\omega' t} W(t) dt \tag{8}$$

is the Fourier transform of the white noise process. If $W(t)$ is Gaussian, the increments are also independent. In general, if the increments in $\beta(\omega)$ are infinitesimal, it is shown in [3] that

$$E\left[d\beta^*(\omega_1)d\beta(\omega_2)\right] = 2\pi\gamma_2^w \delta(\omega_1 - \omega_2)d\omega_1 d\omega_2, \tag{9}$$

where $\delta(\omega)$ is the Dirac delta function and γ_2^w is a constant related to the power of the white noise driving force. Using the spectral representation of $x(t)$, one can derive an expression for the second-order structure function, or generalized correlation

$$\begin{aligned} E[(x(t_1)-x(t))(x(t_2)-x(t))] &= \left(\frac{1}{2\pi}\right)^2 \int_{-\infty}^{\infty}\int_{-\infty}^{\infty} \left[e^{i\omega_1 t_1} - e^{i\omega_1 t}\right]\left(\frac{1}{i\omega_1}\right)^n \\ &\quad \times \left[e^{-i\omega_2 t_2} - e^{-i\omega_2 t}\right]\left(\frac{1}{-i\omega_2}\right)^n E\left[d\beta^*(\omega_1)d\beta(\omega_2)\right] \\ &\quad \times d\omega_1 d\omega_2 \\ &= \gamma_2^w \frac{V_n}{2}\left(|t_1-t|^{2n-1} + |t_2-t|^{2n-1} - |t_2-t_1|^{2n-1}\right), \\ V_n &= \left(\frac{2}{\pi}\right)\Gamma[1-2n]\sin(n\pi), \end{aligned} \tag{10}$$

where $\frac{3}{2} > n > \frac{1}{2}$ for convergence. If $t_2 = t_1 = \tau + t$, then $E\left[|(x(\tau+t)-x(t))|^2\right] = \gamma_2^w V_n |\tau|^{2n-1}$.

Here, we would like to derive the corresponding expression for the fourth-order structure function of an fBm process. We begin by first finding the fourth-order correlation

$$E\left[d\beta^*(\omega_4)d\beta(\omega_1)d\beta(\omega_2)d\beta(\omega_3)\right]$$

of the differential increment $d\beta(\omega) = F(\omega)d\omega$. The fourth-order correlation function of the white noise process was derived in [4] as

$$E\left[W(t_1)W(t_2)W(t_3)W(t_4)\right] = \gamma_4^w \delta(t_1-t_4)\delta(t_2-t_4)\delta(t_3-t_4), \tag{11}$$

where γ_4^w is a constant. Therefore,

$$\begin{aligned} &E\left[d\beta^*(\omega_4)d\beta(\omega_1)d\beta(\omega_2)d\beta(\omega_3)\right] \\ &= \int_{-\infty}^{\infty}\int_{-\infty}^{\infty}\int_{-\infty}^{\infty}\int_{-\infty}^{\infty} e^{i\omega_4 t_4} e^{-i\omega_1 t_1} e^{-i\omega_2 t_2} e^{-i\omega_3 t_3} \end{aligned}$$

$$\times E\left[W(t_1)W(t_2)W(t_3)W(t_4)\right]\mathrm{d}t_1\mathrm{d}t_2\mathrm{d}t_3\mathrm{d}t_4$$
$$\times \mathrm{d}\omega_1\mathrm{d}\omega_2\mathrm{d}\omega_3\mathrm{d}\omega_4 = \gamma_4^w \int_{-\infty}^{\infty}\int_{-\infty}^{\infty}\int_{-\infty}^{\infty}\int_{-\infty}^{\infty} \mathrm{e}^{i\omega_4 t_4}\mathrm{e}^{-i\omega_1 t_1}$$
$$\times \mathrm{e}^{-i\omega_2 t_2}\mathrm{e}^{-i\omega_3 t_3}\delta(t_1-t_4)\delta(t_2-t_4)\delta(t_3-t_4)$$
$$\times \mathrm{d}t_1\mathrm{d}t_2\mathrm{d}t_3\mathrm{d}t_4\mathrm{d}\omega_1\mathrm{d}\omega_2\mathrm{d}\omega_3\mathrm{d}\omega_4$$
$$= \gamma_4^w \int_{-\infty}^{\infty}\int_{-\infty}^{\infty}\int_{-\infty}^{\infty}\int_{-\infty}^{\infty}\int_{-\infty}^{\infty} e^{i(\omega_4-\omega_1-\omega_2-\omega_3)t_4}$$
$$\times \mathrm{d}t_4\mathrm{d}\omega_1\mathrm{d}\omega_2\mathrm{d}\omega_3\mathrm{d}\omega_4$$
$$= \gamma_4^w(2\pi)\int_{-\infty}^{\infty}\int_{-\infty}^{\infty}\int_{-\infty}^{\infty}\int_{-\infty}^{\infty} \delta(\omega_4-\omega_1-\omega_2-\omega_3)$$
$$\times \mathrm{d}\omega_1\mathrm{d}\omega_2\mathrm{d}\omega_3\mathrm{d}\omega_4. \tag{12}$$

By analogy to Eq. (10) one may derive the general correlation of four arbitrary increments in $x(t)$:

$$E\Big[(x(t_1)-x(t))(x(t_2)-x(t))\times(x(t_3)-x(t))(x(t_4)-x(t))\Big]$$
$$= \left(\frac{1}{2\pi}\right)^4 \int_{-\infty}^{\infty}\int_{-\infty}^{\infty}\int_{-\infty}^{\infty}\int_{-\infty}^{\infty} \left[\mathrm{e}^{i\omega_1 t_1}-\mathrm{e}^{i\omega_1 t}\right]\left(\frac{1}{i\omega_1}\right)^n$$
$$\times \left[\mathrm{e}^{i\omega_2 t_2}-\mathrm{e}^{i\omega_2 t}\right]\left(\frac{1}{i\omega_2}\right)^n \left[\mathrm{e}^{i\omega_3 t_3}-\mathrm{e}^{i\omega_3 t}\right]\left(\frac{1}{i\omega_3}\right)^n$$
$$\times \left[\mathrm{e}^{-i\omega_4 t_4}-\mathrm{e}^{-i\omega_4 t}\right]\left(\frac{1}{-i\omega_4}\right)^n$$
$$\times E\left[\mathrm{d}\beta(\omega_1)\mathrm{d}\beta(\omega_2)\mathrm{d}\beta(\omega_3)\mathrm{d}\beta^*(\omega_4)\right]$$
$$= \left(\frac{1}{2\pi}\right)^4 \int_{-\infty}^{\infty}\int_{-\infty}^{\infty}\int_{-\infty}^{\infty}\int_{-\infty}^{\infty} \left[\mathrm{e}^{i\omega_1 t_1}-\mathrm{e}^{i\omega_1 t}\right]\left(\frac{1}{i\omega_1}\right)^n$$
$$\times \left[\mathrm{e}^{i\omega_2 t_2}-\mathrm{e}^{i\omega_2 t}\right]\left(\frac{1}{i\omega_2}\right)^n \left[\mathrm{e}^{i\omega_3 t_3}-\mathrm{e}^{i\omega_3 t}\right]\left(\frac{1}{i\omega_3}\right)^n$$
$$\times \left[\mathrm{e}^{-i\omega_4 t_4}-\mathrm{e}^{-i\omega_4 t}\right]\left(\frac{1}{-i\omega_4}\right)^n (2\pi)\gamma_4^w$$
$$\times \delta(\omega_4-\omega_1-\omega_2-\omega_3)\mathrm{d}\omega_1\mathrm{d}\omega_2\mathrm{d}\omega_3\mathrm{d}\omega_4$$
$$= \left(\frac{1}{2\pi}\right)^3 \gamma_4^w \int_{-\infty}^{\infty}\int_{-\infty}^{\infty}\int_{-\infty}^{\infty} \left[\mathrm{e}^{i\omega_1 t_1}-\mathrm{e}^{i\omega_1 t}\right]$$
$$\times \left[\mathrm{e}^{i\omega_2 t_2}-\mathrm{e}^{i\omega_2 t}\right]\left[\mathrm{e}^{i\omega_3 t_3}-\mathrm{e}^{i\omega_3 t}\right]$$

$$\times \left[\mathrm{e}^{-i(\omega_1+\omega_2+\omega_3)t_4} - \mathrm{e}^{-i(\omega_1+\omega_2+\omega_3)t} \right] \left(\frac{1}{i\omega_1} \right)^n \left(\frac{1}{i\omega_2} \right)^n$$

$$\times \left(\frac{1}{i\omega_3} \right)^n \left(\frac{1}{-i(\omega_1+\omega_2+\omega_3)} \right)^n \mathrm{d}\omega_1 \mathrm{d}\omega_2 \mathrm{d}\omega_3, \tag{13}$$

which is the expression for the fourth-order correlation in terms of a "generalized trispectrum" [5, 6]:

$$\Phi_T(\omega_1, \omega_2, \omega_3) \sim \frac{1}{\omega_1^n \omega_2^n \omega_3^n (-\omega_1 - \omega_2 - \omega_3)^n}. \tag{14}$$

Now, if $t_1 = t_2 = t_3 = t_4 = t + \tau$, Eq. (13) is precisely the fourth-order structure function of $x(t)$. It can easily be seen that the phase factors in t will then cancel out and only phase factors in τ will appear.

Transforming to the variables $\gamma = \left[\omega_1 + \frac{\omega_2+\omega_3}{2}\right]$, $\alpha = [\omega_2 + \omega_3]$, $\beta = \left[\frac{\omega_3}{2} - \frac{\omega_2}{2}\right]$, Eq. (13) becomes

$$E\left[|(x(t+\tau) - x(t))|^4\right])$$
$$= \left(\frac{1}{2\pi}\right)^3 4\gamma_4^w \int_{-\infty}^{\infty} \int_{-\infty}^{\infty} \frac{\left(\cos(\frac{\alpha}{2}\tau) - \cos(\gamma\tau)\right)}{\left(\frac{\alpha^2}{4} - \gamma^2\right)^n} \mathrm{d}\gamma$$
$$\int_{-\infty}^{\infty} \frac{\left(\cos(\frac{\alpha}{2}\tau) - \cos(\beta\tau)\right)}{\left(\frac{\alpha^2}{4} - \beta^2\right)^n} \mathrm{d}\beta \mathrm{d}\alpha. \tag{15}$$

We see that the integrals over β and γ are identical, have even integrands, and for arbitrary n will contain real and imaginary parts. A simultaneous change of sign in both denominators implies that

$$\left[\int_0^{\infty} \frac{\left(\cos(\frac{\alpha}{2}\tau) - \cos(\beta\tau)\right)}{\left(\frac{\alpha^2}{4} - \beta^2\right)^n} \mathrm{d}\beta \right]^2 = \left[\int_0^{\infty} \frac{\left(\cos(\frac{\alpha}{2}\tau) - \cos(\beta\tau)\right)}{\left(\beta^2 - \frac{\alpha^2}{4}\right)^n} \mathrm{d}\beta \right]^2. \tag{16}$$

Equation (16) in turn implies that

$$\int_0^{a/2} \frac{\left(\cos(\frac{\alpha}{2}\tau) - \cos(\beta\tau)\right)}{\left(\frac{\alpha^2}{4} - \beta^2\right)^n} \mathrm{d}\beta = \int_{a/2}^{\infty} \frac{\left(\cos(\frac{\alpha}{2}\tau) - \cos(\beta\tau)\right)}{\left(\beta^2 - \frac{\alpha^2}{4}\right)^n} \mathrm{d}\beta. \tag{17}$$

Now, for $\alpha > 0$, the values of these integrals are

$$\int_0^{a/2} \frac{(\cos(\frac{\alpha}{2}\tau) - \cos(\beta\tau))}{\left(\frac{\alpha^2}{4} - \beta^2\right)^n} \mathrm{d}\beta$$
$$= -\left(\alpha^{\frac{1}{2}-2n}\sqrt{\pi}\Gamma[1-n]\left(-\left(4^n\sqrt{\alpha}\cos\left(\frac{a}{2}\right)\right)\right.\right.$$
$$\left.\left.+2a^n J_{\frac{1}{2}-n}\left(\frac{a}{2}\right)\Gamma\left[\frac{3}{2}-n\right]\right)\right)\left(\frac{1}{4\Gamma\left[\frac{3}{2}-n\right]}\right);$$
$$\int_{a/2}^{\infty} \frac{(\cos(\frac{\alpha}{2}\tau) - \cos(\beta\tau))}{\left(\beta^2 - \frac{\alpha^2}{4}\right)^n} \mathrm{d}\beta$$
$$= \left(\alpha^{\frac{1}{2}-2n}\Gamma[1-n]\left(4^n\sqrt{\alpha}\cos\left(\frac{a}{2}\right)\Gamma\left[-\frac{1}{2}+n\right]\right.\right.$$
$$\left.\left.+2a^n\pi\left(J_{\frac{1}{2}-n}\left(\frac{a}{2}\right)\sec(n\pi) - J_{n-\frac{1}{2}}\left(\frac{a}{2}\right)\tan(n\pi)\right)\right)\right)\left(\frac{1}{4\sqrt{\pi}}\right); \tag{18}$$

([7], p. 427, Eqs. 8 and 9), where $J_\nu(x)$ is the Bessel function of order ν. Simple plots of these results versus α easily demonstrate that the two expressions are in fact not the same. Therefore α cannot be positive but must equal zero in order for the above integrals to exist. For α=0 Eq. (15) then becomes

$$E\left[|(x(t+\tau) - x(t))|^4\right]$$
$$= \left(\frac{1}{2\pi}\right)^3 16\gamma_4^w|\tau|^{m-3}\int_{-\infty}^{\infty} 2\pi\delta(\alpha)$$
$$\int_0^{\infty} \frac{(\cos(\frac{\alpha}{2}) - \cos(\gamma))}{\left(\frac{\alpha^2}{4} - \gamma^2\right)^n}\mathrm{d}\gamma \int_0^{\infty} \frac{(\cos(\frac{\alpha}{2}) - \cos(\beta))}{\left(\frac{\alpha^2}{4} - \beta^2\right)^n}\mathrm{d}\beta\mathrm{d}\alpha$$
$$= \left(\frac{1}{2\pi}\right)^2 16\gamma_4^w|\tau|^{m-3}\left[\int_0^{\infty} \frac{(1-\cos(\beta))}{(\beta^2)^n}\mathrm{d}\beta\right]^2$$
$$= \left(\frac{1}{2\pi}\right)^2 16\gamma_4^w|\tau|^{m-3}\left[\Gamma[1-2n]\sin(n\pi)\right]^2, \tag{19}$$

where we have divided the variables α, β, γ by a scale factor τ, and $m = 4n+1$.[1] The integral $\int_0^\infty \frac{(1-\cos(\beta))}{(\beta^2)^n}\mathrm{d}\beta$ will converge at infinity if $-2n+1 < 0$ or $n > 1/2$

[1] $\delta(\frac{\alpha}{\tau}) = \tau\delta(\alpha)$.

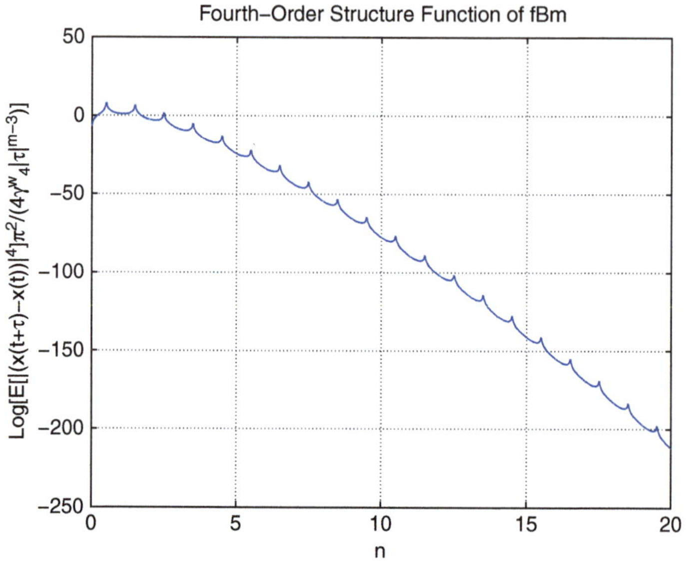

Fig. 1 Behavior of fourth-order correlation versus n

since the numerator remains finite. Near $\beta = 0$, the numerator behaves as $\sim \beta^2$; therefore the integral remains finite if $-2n+3 > 0$ or $n < 3/2$. This implies that $7 > m > 3$ for convergence. The behavior of the fourth-order structure function in Eq. (19) is shown as a function of n in Figs. 1 and 2. Figure 2 is a detailed view of the larger-scale behavior in Fig. 1. The physical region for the fBm process lies between $n = 1/2$ and $n = 3/2$, as it does in the case of the second-order structure function in Eq. (10). It is interesting to note that the large-scale behavior of the logarithm of $E\left[|(x(t+\tau) - x(t)|^4\right]$ versus n resembles a hyperbola.

3 Results

Let us evaluate the integral in Eq. (15) directly for $n = 1$. Since

$$\int_{-\infty}^{\infty} \frac{\left(\cos(\frac{\alpha}{2}) - \cos(\gamma)\right)}{\left(\frac{\alpha^2}{4} - \gamma^2\right)} \mathrm{d}\gamma = 2\frac{\pi}{\alpha} \sin\left(\frac{\alpha}{2}\right), \tag{20}$$

([7], p. 407, Eq. 9), we have

$$\int_{-\infty}^{\infty} 4\pi^2 \frac{\sin^2\left(\frac{\alpha}{2}\right)}{\alpha^2} \mathrm{d}\alpha = 8\pi^2 \frac{\pi}{4} = 2\pi^3, \tag{21}$$

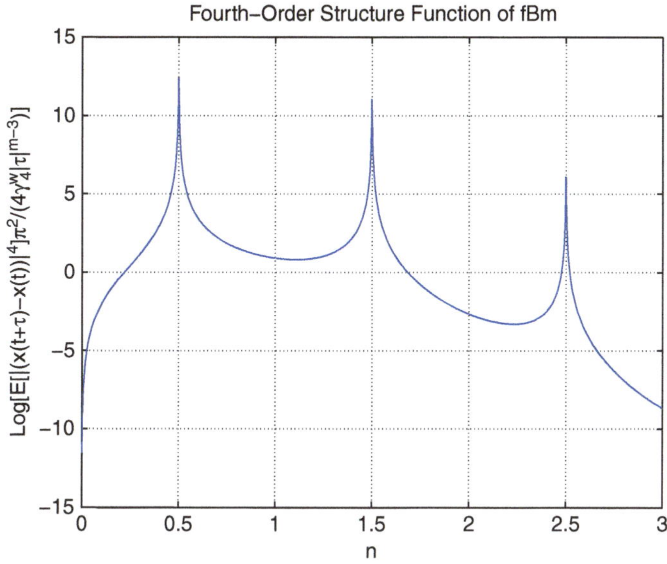

Fig. 2 Behavior of fourth-order correlation versus n near physical region

([7], p. 414, Eq. 3). Alternatively, inserting the delta function in Eq. (15) and setting $\alpha = 0$, we have

$$\int_{-\infty}^{\infty} \frac{(1-\cos(\gamma))}{(\gamma^2)} \mathrm{d}\gamma = 2\frac{\pi}{2}, \tag{22}$$

which when squared and combined with the factor 2π from the delta function gives us $2\pi^3$, the same answer as above. Thus, the fourth-order structure function for fBm is nonzero only along the normal submanifold $\alpha = 0$ of the three-dimensional frequency space.

Dividing Eq. (19) by Eq. (10) squared for $t_1 = t_2 = \tau + t$, we obtain for the flatness function the quantity $\gamma_4^w/(\gamma_2^w)^2$, which, as expected, is a constant for the increments of an fBm. This then implies that the fBm is also not intermittent.

4 Conclusions

In trying to model complex behavior on the battlefield through agent-based simulation, it became necessary to use the theory of complexity, in particular that of multifractal processes. Generalized fractal phenomena are characterized by long-range correlation and intermittency. This work examined the properties of an fBM process, whose increments are stationary. Specifically, in order to determine the intermittency characteristics, it was necessary to calculate the flatness function, the ratio of the fourth-order structure function to the square of the second-order structure function of the increments of an fBm. The focus was on the derivation of the fourth-order structure function through the use of the generalized trispectrum of the process. It was found that the flatness function is independent of the lower edge frequency of the high-pass signal component, which leads to the conclusion that an fBm is not intermittent.

References

1. Frisch, U.: Turbulence, p. 122. Cambridge University Press, Great Britain (1995)
2. Fisher, A., Calvert, L., Mandelbrot, B.: Multifracticality of Deutschmark/US Dollar Exchange Rates. Cowes Foundation Discussion Paper, no. 1165, pp. 3–4 (1997)
3. Reed, I.S., Lee, P.C., Truong, T.K.: Spectral representation of fractional Brownian motion in *n* dimensions and its properties. IEEE Trans. Inform. Theor. **41**(5), 1439–1451 (1995)
4. Nikias, C.L., Petropulu, A.P.: Higher-Order Spectra Analysis: A Nonlinear Signal processing Framework, pp. 24–25. Prentice Hall Signal processing Series, PTR Prentice-Hall, Englewood Cliffs (1993)
5. Mendel, J.M.: Tutorial on higher-order statistics (spectra) in signal processing and system theory: theoretical results and some applications. In: Proceedings of the IEEE **79**(3), 278–305 (1991)
6. Swami, A., Mendel, J.M., Nikias, C.L.: Higher-Order Spectral Analysis Toolbox (For Use With MATLAB), pp. 1–7. The Mathworks, Inc., Version 2 (1998)
7. Gradshteyn, I.S., Ryzhik, I.M.: Table of Integrals, Series, and Products, pp. 420–421, 933. Academic, New York (1980)

Harmonic Analysis of Databases and Matrices

Ronald R. Coifman and Matan Gavish

Abstract We describe methods to organize and process matrices/databases through a bi-multiscale tensor product harmonic analysis on row and column functions. The goal is to reorganize the matrix so that its entries exhibit smoothness or predictability relative to the tensor row column geometry. In particular we show that approximate bi-Holder smoothness follows from simple l^p entropy conditions. We describe various applications both for the analysis of matrices of linear transformations, as well for the extraction of information and structure in document databases.

Keywords Databases • Tensor Haar • Tensor harmonic analysis • Bi-Holder • Partition trees • Machine learning • Diffusion geometry

1 Introduction

In this chapter, we describe a program in analysis, designed to integrate ideas from harmonic analysis and geometry with ideas and methods from machine learning.

Digital data is usually represented as sets of points in high-dimensional Euclidean space. A text document, for example, may be converted to a point, whose coordinates are the list of occurrence frequencies of words in a lexicon. A corpus of documents is then represented as a collection of such points.

Consider now the task of organizing such a dataset. Continuing our example, the documents are to be organized according to their mutual relevance. We can view

R.R. Coifman (✉)
Program in Applied Mathematics, Yale University, New Haven, UK
e-mail: coifman@yale.edu

M. Gavish
Department of Statistics, Stanford University, Stanford, CA 94305, USA
e-mail: gavish@stanford.edu

T.D. Andrews et al. (eds.), *Excursions in Harmonic Analysis, Volume 1*,
Applied and Numerical Harmonic Analysis, DOI 10.1007/978-0-8176-8376-4_15,

this data set as a cloud of points in high-dimensional Euclidean space. Our goal is to build both a geometry and a corresponding harmonic analysis so that various functions on the data are efficiently represented.

We can also view this dataset as a matrix, in which each column is a document whose entries are the frequencies of occurrence of a given word in a lexicon. In a dual fashion, each row is a word whose entries are labeled by the documents. Below, we review a recently introduced organization scheme, which jointly organizes the language (namely, the rows) and the documents (the columns) into a coherent product structure. In this scheme, tensor harmonic analysis of the product set $\{\text{rows}\} \times \{\text{columns}\}$ plays a central role.

The common ground, or underlying geometric fabric, is provided through the construction of various graphs, on which analysis is performed. In mathematics we have seen extensions of tools from harmonic analysis to Riemannian manifolds, or more generally to subsets of $\mathbb{R}^n$. One of the most powerful approaches consisted in the study of the geometry of a manifold through the properties of functions and operators on the manifold. In particular, the Laplace-Beltrami operator, as well as pseudo-differential or singular integral operators, has played a central role in revealing deep connections between geometry and analysis. Here, we describe the interplay of such analytic tools with the geometry and combinatorics of data and information. We will provide a range of illustrations and application to the analysis of operators and, at the same time, to the analysis of documents and questionnaires.

The following topics are interlaced below:

- Geometries of point clouds
- From local to global—the role of eigenfunctions
- Diffusion geometries in original coordinates and organization
- Coupled dual geometries: duality between rows and columns, tensor product geometries
- Sparse tensor product grids and efficient processing of data
- Harmonic analysis, Haar systems, tensor Besov and bi-Hölder functions, Calderon–Zygmund decomposition
- Applications: to mathematics organization of operators the dual geometries of eigenvectors

2 Organization of Point Clouds in $\mathbb{R}^n$ by Eigenvector Embeddings

When representing the dataset as a collection of points in high-dimensional Euclidean space, a naïve organization scheme might be to construct a distance (or similarity) between two documents through the Euclidean distance or the inner product of their corresponding vectors. However, already in moderate dimensions, most points are far away or essentially orthogonal. The distances in high dimensions are informative only when they are quite small. In our document example, if the empirical distributions of the vocabulary in two documents are very close, we can infer that they deal with a similar topic. In this case we can link the two documents and weigh the link by a weight reflecting the probability that the documents are dealing with the same topic. We construct a nondirected, weighted graph whose nodes are the documents, as well as a corresponding random walk (or diffusion process) and a Laplace operator on this graph. This linkage structure provides a local model for the relationship between document, we claim that the eigenvectors of the corresponding diffusion operator enable a global "integration/organization of the data".

This result is remarkable even in the case of Riemannian manifolds, where a local infinitesimal metric defines a Laplace operator or a heat diffusion process and the first few eigenvectors of the Laplace operator provide an explicit global embedding of the manifold in $\mathbb{R}^n$.

We claim that the basic concepts of differential calculus can be restated in terms of eigenvectors of appropriate local linear transformations. Let us illustrate this process of the local-to-global integration in analysis by two concrete examples.

The first example is a nonlinear reformulation of the fundamental theorem of calculus [16]. Consider the following basic problem posed by Cauchy, also known as the *sensor localization* problem: Assume, for example, that you know some of the distances between points in the plane (or in high dimension) and assume also that enough distances are known to determine the system. The problem is to reconstruct the point set, up to a global rigid transformation. If enough local triangles with known length, are given, we can compute a local map which can be assembled bit by bit like a puzzle, in a process analogous to integration. A more powerful equivalent method is obtained by writing each point as the center of mass of its known neighbors, namely,

$$P_i = \sum_{j:j\sim i} A_{ij} P_j \text{ (sum over the neighbors of } i\text{)},$$

where $\sum_{j:j\sim i} A_{ij} = 1$. This equation, which is invariant under rigid motion and scaling, tells us that the vector of x coordinates of all points is an eigenvector corresponding to eigenvalue 1 of the matrix A. Similarly, the vector of y coordinates as well as all of the vectors whose entries equal 1 are in the same space. We thus see that the solution to the rigidity problem is obtained by finding a basis of this space and expressing three points in this basis (using their mutual distances). Indeed, the power iteration algorithm for computing eigenvectors consists of iterating and rescaling matrix A as it acts on the iterates of a random vector. This is effectively a process of integration.

Observe, for example, that if the points are of the form $(i, f(i))$ and we know the differences $|f(i) - f(i-1)|$ and $|f(i) - f(i-2)|$, then we know the distance in the plane between neighbors and therefore can determine f. This is a simple (but nonlinear) version of the fundamental theorem of calculus.

As a second example, let $\Delta = I - A$ where A is an averaging operator, and consider the Poisson-type equation

$$\Delta u = f. \tag{1}$$

We claim that we can find the solution u as an eigenvector

$$Bu = u,$$

where

$$B = 2A - A^2 - \delta A \sigma (I - A)$$

with $\sigma = sgn(f)$ and $\delta = \frac{\Delta f}{A(|f|)}$. Both scalars δ and σ are diagonal multiplications.

This casts the difference equation as a fixed point equation. In other words, the solution of the local discrete infinitesimal model (1) is given in terms of eigenvectors of a corresponding local "affinity" matrix B.

More generally for a cloud of points, similar to the body of text documents mentioned above, we can define a diffusion linking points by building appropriate linear models or affinity kernels, whose eigenvectors are going to map out the data.

The affinity between two points X_p, X_q is defined using the Markov matrix

$$A_{p,q} = \frac{\exp\left(-\left|X_p - X_q\right|^2 / \varepsilon\right)}{\Sigma_q \exp\left(-\left|X_p - X_q\right|^2 / \varepsilon\right)}$$

or, alternatively, as a bi-Markovian version

$$A_{p,q} = \frac{\exp\left(-\left|X_p - X_q\right|^2 / \varepsilon\right)}{\omega(p)\omega(q)},$$

where the weights are selected so that A is Markov in both p and q.

Let us mention two examples where this diffusion operator A arises in classical analysis. First, consider points that are uniformly distributed on a smooth submanifold of Euclidean space, $\Delta = \frac{1}{\varepsilon}(I - A)$ is an approximation of the induced Laplace-Beltrami operator on the manifold, and the eigenvectors of A approximate the eigenvectors of the Laplace operator. Moreover the powers of A correspond to diffusion on the manifold scaled by ε [4, 5, 9].

Second, consider data generated by a stochastic Langevin equation (namely, a stochastic gradient descent differential equation). In this case, $\Delta = \frac{1}{\varepsilon}(I - A)$ approximates the corresponding Fokker–Planck operator [9, 11].

Generally, we have the following organization scheme. We view A as a local model for point affinities (or diffusion probabilities). We diagonalize it and write

$$A_{p,q} = \sum \lambda_\ell^2 \varphi_\ell(X_p) \varphi_\ell(X_q).$$

We then use the eigenvectors φ_ℓ to define a t-dependent family of embeddings $X_p \mapsto \tilde{X}_p^t$. The t-th embedding integrates all local interrelations to range t:

$$\tilde{X}_p^t = \left(\lambda_1^t \varphi_1(X_p), \ldots \lambda_m^t \varphi_m(X_p)\right).$$

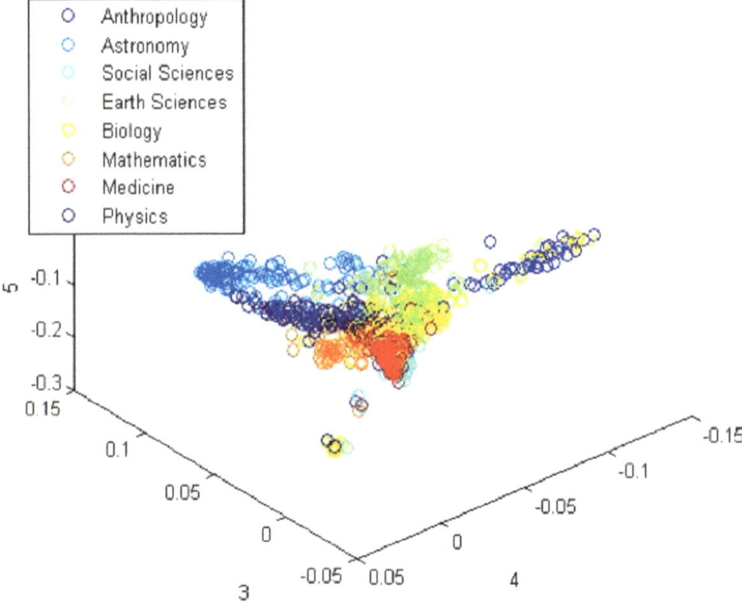

Fig. 1 An embedding of the science news document dataset in $\mathbb{R}^3$

For a given t we determine m so that λ_{m+1}^t is negligible, and embed the data into m-dimensional Euclidean space.

The **diffusion distance** at time t between X_p and X_q is then defined as

$$\begin{aligned} d_t^2(p,q) &= A_{p,p}^t + A_{q,q}^t - 2A_{p,q}^t \\ &= \sum \lambda_\ell^{2t} \left(\varphi_\ell(X_p) - \varphi_\ell(X_q)\right)^2 \\ &= \|\tilde{X}_p^t - \tilde{X}_q^t\|^2. \end{aligned}$$

The diffusion distance at time t is interpreted as the distance in L^2 between the kernels of the diffusion operator at time t starting at p and the one starting at q.

Figure 1 shows a diffusion map embedding in $\mathbb{R}^3$ of the "Science News" database of documents [15], which has been converted to point cloud in high dimension as described above. Note that this low-dimensional embedding already provides a topic organization of the documents: evidently, different disciplines correspond to different colors or different regions in the embedding.

To recap, in this approach, data points are represented as points in high-dimensional Euclidean space; a diffusion operator on this point set is constructed, based on distances between neighboring points; and finally, eigenvectors of a power of this operator that correspond to large eigenvalues are used to embed the data set in a lower-dimensional Euclidean space and to organize it.

3 Organization of Databases Using Coupled Partition Trees

We now turn to an alternative organization method, in which a dataset is viewed as a matrix—a function on a product structure—rather than as a collection of points in Euclidean space. The rows and columns of the matrix are treated on equal footing and are jointly organized.

3.1 *Partition Trees*

Let us consider a multiscale approach to achieve local-to-global integration of local affinities, this is the combinatorial analogue of the diagonalization of a local affinity operator (similar to the relation between wavelets and Fourier series in the classical context). We can collect neighboring points into "folders," or sets, and collect the folders into super-folders, and so on. This yields an organization of the point set through a hierarchical partition tree [7, 8, 12].

A minimal multiscale partition tree organizing a set X at different levels of granularity can be obtained as follows. The first partition, at level $\ell = 0$, consists of singletons. Assume that the partition at level ℓ, denoted by $X = \biguplus x_j^\ell$, has been obtained and that a metric d_ℓ (or an affinity) on the set $\{x_j^\ell\}$ is available.

Let $x_j^{\ell+1}$ be a maximal subcollection of points in x_j^ℓ such that $d_\ell\left(x_j^{\ell+1}, x_i^{\ell+1}\right) > 1/2$. Clearly, each point is at distance less than $1/2$ at scale ℓ from one of the selected key-points. This allows us to create the next partition, at level $\ell+1$, by assigning each point to the nearest point in the key-point grid, that is the point minimizing $d_\ell\left(x, x_i^{\ell+1}\right)$.

This yields a partition tree of disjoint sets at each level ℓ. We refer to this increasingly refined set of partitions as *partition tree* (Fig. 2).

Given a set X equipped with an affinity matrix A, we let d_ℓ be the diffusion distance at time 2^l. We can thus use the local-scaled affinities defined by the powers of A to build a multiscale (in diffusion time) partition tree on X. Observe that this is the usual construction of a vector quantization tree in the Euclidean case.

This construction when applied to text documents (equipped with semantic coordinates) builds an automatic folder structure with corresponding key-documents characterizing the folders or contexts; when applied to the vocabulary, it yields a hierarchy of conceptual word bags. Note that this construction is a well-known stochastic construction used in the past to achieve optimal low-dimensional metric embeddings [2, 3].

For example, Fig. 3 shows an image patch dataset. We consider the dataset of all 8×8 image patches (right panel) taken from a simple image (left panel). The diffusion embedding organizes this dataset as a sphere (lower panel). Using local Euclidean distances in $\mathbb{R}^{64}$ between image patches as affinities, we can alternatively organize this dataset by constructing a partition tree. The folders correspond to

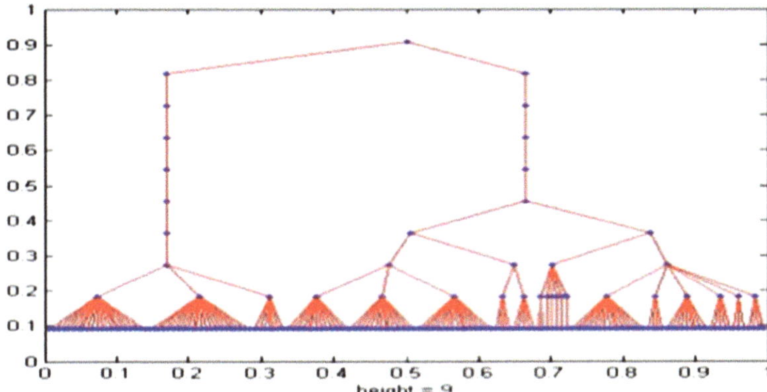

Fig. 2 Hierarchical partition tree

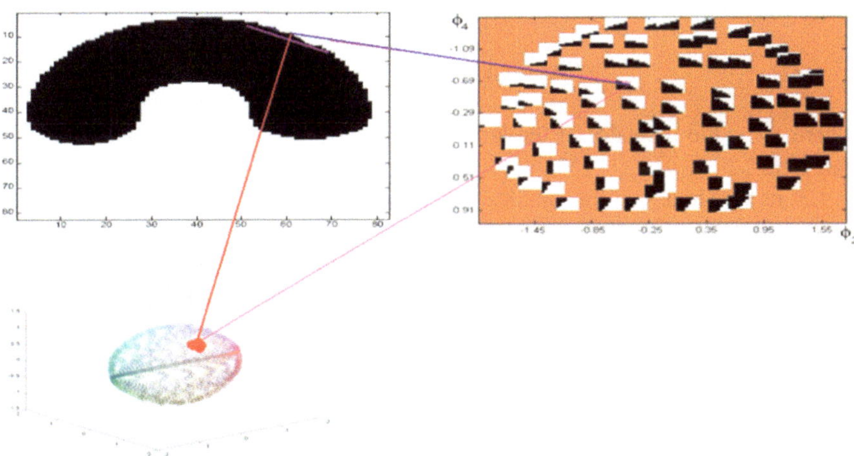

Fig. 3 Organizing an image patch dataset

patches with a similar proportion of black to white and a similar orientation (e.g., the red folder on the lower panel corresponds to the patches indicated on the right panel). Consequently, the tree organization is nothing but a multiscale hierarchy on the sphere.

3.2 *Haar Bases*

A hierarchical partition tree on a dataset X allows harmonic analysis of real-valued functions on X, as it induces special orthonormal bases called *Haar bases* [13].

A Haar basis is obtained from a partition tree as follows. Suppose that a node in the tree has n children, that is, that the set described by the node decomposes into n subsets in the next, more refined, level. Then this node contributes $n-1$ functions to the basis. These functions are all supported on the set described by the node, are piecewise constant on its n subsets, all mutually orthogonal, and are orthogonal to the constant function on the set.

Observe that just like the classical Haar functions, coefficients of an expansion in a Haar basis measure variability of the conditional expectations of the function in sub-nodes of a given node.

Tasks such as compression of functions on the dataset, as well as subsampling, denoising, and learning of such functions, can be performed in Haar coefficient space using methods familiar from Euclidean harmonic analysis and signal processing [13].

Some results for the classical Haar basis on $[0,1]$ extend to generalized Haar bases. Recall that the classical Haar functions are given by

$$h_\ell(x) = \left(|I|^{-\frac{1}{2}}\right)(\chi_- - \chi_+),$$

where χ_- is the indicator of the left half of I and χ_+ is the indicator of the right half of I.

Let us first note that the classical Haar basis on $[0,1]$ is the Haar basis induced by the partition tree of dyadic subintervals of $[0,1]$. This tree defines a natural dyadic distance $d(x,y)$ on $[0,1]$, defined as the length of the smallest dyadic interval containing both points. Hölder classes in the metric d are characterized by the Haar coefficients $a_I = \int f(x)h_I(x)\mathrm{d}x$:

$$|a_I| < c|I|^{\frac{1}{2}+\beta} \Leftrightarrow |f(x) - f(x')| < c \cdot d(x,x')^\beta .$$

This theorem holds for any Haar basis when d is the tree metric induced by the partition tree, and $|I| = \frac{\#I}{\#X}$ is the normalized size of the folder I.

3.3 Coupled Partition Trees

Let us return to the text document example above. Consider a matrix M whose columns are the terms-frequency vectors, so that columns correspond to documents and rows correspond to terms. The partition tree construction can be applied to the columns of M, where the affinity matrix A is obtained from local distances between the columns as vectors in Euclidean space. Each partition in the resulting partition tree can be interpreted as a division of the documents into contexts, or subjects. However, it can also be applied to the set of rows of M, resulting in partitions of the terms into concepts or word bags.

Coupling the construction of the two partition trees—on the columns and the rows—takes us away from the representation of the dataset as a point cloud in Euclidean space toward representation of the dataset as a function on the product set $\{\text{rows}\} \times \{\text{columns}\}$. We now consider data matrix M and assume two partition trees—one on the column set of M and one on the row set of M—have already been constructed. Each tree induces a Haar basis and a tree metric as above.

The tensor product of the Haar bases is an orthonormal basis for the space of matrices of the same dimensions as M. We now consider an analysis of M in this basis.

Denote by $|R| = |I \times J|$ a "rectangle" of entries of M, where I is a folder in the column tree and J is a folder in the row tree. Denote by $|R| = |I||J|$ the volume of the rectangle R. Indexing Haar functions by their support folders, we write $h_I(x)$ for a Haar function on the rows. This allows us to index basis functions in the tensor product basis by rectangles and write $h_R(x,y) = h_I(x)h_J(y)$.

Analysis and synthesis of the matrix M is in the tensor Haar basis is simply

$$a_R = \int M(x,y)h_R(x,y)\mathrm{d}x\mathrm{d}y$$

$$M(x,y) = \sum_R a_R h_R(x,y).$$

The characterization of Hölder functions mentioned above extends to mixed-Hölder matrices [7, 8, 12]:

$$\left|a_R\right| < c\left|R\right|^{1/2+\beta} \Leftrightarrow \left|f(x,y) - f(x',y) - f(x,y') + f(x',y')\right| \leq cd(x,x')^\beta D(y,y')^\beta,$$

where d and D are the tree metrics induced by the partition trees on the rows and columns, respectively.

Simplicity or sparsity of an expansion is quantified by an entropy such as

$$e_\alpha(f) = \left(\sum \left|a_R\right|^\alpha\right)^{1/\alpha}$$

for some $\alpha < 2$.

The relation between this entropy, efficiency of the representation in tensor Haar basis and the mixed-Hölder condition, is given by the following two propositions [7, 8, 12].

Proposition 1. *Assume $e_\alpha(f) = (\sum \left|a_R\right|^\alpha) \leq 1$. Then the number of coefficients needed to approximate the expansion to precision $\varepsilon^{1-\alpha/2}$ does not exceed $\varepsilon^{-\alpha}$, and we need only consider large coefficients corresponding to Haar functions whose support is large. Specifically, we have*

$$\int \left| f - \sum_{|R|>\varepsilon,\, |a_R|>\varepsilon} a_R h_R(x) \right|^\alpha \mathrm{d}x < \varepsilon^{1-\alpha/2}$$

The next proposition shows that $e_\alpha(f)$ estimates the rate at which f can be approximated by Hölder functions outside sets of small measure.

Proposition 2. *Let f be such that $e_\alpha \leq 1$. Then there is a decreasing sequence of sets E_ℓ such that $|E_\ell| \leq 2^{-\ell}$ and a decomposition of Calderon–Zygmund type $f = g_\ell + b_\ell$. Here, b_ℓ is supported on E_ℓ and g_ℓ is Hölder $\beta = 1/\alpha - 1/2$ with constant $2^{(\ell+1)/\alpha}$. Equivalently, g_ℓ has Haar coefficients satisfying $|a_R| \leq 2^{(\ell+1)/\alpha}|R|^{1/\alpha}$.*

To connect the bounded entropy condition with classical Euclidean harmonic analysis, consider the tensor product of smooth wavelet basis on $[0,1]$ with itself, yielding an orthonormal basis for $L_2\left([0,1]^2\right)$. The class of functions, whose wavelet expansion has finite entropy, has been characterized in [10]. There, it is shown that this class is independent of the wavelet used and equals the class of functions having a harmonic extension whose corresponding fractional derivative is integrable in the disk (or bi-disk). The dual spaces are also characterized as Bloch spaces, which in our dyadic structure case are just functions with bounded Haar coefficients. Also observe that for $f : [0,1]^2 \to \mathbb{R}$ $f = \sum |R|^{1/2} a_R |R|^{-1/2} h_R$ is a special atomic decomposition of $\left(\frac{\partial}{\partial x}\right)^{1/2} \left(\frac{\partial}{\partial y}\right)^{1/2} f$ which is therefore in the Hardy space H^1 of the bi-disk. A similar result holds for the other entropies, implying a fractional derivative in the Hardy space.

Proposition 2 decomposes any matrix into a "good," or mixed-Hölder part, and a "bad" part with small support.

Mixed-Hölder matrices indeed deserve to be called "good" matrices, as they can be substantially subsampled. To see this, note that the number of samples needed to recover the functions to a given precision is of the order of the number of tensor Haar coefficients needed for that precision. For balanced partition trees, this is approximately the number of bi-folders R, whose area exceeds the precision ε. This number is of the order of $1/\varepsilon \log(1/\varepsilon)$.

Details of the sampling and reconstruction schemes are available in [12]. The sampling patterns that allow approximate reconstruction of a matrix M from $1/\varepsilon \log(1/\varepsilon)$ samples is analogous to a *sparse grid* [6], a sampling pattern in Euclidean space originally introduced by Smolyak [17].

Propositions 1 and 2 imply that the entropy condition quantifies the compatibility between the pair of partition trees (on the rows and on the columns) and the matrix on which they are constructed. In other words, to construct useful trees, we should seek to minimize the entropy in the induced tensor Haar basis.

For a given matrix M, finding a partition tree pair, which is a global minimum of the entropy, is computationally intractable and not sensible, as the matrix could be the superposition of different structures , corresponding to conflicting organizations. At best we should attempt to peel off organized structured layers.

Let us describe iterative, heuristic procedures for building tree pairs that often perform well in practice.

Observe that the simplest geometry that can be built to fit a given function is to postulate that level sets are points whose proximity is defined by the values of the function, this corresponds to rearranging the functions level sets by decreasing values of the function. In reality we are always confronting a collection of functions such as the coordinates of our cloud of points in high dimensions (or the rows of a matrix viewed as a collection of functions on the columns, representing points in space) and are therefore forced to tune the geometry to render all coordinates as smooth as possible.

If, instead of rearranging scalars, we organize the columns in a quantization tree, as in the construction of partition trees above, then tautologically each coordinate (row) is Hölder relative to the tree metric on the points, enabling us to organize the rescaled (or differentiated) Haar coefficients of the rows in a quantization tree thereby guaranteeing that the matrix is mixed-Hölder. The procedure can then be iterated.

We illustrate a simpler procedure on the example of a questionnaire matrix M, where $M_{i,j}$ is the response of person j to question i. In other words, columns correspond to persons and rows correspond to questions in the questionnaire.

We start by building an affinity graph on the columns (persons) using correlation or any other affinity measure between columns. Bottom-up hierarchical clustering is then performed on the resulting graph, producing a partition tree on the columns. Folders in the different partitions correspond to demographic groups in the population. We append the demographic groups as new pseudo-persons to the dataset. The response of a demographic group "pseudo-person" to question i is the average response to this question along the demographic group. With this extended population at hand, we proceed to build an affinity graph on the questions using correlation or any other affinity measure between the extended rows. Namely, the affinity between rows i_1 and i_2 is the correlation (say) along the answers of both real persons and demographic groups. The process is now iterated, replacing rows and columns: bottom-up hierarchical clustering is performed on the resulting graph on the questions (rows), producing a partition tree on the rows. Folders in this tree correspond to conceptual groups of questions. We append conceptual group as new pseudo-questions to the data set. The response of person j to a conceptual group "pseudo-question" is their average response along that group of questions.

This procedure thus alternates between construction of partition trees on rows and on columns. Empirically, the entropy of the tree pair converges after a few iterations. Figure 4 shows the resulting organization of persons (columns) and questions (rows) in a personality questionnaire. The resulting partition tree pair—a partition tree of demographic groups, or contexts, on the persons, and a partition tree of conceptual groups on the questions is illustrated in Fig. 6.

4 Harmonic Analysis of Operators and Organization of Eigenfunctions

We conclude by relating the methodologies described above to a variety of methods of harmonic analysis.

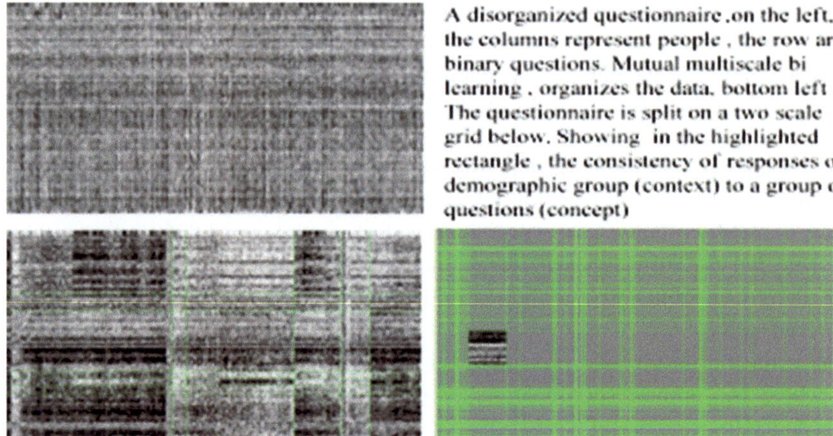

Fig. 4 Organization of a questionnaire

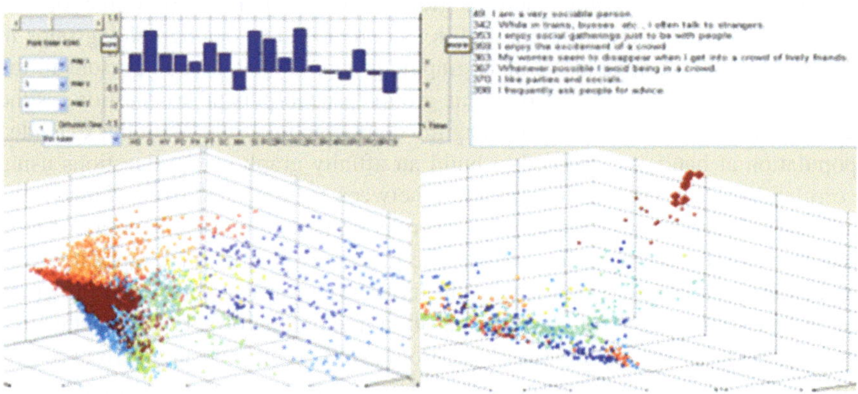

Fig. 5 Diffusion embedding of population and questions

Consider applying the matrix organization described above to an operator matrix, rather than to a dataset matrix. Consider the Hilbert transform or a potential operator that is restricted to discrete sets of sources and receivers. Namely, let $\varphi(x,y)$ be a potential interaction between $x,y \in \mathbb{R}^3$, let $\{y_j\}$ be a set of locations of sources, and let $\{x_i\}$ be a set of locations of receivers. Consider the matrix $M_{i,j} = \varphi(x_i, y_j)$. The matrix columns correspond to sources, and specifically, the j-th columns contains the field generated by source j as sampled by the various receivers. A partition tree on the columns thus organizes the sources by their impact on different receivers, while a partition tree on the rows organizes the receivers by the similarity of the fields that they measure.

When potential function and the spatial layout of the sources and receivers are given, namely, when $\varphi(x,y)$ and the sets $\{x_i\}$ and $\{y_j\}$ are known, the sources and columns can be organized to yield an efficient representation of the matrix M.

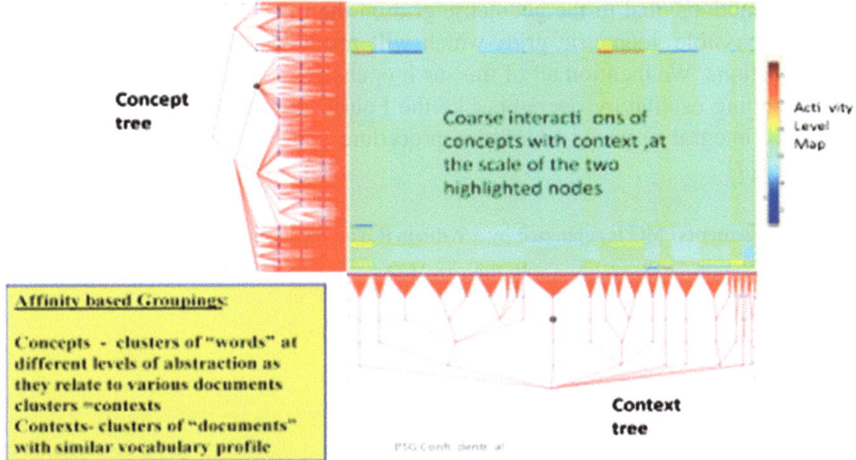

Fig. 6 Concept tree and context tree

The tree of the columns is an adapted Calderon–Zygmund tree organizing the sources, with the corresponding receiver tree. This organization is precisely a generalization of the one introduced by Rokhlin to implement his fast multipole algorithms [14]. The Haar basis provides a low-precision approximation to this algorithm, which can easily be refined by using hierarchical orthogonal polynomials of low degree on each partition as done by Alpert [1]. The point is that these hierarchies are the scaffold of a more interesting and powerful harmonic analysis even for more complex oscillatory transforms.

In fact, for oscillatory systems, the notion of matrix organization may prove to be a powerful harmonic analysis tool. For example, consider a set $\{x_i\} \subset \mathbb{R}^d$ uniformly sampled from a compact manifold embedded in $\mathbb{R}^d$. Let $\{\psi_j\}$ be the eigenfunctions of the Laplace operator on our manifold, and consider the matrix $M_{i,j} = \psi_j(x_i)$.

The geometry of the set $\{\psi_j\}$ is in some sense dual to the geometry of the manifold itself. Traditionally, Laplacian eigenfunctions are only organized by their eigenvalues alone, which measures the frequency of oscillation. This is a one-dimensional organization, regardless of d, that does not reflect more detailed relation (such as direction of oscillation) between the eigenfunctions. Organization of the rows and columns of the matrix M, even without information on the manifold and the sampling points $\{x_i\}$, reveals the geometry of the manifold and the dual geometry of the eigenfunction, respectively.

The matrix organization method we reviewed is needed to organize M. Globally, the eigenfunctions are orthogonal, so they are all equidistant. However, when we correlate them on the various sets of a multiscale partition tree on the manifold, as long as a subset of the tree is such that the product of two eigenfunctions does not oscillate on it, we have a contribution to their affinity. It is a simple exercise to show that for the case of the two-dimensional torus, we recover an affinity corresponding to the distance in the dual grid in two dimensions. This procedure generates a tree of eigenfunctions dual to the geometric grid on the manifold. Of course there are different possible geometric grids which will result in different organization of eigenfunctions. We mention all of this, as it opens up new ways of organizing and understanding oscillatory operators like the Fourier transform, or more generally a Fourier integral operator where this procedure generalizes "microlocalization" geometry [7].

Acknowledgements MG is supported by a William R. and Sara Hart Kimball Stanford Graduate Fellowship.

References

1. Alpert, B., Beylkin, G., Coifman, R., Rokhlin, V.: Wavelet-like bases for the fast solution of second-kind integral equations. SIAM J. Scientific Comput. **14**(1), 159–184 (1993)
2. Bartal, Y.: In: Proceedings of 37th Conference on Foundations of Computer Science, pp. 184–193 (1996)
3. Bartal, Y.: In: Proceedings of the 30th annual ACM Symposium on Theory of computing, ACM, pp. 161–168 (1998)
4. Belkin, M., Niyogi, P.: Laplacian eigenmaps and spectral techniques for embedding and clustering. Advances in Neural Information Processing Systems **14**, 585–591 (2001)
5. Belkin, M., Niyogi, P.: Laplacian eigenmaps for dimensionality reduction and data representation. Neural Comput. **13**, 1373–1397 (2003)
6. Bungartz, H.J., Griebel, M.: Sparse grids. Acta Numerica **13**,147–269 (2004)
7. Coifman, R.R., Gavish, M.: Harmonic analysis of digital data bases. In: Cohen, J., Zayed, A. (eds.) Wavelets and Multiscale Analysis. Birkhäuser, Boston (2011)
8. Coifman, R.R., Lafon, S.: Geometric harmonics: a novel tool for multiscale out-of-sample extension of empirical functions. Appl. Comp. Harmonic Anal. **21**(1), 31–52 (2006)
9. Coifman, R. R., Lafon, S., Lee, A., Maggioni, M.,Nadler, B., Warner, F., Zucker, S.: Geometric diffusions as a tool for harmonic analysis and structure defnition of data, Part I: diffusion maps. Proc. Natl. Acad. Sci. **102**, 7426–7431 (2005)
10. Coifman, R.R., Rochberg, R.: Another characterization of B.M.O. Proc. Am. Math. Soc. **79**, 249–254 (1980)
11. Coifman, R.R., Singer, A.: Non-linear independent component analysis with diffusion maps. Appl. Comput. Harmonic Anal. **25**, 226–239 (2008)
12. Gavish, M., Coifman, R.R.: Sampling, denoising and compression of matrices, by coherent matrix organization. Appl. Comput. Harmonic Anal. (To Appear, 2012)
13. Gavish, M., Nadler, B., Coifman, R.R.: Multiscale wavelets on trees, graphs and high dimensional data: theory and applications to semi supervised learning. In: Proceedings of the 27th International Conference on Machine Learning, ICML (2010)
14. Greengard, L., Rokhlin, V.: A fast algorithm for particle simulations. J. Comput. Phys. **73**, 325–348 (1987)
15. Priebe, C.E., Marchette, D.J., Park, Y., Wegman, E.J., Solka, J.L., Socolinsky, D.A., Karakos, D., Church, K.W., Guglielmi, R., Coifman, R.R., Link, D., Healy, D.M., Jacobs, M.Q., Tsao, A.: Iterative denoising for cross-corpus discovery. In: Proceedings of COMPSTAT 2004, Physica-Verlag/Springer (2004)
16. Singer, A.: Angular synchronization by eigenvectors and semidefinite programming. Appl. Comput. Harmonic Anal. **30**(1), 20–36 (2011)
17. Smolyak, S.A.: Quadrature and interpolation formulas for tensor products of certain classes of functions. Soviet Math. Dokl. **4**, 240–243 (1963) (Russian originalin Dokl. Akad. Nauk SSSR **148**, 1042–1045, 1963)

The Structure of Sidelobe-Preserving Operator Groups

Gregory E. Coxson

Abstract This chapter considers the structure of groups of operators preserving the aperiodic autocorrelation peak sidelobe level of mth root codes. These groups are shown to be helpful for efficient enumeration of codes by peak sidelobe level for a given m and given code length N. In the binary case, it is shown that there is a single Abelian group of order 8 generated by sidelobe-preserving operators. Furthermore, it is shown that shared symmetry in the binary Barker codes can be discovered in a natural way by considering degeneracies of group actions. The group structure for $m = 4$ (the quad-phase case) is shown to have higher complexity; in fact, instead of a single group, there are four groups (two pairs of isomorphic groups), and they are no longer Abelian. Group structure is identified for the cases of odd code lengths N, leaving group structure for even-length cases mostly unresolved. Moving to general mth roots codes, it is shown that results found for the quad-phase case generalize quite well. In particular, it is shown that there are $4m^2$ groups. All m groups are identified for any odd m. When m is even, the structure for odd code lengths N is identified. The group structure for m even and N even is left unresolved.

Keywords Barker code • Skew-symmetry • Autocorrelation sidelobes • Binary code • Polyphase code • Unimodular code • Group action

1 Introduction

In signal processing terminology, a code is a finite sequence of complex scalars, called code elements. A code is called unimodular if each of its elements has modulus 1 (hence unimodularity refers to the elements rather than to the code which,

G.E. Coxson (✉)
Coxson Associates, 17412 Cherokee Lane, Olney, MD 20832, USA
e-mail: gcoxson@ieee.org

T.D. Andrews et al. (eds.), *Excursions in Harmonic Analysis, Volume 1*, Applied and Numerical Harmonic Analysis, DOI 10.1007/978-0-8176-8376-4_16,

if it has N elements, has size $\sqrt{N}$). A subset of the unimodular codes is the set of polyphase codes, for which all elements have elements that are mth roots of unity for some n. Polyphase codes with $n = 2$ are called binary codes; all elements are ± 1.

Binary and polyphase codes that achieve low aperiodic autocorrelation (AAC) average or peak sidelobe levels are valuable for radar and communications applications. This is due to the fact that the autocorrelation function approximates the response for the matched (or North) filter for phase-coded signals [16]. The matched filter is optimal for signal-to-noise ratio, and hence can pull signals out of receiver inputs where the signal is buried in noise.

If it is desired to find the lowest peak sidelobe level for a given code length, or codes which achieve it, the most approach is exhaustive search. Taking the search space as all mth root codes for some m and some length N, it is helpful to consider a partition of this space into equivalence classes relative to a group generated by sidelobe-preserving operators. If a method can be found which involves searching single representatives from each equivalence class, the search may be expedited. Furthermore, listing the best representative (for some measure of sidelobe level of interest) is more efficient than listing the best codes from the search space.

Because search techniques quickly grow computationally costly, even prohibitive-ly so, as code length grows, it is tempting to try and identify patterns in codes that might allow the construction of codes with a good chance of providing low sidelobe levels. Here is possibly another opening for the use of sidelobe-preserving operator groups, (SPGs) may provide some help. For the most notable example of low-sidelobe codes, the binary Barker codes, those of odd length share a skew-symmetric property closely linked to degeneracies in actions of the sidelobe preserving group on these codes. Knowledge of such a symmetry can narrow the search space greatly. For example, for odd-length binary codes of length N, if, rather than searching all the codes, only skew-symmetric codes are searched, the search space is reduced from size 2^N to size $2^{(N-1)/2}$. This computational cost benefit comes at the cost of possibly missing optimal-sidelobe-level codes.

It is natural to ask whether something like skew symmetry, and its connection to a group degeneracy, can be found for non-binary codes. In order to suggest this possibility, this chapter will examine a quad-phase code with Barker-level sidelobes that satisfies a symmetry much like skew symmetry. Furthermore, it will be shown that an operator in the associated group maps this code to itself, meaning that the isometry subgroup for this code has more than one element, and hence its equivalence class degenerates under group action.

The chapter is organized as follows. After an introduction (Sect. 1) and notation and terminology (Sect. 2), Sect. 3 will discuss motivation for examining SPGs. Section 4 will look at the group structure for the binary case. Section 5 will show that consideration of degeneracies in group actions for odd-length binary Barkers leads in a naturalway to the uncovering of their skew-symmetry-property. Section 6 will then consider the group structure for the quad phase case. Finally, Sect. 7 will discuss general mth root, to which findings for the quad-phase case are found to generalize quite well.

2 Basic Notation and Terminology

Let Q_m represent the set of mth roots of unity or the set of m complex numbers z such that $z^m = 1$. For a specified value of $m \geq 2$, let

$$x = [x_1, x_2, \ldots, x_N] \tag{1}$$

denote an N-length code, each of whose elements resides in Q_m. Furthermore, let $(Q_m)_N$ mean the set of codes x with elements in Q_m that is,

$$(Q_m)_N = \{x : |x| = N, x_i \in Q_m, i = 1, \ldots, N\}. \tag{2}$$

Clearly, $|(Q_m)_N| = m^N$. For the special case of $m = 2$, the codes $x \in (Q_2)_N$ will be referred to as binary codes of length N.

The AAC sequence for an $x \in (Q_m)_N$ has length $2N-1$ and is defined by

$$\mathrm{AAC}_x = x * \overline{x^c}, \tag{3}$$

where $*$ means acyclic convolution, $\overline{x}$ means the reversal of a code x, and x^c means elementwise complex conjugation. The elements of the AAC of x may be represented explicitly in terms of sums of pairwise products of elements of x in the following way:

$$\mathrm{AAC}_x(k) = \sum_{i=1}^{N-|k-N|} x_i x^c_{i+|k-N|} \tag{4}$$

for $k = 1, \ldots, 2N-1$. In the binary case, the elements of x are real (either 1 or -1), so the complex conjugation operation can be ignored.

The "peak" of the autocorrelation is $\mathrm{AAC}_x(N)$. The peak is equal to N, since

$$\mathrm{AAC}_x(N) = x_1 x_1^c + \cdots + x_N x_N^c = |x|^2 = N. \tag{5}$$

Elements for indices $k \neq N$ are referred to as "sidelobes" of the autocorrelation. The autocorrelation is symmetric with respect to the peak; that is,

$$\mathrm{AAC}_x(k) = \mathrm{AAC}^c_x(2N-k) \tag{6}$$

for $k = 1, \ldots, 2N-1$.

The "peak sidelobe level" for a code x is defined to be

$$\mathrm{PSL}_x = \max_{k \neq N} |\mathrm{AAC}_x(k)|. \tag{7}$$

The lowest achievable value of PSL_x for $x \in (Q_m)_N$ for any $m \geq 2$ and $N \geq 1$ is 1. This is because when $k = 1$ or $k = 2N - 1$, the sidelobe is a $x_1 x_N$, so its modulus is 1. The binary codes x that achieve $\mathrm{PSL}_x = 1$ are called Barker codes, after the author of an early paper identifying these codes [1]. When $m > 2$, codes $x \in (Q_m)_N$ that achieve $\mathrm{PSL}_x = 1$ are called generalized Barker sequences [7] or polyphase Barker sequences [6].

Finally, some notation is needed for discussing groups and group actions. An expression of the form $< g_1, g_2, \ldots, g_k >$ will mean the group generated by the elements $g_1, \ldots, g_k$. Given a group G and two elements $g, h \in G$, the notation g^h will be shorthand for the conjugation of g by h, that is, hg^{-1} (this is not to be confused with complex conjugation). Given two groups G and H, the notation $G \times H$ will represent the Cartesian product of G with H, and GH will represent a semidirect product of G and H (see, e.g., [2]).

3 PSL-Preserving Operator Groups: Motivation

Codes with low peak sidelobe level are desired in applications such as radar and communications where match filtering is used for detection (see [11, 14, 16]). For a given length, it is useful to know the lowest achievable PSL and some or all the codes which achieve it. Although there exist some well-known construction techniques for codes with low sidelobe levels, often the lowest-PSL codes must be found by random or exhaustive searches. As code length grows, random and exhaustive searches tend to become prohibitively computationally costly.

It can be informative to know how many codes achieve these lowest, or at least relatively low, PSL values. Such enumeration efforts inevitably necessitate a decision about whether to list or enumerate all such codes or to list representatives from code equivalence classes, where the equivalence is defined relative to operations that preserve autocorrelation sidelobe level.

A sidelobe-preserving operator will be understood to mean a transformation that preserves the magnitude of every sidelobe of the autocorrelation AAC_x for each $x \in (Q_m)_N$, for some m and N. Golomb and Win [8] list four sidelobe-preserving operator, for general polyphase codes. They are:

1. Reversal $\overline{x}$
2. Complex conjugation x^c
3. Constant multiple transformation (CMT): given any unit-modulus complex number α, form the product αx
4. Progressive multiplication transformation (PMT): given any unit-modulus complex number ρ, multiply the ith element, x_i, by ρ^i for $i = 1, \ldots, N$

For N-length binary codes x (i.e., $m = 2$), involving only real quantities, the set of four transformations identified by Golomb and Win reduces to a set of three somewhat simpler transformations:

1. Reversal $\overline{x}$
2. Negation $-x$
3. Alternating-sign: multiply element x_i by $(-1)^i$, $i = 1, \ldots, N$

To illustrate the usefulness of these transformations for enumeration, suppose that for $N = 13$, there is a need to determine the lowest achievable PSL for a binary code of length N and the binary codes that achieve it. This length is small enough that an exhaustive search is practical. The simplest, most naive approach would generate each of the 2^{13} codes, compute their PSL values, and keep only those with the lowest PSL. Four codes would be found having the Barker-level PSL of 1, optimal not just for length 13 but for any length. Examination of these codes would lead to the observation that any one of the four could be found by applying various compositions of the three binary transformations listed above. Hence, rather than listing all four, it is enough to list a single representative, say [16]:

$$x = \begin{bmatrix} 1 & 1 & 1 & 1 & 1 & -1 & -1 & 1 & 1 & -1 & 1 & -1 & 1 \end{bmatrix}. \tag{8}$$

Behind the efficiency of this use of representatives are an equivalence relation, and a partition of the search space into equivalence classes. Given that there are three transformations being applied in various orders, these equivalence classes would be expected to hold eight codes, in general, rather than the four found having the optimal PSL for length 13. Indeed, if all eight permutations of the binary transformations are applied to the length-13 Barker code given above, and the set of eight resulting codes are tabulated, this set can be arranged into four sets of twin codes. In other words, the size-8 equivalence class degenerates into one of size 4. This suggests that the code has special structure and the structure is related to "actions" of the three transformations under composition.

Skolnik [16] lists the lowest optimal PSL values for lengths from 3 to 40, which was the best list available in 1990, along with the number of binary codes achieving these values. Skolnik uses the term "allomorphic" for codes transformable into each other by the composition of sidelobe-preserving operations ("allo-" being the Greek root for "other" and "morph" being the Greek root for "form"). The first three columns of Table 1 list these results, along with similar figures for $N = 2$.

Interestingly, the values tabulated in [16] were developed using only two of the three binary code sidelobe-preserving operators (negation and reversal). If the third one is taken into account as well, the result is for most code lengths a reduction in the number of representative codes; the results are listed in the fourth column of Table 1. For most of the lengths, the number of representative codes is reduced by half. However, there is a small set of lengths for which the extra transformation fails to change this number; this means that for these lengths, the third transformation maps the set of minimum-PSL codes into itself. Furthermore, this set of lengths, $\{3,5,7,11,13\}$, is special in that it is the set of odd lengths for which Barker codes exist.

At the least, the behavior of sidelobe-preserving operators is useful for efficient representation of codes of interest for their low peak sidelobe levels. However, it also

Table 1 Adding a third operator changes the number of representatives

N	Best PSL	Number of representatives, for negation and reversal	Number of representatives, for negation, reversal and alternating sign
2	1	2	1
3	1	1	1
4	1	2	1
5	1	1	1
6	2	8	4
7	1	1	1
8	2	16	8
9	2	20	10
10	2	10	5
11	1	1	1
12	2	32	16
13	1	1	1
14	2	18	9
15	2	26	13
16	2	20	10
17	2	8	4
18	2	4	2
19	2	2	1
20	2	6	3
21	2	6	3
22	3	756	378
23	3	1,021	515
24	3	1,716	858
25	2	2	1
26	3	484	242
27	3	774	388
28	2	4	2
29	3	561	283
30	3	172	86
31	3	502	251
32	3	844	422
33	3	278	139
34	3	102	51
35	3	222	111
36	3	322	161
37	3	110	52
38	3	34	17
39	3	60	30
40	3	114	57

appears that degeneracies in the "actions" of compositions of these transformations can uncover structures in codes having low peak sidelobe levels. These ideas will be made more precise in the following sections.

4 Sidelobe-Preserving Operator Groups: The Binary Case

The binary case has the nice property that the sidelobe-preserving transformations can each be effected by matrix operations. Hence, consider defining

1. $g_1 = -xI_N$
2. $g_2 = xJ_N$
3. $g_3 = xA_N$

where I_N is the order-N identity matrix, J_N is the order-N matrix defined by

$$J_N = \begin{pmatrix} 0\,0 \ldots 0\,1 \\ 0\,0 \ldots 1\,0 \\ \vdots\,\vdots\,\vdots\,\vdots\,\vdots \\ 0\,1 \ldots 0\,0 \\ 1\,0 \ldots 0\,0 \end{pmatrix} \tag{9}$$

and A_N is the matrix

$$A_N = \begin{pmatrix} -1 & 0 & \ldots & 0 & 0 \\ 0 & 1 & \ldots & 0 & 0 \\ \vdots & \vdots & \vdots & \vdots & \vdots \\ 0 & 0 & \ldots & (-1)^{N-1} & 0 \\ 0 & 0 & \ldots & 0 & (-1)^N \end{pmatrix}. \tag{10}$$

Then g_1 and g_2 preserve the autocorrelation sequence of any binary code, as can be seen by recalling that $\mathrm{AAC}_x = x * \overline{x}$. The third operator, g_3, which switches the sign of every other element of a code x, has the effect on the autocorrelation of switching the sign of every other sidelobe. However, the magnitude of every sidelobe is preserved.

The three operators g_1, g_2, and g_3 generate a group of order 8. To see this, consider five additional operators:

1. $g_0 = I_N$
2. $g_4 = g_1 \circ g_2$
3. $g_5 = g_1 \circ g_3$
4. $g_6 = g_2 \circ g_3$
5. $g_7 = g_1 \circ g_2 \circ g_3$

where the symbol $\circ$ refers to composition of operations (Table 2). The 8×8 multiplication table is given in Table 2 (where composition is used as the multiplication operator).

These eight operations constitute a group G under composition, as can be checked by showing that the result of composing any two elements lies in the group (i.e., the closure property), that the group includes an identity, that each element

Table 2 Multiplication table for the binary operators

$\circ$	g_0	g_1	g_2	g_3	g_4	g_5	g_6	g_7
g_0	g_0	g_1	g_2	g_3	g_4	g_5	g_6	g_7
g_1	g_1	g_0	g_4	g_5	g_2	g_3	g_7	g_6
g_2	g_2	g_4	g_0	g_6	g_1	g_7	g_3	g_5
g_3	g_3	g_5	g_6	g_0	g_7	g_1	g_2	g_4
g_4	g_4	g_2	g_1	g_7	g_0	g_6	g_5	g_3
g_5	g_5	g_3	g_7	g_1	g_6	g_0	g_4	g_2
g_6	g_6	g_7	g_3	g_2	g_5	g_4	g_0	g_1
g_7	g_7	g_6	g_5	g_4	g_3	g_2	g_1	g_0

has an inverse relative to the identity, and that the associativity property holds [2]. Furthermore, G is Abelian, and isomorphic to $Z_2 \times Z_2 \times Z_2$ (see [3]), and indeed, $G = < g_1 > \times < g_2 > \times < g_3 >$. The group generator relations are simple ones, essentially stating that the three generators are each of order 2 and that the group multiplication is commutative:

1. $g_1 \circ g_1 = g_2 \circ g_2 = g_3 \circ g_3 = g_0$
2. $g_1 \circ g_2 = g_2 \circ g_1$
3. $g_2 \circ g_3 = g_3 \circ g_2$
4. $g_1 \circ g_3 = g_3 \circ g_1$

Note that the only time when it is important to take note of the code length N is when using matrix representation for the operators. It is notable that this same 8×8 group applies to binary codes of all lengths. On the other hand, properties of actions of the group elements on sets of codes can depend on code structure, the parity of N, and on congruence of N modulo 4, as will be shown in the next section.

The next sections will look at group structure for more general mth-root-of-unity codes. A group generated by sidelobe-preserving operators for some m and N will be referred to as a SPG.

5 Equivalence Classes, Group Actions, and the Odd-Length Barker Codes

Consider again the binary case, and the group G defined in the previous section. Furthermore, define two codes $x, y \in (Q_2)_N$ to be equivalent if $y = g_k x$ for some $g_k \in G$. This induces a partition of $(Q_2)_N$ into equivalence classes of size 8 or less.

An interesting question for computational searches is whether it is possible to generate single representatives of each equivalence class by a deterministic algorithm. The answer is that it is possible; one such algorithm was provided in Coxson et al. [4].

As indicated earlier, the odd-length Barker codes provide examples of size-4 equivalence classes. This suggests a shared symmetry that results in degenerate orbits. The theory of group actions suggests that there exists a non-trivial identity (or non-trivial identities, as is actually the case) for the odd-length Barker codes. It is an instructive exercise to find them.

The following candidates can be ruled out quickly:

1. g_1: $g_1 x = -x$ has no fixed points in $(Q_2)_N$ for any $N > 0$.
2. g_3: $g_3 x = xA_N$ has no fixed points in $(Q_2)_N$ for any $N > 0$.
3. g_5: $g_5 x = -xA_N$ has no fixed points in $(Q_2)_N$ for any $N > 0$.

Two more can be ruled out almost as quickly:

1. g_2: $g_2 x = \overline{x}$ fixes symmetric codes x, none of which can achieve $\mathrm{PSL}_x = 1$ for $N > 2$.
2. g_3: $g_4 x = -\overline{x}$ fixes some $x \in (Q_2)_N$, but only when N is even.

That leaves g_6 and g_7 as the only possibilities for nontrivial identities.

Consider first g_7. Matrix representation helps rule out possibilities for solutions to

$$0 = g_7 x - x = -(\overline{xA_N} + x).$$

Indeed, based on simple considerations in the solution of sets of linear equations, it is possible to rule out any solutions when N is even or when $N \equiv 3 \bmod 4$. However, when $N \equiv 1 \bmod 4$, one arrives at the following linear equation (making use of the matrix representation available in the binary case):

$$0 = g_7 x - x = x \begin{pmatrix} -1 & 0 & 0 & \ldots 0 \ldots & 0 & 0 & 1 \\ 0 & -1 & 0 & \ldots 0 \ldots & 0 & -1 & 0 \\ 0 & 0 & -1 & \ldots 0 \ldots & 1 & 0 & 0 \\ \vdots & \vdots & \vdots & \ldots \vdots \ldots & \vdots & \vdots & \vdots \\ 0 & 0 & 0 & \ldots 0 \ldots & 0 & 0 & 0 \\ \vdots & \vdots & \vdots & \ldots \vdots \ldots & \vdots & \vdots & \vdots \\ 0 & 0 & 1 & \ldots 0 \ldots & -1 & 0 & 0 \\ 0 & -1 & 0 & \ldots 0 \ldots & 0 & -1 & 0 \\ -1 & 0 & 0 & \ldots 0 \ldots & 0 & 0 & 1 \end{pmatrix}. \tag{11}$$

Since the matrix on the right-hand side has a zero row, and is hence singular, there exists a solution in R^N. It remains to show that there exists a solution in $(Q_2)_N$. However, the simple form of the set of equations in this case leads in a straightforward way to a set of solutions of the form

$$x = \left[z \; y \; -\overline{zA_{(N-1)/2}} \right], \tag{12}$$

where z can be chosen arbitrarily from $(Q_2)_{(N-1)/2}$ and $y \in \{1, -1\}$.

By a similar process, it is possible to conclude that g_6 has a solution only when $N \equiv 3 \bmod 4$, and the solutions are of the form

$$x = \left[z \; y \; \overline{zA_{(N-1)/2}} \right], \tag{13}$$

where z can be chosen arbitrarily from $(Q_2)_{(N-1)/2}$ and $y \in \{1, -1\}$.

This shared structure of the odd-length Barker codes is well-known (see, for instance, [17]) and is often credited to Golay and referred to as (Golay) skew symmetry (see, e.g., [12]). It is interesting, nonetheless, to rediscover this property using the theory of group actions.

Note that if x has the skew symmetry property, then any code equivalent to it is also skew-symmetric. To see this, let x and y be two members of $(Q_2)_N$ for $N \equiv 3 \bmod 4$ and let $y = g_k x$ for some $g_k \in G$. Then $g_6 x = x$ implies

$$g_6(y) = (g_6 \circ g_k)x = (g_k \circ g_6)x = (g_k)x = y. \tag{14}$$

A similar argument can be made using g_7 for $N \equiv 1 \bmod 4$.

It is easy to check that the odd-length Barker codes are skew-symmetric. Representatives of every odd-length Barker are listed here (see [16]):

1. $N = 3$:
$$\begin{bmatrix} 1 \; 1 \; -1 \end{bmatrix}. \tag{15}$$
2. $N = 5$:
$$\begin{bmatrix} 1 \; 1 \; 1 \; -1 \; 1 \end{bmatrix}. \tag{16}$$
3. $N = 7$:
$$\begin{bmatrix} 1 \; 1 \; 1 \; -1 \; -1 \; 1 \; -1 \end{bmatrix}. \tag{17}$$
4. $N = 11$:
$$\begin{bmatrix} 1 \; 1 \; 1 \; -1 \; -1 \; -1 \; 1 \; -1 \; -1 \; 1 \; -1 \end{bmatrix}. \tag{18}$$
5. $N = 13$:
$$\begin{bmatrix} 1 \; 1 \; 1 \; 1 \; 1 \; -1 \; -1 \; 1 \; 1 \; -1 \; 1 \; -1 \; 1 \end{bmatrix}. \tag{19}$$

It needs to be mentioned that while the odd-length Barker codes are skew-symmetric and achieve the lowest possible PSL, this does not mean that skew-symmetry implies low sidelobe level. If an exhaustive search is done, and a count made of the number of equivalence classes of odd-length binary skew-symmetric codes, for lengths between 3 and 25, the result is the set of tallies given in the table below.

In Table 3, notice that the number of equivalence classes for high PSL values is nearly as high as those for low sidelobe level. The reason that only even values of PSL are listed is that odd-length skew-symmetric binary codes can have only odd PSL (a nice exercise for the reader). This means that for some lengths N, in particular those where the lowest PSL is even, a search over skew-symmetric codes will not be able to find the optimal codes. Nonetheless, such searches will find codes with near-optimal PSL for a considerable savings in computational cost.

Here we see that shared structure in a very special set of codes (those having the lowest achievable peak sidelobe level) can be uncovered by studying degeneracies in group actions for a group generated by sidelobe-preserving operations. A natural question to ask is whether this is a coincidence, and furthermore, if it is not a coincidence, why this connection should exist. These questions are not going to be answered in this chapter. The following sections will pursue the structure of operator groups for a more general set of codes.

Table 3 Number of skew-symmetric binary codes, $N = 3$–25

N	1	3	5	7	9	11	13	15	17	19	21	23
3	1	0	0	0	0	0	0	0	0	0	0	0
5	1	1	0	0	0	0	0	0	0	0	0	0
7	1	2	1	0	0	0	0	0	0	0	0	0
9	0	5	2	1	0	0	0	0	0	0	0	0
11	1	4	8	2	1	0	0	0	0	0	0	0
13	1	9	9	10	2	1	0	0	0	0	0	0
15	0	6	26	24	11	2	1	0	0	0	0	0
17	0	5	45	40	23	12	2	1	0	0	0	0
19	0	4	68	82	59	27	13	2	1	0	0	0
21	0	8	68	195	115	79	30	14	2	1	0	0
23	0	9	107	270	335	154	98	33	15	2	1	0
25	0	3	128	515	552	475	201	119	36	16	2	1

6 Sidelobe-Preserving Operator Group Structure for Quad-Phase Codes

Moving from the binary case to the quad-phase case, the elements of a code $x \in (Q_4)_N$ are chosen not from the set $\{-1,1\}$ but the set $\{-1,1,i,-i\}$, where $i = \sqrt{-1}$. The longest known quad-phase code with PSL $= 1$ (i.e., a generalized Barker sequence) is the length-15 code [13]

$$x = \begin{bmatrix} 1\ 1\ 1\ i\ i\ 1\ -i\ -i\ i\ -1\ -i\ i\ 1\ -1\ 1 \end{bmatrix}. \tag{20}$$

Interestingly, this code satisfies $x = -\overline{xA_{15}}$, where A_{15} is the 15×15 diagonal matrix that effects an alternating-sign transformation on the elements of x; that is, it has diagonal elements $-1, 1, -1, \ldots, (-1)^{15}$. Hence, this code obeys the same symmetry as the binary Barker codes for lengths $N \equiv 1 \bmod 4$.

As will be shown, this means dealing with added complications in the sidelobe-preserving group. One of the complications is that instead of a single group, there are now four, depending on the congruence of code length N modulo 4. Furthermore, the groups have size 64 and are no longer Abelian. Finally, it will no longer be possible to represent transformations in terms of matrix operations.

Before examining this case, it is useful to look at the sidelobe-preserving operations in a more general setting, the general unimodular case where code elements can lie anywhere on the unit circle. For consistency with the notation used previously, let Q_∞ represent the unit circle and let $(Q_\infty)_N$ represent the set of N-length codes whose elements are drawn from the unit circle. Golomb and Win [8] provide a list of the sidelobe-preserving transformations for this quite general case. Let $x \in (Q_\infty)_N$. Then the following operations each preserve the magnitudes of the AAC sequence and hence the peak sidelobe level (using simpler notation than previously, to facilitate the discussions to come):

1. C: elementwise complex conjugation, x^c
2. R: reversal, $\overline{x}$
3. M_μ: multiplication by $\mu \in Q_\infty$ to give μx
4. P_ρ: progressive multiplication (or phase ramp) using $\rho \in Q_\infty$

What is meant by progressive multiplication is that element x_i is multiplied by ρ^i for $i = 1, \ldots, N$. Note that complex conjugation operation cannot be represented using matrix multiplication.

The transformations R and M_μ preserve the autocorrelation sequence, while the operations C and P_ρ preserve the magnitudes of the sidelobes (and hence the peak sidelobe level) but do not preserve the autocorrelation sequence in general.

Moving to the quad-phase case, let x be an arbitrary member of $(Q_4)_N$ for $N > 0$, and consider the following specialization of the generalized list of sidelobe-preserving operations given above:

1. C: elementwise complex conjugation
2. R: reversal, $\overline{x}$
3. M_i: multiplication by $\mu = i$
4. P_i: progressive multiplication by $\rho = i$

No loss of generality results from the particular choice of values for μ and ρ since in each case, the choice of i specifies a generator for the order-4 cyclic group containing every other possibility.

The four operators generate a group of order 64. To see this, first fix $N > 0$. Then $< R, P_i >$ (the group generated by R and P_i) is a dihedral group of order 8. Also, M_i generates a cyclic group of order 4, $< M_i >$. It follows that $< M_i, R, P_i >$ has a normal subgroup, $< M_i >$, modulo in its dihedral-8 subgroup $< R, P_i >$. Hence

$$| < M_i, R, P_i > | = (4)(8) = 32. \tag{21}$$

Now consider the group $< M_i, R, P_i, C >$. Every element may be written $R^a P_i^b C^d M_i^e$ where $a, d \in \{0, 1\}$ and $b, c \in \{0, 1, 2, 3\}$. So

$$| < M_i, R, P_i, C > | \leq (2)(2)(4)(4) = 64. \tag{22}$$

Since C has order 2 and does not belong to $< M_i, R, P_i >$,

$$| < M_i, R, P_i, C > | \geq (2)(32) = 64. \tag{23}$$

Therefore $G = < M_i, R, P_i, C >$ has size 64.

Let g_0 be the group identity. Then with some effort, the list of generator relations is found to be

1. $C^2 = R^2 = g_0$
2. $M_i^4 = P_i^4 = g_0$
3. $RC = CR$
4. $P_i M_i = M_i P_i$

5. $M_iR = RM_i$
6. $CM_i = M_i^{-1}C = -M_iC$
7. $CP_i = P_i^{-1}C$
8. $RP_i = M_i^{N+1}P_i^{-1}R$

Note that the last of these relation, depends on N or, more to the point, the value of N modulo 4. Hence, there are four apparently different sets of relations, yielding four possibly different groups.

To simplify the following discussions, let G_i refer to the group for $N \equiv i \bmod 4$, for $i = 0, \ldots, 3$. When the value of N is not specified, and the discussion applies to all four cases, the notation G will be used.

There exist 267 distinct groups of order 64 (see, e.g., [3]). A first hint at group structure for the four quad-phase groups results from counting the orders of group elements. In the case of G_3, the count of group elements of order 2 is 35. Fortunately, there is a single group of the 267 groups of order 64 having 35 elements of order 2, and that is the Cartesian product of two dihedral-8 groups. Hence G_3 is isomorphic to $D_8 \times D_8$. The count of order-2 elements for G_1 is also 35, suggesting that G_1 and G_3 are isomorphic.

Identification of the group structure for the two remaining cases, G_0 and G_2, is left unresolved for now.

1. 27 elements of order 2
2. 20 elements of order 4
3. 16 elements of order 8

This narrows the possible order-64 group structures to three in these two cases (see [3]).

Element order counts can sometimes be unreliable. Fortunately, it is possible to do better than order tallies. Martin Isaacs, of the Department of Mathematics at University of Wisconsin Madison, has suggested the following approach involving semidirect products and automorphisms on subgroups [10].

Let $A = < M_i, P_i >$. Since M_i and P_i have order 4 and commute, A is Abelian and isomorphic to $Z_4 \times Z_4$, where Z_4 is the integers modulo 4 with respect to addition. Next, let $U = < C, R >$. Since C and R commute and have order 2, U is noncyclic of order 4, and isomorphic to $Z_2 \times Z_2$.

Note that the intersection of U and A contains only the identity. Then $G = AU$, that is, G is the semidirect product of A with U acting on it (in other words, U normalizes A), by the following observations:

1. $CM_iC^{-1} = M_i^C = M_i^{-1}$ and $P_i^C = P_i^{-1}$ imply that C normalizes A.
2. $RM_iR^{-1} = M_i^R = M_i$ and $P_i^R = P_i^{-1}M_i^k$ where $k = N+1$ imply that R normalizes A.

Determination of the structure of G now depends on knowing what automorphisms of A are induced by conjugation by C and R.

Automorphisms of A can be represented as 2×2 matrices over Z_4. Invertibility of a matrix over Z_4 will mean the determinant is ± 1 modulo 4.

Conjugation of P_i and M_i by C gives

1. $P_i^C = CP_iC^{-1} = P_i^{-1}$
2. $M_i^C = CM_iC^{-1} = M_i^{-1}$

These two relationships will be encapsulated in the matrix $-I_2$, the negative of the 2×2 identity matrix.

Similarly, conjugation of P_i and M_i by R gives $P_i^R = RP_i^{-1}R^{-1} = M_i^k P_i^{-1}$ and $M_i^C = CM_iC^{-1} = M_i$, where $k = N+1$. Conjugation of P_i twice by R gives

$$\begin{aligned} R(M_i^k P_i^{-1})R^{-1} &= (RM_i^k R^{-1})(RP_i^{-1}R^{-1} \\ &= M_i^k R(RP_i)^{-1} \\ &= M_i^k R(M_i^k P_i^{-1}R)^{-1} \\ &= M_i^k P_i M_i^{-1} \\ &= P_i. \end{aligned} \tag{24}$$

A 2×2 matrix to represent this is then

$$\begin{pmatrix} 1 & 0 \\ k & -1 \end{pmatrix}, \tag{25}$$

the square of which, modulo 4, is the identity.

The subgroup U is essentially the multiplicative group generated by the two matrices. The easiest case is for $k = N+1 \equiv 0 \bmod 4$, that is, G_3. An equivalent set of generators, then, after setting $k = 0$, is

$$\begin{pmatrix} 1 & 0 \\ 0 & -1 \end{pmatrix} \tag{26}$$

and

$$\begin{pmatrix} 1 & 0 \\ 0 & -1 \end{pmatrix}. \tag{27}$$

It follows that:

1. Conjugation by C leaves M alone but inverts P.
2. Conjugation by R inverts M and leaves P alone.

Together, these imply that G_3 is isomorphic to $D_8 \times D_8$, the same conclusion arrived at from the group element order tally.

Next, if R and C are conjugated by the same invertible matrix, this simply changes the "basis" for A, leaving the group unchanged. Consider using

$$\begin{pmatrix} 1 & 0 \\ 1 & 3 \end{pmatrix}, \tag{28}$$

whose inverse modulo 4 is

$$\begin{pmatrix} 1 & 0 \\ 3 & 1 \end{pmatrix}. \tag{29}$$

Then the matrix for C is unchanged by the conjugation, but the matrix for R becomes

$$\begin{pmatrix} 1 & 0 \\ k-2 & -1 \end{pmatrix}. \tag{30}$$

It follows that the four groups fall into two pairs of isomorphic groups, with G_3 and G_1 isomorphic and G_0 and G_2 isomorphic. Furthermore, G_1 and G_3 are isomorphic to $D_8 \times D_8$, the Cartesian product of the dihedral-8 group with itself. The automorphism argument above means $G_3 =< RC, M_i > \times < R, P_i >$.

To achieve a similar identification of G_1 detailing the generators of the two dihedral-8 groups in the Cartesian product, note that the only generator relation that differs for G_3 and G_1 is the final one. Starting with the form of this last relation that holds for G_1, which is $RP_i = M_i^2 P_i^{-1} R$, observe that it may be rewritten

$$R(M_i^{-1} P_i) = (M_i^{-1} P_i)^{-1} R. \tag{31}$$

Defining a new operator, $\tilde{P}_i = M_i^{-1} P_i$, it is straightforward to check that $\tilde{P}_i$ can replace P_i wherever it appears in the list of generators relations, without affecting the validity of any of the relations. All that has changed is that the "phase ramp" starts at $-i$ rather than i; the element-to-element phase increment remains $\pi/2$. It follows that $G_1 =< RC, M_i > \times < R, M_i^{-1} P_i >$.

Consider again the generalized Barker sequence of length 15:

$$x = \begin{bmatrix} 1 \; 1 \; 1 \; ii \; 1 \; -i \; -ii \; -1 \; -ii \; 1 \; -1 \; 1 \end{bmatrix}. \tag{32}$$

As noted earlier, this code satisfies $x = -\overline{xA_{15}}$, meaning that the composition of operators $(M_i^2) \circ R \circ (P_i^2)$ maps x to itself. Then $(M_i^2) \circ R \circ (P_i^2)$ is a group element of G_3, since the order-4 cyclic group generated by M_i is a subgroup of the dihedral group $< RC, M_i >$ and the order-4 cyclic group generated by P_i is a subgroup of the dihedral group $< R, P_i >$ (and hence $R \circ P_i^2$ is an element of $< R, P_i >$). So, as in the binary Barker case, x is a quad-phase of optimally low peak sidelobe level for which a nonidentity element of the associated SPG fixes x, causing its equivalence class to degenerate. This is due to the fact that the isometry subgroup of x contains an element other than the group identity and therefore has size 2 or greater; then, by Lagrange's orbit-stabilizer theorem [15], the equivalence class of x degenerates to size $32 = 64/2$ or smaller [5,9]. By applying combinations of operators to this code, it is easy to establish that the equivalence class must be at least size 32; hence it is exactly size 32. This example provides further anecdotal support for a link between low-peak-sidelobe codes and degeneracies in SPG group actions.

7 Generalizing from Quad-Phase to *m*th Roots of Unity

Turning from the quad-phase codes $x \in (Q_2)_N$ to the more general *m*th roots codes $x \in (Q_m)_N$ for $m \geq 3$, it will be seen that the approach used in the quad-phase case generalizes well. First, the set of sidelobe preservers becomes:

1. C: elementwise complex conjugation
2. R: reversal, $\overline{x}$
3. M_μ: multiplication by $\mu = \mathrm{e}^{i2\pi/m}$
4. P_μ: progressive multiplication by $\mu = \mathrm{e}^{i2\pi/m}$

No loss of generality results from the particular choice of values for μ. This is because with value $\mathrm{e}^{i2\pi/m}$, μ is a generator for a cyclic group of order m containing the other *m*th roots of unity. Similarly, P_μ is the generator for an order-m cyclic group of "phase ramps" (or progressive multiplication transformations), and hence contains all possible choices for this operator. These cycle groups are subgroups of the SPG or SPGs.

Two of the SPG group generators have order 2 and the other two have order m. The argument for quad-phase group order can be generalized in a natural way to give $(2)(2)(m)(m) = 4m^2$ for group order.

The set of generator relations is:

1. $C^2 = R^2 = g_0$
2. $M_\mu^m = P_\mu^m = g_0$
3. $RC = CR$
4. $P_\mu M_\mu = M_\mu P_\mu$
5. $M_\mu R = RM_\mu$
6. $CM_\mu = M_\mu^{-1} C$
7. $CP_\mu = P_\mu^{-1} C$
8. $RP_\mu = M_\mu^{N+1} P_\mu^{-1} R$

Here, as before, g_0 represents the group identity. The final relation has a different form for each of m powers of μ, implying that there are as many as m groups of order $4m^2$. Let G_i represent the SPG for $N \equiv i \bmod m$, $i = 0, 1, \ldots, m-1$.

Similar arguments as for the quad-phase case ($m = 4$) work here to conclude that $G_{m-1} = < RC, M_\mu > \times < R, P_\mu >$ (i.e., when $N+1 \equiv 0 \bmod m$). Then, turning to the case $N+1 \equiv 2 \bmod m$, it is possible to conclude that by attaching an M_μ^{-1} term to P_μ (as was done in the quad-phase case), leads to $G_1 = < RC, M_\mu > \times < R, M_\mu^{-1} P_\mu >$. This process of incrementing i by 2 and attaching an additional M_μ^{-1} can be repeated as many times as needed, allowing

$$G_{2k-1} = < RC, M_\mu > \times < R, (M_\mu^{-1})^k P_\mu >$$

for any $m-1 \geq k \geq 0$.

Now notice that whenever m is odd, every one of the m SPGs has the structure $< RC, M_\mu > \times < R, (M_\mu^{-1})^k P_\mu >$. This is because when m is odd, every one of the m SPGs is encountered in no more than m jumps, by repeatedly incrementing j by 2, starting with $j = 0$, in $N + 1 \equiv j \bmod m$. Therefore, when m is odd, every SPG is isomorphic to $D_{2m} \times D_{2m}$.

When m is even, it happens that the groups fall into two classes, those for N odd and those for N even. When N is odd,

$$G_{2k-1} =< RC, M_\mu > \times < R, (M_\mu^{-1})^k P_\mu >$$

for any $k \geq 0$, by the same argument used for m odd. Hence, G is isomorphic to $D_{2m} \times D_{2m}$ when m is even and N is odd. The group structure when m is even and N is even is left for others to resolve.

8 Conclusions

This chapter considers the structure of groups of peak-sidelobe-preserving operators for the AAC of mth root codes. These groups are shown to be helpful for efficient enumeration of codes for a given m, by peak sidelobe level. In the binary case, it is shown that the group is an Abelian group of order 8. Furthermore, it is shown that shared symmetry in the binary Barker codes can be discovered in a natural way from considering degeneracies the group actions. The group structure for $m = 4$ (the quad-phase case) is shown to have increased complexity; in fact, instead of a single group, there are four groups (two pairs of isomorphic groups). Group structure is identified for the cases of odd N. Moving to general mth roots codes, it is shown that results found for the quad-phase case generalize quite well. It is shown that there are $4m^2$ groups. All m groups are identified for any odd m. When m is even, the structure for any odd N is identified. The group structure for m even and N even is left unresolved.

Acknowledgments The author would like to acknowledge the help of I. Martin Isaacs of the University of Wisconsin at Madison, in outlining an approach for discovering the sidelobe-preserving operator group structure for quad-phase codes. In addition, Chris Monsour of Travelers Insurance, for the help with in establishing the quad-phase group order. Finally, this chapter is in memory of a friend and advisor, Larry Welch (1956–2003).

References

1. Barker, R.H.: Group synchronization of binary digital systems. In: Jackson, W. (ed.) Communications Theory, pp. 273–287. Academic, London (1953)
2. Carter, N.: Visual Group Theory. MAA Press, Washington DC (2009)

3. Conway, J.H., Curtis, R.T., Norton, S.P., Parker, R.A., Wilson, R.A.: Atlas of Finite Groups. Clarendon Press, Oxford (1985)
4. Coxson, G.E., Cohen, M.N., Hirschel, A.: New results on minimum-PSL binary codes. In: Proceedings of the 2001 IEEE National Radar Conference, pp. 153–156. Atlanta, GA (2001)
5. Dummit, D.S., Foote, R.M.: Abstract Algebra, 3rd edn. Wiley, New York (2004)
6. Friese, M., Zottman, H.: Polyphase Barker sequences up to length 31. Elect. Lett **30**(23), 1930–1931 (1994)
7. Golomb, S., Scholtz, R.: Generalized Barker sequences (transformations with correlation function unaltered, changing generalized Barker sequences. IEEE Trans. Inf. Theor. **11**, 533–537 (1965)
8. Golomb, S., Win, M.Z.: Recent results on polyphase sequences. IEEE Trans. Inf. Theor. **44**, 817–824 (1999)
9. Herstein, I.N.: Topics in Algebra, 2nd edn. Wiley, New York (1975)
10. Isaacs, I.M.: Email communication, 21 June 2006 (2006)
11. Levanon, N., Mozeson, E.: Radar Signals. Wiley, New York (2005)
12. Militzer, B., Zamparelli, M., Beule, D.: Evolutionary search for low autocorrelated binary sequences. IEEE Trans. Evol. Comput. **2**(1), 34–39 (1998)
13. Nunn, C., Coxson, G.E.: Polyphase pulse compression codes with optimal peak and integrated sidelobes. IEEE Trans. Aerosp. Electron. Syst. **45**(2), 41–47 (2009)
14. Pless, V.S., Huffman, V.S., W.C. and Brualdi, R.A., (eds.): Handbook of Coding Theory. Elsevier Publishing, Amsterdam (1998)
15. Roth, R.R.: A history of Lagrange's theorem on groups. Math. Mag. **74**(2), 99–108 (2001)
16. Skolnik, M.: Radar Handbook, 2nd edn . McGraw-Hill, New York (1990)
17. Turyn, R.J., Storer, J.: On binary sequences. Proc. Am. Math. Soc. **12**, 394–399 (1961)

Zeros of Some Self-Reciprocal Polynomials

David Joyner

Abstract We say that a polynomial p of degree n is *self-reciprocal polynomial* if $p(z) = z^n p(1/z)$, i.e., if its coefficients are "symmetric." This chapter surveys the literature on zeros of this family of complex polynomials, with the focus on criteria determining when such polynomials have all their roots on the unit circle. The last section contains a new conjectural criteria which, if true, would have very interesting applications.

Keywords Alexander polynomial • Artin–Weil zeta polynomial (of a curve) • Barker polynomial • Duursma zeta function • Duursma's conjecture • Error-correcting code • Frobenius polynomial (of a curve) • Littlewood polynomial • Littlewood's "two-sided" conjecture • Reciprocal polynomial • Reverse polynomial • Self-reciprocal polynomial

1 Introduction

This talk is about zeros of a certain family of complex polynomials which arise naturally in several areas of mathematics but are also of independent interest. We are especially interested in polynomials which have all their zeros in the unit circle

$$S^1 = \{z \in \mathbb{C} \mid |z| = 1\}.$$

Let p be a polynomial

$$p(z) = a_0 + a_1 z + \cdots + a_n z^n \qquad a_i \in \mathbb{C}, \tag{1}$$

D. Joyner (✉)
Department of Mathematics, US Naval Academy, Annapolis, MD, USA
e-mail: wdj@usna.edu

T.D. Andrews et al. (eds.), *Excursions in Harmonic Analysis, Volume 1*,
Applied and Numerical Harmonic Analysis, DOI 10.1007/978-0-8176-8376-4_17,

and let p^* denote[1] the *reciprocal polynomial* or *reverse polynomial*

$$p^*(z) = a_n + a_{n-1}z + \cdots + a_0 z^n = z^n p(1/z).$$

We say p is *self-reciprocal* if $p = p^*$, i.e., if its coefficients are "symmetric."

The types of polynomials we will be most interested in this talk are self-reciprocal polynomials. The first several sections are surveys. The last section contains a conjecture which is vague enough to probably be new and sufficiently general to hopefully have interesting applications, if true.

2 Where These Self-reciprocal Polynomials Occur

Self-reciprocal polynomials occur in many areas of mathematics—coding theory, algebraic curves over finite fields, knot theory, and linear feedback shift registers, to name several. This section discusses some of these.

2.1 Littlewood Polynomials

This section discusses a very interesting class of polynomials named after the late British mathematician J. E. Littlewood, famous for his collaboration with G. Hardy in the early 1900s. Although the main questions about these polynomials do not involve their zeros, so this section is a bit tangential, there are some aspects related to our main theme. Basically, we present just enough to whet the readers' taste to perhaps pursue the literature further on their own.

This section recalls some relevant facts from Mercer's thesis [16].

A polynomial $p(z)$ as in (1) is a *Littlewood polynomial* if $a_i \in \{\pm 1\}$, for all i, where $a_i = a_i(p)$ is the ith coefficient of the polynomial p. Let L_n denote the set of all Littlewood polynomials of degree n.

Conjecture 1 (Littlewood's "two-sided" conjecture). There are positive constants K_1, K_2 such that, for all $n > 1$, there exists a $p \in L_n$ such that

$$K_1\sqrt{n} \le |p(z)| \le K_2\sqrt{n}. \tag{2}$$

The *autocorrelations* of the sequence $\{a_i\}_{i=0}^n$ are the elements of the sequence given by

$$c_k = c_k(p) = \sum_{j=0}^{n-k} a_j a_{j+k}, \qquad 0 \le k \le n. \tag{3}$$

---

[1]Some authors, such as Chen [2], have p^* denote the complex conjugate of the reverse polynomial. It will not matter for us, since we will eventually assume that the coefficients are real.

One can show that

$$p(z)p^*(z) = c_n + c_{n-1}z + \cdots + c_1 z^{n-1} + c_0 z^n + c_1 z^{n+1} + \cdots + c_n z^{2n}.$$

Littlewood polynomials are studied in an attempt to gain further understanding of pseudorandom sequences of ± 1's. In this connection, one is especially interested in Littlewood polynomials with "small" autocorrelations. A Littlewood polynomial having the property that $|c_k| \leq 1$ is called a *Barker polynomial*. It is an open problem to find a Barker polynomial for $n > 13$ (or show one does not exist). Let

$$b(n) = \min_{p \in L_n} \max_{1 \leq k \leq n} |c_k(p)|,$$

where c_k is as in (3). If $b(n) > 1$ then there is no Barker polynomial of degree n. The asymptotic growth of $b(n)$, as $n \to \infty$, is an open question, although there is a conjecture of Turyn that

$$b(n) \sim K \log(n),$$

for some constant $K > 0$. It is known that $b(n) = O(\sqrt{n \log(n)})$.

The zeros of Littlewood polynomials on the unit circle are of tangential interest to this question. It is known that self-reciprocal Littlewood polynomials have at least one zero on S^1. Such a polynomial would obviously violate (2). A Littlewood polynomial p as in (1) is *skewreciprocal* if, for all j, $a_{d+j} = (-1)^j a_{d-j}$, where $d = m/2$ (m even) or $d = (m-1)/2$ (m odd). A Skewreciprocal Littlewood polynomials have no zeros on S^1. (The Littlewood polynomials having small autocorrelations also tend to be skew-reciprocal.)

We refer to Mercer [16] for more details.

2.2 Algebraic Curves Over a Finite Field

Let X be a smooth projective curve of genus[2] g over a finite field $GF(q)$.

Example 1. The curve

$$y^2 = x^5 - x,$$

over $GF(31)$ is a curve of genus 2.

Suppose X is defined by a polynomial equation $F(x,y) = 0$, where F is a polynomial with coefficients in $GF(q)$. Let N_k denote the number of solutions in $GF(q^k)$ and create the generating function

$$G(z) = N_1 z + N_2 z^2/2 + N_3 z^3/3 + \cdots.$$

[2]These terms will not be defined precisely here. Please see standard texts for a rigorous treatment.

Define the (Artin–Weil) zeta function of X by the formal power series

$$\zeta(z) = \zeta_X(z) = \exp(G(z)) \tag{4}$$

so $\zeta(0) = 1$. The logarithmic derivative of ζ_X is the generating function of the sequence of counting numbers $\{N_1, N_2, \dots\}$. In particular, the logarithmic derivative of $\zeta(z)$ has integral coefficients.

It is known that ζ is a rational function of the form

$$\zeta(z) = \frac{P(z)}{(1-z)(1-qz)},$$

where $P = P_X$ is a polynomial,[3] of degree $2g$ where g is the genus of X. This has a "functional equation" of the form

$$P(z) = q^g z^{2g} P\left(\frac{1}{qz}\right).$$

The Riemann hypothesis (RH) for curves over finite fields states that the roots of P have absolute value $q^{-1/2}$. It is well-known that the RH holds for ζ_X (this is a theorem of André Weil from the 1940s). By a suitable change-of-variable (namely, replacing z by $z/\sqrt{q}$), we thus see that curves over finite fields give rise to a large class of self-reciprocal polynomials having roots on the unit circle.

Example 2. We use Sage to compute an example.

Sage

```
sage: R.<x> = PolynomialRing(GF(31))
sage: H = HyperellipticCurve(x^5 - x)
sage: time H.frobenius_polynomial()
CPU times: user 0.04 s, sys: 0.01 s, total: 0.05 s
Wall time: 0.16 s
x^4 + 62*x^2 + 961
sage: C.<z> = PolynomialRing(CC, "z")
sage: f = z^4+62*z^2+961
sage: rts = f.roots()
sage: [abs(z[0]) for z in rts]
[5.56776436283002, 5.56776436283002]
sage: RR(sqrt(31))
5.56776436283002
```

In other words, the zeta polynomial

$$P_H(z) = 961z^4 + 62z^2 + 1$$

associated to the hyperelliptic curve H defined by $y^2 = x^5 - x$ over $GF(31)$ satisfies the RH. The polynomial $p(z) = P_H(z/\sqrt{31})$ is self-reciprocal, having all its zeros on S^1.

---

[3]Sometimes called the reciprocal of the *Frobenius polynomial*, or the *zeta polynomial*.

It can be shown that if X_1 and X_2 are "isomorphic" curves then the corresponding zeta polynomials are equal. Therefore, these polynomials can be used to help classify curves.

2.3 Error-Correcting Codes

Let $\mathbb{F} = GF(q)$ denote a finite field, for some prime power q.

Definition 1. Fix once and for all a basis for the vector space $V = \mathbb{F}^n$. A subset C of $V = \mathbb{F}^n$ is called a *code of length n*. A subspace of V is called a *linear code of length n*. If $\mathbb{F} = GF(2)$ then C is called a *binary* code. The elements of a code C are called *codewords*.

If $\cdot$ denotes the usual inner product,

$$v \cdot w = v_1 w_1 + \cdots + v_n w_n,$$

where $v = (v_1, \ldots, v_n) \in V$ and $w = (w_1, \ldots, w_n) \in V$, then we define the *dual code* $C^\perp$ by

$$C^\perp = \{v \in V \mid v \cdot c = 0,\ \forall c \in C\}.$$

We say C is *self-dual* if $C = C^\perp$.

For each vector $v \in V$, let

$$\mathrm{supp}(v) = \{i \mid v_i \neq 0\}$$

denote the *support* of the vector. The *weight* of the vector v is $\mathrm{wt}(v) = |\mathrm{supp}(v)|$. The *weight distribution vector* or *spectrum* of a code $C \subset \mathbb{F}^n$ is the vector

$$A(C) = \mathrm{spec}(C) = [A_0, A_1, \ldots, A_n],$$

where $A_i = A_i(C)$ denote the number of codewords in C of weight i, for $0 \leq i \leq n$. Note that for a linear code C, $A_0(C) = 1$, since any vector space contains the zero vector. The *weight enumerator polynomial* A_C is defined by

$$A_C(x,y) = \sum_{i=0}^{n} A_i x^{n-i} y^i = x^n + A_d x^{n-d} y^d + \cdots + A_n y^n.$$

Denote the smallest nonzero weight of any codeword in C by $d = d_C$ (this is the *minimum distance* of C) and the smallest nonzero weight of any codeword in $C^\perp$ by $d^\perp = d_{C^\perp}$.

Example 3. Let $\mathbb{F} = GF(2)$ and

$$C = \{(0,0,0,0), (1,0,0,1), (0,1,1,0), (1,1,1,1)\}.$$

This is a self-dual linear binary code which is a two-dimensional subspace of $V = GF(2)^4$.

The connection between the weight enumerator of C and that of its dual is very close, as the following well-known result shows.

Theorem 1 (MacWilliams' identity). *If C is a linear code over $GF(q)$ then*

$$A_{C^\perp}(x,y) = |C|^{-1} A_C(x+(q-1)y, x-y).$$

2.4 Duursma Zeta Function

Let $C \subset GF(q)^n$ be a linear error-correcting code.

Definition 2. A polynomial $P = P_C$ for which

$$\frac{(xT+(1-T)y)^n}{(1-T)(1-qT)} P(T) = \cdots + \frac{A_C(x,y)-x^n}{q-1} T^{n-d} + \dots .$$

is called a *Duursma zeta polynomial of C* [5]. The *Duursma zeta function* is defined in terms of the zeta polynomial by means of

$$\zeta_C(T) = \frac{P(T)}{(1-T)(1-qT)}.$$

It can be shown that if C_1 and C_2 are "equivalent" codes then the corresponding zeta polynomials are equal. Therefore, these polynomials can be used to help classify codes.

Proposition 1. *The Duursma zeta polynomial $P = P_C$ exists and is unique, provided $d^\perp \geq 2$. In that case, its degree is $n+2-d-d^\perp$.*

This is proven, for example, in Joyner–Kim [9].

It is a consequence of the MacWilliams identity that if C is self-dual (i.e., $C = C^\perp$), then associated Duursma zeta polynomial satisfies a functional equation of the form

$$P(T) = q^g T^{2g} P\left(\frac{1}{qT}\right),$$

where $g = n+1-k-d$. Therefore, after making a suitable change-of-variable (namely, replacing T by $T/\sqrt{q}$), these polynomials are self-reciprocal.

Unfortunately, the analog of the Riemann hypothesis for curves does *not* hold for the Duursma zeta polynomials of self-dual codes. Some counterexamples can be found, for example, in [9].

Example 4. We use Sage to compute an example.

Sage
```
sage: MS = MatrixSpace(GF(2),4,8)
sage: G = MS([[1,1,1,1,0,0,0,0],[0,0,1,1,1,1,0,0],
              [0,0,0,0,1,1,1,1],[1,0,1,0,1,0,1,0]])
sage: C = LinearCode(G)
sage: C == C.dual_code()
True
sage: C.zeta_polynomial()
2/5*T^2 + 2/5*T + 1/5
sage: C.<z> = PolynomialRing(CC, "z")
sage: f = (2*z^2+2*z+1)/5
sage: rts = f.roots()
sage: [abs(z[0]) for z in rts]
[0.707106781186548, 0.707106781186548]
sage: RR(sqrt(2))
1.41421356237310
sage: RR(1/sqrt(2))
0.707106781186548
```

In other words, the Duursma zeta polynomial

$$P_C(T) = (2T^2 + 2T + 1)/5$$

associated to "the" binary self-dual code of length 8 satisfies the analog of the RH. The polynomial $p(z) = P(z/\sqrt{2})$ is self-reciprocal, with all roots on S^1.

2.4.1 Duursma's Conjecture

There is an infinite family of Duursma zeta functions for which Duursma has conjecture that the analog of the Riemann hypothesis always holds. The linear codes used to construct these zeta functions are so-called "extremal self-dual codes."

To be more precise, we must take a more algebraic approach and replace "codes" by "weight enumerators." This tactic avoids some constraints which hold for codes and not for weight enumerators. We briefly describe how to do this. (For details, see Joyner–Kim [9], Chapter 2.) If $F(x,y) = x^n + \sum_{i=d}^{n} A_i x^{n-i} y^i \in \mathbb{Z}[x,y]$ is a homogeneous polynomial with $A_d \neq 0$ then we call n the *length* of F and d the *minimum distance* of F. We say F is *virtually self-dual weight enumerator* (over $GF(q)$) if and only if F satisfies the invariance condition

$$F(x,y) = F\left(\frac{x+(q-1)y}{\sqrt{q}}, \frac{x-y}{\sqrt{q}}\right). \tag{5}$$

Assume F is a virtually self-dual weight enumerator. We say F is *extremal, Type I* if $q = 2$, n is even, and $d = 2[n/8]+2$. We say F is *extremal, Type II* if $q = 2$, $8|n$, and $d = 4[n/24]+8$. We say F is *extremal, Type III* if $q = 3$, $4|n$, and $d = 3[n/12]+3$. We say F is *extremal, Type IV* if $q = 4$, n is even, and $d = 2[n/6]+2$.

If F is an extremal virtually self-dual weight enumerator then the zeta function $Z = Z_F$ can be explicitly computed. First, some notation. If F is a virtually self-dual weight enumerator of minimum distance d and $P = P_F$ is its zeta polynomial then define

$$Q(T) = \begin{cases} P(T), & \text{Type I,} \\ P(T)(1-2T+2T^2), & \text{Type II,} \\ P(T)(1+3T^2), & \text{Type III,} \\ P(T)(1+2T), & \text{Type IV.} \end{cases} \tag{6}$$

Let $(a)_m = a(a+1)\dots(a+m-1)$ denote the *rising generalized factorial* and write $Q(T) = \sum_j q_j T^j$, for some $q_j \in \mathbb{Q}$. Let

$$\gamma_1(n,d,b) = (n-d)(d-b)_{b+1}A_d/(n-b-1)_{b+2},$$

and

$$\gamma_2(n,d,b,q) = (d-b)_{b+1}\frac{A_d}{(q-1)(n-b)_{b+1}},$$

where recall A_d denoted the coefficient of $x^{n-d}y^d$ in the virtual weight enumerator $F(x,y)$.

Theorem 2 (Duursma [6]). *If F is an extremal virtually self-dual weight enumerator then the coefficients of $Q(T)$ satisfy the following conditions*

(a) *If F is Type I then*

$$\sum_{i=0}^{2m+2\nu} \binom{4m+2\nu}{m+i} q_i T^i = \gamma_1(n,d,2)\cdot(1+T)^m(1+2T)^m(1+2T+2T^2)^\nu,$$

where $m = d-3$, $4m+2\nu = n-4$, $b = q = 2$, $0 \le \nu \le 3$.

(b) *If F is Type II then*

$$\begin{aligned} &\textstyle\sum_{i=0}^{4m+8\nu} \binom{6m+8\nu}{m+i} q_i T^i \\ &= \gamma_1(n,d,2)\cdot(1+T)^m(1+2T)^m(1+2T+2T^2)^m B(T)^\nu, \end{aligned}$$

where $m = d-5$, $6m+8\nu = n-6$, $b = 4$, $q = 2$, $0 \le \nu \le 2$, and $B(T) = W_5(1+T,T)$, where $W_5(x,y) = x^8 + 14x^4y^4 + y^8$ is the weight enumerator of the Type II $[8,4,4]$ self-dual code.

(c) *If F is Type III then*

$$\sum_{i=0}^{2m+4\nu} \binom{4m+4\nu}{m+i} q_i T^i = \gamma_2(n,d,3,3)\cdot(1+3T+3T^2)^m B(T)^\nu,$$

where $m = d-4$, $4m+4\nu = n-4$, $b = q = 3$, $0 \le \nu \le 2$, *and* $B(T) = W_9(1+T,T)$, *where* $W_9(x,y) = x^4+8xy^3$ *is the weight enumerator of the Type III self-dual ternary code.*

(d) If F is Type VI then

$$\sum_{i=0}^{m+2\nu} \binom{3m+2\nu}{m+i} q_i T^i = \gamma_2(n,d,2,4)\cdot(1+2T)^m(1+2T+4T^2)^\nu,$$

where $m = d-3$, $3m+2\nu = n-3$, $b = 2$, $q = 4$, *and* $0 \le \nu \le 2$.

Although the construction of these codes is fairly technical (see [9] for an expository treatment), we can give some examples.

Example 5. Let P be a Duursma zeta polynomial as above, and let

$$p(z) = a_0 + a_1 z + \cdots + a_N z^N$$

denote the normalized Duursma zeta polynomial, $p(z) = P(z/\sqrt{q})$. By the functional equation for P, p is self-reciprocal. Some examples of the lists of coefficients $a_0, a_1, \ldots$, computed using Sage, are given below. We have normalized the coefficients so that they sum to 10 and represented the rational coefficients as decimal approximations to give a feeling for their relative sizes. The notation for m below is that in (6) and Theorem 2:

- Case Type I
  $m = 2$: [1.1309, 2.3990, 2.9403, 2.3990, 1.1309]
  $m = 3$: [0.45194, 1.2783, 2.0714, 2.3968, 2.0714, 1.2783, 0.45194]
  $m = 4$: [0.18262, 0.64565, 1.2866, 1.8489, 2.0724, 1.8489, 1.2866, 0.64565, 0.18262]
- Case Type II
  $m = 2$: [0.43425, 0.92119, 1.3028, 1.5353, 1.6129, 1.5353, 1.3028, 0.92119, 0.43425]
  $m = 3$: [0.12659, 0.35805, 0.63295, 0.89512, 1.1052, 1.2394, 1.2854, 1.2394, 1.1052, 0.89512, 0.63295, 0.35805, 0.12659]
  $m = 4$: [0.037621, 0.13301, 0.28216, 0.46554, 0.65783, 0.83451, 0.97533, 1.0656, 1.0967, 1.0656, 0.97533, 0.83451, 0.65783, 0.46554, 0.28216, 0.13301, 0.037621]
- Case Type III
  $m = 2$: [1.3397, 2.3205, 2.6795, 2.3205, 1.3397]
  $m = 3$: [0.58834, 1.3587, 1.9611, 2.1836, 1.9611, 1.3587, 0.58834]
  $m = 4$: [0.26170, 0.75545, 1.3085, 1.7307, 1.8874, 1.7307, 1.3085, 0.75545, 0.26170]
- Case Type IV
  $m = 2$: [2.8571, 4.2857, 2.8571]
  $m = 3$: [1.6667, 3.3333, 3.3333, 1.6667]
  $m = 4$: [0.97902, 2.4476, 3.1469, 2.4476, 0.97902]

Hopefully it is clear that, at least in these examples, these "normalized, extremal" Duursma zeta functions have coefficients which have "increasing symmetric form." We discuss this further in Sect. 6.

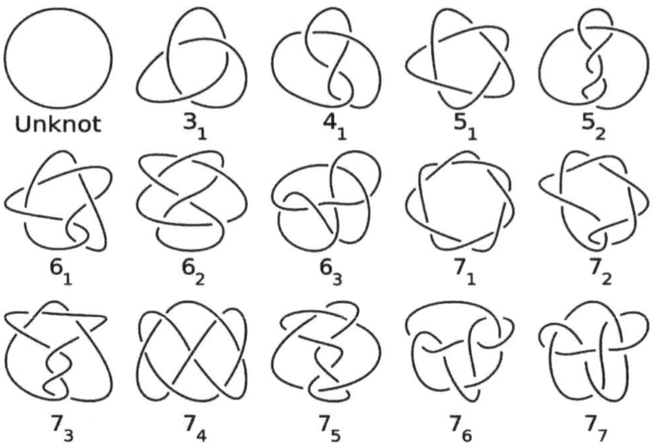

Fig. 1 Examples of knots

2.5 Knots

A *knot* is an embedding of S^1 into $\mathbb{R}^3$. If K is a knot then the *Alexander polynomial* is a polynomial $\Delta_K(t) \in \mathbb{Z}[t,t^{-1}]$ which is a topological invariant of the knot. For the definition, we refer, for example, to [1]. One of the key properties is the fact that

$$\Delta_K(t^{-1}) = \Delta_K(t).$$

If

$$\Delta_K(t) = \sum_{-d}^{d} a_i t^i,$$

then the polynomial $p(t) = t^d \Delta_K(t)$ is a self-reciprocal polynomial in $\mathbb{Z}[t]$. There is a special class of knots ("special alternating knots") which have the property that all its roots lie on the unit circle (see [17, 18]).

Kulikov [10] constructed an analogous Alexander polynomial Δ associated to a complex plane algebraic curve. Under certain technical conditions, such a Δ is a self-reciprocal polynomial in $\mathbb{Z}[t]$, all of whose roots lie on the unit circle.

Example 6. In Fig. 1, we give several examples of knots. These figures can be found in several places, for example, from [20].

The Alexander polynomial of the "unknot" is the constant function $\Delta_{S^1}(t) = 1$. The Alexander polynomials of the other knots in Fig. 1 are:

$$\begin{aligned}
\Delta_{3_1}(t) &= t^{-1} - 1 + t, & \Delta_{7_1}(t) &= t^{-3} - t^{-2} + t^{-1} - 1 + t - t^2 + t^3,\\
\Delta_{4_1}(t) &= -t^{-1} + 3 - t, & \Delta_{7_2}(t) &= 3t^{-1} - 5 + 3t,\\
\Delta_{5_1}(t) &= t^{-2} - t^{-1} + 1 - t + t^2, & \Delta_{7_3}(t) &= 2t^{-2} - 3t^{-1} + 3 + 3t + 2t^2,\\
\Delta_{5_2}(t) &= 2t^{-1} - 3 + 2t, & \Delta_{7_4}(t) &= 4t^{-1} - 7 + 4t,\\
\Delta_{6_1}(t) &= -2t^{-1} + 5 - 2t, & \Delta_{7_5}(t) &= t^{-2} - 4t^{-1} + 5 - 4t + 2t^2,\\
\Delta_{6_2}(t) &= -t^{-2} + 3t^{-1} - 3 + 3t - t^2, & \Delta_{7_6}(t) &= -t^{-2} + 5t^{-1} - 7 + 5t - t^2,\\
\Delta_{6_3}(t) &= t^{-2} - 3t^{-1} + 5 - 3t + t^2, & \Delta_{7_6}(t) &= t^{-2} - 5t^{-1} + 9 - 5t + t^2.
\end{aligned}$$

We use Sage [19] to compute their roots in several examples.

Sage

```
sage: t = var('t')
sage: RC.<z> = PolynomialRing(CC,"z")
sage: z = RC.gen()

sage: Delta51 = (t^(-2)-t^(-1)+1-t+t^2)*t^2
sage: f = RC(expand(Delta51)(t=z))
sage: [r.abs() for r in f.complex_roots()]
[1.00000000000000, 1.00000000000000, 1.00000000000000,
 1.00000000000000]

sage: Delta63 = (t^(-2)-3*t^(-1)+5-3*t+t^2)*t^2
sage: f = RC(expand(Delta63)(t=z))
sage: [r.abs() for r in f.complex_roots()]
[0.580691831992952, 0.580691831992952, 1.72208380573904,
 1.72208380573904]

sage: Delta71 = (t^(-3)-t^(-2)+t^(-1)-1+t-t^2+t^3)*t^3
sage: f = RC(expand(Delta71)(t=z))
sage: [r.abs() for r in f.complex_roots()]
[1.00000000000000, 1.00000000000000, 1.00000000000000,
 1.00000000000000, 1.00000000000000, 1.00000000000000]

sage: Delta75 = (2*t^(-2)-4*t^(-1)+5-4*t+2*t^2)*t^2
sage: f = RC(expand(Delta75)(t=z))
sage: [r.abs() for r in f.complex_roots()]
[1.00000000000000, 1.00000000000000, 1.00000000000000,
 1.00000000000000]

sage: Delta77 = (t^(-2)-5*t^(-1)+9-5*t+t^2)*t^2
sage: f = RC(expand(Delta77)(t=z))
sage: [r.abs() for r in f.complex_roots()]
[0.422082440385454, 0.422082440385454, 2.36920540709247,
 2.36920540709247]
```

2.6 Cryptography and Related Fields

The class of self-reciprocal polynomials also arise naturally in the field of cryptography (e.g., in the construction of "symmetric" linear feedback shift registers) and coding theory (e.g., in the construction of "symmetric" cyclic codes). However, such polynomials have coefficients in a finite field, so would take us away from our main topic. We refer the interested reader, for example, to Gulliver [8] and Massey [15].

3 Characterizing Self-reciprocal Polynomials

Let

$$\mathbb{R}[z]_m = \{p \in \mathbb{R}[z] \mid \deg(p) \le m\}$$

denote the real vector space of polynomials of degree m or less. Let

$$R_m = \{p \in \mathbb{R}[z]_m \mid p = p^*\}$$

denote the subspace of self-reciprocal ones.

Here is a basic fact about even degree self-reciprocal polynomials. Let

$$p(z) = a_0 + a_1 z + \cdots + a_{2n} z^{2n}, \qquad a_i \in \mathbb{R}.$$

Lemma 1 ([4], Section 2.1; see also [12]). *The polynomial $p \in \mathbb{R}[z]_{2n}$ is self-reciprocal if and only if it can be written*

$$p(z) = z^n \cdot (a_n + a_{n+1} \cdot (z + z^{-1}) + \cdots + a_{2n} \cdot (z^n + z^{-n})),$$

if and only if it can be written

$$p(z) = a_{2n} \cdot \prod_{k=1}^{n} (1 - \alpha_k z + z^2), \tag{7}$$

for some real $\alpha_k \in \mathbb{R}$.

Example 7. Note

$$1 + z + z^2 + z^3 + z^4 = (1 + \phi \cdot z + z^2)(1 + \overline{\phi} \cdot z + z^2),$$

where $\phi = \frac{1+\sqrt{5}}{2} = 1.618\ldots$ is the "golden ratio," and $\overline{\phi} = \frac{1-\sqrt{5}}{2} = -0.618\ldots$ is its "conjugate."

The *Chebyshev transformation* $T : R_{2n} \to \mathbb{R}[x]_n$ is defined on the subset[4] of polynomials of degree $2n$ by

$$T_p(x) = a_{2n} \prod_{k=1}^{n} (x - \alpha_k),$$

where $x = z + z^{-1}$, and p and the α_i's are as in (7).

The following statement is proven in Lakatos [12].

Lemma 2. *The Chebyshev transformation* $T : R_{2n} \to \mathbb{R}[x]_n$ *is a vector space isomorphism.*

For any $X_i \in \mathbb{C}$ $(1 \le i \le n)$, let

$$\begin{aligned} e_0(X_1, X_2, \ldots, X_n) &= 1, \\ e_1(X_1, X_2, \ldots, X_n) &= \textstyle\sum_{1 \le j \le n} X_j, \\ e_2(X_1, X_2, \ldots, X_n) &= \textstyle\sum_{1 \le j < k \le n} X_j X_k, \\ e_3(X_1, X_2, \ldots, X_n) &= \textstyle\sum_{1 \le j < k < l \le n} X_j X_k X_l, \\ &\vdots \\ e_n(X_1, X_2, \ldots, X_n) &= X_1 X_2 \cdots X_n. \end{aligned}$$

It is possible to describe explicitly how the α_k's determine the a_j's in (7). The following result is proven in Losonczi [13].

Lemma 3. *For each* $n \ge 1$ *and* $\alpha_i \in \mathbb{C}$, *we have*

$$\prod_{k=1}^{n} (z^2 - \alpha_k z + 1) = \sum_{k=1}^{2n} c_{2n,k} z^k,$$

where $c_{2n,k} = c_{2n,2n-k}$ *and*

$$c_{2n,k} = (-1)^k \sum_{\ell=1}^{[k/2]} \binom{n-k+2\ell}{\ell} e_{k-2\ell}(\alpha_1, \ldots, \alpha_n),$$

for $0 \le k \le n$.

4 Those with All Roots on S^1

There are several results concerning the set of self-reciprocal polynomials all of whose roots lie on S^1.

[4]In this definition, we assume for simplicity $a_{2n} \neq 0$; see [12] for the general definition of $p \mapsto T_p$.

Remark 1. Note that if $p \in R_m$ is a real self-reciprocal polynomial of degree m then $f(z) = z^{-m/2}p(z)$ is invariant under $z \mapsto z^{-1}$. Therefore, $f(z)$ is a real-valued on S^1, which implies that it is a cosine transform of its coefficients. Saying $p(z)$ has all its roots on S^1 is equivalent to saying $f(e^{i\theta})$ has n zeros on $[0, 2\pi)$.

One of the simplest examples of a polynomial in R_m with all its zeros on S^1 is

$$c_m(z) = 1 + z + \cdots + z^m.$$

If m is even then c_m does not have ± 1 as roots. Many results in the theory fall into the following category.

Metatheorem: If $p \in R_m$ is "close" to c_m then p has all its roots in the unit circle S^1. For example, the polynomials c_m above satisfy this.

Theorem 3 (Lakatos [11]). *Take the notation as in Lemma 1. The polynomial $p \in R_{2n}$ has all its roots in S^1 if and only if $-2 \le \alpha_k \le 2$ for all k.*

Here's another one of those metatheorem-type results.

Theorem 4 (Lakatos [11]). *The polynomial $p \in R_m$ given by*

$$p(z) = \sum_{j=0}^{m} a_j z^j$$

has all its roots on S^1, provided the coefficients satisfy the following condition:

$$|a_m| \ge \sum_{j=0}^{m} |a_j - a_m|.$$

Example 8. Let $p(z) = p(t, z) = c_2(z) + t \cdot z$. The theorem above says, in this case, $|t| \le 1$ implies all roots of $p(z)$ belong to S^1.

There are several other characterizations of self-reciprocal polynomials all of whose roots lie on S^1. Those due to Cohn, Chinen, Chen, and Fell, are discussed next.

Theorem 5 (Schur-Cohn[5]). *Let $p \in \mathbb{C}[z]_n$ be as in (1). The polynomial p has all its zeros on S^1 if and only if:*

(a) There is a $\mu \in S^1$ such that, for all k with $0 \le k \le n$, we have $a_{n-k} = \mu \cdot \overline{a_k}$.
(b) All the zeros of p' lie inside or on S^1.

[5] See, for example, Chen [2], Section 1.

According to Chen [2], this result of Cohn, published in 1922, is closely related[6] to a result of Schur, published in 1918. The following result is an immediate corollary of this theorem.

Corollary 1. *$p \in R_m$ has all its zeros on S^1 if and only if all the zeros of p' lie inside or on S^1.*

Remark 2.

- As a corollary to the corollary, by "version 2" of the Eneström-Kakeya theorem (see Remark 3), if $p \in R_m$ is "near" c_m, then the coefficients of p' are increasing and positive, so all roots of p' are inside S^1.
- For example, if $|a_i - a_{i-1}| < a_{i-1}/i$ for all i, then the corollary above and the Eneström-Kakeya theorem imply that (all roots of p' are inside S^1 and so) $p(z)$ has all its roots on S^1. However, this rate of growth is not sufficient for application to the Duursma zeta polynomials of extremal type.

The next result was proven by Chen [2] and later independently by Chinen[7] [3]. It provides a very large class of self-reciprocal polynomials having roots on the unit circle.

Theorem 6 (Chen–Chinen). *If $p \in R_m$ has "decreasing symmetric form"*

$$p(z) = a_0 + a_1 z + \cdots + a_k z^k + a_k z^{m-k} + a_{k-1} z^{m-k+1} + \cdots + a_0 z^m,$$

with $a_0 > a_1 > \cdots > a_k > 0$, then all roots of $p(z)$ lie on S^1, provided $m \geq k$.

We prove the following more general version of this.

Theorem 7. *If $g(z) = a_0 + a_1 z + \cdots + a_k z^k$ and $0 < a_0 < \cdots < a_{k-1} < a_k$, then, for each $r \geq 0$, the roots of $z^r g(z) + g^*(z)$ all lie on the unit circle.*

Proof. We shall adapt some ideas from Chinen [3] for our argument. □

The proof requires recalling the following well-known theorem, discovered independently by Eneström (in the late 1800s) and Kakeya (in the early 1900s).

Theorem 8 (Eneström–Kakeya, version 1). *Let $f(z) = a_0 + a_1 z + \cdots + a_k z^k$ satisfy $a_0 > a_1 > \cdots > a_k > 0$. Then $f(z)$ has no roots in $|z| \leq 1$.*

Remark 3. Replacing the polynomial by its reverse, here is "version 2" of the Eneström–Kakeya theorem: Let $f(z) = a_0 + a_1 z + \cdots + a_k z^k$ satisfy $0 < a_0 < a_1 < \cdots < a_k$. Then $f(z)$ has no roots in $|z| \geq 1$.

Back to the proof of Theorem 7.

Claim. $g^*(z)$ has no roots in $|z| \leq 1$.

---

[6] In fact, both are exercises in Marden [14].

[7] In one sense, Chinen's version is slightly stronger, and it is that version which we are stating.

Proof. This is equivalent to the statement of the Eneström–Kakeya theorem (Theorem 8). □

Claim. $g(z)$ has no roots in $|z| \geq 1$.

Proof. This follows from the previous claim and the observation that the roots of $g(z)$ correspond to the inverse of the roots of $g^*(z)$. □

Claim. $|g(z)| < |g^*(z)|$ on $|z| < 1$.

Proof. By the above claims, the function $\phi(z) = g(z)/g^*(z)$ is holomorphic on $|z| \leq 1$. Since $g(z^{-1}) = \overline{g(z)}$ on $|z| = 1$, we have $|g(z)| = |g^*(z)|$ on $|z| = 1$. The claim follows from the maximum modulus principle. □

Claim. The roots of $z^r g(z) + g^*(z)$ all lie on the unit circle, $r \geq 0$.

Proof. By the previous claim, $z^r g(z) + g^*(z)$ has the same number of zeros as $g^*(z)$ in the unit disc $|z| < 1$ (indeed, the function $\frac{z^r g(z)+g^*(z)}{g^*(z)} = 1 + \frac{z^r g(z)}{g^*(z)}$ has no zeros). Since $g^*(z)$ has no roots in $|z| < 1$, neither does $z^r g(z) + g^*(z)$. But since $z^r g(z) + g^*(z)$ is self-reciprocal, it has no zeros in $|z| > 1$ either. □

This proves Theorem 7. If $P_0(z)$ and $P_1(z)$ are polynomials, let

$$P_a(z) = (1-a)P_0(z) + aP_1(z),$$

for $0 \leq a \leq 1$. Next, we recall an interesting characterization of polynomials (not necessarily self-reciprocal ones) with roots on S^1, due to Fell [7].

Theorem 9 (Fell). *Let $P_0(z)$ and $P_1(z)$ be real monic polynomials of degree n having zeros on $S^1 - \{1,-1\}$. Denote the zeros of $P_0(z)$ by $w_1, w_2, \ldots, w_n$ and of $P_1(z)$ by $z_1, z_2, \ldots, z_n$. Assume*

$$w_i \neq z_j,$$

for $1 \leq i, j \leq n$. Assume also that

$$0 < \arg(w_i) \leq \arg(w_j) < 2\pi,$$
$$0 < \arg(z_i) \leq \arg(z_j) < 2\pi,$$

for $1 \leq i, j \leq n$. Let A_i be the smaller open arc of S^1 bounded by w_i and z_i, for $1 \leq i \leq n$. Then the locus of $P_a(z)$, $0 \leq a \leq 1$ is contained on S^1 if and only if the arcs A_i are all disjoint.

This theorem is used in the "heuristic argument" given in Sect. 6.

5 Smoothness of Roots

A natural question to ask about zeros of polynomials is how "smoothly" do they vary as a function of the coefficients of the polynomial?

To address this, suppose that the coefficients a_i of the polynomial p are functions of a real parameter t. Abusing notation slightly, identify $p(z) = p(t,z)$ with a function of two variables ($t \in \mathbb{R}$, $z \in \mathbb{C}$). Let $r = r(t)$ denote a root of this polynomial, regarded as a function of t:

$$p(t,r(t)) = 0.$$

Using the two-dimensional chain rule,

$$0 = \frac{\mathrm{d}}{\mathrm{d}t} p(t,r(t)) = p_t(t,r(t)) + r'(t) \cdot p_z(t,r(t)),$$

so $r'(t) = -p_t(t,r(t))/p_z(t,r(t))$. Since $p_z(t,r(t)) = p'(r)$, the denominator of this expression for $r'(t)$ is zero if and only if r is a double root of p (i.e., a root of multiplicity 2 or more).

In answer to the above question, we have proven the following result on the "smoothness of roots."

Lemma 4. *$r = r(t)$ is smooth (i.e., continuously differentiable) as a function of t, provided t is restricted to an interval on which $p(t,z)$ has no double roots.*

Example 9. Let

$$p(z) = 1 + (1+t) \cdot z + z^2,$$

so we may take

$$r(t) = \frac{-1 - t + \sqrt{(1+t)^2 - 4}}{2}.$$

Note that $r(t)$ is smooth provided t lies in an interval which does not contain 1 or -3. We can directly verify the lemma holds in this case. Observe (for later) that if $-3 < t < 1$ then $|r(t)| = 1$.

Let $p(z) = p(t,z)$ and $r = r(t)$ be as before. Consider the distance function

$$d(t) = |r(t)|$$

of the root r. Another natural question is how smooth is the distance function of a root as a function of the coefficients of the polynomial p?

The analog to Lemma 4 holds, with one extra condition.

Lemma 5. *$d(t) = |r(t)|$ is smooth (i.e., continuously differentiable) as a function of t, provided t is restricted to an interval one which $p(t,z)$ has no double roots and $r(t) \neq 0$.*

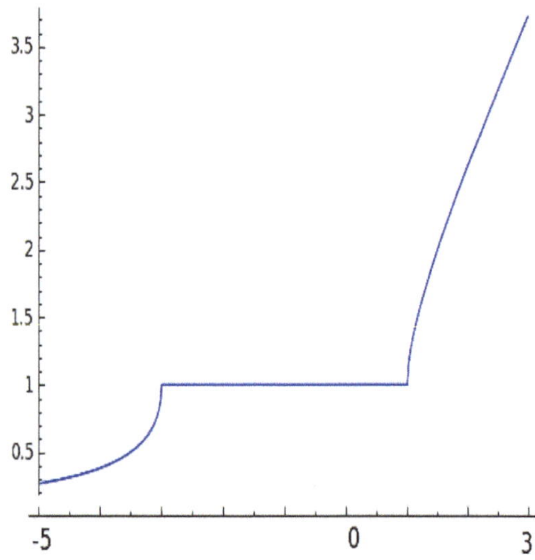

Fig. 2 Size of largest root of the polynomial $1+(1+t)z+z^2$, $-5<t<3$. The plot was created using Sage's `list plot` command, though the axes labels were modified using `GIMP` for ease of reading

Proof. This is basically an immediate consequence of the above lemma and the chain rule,

$$\frac{\mathrm{d}}{\mathrm{d}t}|r(t)| = r'(t) \cdot \left(\frac{\mathrm{d}|x|}{\mathrm{d}x}|_{x=r(t)}\right).$$

□

Example 10. This is a continuation of the previous Example 9. Figure 2 is a plot of $d(t)$ in the range $-5<t<3$.

In the next section, we will find this "smoothness" useful.

6 A Conjecture

Are there conditions under which self-reciprocal polynomials with in "increasing symmetric form" have all their zeros on S^1?

We know that self-reciprocal polynomial with "decreasing symmetric form" have all their roots on S^1. Under what conditions is the analogous statement true for functions with "increasing symmetric form?" The remainder of this section considers this question for polynomials of even degree.

Let d be an odd integer and let $f(z) = f_0 + f_1 z + \cdots f_{d-1} z^{d-1} \in R_{d-1}$ be a self-reciprocal polynomial with "increasing symmetric form"

$$0 < f_0 < f_1 < \cdots < f_{\frac{d-1}{2}}.$$

For each $c \geq f_{\frac{d-1}{2}}$, the polynomial

$$g(z) = c \cdot (1 + z + \cdots + z^{d-1}) - f(z) = g_0 + g_1 z + \cdots g_{d-1} z^{d-1} \in R_{d-1},$$

is a self-reciprocal polynomial having nonnegative coefficients with "decreasing symmetric form." If $c > f_{\frac{d-1}{2}}$, the Chen–Chinen theorem (Theorem 7) implies, all the zeros of $g(z)$ are on S^1. Let

$$P_0(z) = \frac{g(z)}{g_{d-1}}, \quad P_1(z) = \frac{f(z)}{f_{d-1}}, \quad P_a(z) = (1-a)P_0(z) + aP_1(z),$$

for $0 \leq a \leq 1$. By the Chen–Chinen theorem, there is a $t_0 \in (0,1)$ such that all zeros of $P_t(z)$ are on S^1 for $0 \leq t < t_0$. In fact, if

$$t = \frac{f_{\frac{d-1}{2}} - f_{d-1}}{f_{\frac{d-1}{2}}},$$

then $P_t(z)$ is a multiple of $1 + z + \cdots + z^{d-1}$.

Do any of the polynomials $P_t(z)$ have multiple roots $(0 < t < 1)$? Using the notation of Sect. 5, in the case $p(t,z) = P_t(z)$, we have

$$r'(t) = -p_t(t, r(t))/p_z(t, r(t)) = \frac{P_1(r(t)) - P_0(r(t))}{P_t'(r(t))}.$$

If no $P_t(z)$ has a multiple root, then by the second "smoothness of roots lemma" (Lemma 5), all the roots of $f(z)$ are also on S^1. This heuristic argument supports the hope expressed in the following statement.

Conjecture 2. Let $s : \mathbb{Z}_{>0} \to \mathbb{R}_{>0}$ be a "slowly increasing" function.

- Odd degree case. If $g(z) = a_0 + a_1 z + \cdots + a_d z^d$, where $a_i = s(i)$, then the roots of $p(z) = g(z) + z^{d+1} g^*(z)$ all lie on the unit circle.
- Even degree case. The roots of

$$p(z) = a_0 + a_1 z + \cdots + a_{d-1} z^{d-1} + a_d z^d + a_{d-1} z^{d+1} + \cdots + a_1 z^{2d-1} + a_0 z^{2d}$$

all lie on the unit circle.

Remark 4.

- Though this is supported by some numerical evidence, I don't know what "slowly increasing" should be here.[8] In any case, the correct statement of this form, whatever it is, would hopefully allow for the inclusion of the extremal type Duursma polynomials!

[8] For example, numerical experiments suggest "linear growth" seems too fast but "logarithmic growth" seems sufficient.

- Note that if $p(z)$ is as above and m denotes the degree then $f(z) = z^{-m/2}p(z)$ is a real-valued function on S^1. Therefore, the above conjecture can be reformulated as a statement about zeros of cosine transforms.

Acknowledgment I thank Mark Kidwell for discussions of knots and the references in Sect. 2.5, Geoff Price for pointing out the applications in Sect. 2.6, and George Benke for helpful suggestions.

References

1. Adams, C.: The Knot Book: An Elementary Introduction to the Mathematical Theory of Knots. American Mathematical Society, Providence, RI (2004)
2. Chen, W.: On the polynomials with all there zeros on the unit circle. J. Math. Anal. Appl. **190**, 714–724 (1995)
3. Chinen, K.: An abundance of invariant polynomials satisfying the Riemann hypothesis. Discrete Math. **308**, 6426–6440 (2008). Available: http://arxiv.org/abs/0704.3903
4. DiPippo, S., Howe, E.: Real polynomials with all roots on the unit circle and abelian varieties over finite fields. J. Num. Theor. **78**, 426–450 (1998). Available: http://arxiv.org/abs/math/9803097
5. Duursma, I.: Weight distributions of geometric Goppa codes.Trans. Am. Math. Soc. **351**, 3609–3639 (1999). Available: http://www.math.uiuc.edu/~duursma/pub/
6. Duursma, I.: Extremal weight enumerators and ultraspherical polynomials. Discrete Math. **268**(1–3), 103–127 (2003)
7. Fell, H.J.: On the zeros of convex combinations of polynomials. Pac. J. Math. **89**, 43–50 (1980)
8. Gulliver, T.A.: Self-reciprocal polynomials and generalized Fermat numbers. IEEE Trans. Inf. Theor. **38**, 1149–1154 (1992)
9. Joyner, D., Kim, J.L.: Selected Unsolved Problems in Coding Theory. Birkhäuser, Boston (2011)
10. Kulikov, V.: Alexander polynomials of plane algebraic curves. Russian Acad. Sci. Izv. Math. **42**, 67–88 (1994)
11. Lakatos, P.: On polynomials having zeros on the unit circle. C. R. Math. Acad. Sci. Soc. R. Can. **24**(2), 9196 (2002)
12. Lakatos, P.: On zeros of reciprocal polynomials, vol. 61, pp. 645–661. Publicationes Mathematicae, Debrecen (2002)
13. Losonczi, L.: On reciprocal polynomials with zeros of modulus one. Math. Inequalities Appl. **9**, 286–298 (2006). http://mia.ele-math.com/09-29/
14. Marden, M.: Geometry of Polynomials. American Mathematical Society, Providence, RI (1970)
15. Massey, J.L.: Reversible codes. Inf. Control **7**, 369–380 (1964) Available at: http://www.isiweb.ee.ethz.ch/archive/massey_pub/dir/pdf.html
16. Mercer, I.D.: Autocorrelation and Flatness of Height One Polynomials. PhD thesis, Simon Fraser University (2005) http://www.idmercer.com/publications.html
17. Murasugi, K.: On alternating knots. Osaka Math. J. **12**, 227–303 (1960)
18. Riley, R.: A finiteness theorem for alternating links. J. London Math. Soc. **5**, 263–266 (1972)
19. Stein, W., Sage group, Sage -Mathematical Software, version 4.5 (2010) http://www.sagemath.org/.
20. Knot theory. Wikipedia http://en.wikipedia.org/wiki/Knot_theory

Part IV
Applications of Data Processing

Several applications of data processing are included in this part. The five chapters cover areas of signal processing and information theory. Specifically the papers by Claussen et al., Patel et al., and Teolis et al. deal with audio signal processing, image signal processing, and eddy current signal processing, respectively. The paper by Tyagi et al. addresses a problem of channel coding.

The first chapter of this part is by HEIKO CLAUSSEN, JUSTINIAN ROSCA, VISWANATHAN RAMASUBRAMANIAN, and SUBRAMANI THIYAGARAJAN. The authors consider the signal detection problem where the signal of interest is corrupted additively by structured perturbations. Specifically, a set of measurements is collected as columns of a $D \times N$ matrix X. The authors estimate the signal of interest as the vector given by $w = X(X^T X + \lambda I)^{-1}\underline{\mathbf{1}}$, where λ is a scalar that parameterizes the generalized mutual interdependence analysis (GMIA) solution. Next the authors apply this estimate to the problem of voice activity detection for hearing-aid devices. They compare performance of GMIA with MFCC and cepstral-mean features.

VISHAL M. PATEL and RAMA CHELLAPPA study shape and image reconstruction from gradient field information. In particular they discuss four methods: a Poisson-based method, a Fourier-based method (the Frankot–Chellappa algorithm), the shapelet approach, and a wavelet-based algorithm. In the second part of this chapter, the authors discuss image recovery from partial Fourier measurements using inversion of the estimated gradient field. Numerical comparisons with the inversion algorithm using total variation are also presented.

ANTHONY TEOLIS analyzes the linear system response to analytic signals, which are functions in the Hardy space H^2. Specifically, he considers input signals of the form $p(t) = A(t)\mathrm{e}^{j\Phi(t)}$, where $\Phi(t) = 2\pi \int_{t_0}^{t} f(s)\mathrm{d}s$. Here $f(s)$ denotes the instantaneous frequency. This chapter considers linear systems which are small perturbations of the identity; "small" is with respect to the operator norm on $L^2(\mathbb{R})$. The instantaneous frequency of the output signal is computed in terms of the input signal and the linear system impulse response. In the last part of the chapter, the author applies these results to two classes of signals: chirps and FM chirps.

The fourth chapter of this part is by CAROLE TEOLIS, DAVID GENT, CHRISTINE KIM, ANTHONY TEOLIS, JAMES PADUANO, and MICHELLE BRIGHT. The authors present an algorithm for gas turbine monitoring and stall detection using eddy current sensors. First, they analyze the eddy current sensor signal using a combination of band-pass filter center around the second harmonic, followed by a narrowband wavelet filterbank. The engine-dependent signature corresponds to the narrowband channel output where the maximum occurs. A simplified implementation of this algorithm uses the second derivative of the instantaneous phase, which is also the first derivative of the instantaneous frequency. The authors test these algorithms on real data obtained at the NASA Glenn W8 compressor test facility.

The last chapter of this part is by HIMANSHU TYAGI and PRAKASH NARAYAN. The authors discuss the fundamental problem of reliable channel coding in information theory. They consider the case of a state-dependent discrete memoryless channel with known underlying state process distribution. Additionally it is assumed that the transmitter knows the channel state. Two classical results on state-dependent channel capacity are due to Shannon (when the encoder is causal) and Gelfand and Pinsker (when the encoder is noncausal). It was known that Shannon's result admits a strong converse. The authors prove a strong converse for the Gelfand–Pinsker theorem. During this exposition they also obtain upper bounds on the reliability function (the exponent for which transmission error decays to zero) for both channel models.

Generalized Mutual Interdependence Analysis of Noisy Channels

Heiko Claussen, Justinian Rosca, Viswanathan Ramasubramanian, and Subramani Thiyagarajan

Abstract The main motivation for our present work is to reliably perform voice (or signal) detection for a source of interest from a single microphone recording. We rely on the assumption that the input signal contains invariant information about the channel, or transfer function from each source to the microphone, which could be reliably exploited for signal detection and classification. In this chapter we employ a nonconventional method called generalized mutual interdependence analysis (GMIA) that proposes a model for the computation of this hidden invariant information present across multiple measurements. Such information turns out to be a good characteristic feature of a signal source, transformation, or composition that fits the model. This chapter introduces a unitary and succinct description of the underlying model of GMIA, and the formulation and solution of the corresponding optimization problem. We apply GMIA for feature extraction in the problem of own-voice activity detection, which aims at classification of a near-field channel based on access to prior information about GMIA features of the channel. It is extremely challenging to recognize the presence of voice in noisy scenarios with interference from music, car noise, or street noise. We compare GMIA with MFCC and cepstral-mean features. For example, GMIA performs with equal error rates below 10 % for music interference of SNRs down to −20 dB.

Keywords Acoustic signal classification • Voice detection • Feature extraction • Near-field channels • Speaker verification • Head related transfer functions • Mutual interdependence analysis • Hearing aids

---

H. Claussen • J. Rosca (✉)
Siemens Corporation, Corporate Research, 755 College Road East, Princeton, NJ 08540, USA
e-mail: heiko.claussen@siemens.com; justinian.rosca@siemens.com

V. Ramasubramanian • S. Thiyagarajan
Siemens Corporate Research and Technologies-India, Bangalore, India
e-mail: V.Ramasubramanian@siemens.com; thiyagarajan.s@siemens.com

T.D. Andrews et al. (eds.), *Excursions in Harmonic Analysis, Volume 1*, Applied and Numerical Harmonic Analysis, DOI 10.1007/978-0-8176-8376-4_18,

1 Introduction

Our goal is to compute a simplified statistical data representation that retains invariant information that is necessary for subsequent tasks such as classification or prediction. Methods such as Fisher's linear discriminant analysis (FLDA) [10], canonical correlation analysis (CCA) [14], or ridge regression [23] extract "optimal" representations of a dataset. For instance, FLDA defines a projection space that maximizes the ratio of the between- and within-class scatter of the training data to reduce the dimensionality of the input. CCA assumes one common source in two datasets. The dimensionality of the data is reduced by retaining the space that is spanned by pairs of projecting directions in which the datasets are maximally correlated. In contrast, ridge regression finds a linear combination of the inputs that best fits a desired response. In this chapter, we review an alternative second-order statistical criterion to find an "optimal" dataset representation, called GMIA. We aim to define an invariant computation or feature of high dimensional instances of a single class, which does not change within its class, where the number of input instances N is smaller than their dimensionality D.

We further consider the application of GMIA to the system identification problem of an acoustical channel, as follows. Multiple people (representing the multiple inputs of a linear acoustic system) could be engaged in conversational speech. Audio could be captured using multiple microphones, which are the system outputs available for identification of the linear time invariant system representing the channels. Each transfer function input to output can be modeled as an FIR filter, and the system can be modeled as a MIMO FIR acoustic system. Such a scenario, encountered not just in acoustics but also in communications and other areas, is conventionally addressed by blind source separation (for source estimation) and blind channel identification techniques (for channel identification).

In this section we are interested in one sensor only, and we aim to exploit partial additional information about the channel or source in order to recognize if a particular channel, and consequently its source, is active. For example, practical problems abstracted by this scenario are the own-voice activity detection (OVAD) for hearing aids and headsets. The channel of interest corresponds to the invariant channel of the owner's voice to a single microphone. Detecting when the owner's voice is active, in contrast to external active speakers or noises, is of importance for automatic processing (e.g., in the hearing aid). We are interested in a semi-blind solution to OVAD, which exploits training information about the owner's channel (and possibly the owner's voice) to assess if the currently identified active channel fits the owner in contrast to external sources of sound.

Methods to blindly or semi-blindly identify the channel include second order and higher-order statistical approaches. The latter require large amounts of data to achieve good recognition performance, while second-order methods promise speed and efficiency. We will apply GMIA, a second-order method, to effectively capture the invariant own-voice channel information in noisy scenarios. Other applications, in addition to OVAD for hearing aids and headsets, are the detection of the owner's voice in videoconferencing, the detection and tracking of slowly varying dynamic speech channels in interactive speech gaming, or the detection of active speech channels in hands free communication. All could exploit a GMIA-based approach to the corresponding single-input single-output (SISO) problem to address more complex MIMO channel detection solutions.

The outline of this chapter is as follows. In Sect. 2 we discuss the importance of voice detection applications and present related work. Section 3 revisits the generalized mutual interdependence analysis (GMIA) method [4–7]. In Sect. 4 we bring in a generative model for GMIA(λ) parameterized by λ and demonstrate the effect of noise on the extracted features. Section 5 analyzes the applicability of GMIA for channel extraction and classification from monaural speech. In Sect. 6 we evaluate the performance of GMIA for OVAD and compare these results with mel-frequency cepstral coefficients (MFCC) and cepstral-mean (CM)-based approaches. We draw conclusions in Sect. 7.

2 Motivation and Related Work

Signal detection in continuous or discrete time is a cornerstone problem in signal processing. One particularly well-studied instance in speech and acoustic processing is voice detection, which subsumes a solution to the problem of distinguishing the most likely hypothesis between one assuming speech presence and a second assuming the presence of noise. Furthermore, when multiple people are speaking, it is difficult to determine if the captured audio signal is from a speaker of interest or from other people. Speech coding, speech/signal processing in noisy conditions, and speech recognition are important applications where a good voice/signal detection algorithm can substantially increase the performance of the respective system.

Traditionally, voice detection approaches used energy criteria such as short-time SNR estimation based on long-term noise estimation [22], likelihood ratio test of the signal and exploiting a statistical model of the signal [3], or attempted to extract robust features (e.g., the presence of a pitch [9], the formant shape [15], or the cepstrum [13]) and compare them to a speech model. Diffuse, nonstationary noise, with a time-varying spectral coherence, plus the presence of a superposition of spatially localized but simultaneous sources make this problem extremely challenging when using a single sensor (microphone).

Not surprisingly, during the last decade, researchers have focused on multimodality sensing to make this problem tractable. Multiple channel voice detection algorithms take advantage of the extra information provided by additional sensors. For example, [21] blindly identify the mixing model and estimates a signal with maximal signal-to-interference-ratio (SIR) obtainable through linear filtering. Although the filtered signal contains large artifacts and is unsuitable for signal estimation it was proven ideal for signal detection. Another example, is the WITTY (Who is Talking to You) project from Microsoft [24], which deals with the voice detection problem by means of integrated heterogeneous sensors (e.g., a combination of a close-talk microphone and a bone-conductive microphone). Even further, multimodal systems using both microphones and cameras have been studied [17].

The main motivation for our present work is to perform voice (or signal) detection for the source of interest with the reliability of multimodal approaches such as WITTY but in the absence of additional sensors such as a bone-conducting microphone. We will demonstrate that a single microphone signal contains invariant information about what may be the channel, or transfer function from each source to the microphone, which could be reliably exploited for signal detection and classification (e.g., OVAD). We use GMIA [6] to extract this invariant information for both reference (training) and testing, and further to compare classification performance on the OVAD problem to MFCC and CM-based approaches.

Mutual interdependence analysis (MIA) was first introduced by Claussen et al. [4] to extract a representation, also called common or mutual component, which is equally correlated with all the inputs. After successfully applying MIA to text-independent speaker verification and illumination-robust face recognition [5], the method was generalized to GMIA [6] to account for different noise levels and to relax the requirement for equal correlation of the common component with each input. A conclusive up-to-date statement of GMIA is presented in [7]. In the next section we review GMIA and some of its properties.

3 Generalized Mutual Interdependence Analysis

In the following let $\mathbf{x}_i \in \mathbb{R}^D$ denote the ith input vector $i = 1 \dots N$ and a column of the input matrix $\mathbf{X}$. Moreover, $\mu = \frac{1}{N}\sum_{i=1}^{N} \mathbf{x}_i$, $\underline{\mathbf{1}}$ is a vector of ones and $\mathbf{I}$ represents the identity matrix.

Extracting a common component $\mathbf{s} \in \mathbb{R}^D$ in the inputs $\mathbf{X}$ can be defined as finding a direction in $\mathbb{R}^D$ that is equally correlated with the inputs. That is:

$$\zeta \underline{\mathbf{1}} = \mathbf{X}^T \cdot \mathbf{s} \quad \text{where } \zeta \text{ is a constant.} \tag{1}$$

This is an underdetermined problem if $D \geq N$. MIA finds an estimate of $\mathbf{s}$, i.e., a direction denoted by $\mathbf{w}_{\text{MIA}} \in \mathbb{R}^D$ that minimizes the projection scatter of the inputs $\mathbf{x}_i$, under the linearity constraint to be in the span of $\mathbf{X}$. That is, $\mathbf{w} = \mathbf{X} \cdot \mathbf{c}$. Generally, MIA is used to extract a common component from high-dimensional data $D \geq N$. Its cost function is given as:

$$\mathbf{w}_{\text{MIA}} = \underset{\mathbf{w}, \mathbf{w}=\mathbf{X}\cdot\mathbf{c}}{\arg\min} \left(\mathbf{w}^T \cdot (\mathbf{X} - \mu \cdot \underline{\mathbf{1}}^T) \cdot (\mathbf{X} - \mu \cdot \underline{\mathbf{1}}^T)^T \cdot \mathbf{w} \right). \tag{2}$$

By solving Eq. (2) in the span of the original inputs rather than mean subtracted inputs, a closed-form solution can be found [4]:

$$\mathbf{w}_{\text{MIA}} = \zeta \mathbf{X} \cdot \left(\mathbf{X}^T \cdot \mathbf{X}\right)^{-1} \cdot \underline{\mathbf{1}}. \tag{3}$$

The properties of MIA are captured in the following theorems:

Theorem 1. *The minimum of the criterion in Eq. (2) is zero if the inputs* $\mathbf{x}_i$ *are linearly independent.*

If inputs are linearly independent and span a space of dimensionality $N \leq D$, then the subspace of the mean subtracted inputs in Eq. (2) has dimensionality $N-1$. There exists an additional dimension in $\mathbb{R}^N$, orthogonal to this subspace. Thus, the scatter of the mean subtracted inputs can be made zero. The existence of a solution where the criterion in Eq. (2) becomes zero is indicative of an invariance property of the data.

Theorem 2. *The solution of Eq. (2) is unique (up to scaling) if the inputs* $\mathbf{x}_i$ *are linearly independent.*

This is shown by the existence of the closed-form solution in Eq. (3). However, it is important to note that, if $\mathbf{w}$ is not constrained to the span of the inputs, any combination $\hat{\mathbf{w}}_{\text{MIA}} + \mathbf{b}$ with $\mathbf{b}$ in the nullspace of $\mathbf{X}$ is also a solution. Also, the MIA problem has no defined solution if the inputs are zero mean, that is, if $\mathbf{X} \cdot \underline{\mathbf{1}} = \underline{\mathbf{0}}$. The reason is that there exists $\mathbf{w} = \underline{\mathbf{0}}$ in the span of the inputs as a trivial solution to Eq. (2).

The MIA data model in Eq. (1) is extended in [6] to incorporate measurement noise $\mathbf{n} \sim \mathscr{N}(\underline{\mathbf{0}}, \mathbf{C_n})$ and to relax the equal correlation constraint from $\zeta\underline{\mathbf{1}}$ to $\mathbf{r}$:

$$\mathbf{r} = \mathbf{X}^T \cdot \mathbf{w} + \mathbf{n}. \tag{4}$$

We assume $\mathbf{w}$ to be a random variable. Our goal is to estimate $\mathbf{w} \sim \mathscr{N}(\mu_{\mathbf{w}}, \mathbf{C_w})$ assuming that $\mathbf{w}$ and $\mathbf{n}$ are statistically independent. Given the model in Eq. (4), the generalized MIA criterion (GMIA) is defined as:

$$\mathbf{w}_{\text{GMIA}} = \mu_{\mathbf{w}} + \mathbf{C_w} \cdot \mathbf{X} \cdot \left(\mathbf{X}^T \cdot \mathbf{C_w} \cdot \mathbf{X} + \mathbf{C_n}\right)^{-1} \cdot \left(\mathbf{r} - \mathbf{X}^T \cdot \mu_{\mathbf{w}}\right) \tag{5}$$

$$= \mu_{\mathbf{w}} + \left(\mathbf{X} \cdot \mathbf{C_n}^{-1} \cdot \mathbf{X}^T + \mathbf{C_w}^{-1}\right)^{-1} \cdot \mathbf{X} \cdot \mathbf{C_n}^{-1} \cdot \left(\mathbf{r} - \mathbf{X}^T \cdot \mu_{\mathbf{w}}\right). \tag{6}$$

Throughout the remainder of the document, the GMIA parameters are $\mathbf{C_w} = \mathbf{I}$, $\mathbf{C_n} = \lambda\,\mathbf{I}$, $\mathbf{r} = \zeta\underline{\mathbf{1}}$ and $\mu_{\mathbf{w}} = \underline{\mathbf{0}}$. We refer to this parameterization by

$$\text{GMIA}(\lambda) = \zeta\mathbf{X} \cdot \left(\mathbf{X}^T \cdot \mathbf{X} + \lambda\,\mathbf{I}\right)^{-1} \cdot \underline{\mathbf{1}}. \tag{7}$$

When $\lambda \to \infty$, the GMIA solution represents the mean of the inputs. Indeed, the inverse $\left(\mathbf{X}^T \cdot \mathbf{X} + \lambda\,\mathbf{I}\right)^{-1} \to \frac{1}{\lambda}\mathbf{I}$ simplifying the solution to $\mathbf{w}_{\text{GMIA}} \to \frac{\zeta}{\lambda}\mathbf{X} \cdot \underline{\mathbf{1}}$. Furthermore, MIA [solution to Eq. (3)] is equivalent to GMIA(λ) when $\lambda = 0$. In the rest of this chapter, we denote MIA by GMIA(0) to emphasize their common theoretical foundation.

4 Generative Signal Model for GMIA

This section evaluates the behavior of GMIA(λ) for different types and intensities of additive distortions. In particular, we evaluate the effect of noise components that are either recurring uncorrelated components or Gaussian noise. We use the generative signal model in [7] to generate synthetic data with various properties. In contrast to published work we show a gradual change in the intensities of the different noise types and compare the feature extraction result to the true feature desired. This allows an interpretation of GMIA(λ) and analysis of its performance on data with unknown noise conditions from the field.

Assume the following generative model for input data $\mathbf{x}$:

$$\begin{aligned} \mathbf{x}_1 &= \alpha_1 \mathbf{s} + \mathbf{f}_1 + \mathbf{n}_1 \\ \mathbf{x}_2 &= \alpha_2 \mathbf{s} + \mathbf{f}_2 + \mathbf{n}_2 \\ &\vdots \\ \mathbf{x}_N &= \alpha_N \mathbf{s} + \mathbf{f}_N + \mathbf{n}_N, \end{aligned} \tag{8}$$

where $\mathbf{s}$ is a common, invariant component or feature we aim to extract from the inputs, α_i, $i = 1,\ldots,N$ are scalars (typically all close to 1), $\mathbf{f}_i$, $i = 1,\ldots,N$ are combinations of basis functions from a given orthogonal dictionary such that any two are orthogonal, and $\mathbf{n}_i$, $i = 1,\ldots,N$ are Gaussian noises. We will show that GMIA estimates the invariant component $\mathbf{s}$, inherent in the inputs $\mathbf{x}$.

Let us make this model precise. As before, D and N denote the dimensionality and the number of observations. Additionally, K is the size of a dictionary $\mathbf{B}$ of orthogonal basis functions. Let $\mathbf{B} = [\mathbf{b}_1,\ldots,\mathbf{b}_K]$ with $\mathbf{b}_k \in \mathbb{R}^D$. Each basis vector $\mathbf{b}_k$ is generated as a weighted mixture of maximally J elements of the Fourier basis which are not reused ensuring orthogonality of $\mathbf{B}$. The actual number of mixed elements is chosen uniformly at random, $J_k \in \mathbb{N}$ and $J_k \sim \mathscr{U}(1,J)$. For $\mathbf{b}_k$, the weights of each Fourier basis element i are given by $w_{jk} \sim \mathscr{N}(0,1)$, $j = 1,\ldots,J_k$. For $i = 1,\ldots,D$ (analogous to a time dimension) the basis functions are generated as:

$$b_k(i) = \frac{\sum_{j=1}^{J_k} w_{jk} \sin\left(\frac{2\pi i \alpha_{jk}}{D} + \beta_{jk}\frac{\pi}{2}\right)}{\sqrt{\frac{D}{2}\sum_{j=1}^{J_k} w_{jk}^2}}$$

with

$$\alpha_{jk} \in \left[1,\ldots,\frac{D}{2}\right]; \beta_{jk} \in [0,1]; \left[\alpha_{jk},\beta_{jk}\right] \neq \left[\alpha_{lp},\beta_{lp}\right] \forall j \neq l \text{ or } k \neq p.$$

In the following, one of the basis functions $\mathbf{b}_k$ is randomly selected to be the common component $\mathbf{s} \in [\mathbf{b}_1,\ldots,\mathbf{b}_K]$. The common component is excluded from the basis used to generate uncorrelated additive functions $\mathbf{f}_n$, $n = 1,\ldots,N$. Thus only $K-1$ basis functions can be combined to generate the additive functions $\mathbf{f}_n \in \mathbb{R}^D$. The actual number of basis functions J_n is randomly chosen, similarly to J_k, with $J = K-1$. The randomly correlated additive components are given by:

$$f_n(i) = \frac{\sum_{j=1}^{J_n} w_{jn} c_{jn}(i)}{\sqrt{\sum_{j=1}^{J_n} w_{jn}^2}}$$

with

$$\mathbf{c}_{jn} \in [\mathbf{b}_1, \ldots, \mathbf{b}_K]; \mathbf{c}_{jn} \neq \mathbf{s}, \forall j, n; \mathbf{c}_{jn} \neq \mathbf{c}_{lp}, \forall j \neq l \text{ and } n = p.$$

Note that $\|\mathbf{s}\| = \|\mathbf{f}_n\| = \|\mathbf{n}_n\| = 1, \forall n = 1, \ldots, N$. To control the mean and variance of the norms of common, additive, and noise components in the inputs, each component is multiplied by the random variable $a_1 \sim \mathcal{N}(m_1, \sigma_1^2)$, $a_2 \sim \mathcal{N}(m_2, \sigma_2^2)$ and $a_3 \sim \mathcal{N}(m_3, \sigma_3^2)$, respectively. Finally, the synthetic inputs are generated as:

$$\mathbf{x}_n = a_1 \mathbf{s} + a_2 \mathbf{f}_n + a_3 \mathbf{n}_n \tag{9}$$

with $\sum_{i=1}^{D} x_n(i) \approx 0$. The parameters of the artificial data generation model are chosen as $D = 1{,}000$, $K = 10$, $J = 10$, and $N = 20$.

Throughout the experiments we keep the parameters $m_1 = 1$, $\sigma_1 = 0.1$, $\sigma_2 = 0.1$ and $m_3 = 0$ of the distributions for a_1, a_2, and a_3 constant. We vary the mean amplitude m_2 of the recurring uncorrelated components and the variance σ_3 of the Gaussian noise and illustrate its effect on GMIA(0), GMIA(λ), and the sample mean in Fig. 1. The figure shows a matrix of 3D histograms for different parameters m_2 and σ_3. Each point in a histogram represents an experiment for a given value of λ (x-axis). The y-axis indicates the correlation of the GMIA solution with $\mathbf{s}$, the true common component. The intensity (z-axis) of the point represents the number of experiments, in a series of random experiments, where we obtain this specific correlation value for the given λ. Overall, we performed $1{,}000$ random experiments with randomly generated inputs using various values of λ per histogram.

Results show that a change in the mean amplitude m_2 of the recurring uncorrelated components $\mathbf{f}_i$ has a minimal effect on GMIA(0) but greatly affects the correlation coefficient of $\mathbf{s}$ with the sample mean. That is, the sample mean results is a good representation of $\mathbf{s}$ only if m_2 is low and the common component $\mathbf{s}$ is dominant in the data. Moreover, this indicates that GMIA(0) succeeds in finding a good representation of $\mathbf{s}$.

The second row of Fig. 1 shows that an increased variance σ_3 of the noise can improve the GMIA(0) result. The increased noise level appears to acts as a regularization in the matrix inversion when computing GMIA. This has the same effect as an increased value of the regularization parameter λ.

Moreover, the experiments show that the results for all λ suffer for high noise variances σ_3, but that the spectral mean is affected the most. In all experiments, GMIA(λ) performs equally or outperforms GMIA(0) and the spectral mean. This demonstrates that GMIA is more versatile than the spectral mean in extracting a common component from data with an unknown and possibly varying distortion. In the following section we evaluate how the extraction results are affected for nonstationary, real-world data such as speech.

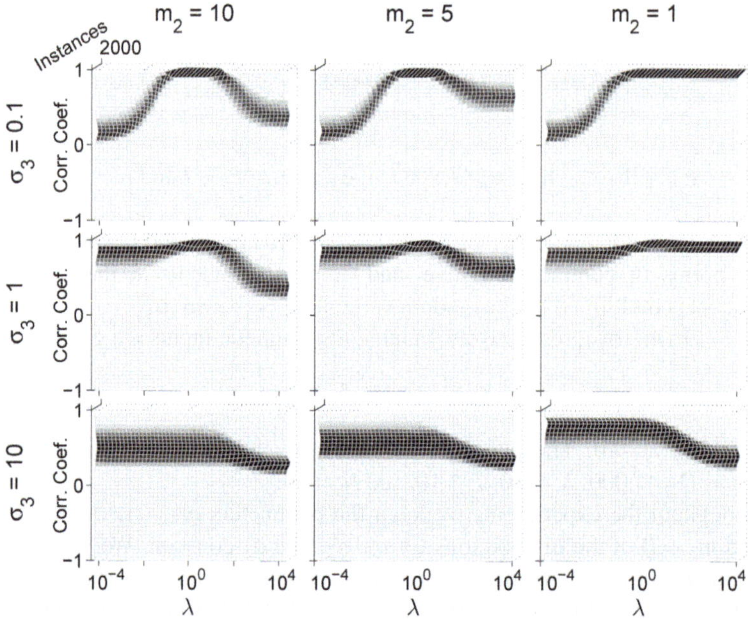

Fig. 1 Histograms of GMIA(λ) extraction performance for different levels of additive Gaussian noise and uncorrelated additive components $\mathbf{f}_i$. The mean of the inputs extracts the common component **s** well for low energy contributions of $\mathbf{f}_i$. Small levels of Gaussian noise result in a drop of the GMIA(0) performance. Larger amount of Gaussian noise results first in an improved GMIA(0) performance and later in a reduced extraction result overall λ. High levels of noise are better addressed by GMIA(0) then the mean

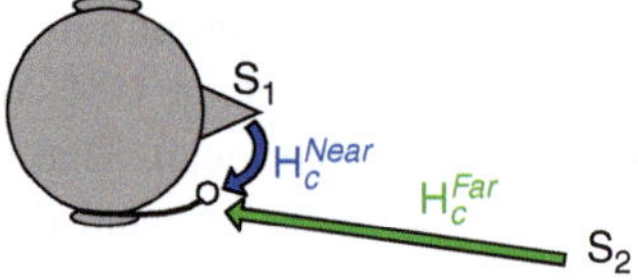

Fig. 2 Own-voice activity detection (OVAD) scenario: person is wearing a headset; speech from near-field (own), S_1, or far-field (external sources), S_2, is recorded on a single nearby microphone. The signal incorporates channel information, e.g., $\mathbf{H}_c^{\mathrm{Near}}$ or $\mathbf{H}_c^{\mathrm{Far}}$, respectively

5 Channel Extraction from Mono Speech Recordings

Lets us consider a single microphone recording of near-field and far-field nonoverlapping conversational speech as in Fig. 2. As noted in Sect. 2, a potential application of GMIA is to extract channel features in the context of owner speech detection for hearing aids. This problem has been referred to as OVAD, to imply the recognition of when the wearer of the hearing-aid (owner) is talking and when an external speaker is talking in a conversation (between the owner and the external speaker).

Such a detection facilitates, e.g., the hearing aid signal processing to be adapted dynamically to own-voice (OV) or external-speaker (EXT) characteristics.

We aim for an understanding of the domain and timescales where real-world acoustic data (e.g., conversational speech) fits the generative model studied in Eq. (9). As a first step, in this section, we review the model for the recorded signal and its dependence on speaker and channel characteristics. We use data from one or more speakers for fixed positions (i.e., exhibiting common channels), as in Fig. 2, to extract channel information using GMIA. Later, in Sect. 6, we address the OVAD problem.

5.1 Speech and Channel Models

A speech signal can be modeled as an excitation that is convolved with a linear dynamic filter, which represents the channel including the microphone characteristic, the channel impulse response of the environment, and the vocal tract. The excitation signal can be modeled for voiced speech as a periodic signal and for unvoiced speech as random noise [8, p. 50]. Let $\mathbf{E}^{(p)}$, $\mathbf{H}_v^{(p)}$, $\mathbf{H}_c$, and $\mathbf{S}^{(p)}$ be the spectral representations of the excitation or pitch signal (covering the lungs and vocal chords), the vocal tract filter (covering the mouth, tongue, teeth, lips, and nasal cavity), the external channel impulse response, and the speech signal parts of person p, respectively. Note that the channel impulse response implicitly depends on the spatial location of the receiver. This can vary substantially from near-field to far-field, or even over different far-field only or near-field only locations. If the environment of the speaker is invariant (e.g., the speaker does not move significantly) and we make simplifying assumptions to idealize the spectrum and capture important features at the timescale of interest, assume the data can be modeled as: $\mathbf{S}^{(p)} = \mathbf{E}^{(p)} \cdot \mathbf{H}_v^{(p)} \cdot \mathbf{H}_c$. For person p and instance[1] i, we obtain:

$$\log \mathbf{S}_i^{(p)} = \log \mathbf{E}_i^{(p)} + \log \mathbf{H}_v^{(p)} + \log \mathbf{H}_c. \tag{10}$$

$\mathbf{E}_i^{(p)}$ is nonstationary in general for timescales larger than the pitch period.[2] $\mathbf{H}_v^{(p)}$ may capture invariant characteristics of the speaker's vocal tract as well as phoneme-specific characteristics (and underlying speech neural control) that can be considered stationary and hence invariant within phonetic timescales, in keeping with the quasistationary assumptions of the speech process.[3] This fundamental

[1]The instance i implicitly represents the timescale of interest, e.g., a timescale of the order of the pitch period (10–20 ms) or of the order of the average word period (500 ms).

[2]The spectrum of the excitation changes slowly for voiced sounds and appears unchanged although radically different over the duration of a consonant, at the phonetic timescale.

[3]A detailed analysis of these components of the speech production model is beyond present scope.

model of speech production extended with the external channel transfer function is the basis for defining inputs $\mathbf{x}_i$ and the corresponding timescales where various components play the role of $\mathbf{s}$ and $\mathbf{f}_n$ from Eq. (9).

For example, [7] use training data from different nonlinearly distorted channels for each person from various portions of the NTIMIT database [11]. The intuition was that the channel variation results in a low contribution of the channel in the GMIA extract while the vocal tract characteristic $\log \mathbf{H}_v^{(p)}$ is retained. In contrast, in this chapter, we considered training instances $\mathbf{x}_i$ from multiple people exploring an identical external channel $\mathbf{H}_c$ (e.g., from the same external position and using the same microphone, which is the case for own-voice recordings in OVR). In this case the $\log \mathbf{E}_i^{(p)}$ and $\log \mathbf{H}_v^{(p)}$ components in Eq. (10) play the role of the orthogonal components $\mathbf{f}_n$ in our synthetic model (Eq. (9)), while $\log \mathbf{H}_c$ is the invariant. In such a setup, GMIA can be used to identify invariant characteristics of the channel (e.g., near-field channel for OVR).

We use various portions of the TIMIT database [12] for our experiments in this section. TIMIT contains speech from 630 speakers that is recorded with a high quality microphone in a recording studio like environment. Each speaker is represented by 10 utterances. We convolve the TIMIT speech with a head-related transfer function (HRTF) to simulate various invariant channels. The output of an algorithm for channel identification can thus be compared directly with the true HRTF used to generate the data.

We chose a HRTF from a position on the right side of a dummy head with a source distance of 20 cm, azimuth of 0° and at an elevation of −30° as invariant channel, and a HRTF for the right side of the dummy head with a source distance of 160 cm, azimuth of 0° and at an elevation of 0° as external channel. The HRTF data has been obtained from [18]. Thereafter, the data is windowed with half overlapping Hann windows of 0.2 s length and transferred into the power spectral domain.

Our goal is to apply GMIA to extract channel information and evaluate if GMIA representations can be used to distinguish different channels. Person-dependent information is minimized by introducing variation in the excitation $\mathbf{E}_i^{(p)}$ using speech from both voiced and unvoiced signals. Note that speech signals contain silence periods where no channel information is present. Furthermore, voiced speech is sparse in the spectral domain. Therefore, not all parts of the channel characteristic are fully represented at all times. Clearly, the channel does not equally correlate with the spectral information of the speech from different time windows. A GMIA representation will be computed separately from speech of the same or multiple speakers.

5.2 *Speaker Model*

For one person p_0, consider the vector $\mathbf{x}_i$ obtained from a speech clip i:

$$\mathbf{x}_i = \log \mathbf{S}_i^{(p_0)} = \left(\log \mathbf{H}_c + \log \mathbf{H}_v^{(p_0)}\right) + \left(\log \mathbf{E}_i^{(p_0)}\right) \approx \mathbf{s} + \mathbf{f}_i. \tag{11}$$

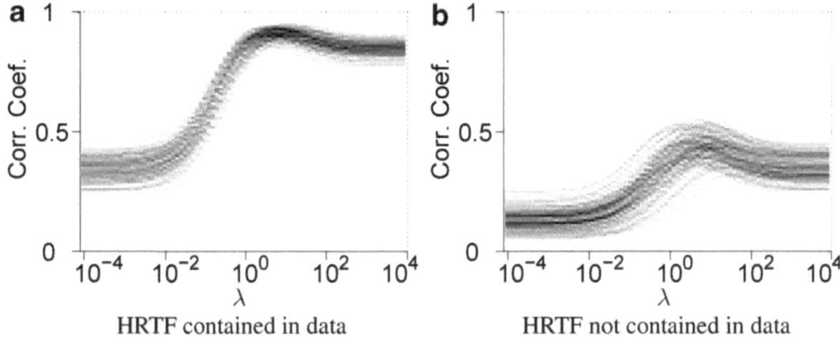

Fig. 3 Histograms as vertical slices of the plot of the correlation coefficients between GMIA(λ), for a fixed value of λ, from single-speaker data and two different HRTF's. *Dark gray bins* represent a large number and *light gray bins* a low number of instances with a particular correlation coefficient. (**a**) The HRTF used to generate the speech data is well represented by the GMIA(λ) result for $\lambda = 10^1$, resulting in a mean correlation coefficient of 0.9. (**b**) An HRTF that is not contained in the speech data minimally correlates with the GMIA extract

We use data as above for one single person and with channels for near- and far-field given by the HRTFs to the right side of the dummy head. According to the data model in Eq. (11) we expect that GMIA computes a common component capturing information about both the channel and the speaker characteristics. Indeed, $\log \mathbf{H}_c + \log \mathbf{H}_v^{(p_0)}$ is invariant to the actual clip i used as input. Next we compute GMIA and correlate the result with known channel information (HRTF) to verify our hypothesis.

All experiments are repeated for 100 speakers and various values of λ. Figure 3a illustrates the histogram of the correlation coefficients of the GMIA extract from the near-field speech with the ground truth near-field HRTF for a 20 cm source/receiver distance. Note that both $\mathbf{w}_{\text{GMIA}(10^{-4})} \approx \mathbf{w}_{\text{MIA}}$ and $\mathbf{w}_{\text{GMIA}(10^4)} \approx \mu$ (the mean of the inputs) do not compute maximal correlation coefficients. The median correlation value at $\lambda = 10^1$ is 0.9, demonstrating that GMIA can extract good representations of the original HRTF. In contrast, Fig. 3b shows histograms of the correlation coefficients with the HRTF from a far-field position (160 cm source/receiver distance) that was not used in the data generation. The low correlation coefficients indicate that channel characteristics are well separable with the extracted GMIA representations.

Note that Fig. 3a is similar to Fig. 1 for $\sigma_3 = 0.1$ and $m_2 = 5$, which represents the case where the common component intensity varies over different training instances. This confirms that for speech the channel is not equally correlated with the spectral information from different time windows.

5.3 Channel Model

The previous subsection shows that the GMIA projection correlates well with the channel and that it can be used as feature for channel detection or as classifier of the

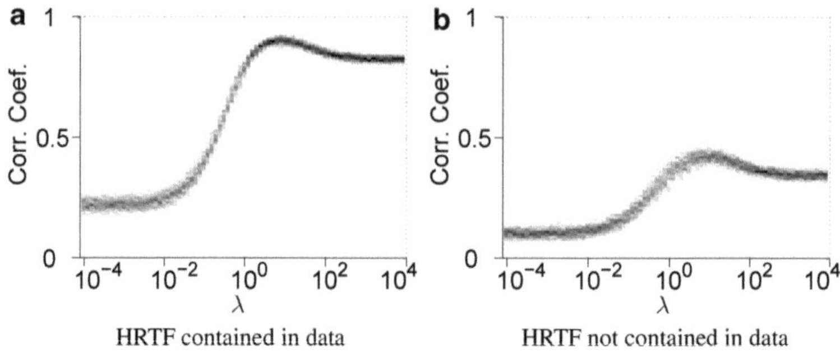

Fig. 4 Histograms (vertical slices of the plot) of the correlation coefficients between GMIA(λ) from multiple-speaker data and two different HRTF's. *Dark gray bins* represent a large number and *light gray bins* a low number of instances with a particular correlation coefficient. (**a**) The HRTF that is convolved with the data is well extracted for GMIA(λ) with $\lambda = 10^1$ resulting in a mean correlation coefficient of 0.9. The variance of the result is lower than for GMIA(λ) from single-speaker data (see Fig. 3) (**b**) The HRTF that is not contained in the data only minimally correlates with the GMIA extract

channel. We would like to make the model in Eq. (11) more precise and eliminate the speaker dependence as much as possible. For this we use data from multiple speakers p_i with $i = 1 \dots N$ as follows:

$$\mathbf{x}_i = \log \mathbf{S}_i^{(p_i)} = (\log \mathbf{H}_c) + \left(\log \mathbf{E}_i^{(p_i)} + \log \mathbf{H}_v^{(p_i)}\right) \approx \mathbf{s} + \mathbf{f}_i. \tag{12}$$

We expect to compute a common component that better captures the channel. The experiment is performed as follows. First, a number of speakers, corresponding to the number of training instances N in Sect. 5.2, are selected randomly from the TIMIT database. One of their 10 utterances is randomly selected and convoluted with the previously chosen near-field HRTF. Thereafter, one speech segment (e.g., 0.2 s long) is randomly selected from each speaker. These segments are thereafter used to extract a GMIA representation in the log-spectral domain. The experiment is repeated for 100 randomly selected sets of speakers and various values of λ. Figure 4a shows a histogram of the correlation coefficients of the GMIA result and the ground truth for the channel, the near-field HRTF. Figure 4b illustrates the correlation coefficients between the GMIA extract and the HRTF from the external channel (160 cm source/receiver distance) that was not used in the data generation.

Indeed, Fig. 4 shows a reduced variance of the correlation coefficients for different speakers compared to Fig. 3 and thus a more reliable extraction of the channel. GMIA will be further used for channel estimation in the OVAD problem.

6 Own Voice Activity Detection

Section 5 demonstrated the efficacy of GMIA to extract channel features using a known HRTF as the channel convolved with the TIMIT speaker data under both speaker model and channel model formulations. In this section, we extend this further to a realistic scenario of OVAD using the same large speaker database convolved with near-field and far-field HRTFs to closely approximate own-voice speakers and external speakers.

In the experimental scenarios used here with such data for OVAD, though the underlying HRTF information is available (as was used in Sect. 5 for measuring the correlation coefficients between extracted MIA features and the reference HRTF), we assume the underlying HRTF information to be unknown and unavailable, thereby treating them as implicit in the speech data (as is the case with real recorded OV and EXT speaker data at an hearing aid); for this purpose, the underlying OV and EXT channel information are equivalently considered only in the form as available by means of estimates of channel information from held-out reference data, such as by the GMIA extraction proposed here. Thus, in this scenario, how well the GMIA-based features offer a good own-voice recognition performance when set in a channel detection framework will serve to demonstrate the effectiveness of GMIA to extract the underlying channel information from the actual OV and EXT speech data.

Toward this, we demonstrate in this section the use of GMIA-based channel features for OVAD in a verification framework posed as an hypotheses testing problem. Further, in order to provide a comparative reference for the GMIA-based approach, we consider two alternate approaches: one using cepstral mean as an alternate channel feature and set in the same verification framework, and the other using the conventional speech feature, namely, MFCC, set in a speaker verification framework. We work with a 100-speaker database convolved with near-field and far-field HRTFs to closely represent own-voice and external speakers. The performance of these three verification systems are given and compared in terms of the equal-error-rate (EER) measure. Additionally, given that GMIA is specifically formulated to handle real-world data with additive noise, we also demonstrate the effectiveness of GMIA for noisy data by considering three noise conditions, namely, street, car, and music noises, at different SNRs (clean, 20 dB, 10 dB, 0 dB, −10 dB and −20 dB) and show how its parameterization (in terms of λ—the assumed noise variance) allows a superior performance at a range of optimal λ, in comparison to the other two approaches (cepstral-mean- and MFCC-based speaker-verification).

6.1 GMIA Based Verification Framework for OVAD

Given the conversational speech signal, the OVAD problem can be reduced to that of detecting the underlying channel. This in turn involves extracting the channel feature from the speech signal and classifying it as own-voice or external-speaker channel, thereby comprising a 2-class problem. Alternately, this can also be viewed as a "own-voice verification" problem (e.g., as in speaker-verification), set in a hypothesis testing framework of deciding between the two hypotheses:

H_0: Input speech segment is own voice.
H_1: Input speech segment is not own voice (i.e., external speaker).

The verification framework adopted here is essentially as in speaker-verification, which is well established [2, 19]. We outline this here as adopted to the channel verification task: Given a set of OV and EXT speakers, a test OV speaker is verified as OV speaker with respect to a set of OV channel features extracted from another set of OV speakers. The latter is referred to as "reference" OV channel features, and serves to represent the underlying unknown OV channel, as extracted by GMIA; such a channel information, by virtue of being speech- and speaker independent, can be expected to be invariant across a set of OV speakers and to generalize to an unseen test OV speaker. Likewise, a test external (EXT) speaker can be verified as "not OV" speaker against the same set of reference OV channel features. In general, a set of test OV/EXT speakers represented in terms of their channel features are verified in this manner with respect to another set of reference OV channel features, thus constituting a statistically robust channel verification task.

Ideally, the OV test GMIA features ought to yield high correlation scores (or alternately, low distance scores) with OV reference channel features, while the EXT test GMIA features yield low correlation scores with the OV reference channel features. If the features represent the OV and EXT channels well and offer good separability in the GMIA feature space, the corresponding OV and EXT score distributions are also well separated. An optimal threshold is determined on the OV and EXT score distributions which minimizes false rejections (fr, which is the number of true OV features rejected as "not OV") and false acceptances (fa, which is the number of true EXT features accepted as "OV"). The corresponding EER of (Prob(fr), Prob(fa)) is reported as the OVR system performance, with lower EER implying a better performance.

6.2 Alternate Approaches

In order to provide a baseline reference to the OVAD by GMIA-based channel features as discussed above, we also consider two other alternatives to OVAD: one using an alternate channel feature extraction, namely, the "cepstral mean," and another using a speaker-verification approach wherein OVR is carried out in terms of verifying whether the input speaker is the wearer or not.

6.2.1 Cepstral-Mean-Based OVAD

The mean vector obtained from GMIA for large λ ($\lambda \to \infty$) corresponds to the mean of the log-spectral vectors in a clip (analysis window for extracting a GMIA vector). Alternately, one can consider the mean of the cepstral vectors derived by an inverse FFT or DCT of the log-spectral vectors, as is done for deriving cepstral coefficients or MFCCs in speech recognition [16]. This mean vector, referred to as "cepstral-mean" (CM) in speech recognition, is popularly used in the context of cepstral mean normalization (CMN) for channel compensation [1, 16]. Here, it is already a well established concept that the cepstral mean of the log spectra of long speech intervals approximates the channel cepstra and that subtraction of this long-term averaged cepstral-mean from the individual frames of cepstral features removes the channel effect, thereby rendering the resultant cepstral vectors robust to channel variability (such as arising from channel differences in telephony speech recognition due to differences in handset, physical channel media, wireless network channels, etc., particularly between training and test conditions).

6.2.2 Speaker-Verification-Based OVAD

In a OVAD task, the OV speaker is fixed and given and can be made to provide training data to define OV models that characterize the OV speaker. By this, the OVAD task can be alternately defined as a conventional speaker-verification task of treating the OV speaker as the target speaker and EXT speakers as the impostor speakers. For this, it becomes necessary to use conventional "speaker" feature representations, such as MFCC [2, 19]. In this case, the OV speaker is represented by a statistical model (GMM) or a nonparametric model (VQ) in the MFCC feature space.

The distribution of the MFCC vectors (and the GMM- or VQ-based representation of this distribution) of a speaker characterizes the unique acoustic signature or footprint of that speaker in the MFCC feature space as manifesting in the unique spectral characteristics of his voice, manner of articulation of the different sounds of the language (phonemes), and spectral dynamics (which can be potentially captured in the delta and delta-delta MFCCs). The OV and EXT speaker data occupy different regions in the feature space, by virtue of the fact that the spectral characteristics of each of these speech is a result of convolution with different channels (here, HRTF). An OV speaker model thereby offers a better match with OV test speaker data than with EXT test speaker data, which then becomes the primary basis of OVAD by MFCC-based speaker verification. The verification task is thus essentially as described in Sect. 6.1, but constituting a "speaker" verification (as against "channel" verification, since the MFCC features here serve as "speaker" features) in this case taking the form of computing OV scores between OV test MFCC vectors and the OV models and EXT scores between EXT test MFCC vectors and the OV models, subsequently forming the OV and EXT score distributions and then determining the EER.

6.3 Experimental Setup

Here, we present the experimental details of the three OVAD tasks, namely, GMIA-based channel verification, cepstral-mean (CM)-based channel verification, and MFCC-based speaker-verification. These three frameworks are as described generically earlier in Sects. 6.1 and 6.2. While the three tasks have specific differences due to their underlying idiosyncratic frameworks, they share an overall experimental scenario, comprising the following common aspects.

All the OVAD experiments use a randomly selected (but fixed) subset of 100 speakers from the TIMIT database (of 630 speakers) as the test set of OV and EXT speakers, with each speaker having 10 sentences, each 3 to 4 s duration. The fixed subset of 100 test speakers is convolved with single fixed near-field and far-field HRTFs to generate the own voice and external type of speakers, respectively (referred to as OV and EXT henceforth); the HRTFs used here are as described in Sect. 5. In order to examine the noise robustness of GMIA and the two alternate approaches, we consider three different noise conditions, namely, street, car, and music, and five SNRs for each of these noise conditions (20 dB, 10 dB, 0 dB, −10 dB, and 20 dB), in addition to the clean case. The specific noise data is added to the original clean TIMIT sentences at the desired SNR subsequent to the HRTF convolutions, i.e., to the OV and EXT data.

We now describe the specific variations in the experiments for each of the three OVAD tasks.

6.3.1 GMIA-Based OVAD

While the 100 speakers as defined above constitutes the test data, GMIA experiments use a set of 300 speakers (different from the 100 test speakers) to define the "reference" OV channel feature. This is motivated by the channel model formulation in Sect. 5.3, where a GMIA vector is extracted in a speaker-independent manner. Here, a single GMIA reference vector is extracted from the 300-speaker clean data, i.e., with $N = 300$, as defined in Sect. 5.3.

For the noise-added experiments, only the test data is made noisy, while the above reference GMIA vector is extracted and kept fixed from clean 300 speaker data. For the purposes of examining and establishing the noise-robust advantage intrinsic to GMIA through its parameter λ, the GMIA-based channel verification experiments are conducted for λ varying over the range of $[10^{-4}$ to $10^4]$. One such experiment (for a given λ) consists of using 100 test OV and EXT speaker data and computing 1 GMIA vector for each speaker (from the entire duration of 30 to 40 s of that speaker, corresponding to $N = 300$–400 in $\mathbf{X}$ of Eq. (7)). The test database of 100 speakers thus yields 100 OV and EXT scores, from which the EER corresponding to the given λ is obtained. For a given noise-type and SNR, EER is obtained as a function of λ over the range $[10^{-4}$ to $10^4]$. Such an EER-vs-λ curve is obtained for all the 6 SNRs (clean, 20 dB, 10 dB, 0 dB, −10 dB, and 20 dB), for each noise type (street, car, and music).

6.3.2 Cepstral-Mean-Based OVAD

The experimental framework for this task uses the identical test set of 100 speakers as above, while differing only in the way the reference cepstral-mean channel feature vector is derived and in how the test set scores are computed in a leave-one-out framework, in order to offer a statistically robust verification task; this is outlined below.

For a given speaker (OV or EXT), a cepstral-mean vector is computed from the entire duration of that speaker (30 to 40 s, yielding 300–400 cepstral vectors, each obtained using framesize of 200 ms and overlap of 100 ms). The cepstral vector for each frame is obtained by a DCT of the log-spectral vector.

For a given test OV speaker (among 100 test speakers), the remaining 99 OV speakers are defined as the reference channel speakers. 1 cepstral-mean vector is computed for each of these 99 speakers (from clean data), thereby providing 99 clean reference channel vectors (for that test OV speaker). One score is computed between the test cepstral-mean vector (from the entire duration of that test speaker) and the reference cepstral-mean vector (from among the 99 reference vectors) which has the highest correlation with the test cepstral-mean vector. For the given test OV speaker, the corresponding EXT speaker (the same speaker in the 100 speaker database, but now from the EXT set) is used to compute the EXT score with respect to the same OV reference channel vectors.

The above is repeated for each of the 100 test OV speakers as the test speaker (with the remaining 99 speakers forming the reference channel set), thereby yielding 100 OV and EXT scores, from which the score distribution is formed and EER determined; this corresponds to a specific noise type and SNR. EERs are obtained for all 5 SNRs and clean cases for the 3 noise types (street, car, and music).

6.3.3 MFCC-Based OVAD

This OVAD task differs in several respects from the above two channel verification tasks, in that it is essentially a speaker verification task and therefore has a fairly different experimental setup, though sharing the broad parameters with the above tasks to allow for a fair comparison.

The primary feature for this task is the MFCC vector computed with a framesize of 20 ms and overlap of 10 ms, constituting quasistationary timescales as required to derive spectral information of speech data. This yields 100 MFCC vectors per second of speech data, and each TIMIT speaker (of duration 30–40 s) has about 3000–4000 vectors. The MFCC feature vector used here is derived with a set of 40 triangular filters applied on the log spectra of a frame followed by DCT on the filter energy outputs to yield the cepstral coefficients; the MFCC vector used is of dimension 36, consisting of 12 cepstral coefficients (coefficients 2 to 13, with the first energy co-efficient not used, thereby making the feature insensitive to signal energy), 12 delta and 12 delta-delta coefficients.

The verification task here is set in the leave-one-out framework (as defined for the cepstral-mean task). For a given test speaker, the remaining 99 speakers are used to define the reference OV speakers against which the test speaker MFCCs are scored. Each of these 99 speakers is represented by a VQ codebook of size 64, considered adequate from established speaker-identification tasks [20].

A scoring window is defined for the test data for deriving a score with respect to the reference VQ codebooks. The scoring windows used here are 1, 2, 4, 8, 16 and 30 s. For a specific scoring window duration, an accumulated dissimilarity (distance) score is computed for the window with respect to each of the 99 VQ codebooks. The accumulated score for a VQ codebook is the sum of the individual scores of the MFCC vectors in the window, the individual score of a vector being the distance between the vector and the nearest codevector in the VQ codebook. The final score of the test window is determined as the minimum across the 99 VQ codebooks, i.e., a window of test vectors has a single score with respect to the best scoring reference VQ codebook.

For a given test window duration, OV and EXT scores are computed over the test data duration of a speaker and score distributions formed from such scores from all test speakers in the above leave-one-out framework; an EER is obtained for each test window duration for a given noise type and SNR. For the different noise types and SNRs, only the test data is subjected to noise, while the reference VQ codebooks are maintained as derived from clean data.

6.4 OVAD Results Analysis

In this section, we present results of the above three OVAD tasks (GMIA based channel verification, CM-based channel verification, and MFCC-based speaker-verification) for different noise types and SNRs. The performance of the three verification approaches are given in terms of EER, as defined earlier in Sect. 6.1, in street, car, and music noises, respectively, for different SNRs.

6.4.1 OVAD for GMIA, CM, and MFCC in Noisy Conditions

Figure 5a–c show EER as a function of λ for GMIA. As expected, the EER shows a pronounced dependence on λ, consequently offering the best performance at $\lambda = 10^0$ consistently for both clean and noisy cases. This is in agreement with the similar dependence and optimality shown by the correlation coefficients for the experiments reported in Sect. 5.3. Optimal results are obtained similarly for values of $\lambda = 0.1$–10. This validates the importance of the parameterization of GMIA in terms of λ to handle real-world noisy data.

Fig. 5 Own-voice activity detection with GMIA(λ), MFCC, and CM for various noise types and levels. (**a**) Street noise above 0, dB SNRs enables GMIA-based own-voice activity detection with EERs below 10 %. GMIA(10^0) achieves best results. (**b**) Car noise above -10 dB SNRs enables GMIA-based own-voice activity detection with EERs below 5 %. There is a clear improvement for $\lambda = 10^0$ over the spectral mean. (**c**) Music noise is least affecting the GMIA-based own-voice activity detection. (**d**) CM performs mostly below the spectral mean and by a large margin below GMIA(10^0). MFCC performs below GMIA(10^0) for high SNRs and at level for low SNRs

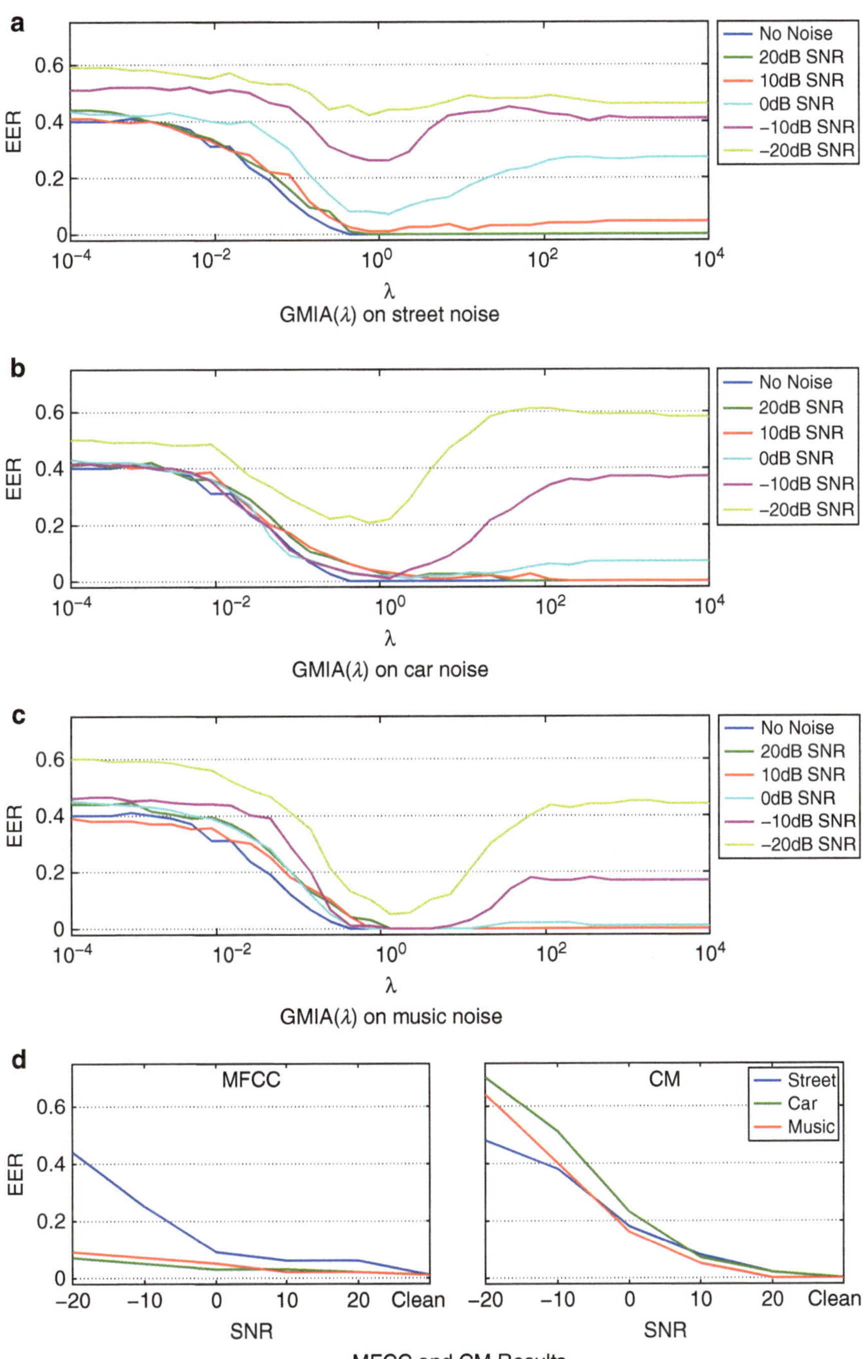

Fig. 5 (continued)

More specifically, it can be noted that for low SNRs and for all noise types, the optimal λ results in a prominent dip in EER, when compared to MIA (for $\lambda = 10^{-4}$) and the spectral mean (for $\lambda = 10^4$). This is in line with the basis of the channel model formulation in Sect. 5.3, indicating the ability of GMIA (at optimal λ) to extract an invariant component in the presence of a higher degree of uncorrelated additive components $\mathbf{f}_i$ [in Eq. (12)], in this case corresponding to large variability in log-spectral components corrupted with higher levels of noise (lower SNRs).

With regard to MFCCs, Fig. 5d shows that MFCC offers competitive performance to GMIA (comparable or even lower EERs at times, such as for street noise at -20 dB and -10 dB and car noise at -20 dB) for lower SNRs, while the optimal GMIA performances are better than MFCC for high SNRs. The better performance of GMIA over MFCCs (particularly for high SNR cases) is accounted for as follows. MFCC-based speaker-verification approach attempts to model the OV (or EXT) space as the feature space spanned by the speech of the owner (or external) speaker (i.e., spanned by all the phonetic realizations as is unique to a speaker) and hence implicitly captures both the channel and speaker information. This in turn makes the feature space occupied by the OV and EXT speaker data to be large and diffuse, leading to potentially higher overlap of their feature spaces and a consequent higher overlap of the OV and EXT score distributions with associated higher EERs. In contrast, the GMIA features represent the channel information directly with minimal associated speaker information (as was evident from the results in Fig. 4, where the channel model, being extracted in a speaker-independent manner, offers lower variance of the correlation coefficients) and consequently better separability between the OV and EXT spaces and associated lower EERs.

Within the channel modeling framework, the alternative cepstral-mean features (Fig. 5d) have higher EERs than the "spectral mean" of GMIA at $\lambda = 10^4$ (i.e., the asymptotic performance for GMIA for $\lambda \to \infty$), particularly for lower SNRs. Moreover, the EERs for cepstral mean are significantly higher than the best GMIA EERs for all noise types and SNRs. In general, while CM offers reasonably comparable performance at clean conditions, it degrades severely with increase in noise levels and has poor noise robustness. When compared to MFCC, MFCC clearly outperforms CM for all cases.

6.4.2 OVAD for GMIA, CM, and MFCC for Varying Test Durations

Figure 6 shows an important computational aspect of GMIA—the duration over which a single GMIA vector is computed. In this figure, EER-vs-λ is shown for varying durations (1, 2, 4, 8, and 16 s) over which the GMIA vector is computed in the test data. GMIA exhibits no particular sensitivity to this duration (at the optimal λ) for clean case (Fig. 6a). Even 1 s of data is sufficient to realize a 0 % EER for the clean case at the optimal λ.

However, for the noisy case (car noise at 0 dB) in Fig. 6b, the EER curve worsens with decrease in the duration (from 16 s to 1 s). For 1 s data, even the EER at optimal λ is as high as $\sim$30 % and it needs 4 s of data to enable EERs $\sim$8 %.

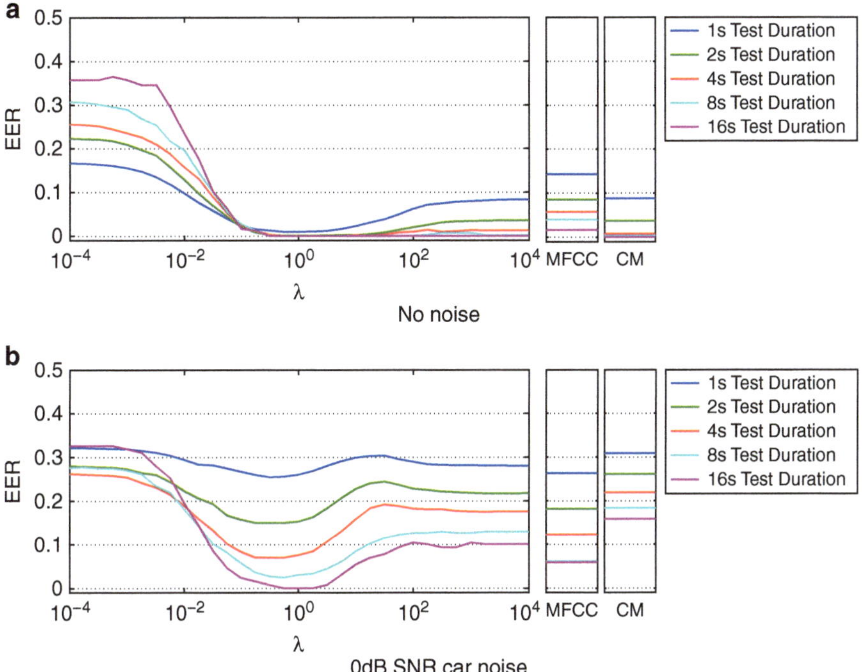

Fig. 6 Own-voice activity detection with GMIA(λ) and MFCC for various test durations. (**a**) GMIA(10^0) and the spectral mean both outperform the MFCC in case of no noise. (**b**) GMIA(10^0) outperforms MFCC for car noise with long test durations and achieves similar results for short test durations. MFCC performs better then the spectral mean

This shows that channel extraction with GMIA requires large amounts of data to enable noise-robust extraction, i.e., larger data implying sufficient uncorrelated components [**f** in Eq. (12)] to enable their cancellation and reliable extraction of the common channel component. This will impact online applications, where shorter durations (over which an OVAD decision is reported) will be clearly preferred.

Considering MFCC, GMIA(10^0) offers better performance than MFCC for the clean case. For the noisy case (Fig. 6b), GMIA(10^0) is again better than MFCC for longer durations, but comparable for shorter durations. The dependence of MFCC on longer durations is consistent with previously reported results on MFCC-based speaker verification where it is known that test durations of the order of 5–10 s are necessary to achieve optimal performance [20]; this is primarily due to the fact that such speaker verification relies on having long acoustic signature of the speaker to yield a sufficiently discriminating accumulated score.

Considering CM, for clean cases, CM has comparable performance to the spectral mean (GMIA(10^4)); however, for the noisy case, CM is worse than MFCC and also the spectral mean (GMIA(10^4)), indicating that CM is more sensitive to noise than GMIA, though it can offer comparable performance to the spectral mean for clean conditions.

7 Conclusion

GMIA is a low-complexity second-order statistical method for projecting data in a subspace that captures invariant properties of the data. This chapter summarizes the theory behind GMIA in a unitary presentation and most importantly carries the reader through a succession of increasingly difficult application examples. The examples come from a conspicuous albeit well-studied signal processing problem: voice (signal) activity detection and classification. We show how real-world conversational speech data should be modeled to fit the GMIA assumptions. From there, low-complexity GMIA computations can induce reliable features that are used for classification under noisy conditions and operate with small amounts of data. Furthermore, our results push the state of the art and are intriguing. For example, GMIA features perform better than cepstral power and mel-frequency cepstral coefficient features, particularly in noisy conditions, and are amenable to online (real-time) detection algorithms. More significantly, the approach opens the door for a large number of possible applications where a signal source (e.g., a speaker), characterized by a slow varying or invariant channel that is learned can be tracked from single channel data. The GMIA approach derived and applied in this chapter resonates with the principle of doing more with less, which will certainly find new applications in discrete time signal processing in the near future.

References

1. Benesty, J., Sondhi, M.M., Huang, Y.: Handbook of Speech Processing. Springer, Berlin (2008)
2. Bimbot, F., Bonastre, J.F., Fredouille, C., Gravier, G., Magrin-Chagnolleau, I., Meignier, S., Merlin, T., Ortega-Garcia, J., Petrovska-Delacretaz, D., Reynolds, D.A.: A tutorial on text-independent speaker verification. EURASIP J. Appl. Signal Process. **4**, 430–451 (2004)
3. Cho, Y., Al-Naimi, K., Kondoz, A.: Improved voice activity detection based on a smoothed statistical likelihood ratio. In: International Conference on Acoustics, Speech and Signal Processing, vol. 2, pp. 737–740. IEEE, Los Alamitos, CA (2001)
4. Claussen, H., Rosca, J., Damper, R.: Mutual interdependence analysis. In: Independent Component Analysis and Blind Signal Separation, pp. 446–453. Springer, Heidelberg (2007)
5. Claussen, H., Rosca, J., Damper, R.: Mutual features for robust identification and verification. In: International Conference on Acoustics, Speech and Signal Processing, pp. 1849–1852. Las Vegas, NV (2008)
6. Claussen, H., Rosca, J., Damper, R.: Generalized mutual interdependence analysis. In: International Conference on Acoustics, Speech and Signal Processing, pp. 3317–3320. Taipei, Taiwan (2009)
7. Claussen, H., Rosca, J., Damper, R.I.: Signature extraction using mutual interdependencies. Pattern Recognit. **44**, 650–661 (2011)
8. Deng, L., O'Shaughnessy, D.: Speech Processing: A Dynamic and Optimization-Oriented Approach. Signal Process. Commun. Dekker, New York (2003)
9. ETSI: Digital cellular telecommunication system (phase 2+); voice activity detector VAD for adaptive multi rate (AMR) speech traffic channels; general description. Technical Report V.7.0.0, ETSI (1999)

10. Fisher, R.A.: The use of multiple measurements in taxonomic problems. Ann. Eugenics **7**, 179–188 (1936)
11. Fisher, W.M., Doddington, G.R., Goudie-Marshall, K.M., Jankowski, C., Kalyanswamy, A., Basson, S., Spitz, J.: NTIMIT. Linguistic Data Consortium, Philadelphia CDROM (1993). http://www.ldc.upenn.edu/Catalog/CatalogEntry.jsp?catalogId=LDC93S2
12. Garofolo, J.S., Lamel, L.F., Fisher, W.M., Fiscus, J.G., Pallett, D.S., Dahlgren, N.L., Zue, V.: TIMIT acoustic-phonetic continuous speech corpus. Linguistic Data Consortium, Philadelphia CDROM (1993). http://www.ldc.upenn.edu/Catalog/CatalogEntry.jsp?catalogId=LDC93S1
13. Haigh, J., Mason, J.: Robust voice activity detection using cepstral features. In: IEEE Region 10 Conference TENCON, vol. 3, pp. 321–324. IEEE (1993)
14. Hotelling, H.: Relation between two sets of variates. Biometrika **28**, 322–377 (1936)
15. Hoyt, J.D., Wechsler, H.: Detection of human speech in structured noise. In: International Conference on Acoustics, Speech and Signal Processing, vol. 2, pp. 237–240. IEEE (1994)
16. Huang, X., Acero, A., Hon, H.W.: Spoken Language Processing: A guide to Theory, Algorithm, and System Development. Prentice Hall, New York (2001)
17. Liu, P., Wang, Z.: Voice activity detection using visual information. In: International Conference on Acoustics, Speech and Signal Processing, vol. 1, pp. 609–612. Montreal, Canada (2004)
18. Qu, T., Xiao, Z., Gong, M., Huang, Y., Li, X., Wu, X.: Distance-dependent head-related transfer functions measured with high spatial resolution using a spark gap. IEEE Trans. Audio, Speech Lang. Process. **17**(6), 1124–1132 (2009)
19. Reynolds, D.A., Campbell, W.M.: Text-independent speaker recognition. In: Benesty, J., Sondhi, M., Huang, Y. (eds.) Handbook of Speech Processing and Communication, pp. 763–781. Springer GMBH, New York (2007)
20. Reynolds, D.A., Rose, R.C.: Robust text-independent speaker identification using Gaussian mixture speaker models. IEEE Trans. Speech, Audio Process. **3**(1), 72–83 (1995)
21. Rosca, J., Balan, R., Fan, N., Beaugeant, C., Gilg, V.: Multichannel voice detection in adverse environments. In: European Signal Processing Conference (2002)
22. Srinivasan, K., Gersho, A.: Voice activity detection for cellular networks. In: IEEE Speech Coding Workshop, pp. 85–86 (1993)
23. Tikhonov, A.: On the stability of inverse problems. Doklady Akademii Nauk SSSR **39**(5), 195–198 (1943)
24. Zhang, Z., Liu, Z., Sinclair, M., Acero, A., Deng, L., Huang, X., Zheng, Y.: Multi-sensory microphones for robust speech detection, enhancement and recognition. In: International Conference on Acoustics, Speech and Signal Processing, pp. 781–784. IEEE (2004)

Approximation Methods for the Recovery of Shapes and Images from Gradients

Vishal M. Patel and Rama Chellappa

Abstract Recovery of shapes and images from gradients is an important problem in many fields such as computer vision, computational photography, and remote sensing. For instance, techniques such as photometric stereo and shape from Shading recover the underlying 3D shape by integrating an estimated surface gradient field or surface normals. In applications such as image stitching and image editing, gradients of given images are first manipulated. The final image is then reconstructed from the modified gradient field. The estimated or modified gradient field is usually nonintegrable due to the presence of noise, outliers in the estimation process, and inherent ambiguities. This chapter reviews some approximation-based methods for surface reconstruction from the given nonintegrable gradient field with applications to 3D modeling and image reconstruction.

Keywords Photometric stereo • Shape from shading • Image gradients • Image recovery • Compressive sampling • Poisson solver • Shapelets • Shape recovery • Surface reconstruction • Sparsity

1 Introduction

Reconstruction of images from gradient fields is an important problem in many applications such as shape from shading (SfS) [9, 10, 12, 29], photometric stereo (PS) [28], compressive sampling (CS) [19], computational photography [26], phase unwrapping [8], high dynamic range compression [5], image editing [20], and image matting [24]. Given an estimate of gradient field, the final surface or image is obtained by integrating the available gradient field.

V.M. Patel • R. Chellappa (✉)
Department of Electrical and Computer Engineering, Center for Automation Research, UMIACS, University of Maryland, College Park, MD 20742, USA
e-mail: pvishalm@umiacs.umd.edu; rama@umiacs.umd.edu

T.D. Andrews et al. (eds.), *Excursions in Harmonic Analysis, Volume 1,* Applied and Numerical Harmonic Analysis, DOI 10.1007/978-0-8176-8376-4_19,

For instance, in image editing [20], the gradient fields of one or more images are manipulated to obtain the desired goal and the final image is recovered by integrating the modified gradient field. In PS [28], multiple images of an object under varying illumination are captured to first estimate a surface gradient field. The estimated surface gradient field is then integrated to obtain the final shape of an object. In [19], a method for recovering images from their partial Fourier measurements is presented. Given partial Fourier measurements of an image and using the fact that the Fourier transform of the gradients of an image are precisely equal to a diagonal transformation of the Fourier transform of the original image, CS methods are utilized to directly recover the horizontal and vertical differences of the desired image. The final image is then reconstructed by integrating the estimated gradient field.

The gradient field of a scalar surface or an image should be integrable (conservative). That is, the integral along any closed curve should be equal to zero and the reconstruction should not depend on the path of the integration. However, this is often not the case when inherent noise during the estimation process contaminates the gradient field. In this chapter, we review a few approaches for enforcing integrability. We first briefly describe the idea behind SfS and photometric stereo.

1.1 Shape from Shading

SfS attempts to recover 3D shape from a single image. In this method, the reflectance map is assumed to be known or given. Let **s** denote the direction of the light source

$$\mathbf{s} = \frac{1}{\sqrt{1+p_s^2+q_s^2}}(-p_s,-q_s,1)^T,$$

for some p_s and q_s. Let Z be the surface height and $p = \frac{\partial Z(x,y)}{\partial x}$, $q = \frac{\partial Z(x,y)}{\partial y}$. The surface normal **n** is then given by

$$\mathbf{n} = \frac{1}{\sqrt{1+p^2+q^2}}(-p,-q,1)^T.$$

For a Lambertian surface, the image intensity $I(x,y)$ is modeled as

$$\begin{aligned} I(x,y) &= \rho(x,y)\mathbf{n}(x,y).\mathbf{s} \\ &= \rho\frac{1+pp_s+qq_s}{\sqrt{1+p_s^2+q_s^2}\sqrt{1+p^2+q^2}}, \end{aligned}$$

where ρ is the albedo of the surface and. denotes dot product. Often times, ρ is assumed to be constant over the surface. In that case, one can also write the image irradiance equation as

$$I(x,y) = R(p,q)$$
$$= \frac{1+pp_s+qq_s}{\sqrt{1+p_s^2+q_s^2}\sqrt{1+p^2+q^2}},$$

where $R(p,q)$ is called the reflectance map. Given an estimate of the light source direction, the general SfS problem then reduces to estimating the surface gradients p and q from the image intensity I.

One can use the algorithm by Brooks and Horn for SfS [9]. The algorithm assumes a Lambertian reflectance model for the surfaces. At each iteration, new estimates of the surface gradients $\begin{bmatrix}\hat{p}\\ \hat{q}\end{bmatrix}_{k+1}$ are obtained from the previous estimates $\begin{bmatrix}\hat{p}\\ \hat{q}\end{bmatrix}_{k}$ as

$$\begin{bmatrix}\hat{p}\\ \hat{q}\end{bmatrix}_{k+1} = \begin{bmatrix}\hat{\hat{p}}\\ \hat{\hat{q}}\end{bmatrix}_{k} + \lambda(I-R)\begin{bmatrix}R_x\\ R_y\end{bmatrix},$$

where $\hat{\hat{p}}_k$ and $\hat{\hat{q}}_k$ denote the smoothed values of $\hat{p}_k$ and $\hat{q}_k$, respectively, I is the input image, R is the reflectance map, and R_x, R_y are the corresponding derivatives. In the next section, we present more sophisticated algorithms that can be used to incorporate integrability at each iteration of this algorithm by projecting the nonintegrable gradient fields onto the basis functions such as Fourier, wavelet, and shapelet.

1.2 Photometric Stereo

Photometric stereo attempts to estimate the 3D shape of an object from images taken from the same viewpoint but illuminated from distant point light sources from multiple directions. In this method, a minimum of three images are required to estimate the shape with the constraint that the direction of light sources should not be coplanar. If $I_1,\ldots,I_n$ are n images captured under distant point light sources whose directions are given by $\mathbf{s}_1,\ldots,\mathbf{s}_n$, then under the Lambertian reflectance assumption, one can write

$$I_i(x,y) = \rho(x,y)\mathbf{n}(x,y).\mathbf{s}_i \quad i = 1,\ldots,n.$$

If the light source directions are known, then for each pixel, one can write a linear system for the scaled albedo $\mathbf{a}(x,y) = \rho(x,y)\mathbf{n}(x,y)$ by stacking the image intensities

$$\begin{bmatrix}\mathbf{s}_1^T\\ \mathbf{s}_2^T\\ \vdots\\ \mathbf{s}_n^T\end{bmatrix}\mathbf{a}(x,y) = \begin{bmatrix}I_1(x,y)\\ I_2(x,y)\\ \vdots\\ I_n(x,y)\end{bmatrix}.$$

This linear system is solved to estimate the scaled albedo $\mathbf{a}$. The albedo, ρ, can be obtained as the norm of $\mathbf{a}$

$$\rho(x,y) = |\mathbf{a}(x,y)|.$$

The surface normal $\mathbf{n}(x,y)$ is given by

$$\mathbf{n}(x,y)\frac{\mathbf{a}(x,y)}{|\mathbf{a}(x,y)|}.$$

Given $\mathbf{n}(x,y) = [n_x, n_y, n_z]^T$, the surface gradients can be obtained as $p = -\frac{n_x}{n_z}$ and $q = -\frac{n_y}{n_z}$. The problem then reduces to estimating Z from the estimated gradient field (p,q).

The rest of this chapter is organized as follows: Sect. 2 presents various methods for reconstructing images and surfaces from gradient fields. In particular, Poisson-based, Fourier-based, wavelet-based, and shapelet-based methods are described in detail. Finally, in Sect. 3, a method for reconstructing images from their partial Fourier measurements that use gradient information is presented.

2 Shape Recovery from Gradients

In this section, we present various methods for recovering shapes from the estimated gradients. In particular, we discuss a Poisson-based method [23], Fourier-based method [6], shapelet approach [15], and a wavelet-based algorithm [27] in detail.

2.1 *Poisson Solver*

Let $Z(x,y)$ be a 2D real-valued scalar function on a $H \times W$ rectangular grid (x,y) of image pixels. Let $\{p(y,x), q(y,x)\}$ denote the given nonintegrable gradient field over the grid. Define the curl and divergence operators as

$$\texttt{curl}(p,q) = \frac{\partial p}{\partial y} - \frac{\partial q}{\partial x}$$

$$\texttt{div}(p,q) = \frac{\partial p}{\partial x} - \frac{\partial q}{\partial y}.$$

Given (p,q), the objective is to obtain a surface Z. Let (Z_x, Z_y) denote the gradient field of Z. A common approach is to minimize the least square error functional given by

$$J(Z) = \int\int \left((Z_x - p)^2 - (Z_y - q)^2\right)\, \mathrm{d}x\, \mathrm{d}y. \tag{1}$$

The corresponding Euler–Lagrange equation gives

$$\frac{\partial J}{\partial Z}-\frac{\partial}{\partial x}\frac{\partial J}{\partial Z_x}-\frac{\partial}{\partial y}\frac{\partial J}{\partial Z_y}=0$$

$$\frac{\partial}{\partial x}(Z_x-p)+\frac{\partial}{\partial y}(Z_y-q)=0$$

$$\frac{\partial^2 Z}{\partial x^2}+\frac{\partial^2 Z}{\partial y^2}=\frac{\partial p}{\partial x}+\frac{\partial q}{\partial y}$$

$$\nabla^2 Z=\mathtt{div}(p,q).$$

This is the Poisson equation [10], where $\nabla^2=\frac{\partial^2}{\partial x^2}+\frac{\partial^2}{\partial y^2}$ is the Laplacian operator. This method is often referred to as the Poisson solver [23]. The integrable gradient field is found by differentiating the estimated surface Z. Thus, (Z_x,Z_y) is the integrable gradient field corresponding to the given nonintegrable gradient field (p,q). Note that

$$\mathtt{div}(Z_x,Z_y)=\frac{\partial Z_x}{\partial x}+\frac{\partial Z_y}{\partial y}=\nabla^2 Z=\mathtt{div}(p,q).$$

In other words, the Poisson solver enforces integrability by finding a zero curl gradient field which has the same divergence as the given non-integrable gradient field.

Assuming Neumann boundary conditions given by $\nabla Z.\hat{\mathbf{n}}=0$, one can discretize the Poisson equation as

$$\mathbf{LZ}=\mathbf{u},$$

where $u=\mathtt{div}(p,q)$, $\mathbf{u}=[u(1,1),\ldots,u(H,W)]^T$, and $\mathbf{L}$ is the sparse Laplacian matrix of size $HW\times HW$. Each row of $\mathbf{L}$ has -4 at the diagonal entry and four 1's corresponding to the isotropic Laplacian kernel ∇^2. With this, Z can be obtained as $\mathbf{Z}=\mathbf{L}^{-1}\mathbf{u}$. See [23] for more details.

2.2 *Frankot–Chellappa Algorithm*

In [6], Frankot and Chellappa present a method for projecting a gradient field to the nearest integrable solution. They suggested to use a set of integrable basis functions to represent the surface slopes so as to minimize the distance between the integrable gradient field and a nonintegrable one. Suppose that $Z(x,y)$ denotes the reconstructed height at the image location with coordinates (x,y). As discussed earlier, the integrability condition for the surface requires that the height function does not depend on the integration path. This implies that the surface must satisfy the following condition:

$$Z_{xy} = Z_{yx}, \tag{2}$$

where $Z_{xy} = \frac{\partial^2 Z(x,y)}{\partial x \partial y}$ and $Z_{yx} = \frac{\partial^2 Z(x,y)}{\partial y \partial x}$. This condition can also be regarded as a smoothness constraint, since the partial derivative of the surface needs to be continuous in order that it can be integrable.

An integrable surface Z can be represented by the basis expansion

$$\tilde{Z}(x,y) = \sum_{\boldsymbol{\omega} \in \Omega} \tilde{C}(\boldsymbol{\omega}) \varphi(x,y,\boldsymbol{\omega}), \tag{3}$$

where $\varphi(x,y,\boldsymbol{\omega})$ are the basis functions, $\boldsymbol{\omega} = (\omega_x, \omega_y)$ is a two-dimensional index, Ω is a finite set of indexes, and the members of $\varphi(x,y,\boldsymbol{\omega})$ are not necessarily mutually orthogonal. If each $\varphi(\boldsymbol{\omega})$ satisfies (2), then it follows that Z does as well. The first partial derivatives of $\tilde{Z}$ can be expressed in terms of this set of basis functions as

$$\frac{\partial \tilde{Z}(x,y)}{\partial x} = \sum_{\boldsymbol{\omega} \in \Omega} \tilde{C}(\boldsymbol{\omega}) \varphi_x(x,y,\boldsymbol{\omega}) \tag{4}$$

and

$$\frac{\partial \tilde{Z}(x,y)}{\partial y} = \sum_{\boldsymbol{\omega} \in \Omega} \tilde{C}(\boldsymbol{\omega}) \varphi_y(x,y,\boldsymbol{\omega}), \tag{5}$$

where $\varphi_x = \frac{\partial \varphi}{\partial x}$ and $\varphi_y = \frac{\partial \varphi}{\partial y}$. Since these are the first partial derivatives of an integrable surface, they share the same set of coefficients $\tilde{C}(\boldsymbol{\omega})$. Similarly, the non-integrable gradient field can be represented as

$$\frac{\partial \hat{Z}(x,y)}{\partial x} = \sum_{\boldsymbol{\omega} \in \Omega} \hat{C}_1(\boldsymbol{\omega}) \varphi_x(x,y,\boldsymbol{\omega})$$

and

$$\frac{\partial \hat{Z}(x,y)}{\partial y} = \sum_{\boldsymbol{\omega} \in \Omega} \hat{C}_2(\boldsymbol{\omega}) \varphi_y(x,y,\boldsymbol{\omega}).$$

Since this set of first partial derivatives is not integrable, their corresponding transform coefficients are not the same. That is $\hat{C}_1(\boldsymbol{\omega}) \neq \hat{C}_2(\boldsymbol{\omega})$. One can minimize the distance between the transform coefficients of the non-integrable and the integrable gradient fields by making $\hat{C}_1(\boldsymbol{\omega}) = \hat{C}(\boldsymbol{\omega}) = \hat{C}_2(\boldsymbol{\omega})$. This requires the minimization of the following quantity:

$$d\left[(\hat{Z}_x,\hat{Z}_y),(\tilde{Z}_x,\tilde{Z}_y)\right] = \int\int \|\tilde{Z}_x - \hat{Z}_x\|^2 + \|\tilde{Z}_y - \hat{Z}_y\|^2 \, \mathrm{d}x \, \mathrm{d}y. \tag{6}$$

The following proposition is proven in [6].

Proposition 1. *The expansion coefficients $C(\boldsymbol{\omega})$ in (3) that minimize (6) given a possibly non-integrable estimate of surface slopes $\hat{Z}_x(x,y), \hat{Z}_y(x,y)$ are given by*

$$\tilde{C}(\boldsymbol{\omega}) = \frac{P_x(\boldsymbol{\omega})\hat{C}_1(\boldsymbol{\omega}) + P_y(\boldsymbol{\omega})\hat{C}_2(\boldsymbol{\omega})}{P_x(\boldsymbol{\omega}) + P_y(\boldsymbol{\omega})} \tag{7}$$

for each $\boldsymbol{\omega} \in \Omega$ where $P_x(\boldsymbol{\omega}) = \int\int |\varphi_x(x,y,\boldsymbol{\omega})|^2 \ \mathrm{d}x \ \mathrm{d}y$ and $P_y(\boldsymbol{\omega}) = \int\int |\varphi_y(x,y,\boldsymbol{\omega})|^2 \ \mathrm{d}x\,\mathrm{d}y$. The integrated surface $\tilde{Z}(x,y)$ and integrable surface slopes $\tilde{Z}_x(x,y), \tilde{Z}_y(x,y)$ are then obtained by substituting $\tilde{C}(\boldsymbol{\omega})$ into the expansions (3), (4), and (5).

In the case when Fourier basis functions are used, $\varphi(x,y,\omega) = \exp(2\pi i\omega_x x + 2\pi i\omega_y y)$, one can show that (6) is minimized by taking

$$\tilde{C}(\boldsymbol{\omega}) = \frac{-i\omega_x\hat{C}_x(\boldsymbol{\omega}) - i\omega_y\hat{C}_y(\boldsymbol{\omega})}{\omega_x^2 + \omega_y^2},$$

with the Fourier coefficients of the constrained surface slopes given by

$$\tilde{C}_x(\boldsymbol{\omega}) = i\omega_x\tilde{C}(\boldsymbol{\omega}),$$
$$\tilde{C}_y(\boldsymbol{\omega}) = i\omega_y\tilde{C}(\boldsymbol{\omega}).$$

In this setting, one can recover the surface $\tilde{Z}$ up to an unknown scaling factor. This method is also robust to noise. In the following sections, we show how this method can be generalized using cosine basis functions as well as by using well-localized basis functions such as those found in wavelets [11, 13], and shapelets [15].

2.3 Wavelet-Based Methods

The method proposed in the previous section can be extended using wavelet basis functions. One such method for SfS was proposed by Hsieh et al. in [11]. Karacali and Snyder use a reconstruction approach based on constructing an orthonormal set of gradient field that span a feasible subspace of the gradient space using wavelets [13]. They adapt wavelet shrinkage methods to successfully reduce the influence of noise during the reconstruction process. Another approach based on using Daubechies wavelet basis and connection coefficients was proposed by Wei and Klette in [27]. In what follows, we briefly describe this method.

2.4 Connection Coefficients

Let $\phi(x)$ and $\psi(x)$ be the Daubechies scaling function and wavelet, respectively, defined as

$$\phi(x) = \sum_{k\in\mathbb{Z}} a_k \phi(2x-k),$$

$$\psi(x) = \sum_{k\in\mathbb{Z}} (-1)^k a_{1-k} \phi(2x-k),$$

where a_k are the wavelet filter coefficients. Let $L^2(\mathbb{R})$ be the space of square integrable functions on the real line. Let V_j be the closure of this space spanned by $\phi_{j,k}(x) = 2^{j/2}\phi(2^j x - k)$, $j,k \in \mathbb{Z}$, and suppose that W_j, the orthogonal complementary of V_j in V_{j+1}, be the closure of the function subspace generated by $\psi_{j,k}(x) = 2^{j/1}\psi(2^j x - k)$, $k \in \mathbb{Z}$. Then the function subspaces V_j and W_j have the following properties: $V_j \subseteq V_{j+1}$ for all $j \in \mathbb{Z}$; $\cap_{j\in\mathbb{Z}} V_j = \{0\}$; $\cup_{j\in\mathbb{Z}} V_j = L^2(\mathbb{R})$; $V_{j+1} = V_0 \oplus W_0 \oplus W_1 \oplus \cdots \oplus W_j$, where $\oplus$ denotes the orthogonal direct sum. The scaling functions $\{\phi_{j,k}(x), k \in \mathbb{Z}\}$ form an orthonormal basis for V_j for each fixed scale j. Similarly, the wavelets $\{\psi_{j,k}(x), k \in \mathbb{Z}\}$ form an orthonormal basis for W_j. The set of subspaces V_j is often known as a multiresolution analysis of $L^2(\mathbb{R})$. Let J be a positive integer. A function $f(x) \in V_J$ can be represented as

$$f(x) = \sum_{k\in\mathbb{Z}} c_{J,k} \phi_{J,k}(x),$$

where the coefficients $c_{J,k}$ are found by $c_{J,k} = \int f(x)\phi_{J,k}(x)\,\mathrm{d}x$. Alternatively, f can be represented as

$$f(x) = \sum_{k\in\mathbb{Z}} c_{0,k}\phi_{0,k}(x) + \sum_{j=0}^{J-1}\sum_{k\in\mathbb{Z}} d_{j,k}\psi_{j,k}(x).$$

Assume that the scaling function $\phi(x)$ has N vanishing moments. For $k \in \mathbb{Z}$, we define that

$$\Gamma_k^0 = \int \phi(x)\phi(x-k)\;\mathrm{d}x,$$

$$\Gamma_k^1 = \int \phi^{(x)}(x)\phi(x-k)\;\mathrm{d}x,$$

$$\Gamma_k^2 = \int \phi^{(x)}(x)\phi^{(x)}(x-k)\;\mathrm{d}x,$$

where $\phi^{(x)}$ denotes the derivative of ϕ with respect to x. The connection coefficients have the following properties: For a scaling function $\phi(x)$ with N vanishing moments, $\Gamma_0^1 = 0$, $\Gamma_k^1 = \Gamma_k^2 = 0$, $k \notin [-2N+2, 2N-2]$; and $\Gamma_k^0 = 1$ for $k = 0$ and 0 otherwise.

2.5 Wavelet-Based Reconstruction from Gradients

Assume that the size of the domain of the surface $Z(x,y)$ is $M \times M$ and the surface $Z(x,y)$ can be represented by a linear combination of a set of the Daubechies scaling functions as

$$Z(x,y) = \sum_{m=0}^{M-1}\sum_{n=0}^{M-1} z_{m,n}\phi_{m,n}(x,y), \tag{8}$$

where $z_{m,n}$ are the coefficients and $\phi_{m,n}(x,y) = \phi(x-m)\phi(y-n)$. For the given gradient values $p(x,y)$ and $q(x,y)$, we assume that

$$p(x,y) = \sum_{m=0}^{M-1}\sum_{n=0}^{M-1} p_{m,n}\phi_{m,n}(x,y), \tag{9}$$

$$q(x,y) = \sum_{m=0}^{M-1}\sum_{n=0}^{M-1} q_{m,n}\phi_{m,n}(x,y), \tag{10}$$

where the coefficients $p_{m,n}$ and $q_{m,n}$ can be found by

$$p_{m,n} = \int\int p(x,y)\phi_{m,n}(x,y)\;\mathrm{d}x\,\mathrm{d}y,$$

$$q_{m,n} = \int\int q(x,y)\phi_{m,n}(x,y)\;\mathrm{d}x\,\mathrm{d}y.$$

Substituting (8), (9), and (9) into (1), we have

$$\begin{aligned} d &= \int\int \Big(\sum_m\sum_n z_{m,n}\phi_{m,n}^{(x)}(x,y) - \sum_m\sum_n p_{m,n}\phi_{m,n}(x,y)\Big)^2 \\ &\quad + \Big(\sum_m\sum_n z_{m,n}\phi_{m,n}^{(y)}(x,y) - \sum_m\sum_n q_{m,n}\phi_{m,n}(x,y)\Big)^2 \;\mathrm{d}x\,\mathrm{d}y \\ &= d_1 + d_2, \end{aligned}$$

where $\phi_{m,n}^{(x)}(x,y) = \frac{\partial\phi_{m,n}(x,y)}{\partial x}$ and $\phi_{m,n}^{(y)}(x,y) = \frac{\partial\phi_{m,n}(x,y)}{\partial y}$.

To derive the iterative scheme for Z, let $\Delta z_{i,j}$ represent the updating amount of $z_{i,j}$ in the iterative equation, $z'_{i,j}$ be the value after update. Then, $z'_{i,j} = z_{i,j} + \Delta z_{i,j}$. Substituting $z'_{i,j}$ into d_1, it will be changed by the following amount $d'_1 = d_1 + \Delta d_1$. Similarly, using the d_2 will be changed as $d'_2 = d_2 + \Delta d_2$. Using d'_1 and d'_2, it can be shown that

$$\begin{aligned} \Delta d &= \Delta d_1 + \Delta d_2 \\ &= 2\Delta z_{i,j}\sum_{m,n} z_{m,n}(\Gamma_{i-m}^2\Gamma_{j-n}^0 + \Gamma_{i-m}^0\Gamma_{j-n}^2) \end{aligned}$$

$$-2\Delta z_{i,j}\sum_{m,n} p_{m,n}\Gamma^1_{i-m}\Gamma^0_{j-n}$$

$$-2\Delta z_{i,j}\sum_{m,n} q_{m,n}\Gamma^0_{i-m}\Gamma^1_{j-n} + 2\Delta z^2_{i,j}\Gamma^2_0.$$

In order to make the cost function decrease as fast as possible, Δd must be maximized. From $\frac{\partial \Delta d}{\partial \Delta z_{i,j}} = 0$, we have

$$\Delta z_{i,j} = \frac{1}{2\Gamma^2_0}\sum_{k=-2N+2}^{2N-2}\left[(p_{i-k,j}+q_{i,j-k})\Gamma^1_k - (z_{i-k,j}+z_{i,j-k})\Gamma^2_k\right].$$

With the initial values set to zero, we get the following iterative equation:

$$z^{[t+1]}_{i,j} = z^{[t]}_{i,j} + \Delta z_{i,j}.$$

2.6 Shapelet-Based Method

The motivation behind this method is that correlation between the gradients of a signal and the gradients of a basis function can provide information equivalent to direct correlation between the signal and basis function up to a signal offset because differentiation is linear [15]. If the surface gradient information is correlated with the gradients of a bank of shapelet[1] basis functions, one can reconstruct the surface shape, up to an offset, by simply adding the correlation results. Summation of the basis correlations automatically imposes a continuity constraint and performs an implicit integration of the surface from its gradients. However, a shapelet function must satisfy certain properties [15]. With a properly chosen shapelet and assuming that we know the slant and tilt values over the surface, one can recover a shape from its gradients as follows.

Correlation of the surface and shapelet slants is calculated by

$$|\nabla| = \tan(\sigma),$$

where σ is the slant and $\nabla = (p,q)$ are the gradients. The gradient correlation is then calculated as

$$A_{\nabla i} = |\nabla_Z| \star |\nabla_{si}|,$$

where ∇_Z and ∇_{si} are the gradients of the surface and shapelet at scale i. Without tilt information, this correlation matches positive and negative gradients equally because only gradient magnitudes are considered. Hence, shape tilt information

[1]The term shapelet is used to describe any basis function of finite support used for representing a shape.

needs to be considered. To form the tilt correlation between a shapelet at scale i and the surface, one can use the cosine of the tilt differences between points on the surface and shapelet and use the standard trigonometric difference equation to overcome any angle wraparound problem at the origin. Thus,

$$A_{\tau i} = \cos(\tau_Z) \star \cos(\tau_{si}) + \sin(\tau_Z) \star \sin(\tau_{si}),$$

where τ_Z and τ_{si} denote the tilts of the surface and shapelet at scale i, respectively. With this, one can obtain the overall correlation between surface and shapelet at scale i by taking the point-wise product of the gradient and tilt correlations

$$\begin{aligned} \mathbf{A}_i &= A_{\nabla i}.A_{\tau i} \\ &= [|\nabla_Z| \star |\nabla_{si}|] . [\cos(\tau_Z) \star \cos(\tau_{si}) + \sin(\tau_Z) \star \sin(\tau_{si})] \\ &= [|\nabla_Z| . \cos(\tau_Z)] \star [|\nabla_{si}| . \cos(\tau_{si})] + [|\nabla_Z| . \sin(\tau_Z)] \star [|\nabla_{si}| . \sin(\tau_{si})], \end{aligned}$$

where denotes point-wise multiplication. This process can be performed over multiple scales, and the results are summed to form the reconstruction

$$\tilde{Z} = \sum_i \mathbf{A}_i.$$

This method was also shown to be robust to noise [15].

2.7 *Other Methods*

Many other methods have been proposed for recovering shapes from their gradients. For instance, a discrete cosine transform-based method was proposed in [7]. A Gaussian kernel-based method was recently proposed in [17]. Motivated by the recent advances in sparse representation and compressed sensing, a robust method for enforcing integrability was proposed in [21]. Some of the other methods include [1, 2]. See [14, 29], and the references therein for more details on surface reconstruction from gradients.

2.8 *Numerical Examples*

In this section, we show some examples of reconstruction obtained by applying various methods to the problem of shape recovery from gradients. In particular, using a synthetic surface, we compare the performance of different algorithms such as a Poisson solver-based method [23], shapelet-based approach [15], Frankot–Chellappa algorithm [6], and an ℓ_1-minimization approach [21]. We used a synthetic surface shown in Fig. 1a to generate the gradient field. We then contaminated

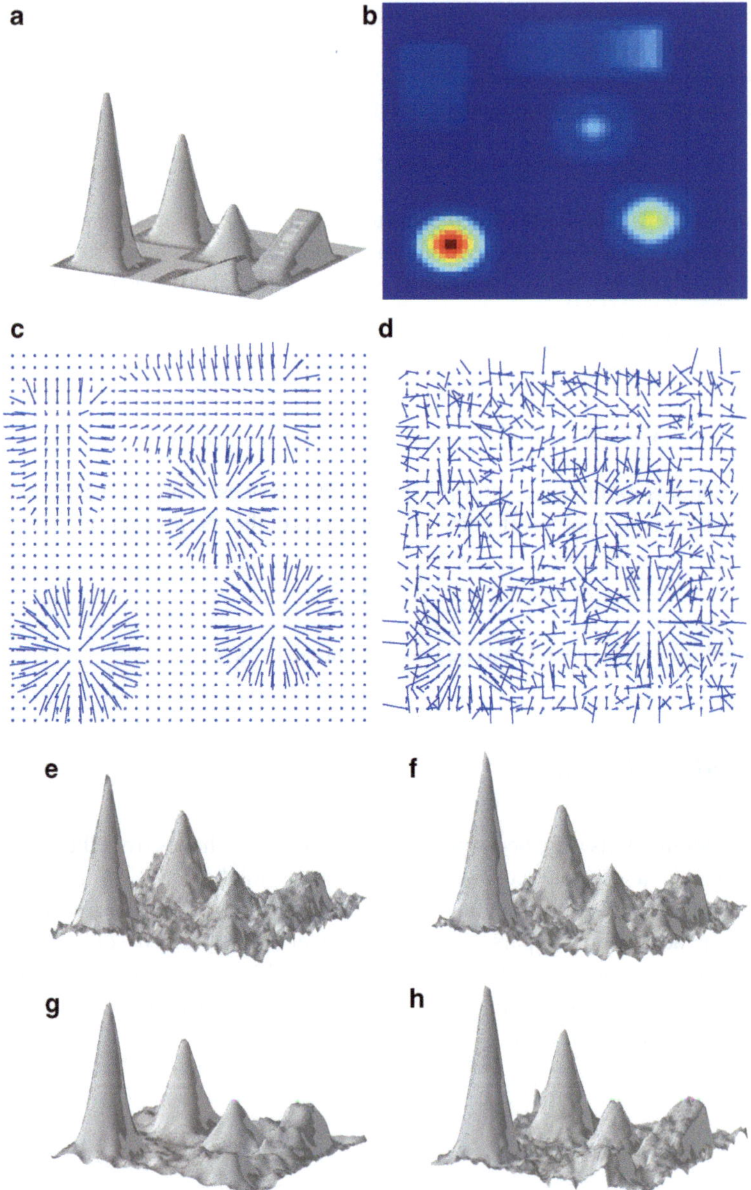

Fig. 1 Reconstructed surface when the gradient field is corrupted by both outliers (at 3 % locations) and noise (Gaussian with $\sigma = 7\,\%$). (**a**) Original surface. (**b**) Image plot of the surface shown in (a). (**c**) Surface normal needle plot of the original surface. (**d**) Surface normal needle plot corresponding to the noisy gradients. (**e**) Reconstruction using Poisson solver. (**f**) Reconstruction using Frankot–Chellappa method. (**g**) Reconstruction using shapelets. (**h**) Reconstruction using ℓ_1-minimization method

the gradient field by adding Gaussian noise and outliers. The noisy gradient field was then used as input to different integration methods. Figure 1 shows the reconstructions from different methods. As can be seen from the figure, the Poisson-based method suffers the most from outliers and noise producing very poor reconstruction. Both shapelet-based method and an ℓ_1-minimization approach produce good results. Frankot–Chellappa method performs significantly better compared to the Poisson-based method.

3 Image Recovery from Partial Fourier Measurements

Reconstruction of imagery from an incomplete set of samples from a Fourier representation is an important goal to improve several scanning technologies such as magnetic resonance imaging (MRI) [16] and synthetic aperture radar (SAR) [18]. Solutions to such a problem would allow significant reductions in collection times and improve the capacity to collect very time sensitive events.

The images of interest tend to enjoy the property of being sparse or compressible in some transform domain (e.g., wavelet, gradient, Fourier). Images such as angiograms are inherently sparse in the pixel domain or gradient domain. For instance, if the image is piecewise constant, then a gradient representation would only contain nonzero values near boundary positions. The redundancies in these images suggest a possible dimensionality reduction that should allow for Fourier sampling at sub-Nyquist rates. An example of a sparse-gradient image along with an image of its edges is shown in Fig. 2.

Let $X \in \mathbb{C}^{N\times N}$ denote an image. Any particular pixel of X will be denoted as $X_{n,m}$. The discrete directional derivatives on X are defined pixel-wise as

$$\begin{aligned}(X_x)_{n,m} &= X_{n,m} - X_{n-1,m},\\ (X_y)_{n,m} &= X_{n,m} - X_{n,m-1}.\end{aligned} \tag{11}$$

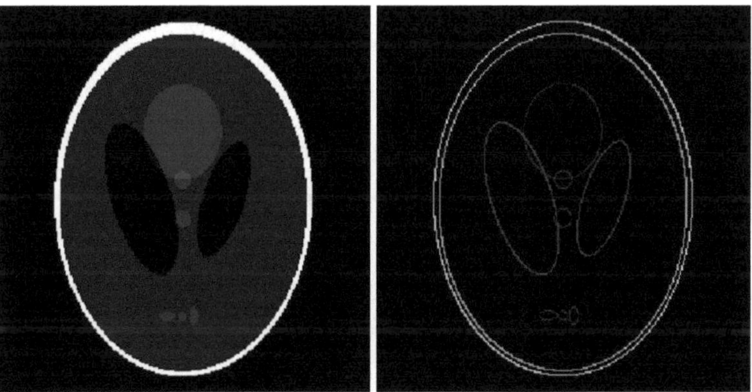

Fig. 2 512×512 Shepp-Logan phantom image and its edges

Based on these, the discrete gradient operator ∇ where $\nabla X \in \mathbb{C}^{N\times N\times 2}$ is defined as

$$(\nabla X)_{n,m} = ((X_x)_{n,m}, (X_y)_{n,m}).$$

From these operators, one can define the discrete total-variational operator TV or $|\nabla|$ on X as

$$\begin{aligned}(TV[X])_{n,m} &= (|\nabla|(X))_{n,m} \\ &= \sqrt{|(X_x)_{n,m}|^2 + |(X_y)_{n,m}|^2},\end{aligned} \tag{12}$$

from which one can also define the total-variation seminorm of X as

$$\|X\|_{TV} = \|TV(X)\|_1, \tag{13}$$

where $\|\cdot\|_p$ for $0 < p < \infty$ is the ℓ_p norm defined as

$$\|X\|_p = \left(\sum_{n=1}^{N}\sum_{m=1}^{N} |X_{n,m}|^p\right)^{\frac{1}{p}}.$$

For $p = 0$, we have the quasi-norm

$$\|X\|_0 = \{\#(n,m) : X_{n,m} \neq 0\},$$

i.e., $\|X\|_0$ is the number of non-zero pixels in X. It is said that an image X is K-sparse if $\|X\|_0 = K$. It is also said that X is K-sparse in gradient (or in the total-variational sense) if $\||\nabla|(X)\|_0 = K$.

The objective is to recover an image X that is K-sparse in gradients from a set of $M \ll N^2$ Fourier measurements. To that end, define a set Ω of M two-dimensional frequencies $\underline{\omega}_k = (\omega_{x,k}, \omega_{y,k})$, $1 \le k \le M$ chosen according to a particular sampling pattern from $\{0, 1, \ldots, N-1\}^2$. Let $\mathscr{F}$ denote the two-dimensional DFT of X

$$\mathscr{F}(\omega_x, \omega_y) = \sum_{n=0}^{N-1}\sum_{m=0}^{N-1} X(n,m)\exp\left(-2\pi i\left(\frac{n\omega_x}{N}, \frac{m\omega_y}{N}\right)\right)$$

and $\mathscr{F}^{-1}$ its inverse

$$\mathscr{F}^{-1}\{\mathscr{F}(\omega_x, \omega_y)\} = X(n,m) = \frac{1}{N^2}\sum_{\omega_x=0}^{N-1}\sum_{\omega_y=0}^{N-1} \mathscr{F}(\omega_x, \omega_y)\exp\left(2\pi i\left(\frac{n\omega_x}{N}, \frac{m\omega_y}{N}\right)\right).$$

Next define the operator $\mathscr{F}_\Omega : \mathbb{C}^{N\times N} \to \mathbb{C}^M$ as

$$(\mathscr{F}_\Omega X)_k = (\mathscr{F}X)_{\underline{\omega}_k},$$

i.e., Fourier transform operator restricted to Ω. $\mathscr{F}_{\Omega}^{*}$ will represent its conjugate adjoint. Equipped with the above notation, the main problem can be formally stated as follows:

Problem 1. Given a set Ω of $M \ll N^2$ frequencies and Fourier observations of a K-sparse in gradient image X given by $\mathscr{F}_{\Omega}X$, how can one estimate X accurately and efficiently?

The most popular method of solving this problem is to find the image of the least total variation that satisfies the given Fourier constraints. This corresponds to solving the following convex optimization problem:

$$\tilde{X} = \arg\min_{Y} \|Y\|_{TV} \text{ s.t. } \mathscr{F}_{\Omega}Y = \mathscr{F}_{\Omega}X. \tag{14}$$

Based on an extension of Theorem 1.5 in [4] and the result in [22] regarding Fourier measurements, one can prove the following proposition:

Proposition 2. *Let X be a real-valued K-sparse in gradient image. If $M = \mathscr{O}(K\log^4 N)$, then the solution $\tilde{X}$ of (14) is unique and equal to X with probability at least* $1 - \mathscr{O}(N^{-M})$.

In the case of an image corrupted by noise, the measurements take the form

$$b = \mathscr{F}_{\Omega}X + \eta,$$

where η is the measurement noise with $\|\eta\|_2 = \varepsilon$.

It was shown in [19] that instead of reconstructing an image by TV minimization, one can reconstruct the image by separately reconstructing the gradients and then solving for the image. This allows one to reconstruct the image with a far fewer number of measurements than required by the TV0-minimization method. Figure 3 presents an important comparison in the sparsity of X_x, X_y, and the TV measure. The plots of the sorted absolute values of the coefficients of the gradients X_x, X_y, and the TV measure for the Shepp-Logan phantom image (Fig. 2) indicate that X_x and X_y decay much faster than the TV measure. In fact, it is easy to see from the expression of TV [Eq. (12)] that the coefficients of X_x and X_y will always decay faster than the coefficients of TV. This means, one can take advantage of this and be able to reconstruct an image with far fewer measurements than that required by using the TV-based method.

4 Image Gradient Estimation

Given Fourier observations $\mathscr{F}_{\Omega}X$ over some set of frequencies Ω, one can obtain the Fourier observations of X_x and X_y over Ω via the equations

$$(\mathscr{F}_{\Omega}X_x)_k = (1 - \mathrm{e}^{-2\pi i\omega_{x,k}/N})(\mathscr{F}_{\Omega}X)_k, \tag{15}$$

$$(\mathscr{F}_{\Omega}X_y)_k = (1 - \mathrm{e}^{-2\pi i\omega_{y,k}/N})(\mathscr{F}_{\Omega}X)_k. \tag{16}$$

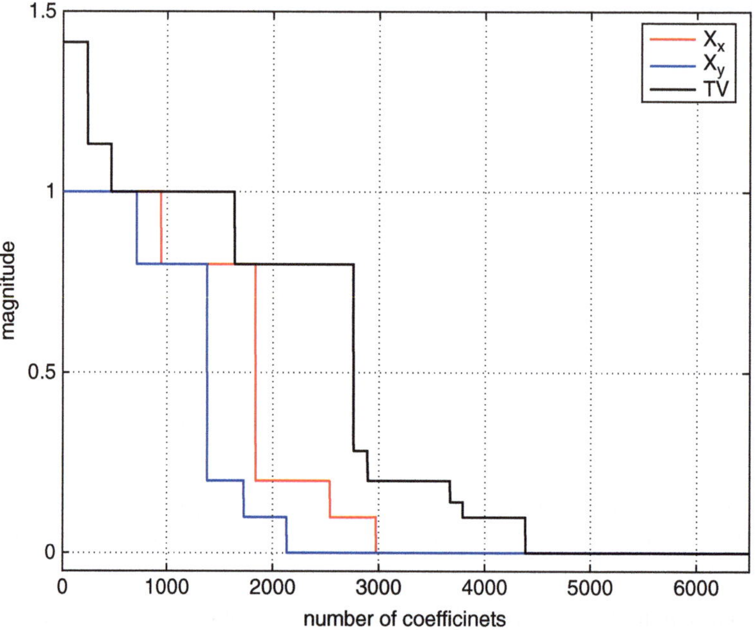

Fig. 3 The magnitude of *TV* (*black*) and X_x (*red*) and X_y (*blue*) coefficients in decreasing order for the Shepp-Logan phantom image (see Fig. 2)

After this is done, any one of many CS recovery algorithms can be used to recover X_x and X_y from their respective Fourier observations. For instance, taking into account the presence of additive noise during the measurement process, gradients can be estimated by solving the following two optimization problems:

$$\widetilde{X}_x = \arg\min_{X'_x} \|X'_x\|_1 \;\; \text{s. t.} \;\; \|\mathscr{F}_\Omega X'_x - b_x\|_2 \leq \varepsilon_x, \tag{17}$$

$$\widetilde{X}_y = \arg\min_{X'_y} \|X'_y\|_1 \;\; \text{s. t.} \;\; \|\mathscr{F}_\Omega X'_y - b_y\|_2 \leq \varepsilon_y, \tag{18}$$

where we have assumed that the measurements are of the following form:

$$b_x = \mathscr{F}_\Omega X_x + \eta_x,$$
$$b_y = \mathscr{F}_\Omega X_y + \eta_y$$

with $(\eta_x)_k = (1 - \mathrm{e}^{-2\pi i \omega_{x,k}/N})(\eta)_k$, $\varepsilon_x = \|\eta_x\|_2$, $(\eta_y)_k = (1 - \mathrm{e}^{-2\pi i \omega_{y,k}/N})(\eta)_k$, and $\varepsilon_y = \|\eta_y\|_2$.

Note that the recovery of sparse gradients from their respective Fourier measurements will depend on the RIP of the resulting sensing matrix. It is very difficult to prove any general claim that the resulting CS matrix satisfies the RIP for any particular restriction set Ω. However, empirical studies have shown that RIP holds for many practical measurement schemes arising in medical imaging [4,16,25].

5 Image Reconstruction from Gradients

After obtaining estimates $\widetilde{X}_x$ and $\widetilde{X}_y$ of X_x and X_y, respectively, some kind of integration must be performed to recover an estimate $\tilde{X}$ of X. To obtain X from $\widetilde{X}_x$ and $\widetilde{X}_y$, the following optimization problem was proposed in [19]:

$$\tilde{X} = \arg\min_{Y} \left\| Y_x - \widetilde{X}_x \right\|_2^2 + \left\| Y_y - \widetilde{X}_y \right\|_2^2 + \beta \left\| Y_x \right\|_2^2 + \beta \left\| Y_y \right\|_2^2 + \lambda \left\| \mathscr{F}_\Omega Y - b \right\|_2^2,$$

where β and λ are penalty parameters that determine the degrees to which the TV minimization and Fourier constraints are enforced.

Now observe that if hats are used to denote the Fourier transform operator, it is possible to use the Parseval's theorem to rewrite (19) as the following equivalent problem in the Fourier domain:

$$\begin{aligned}\hat{\tilde{X}} = \arg\min_{\hat{Y}} & \left\| \left(1 - \mathrm{e}^{-2\pi i \omega_1/N}\right) \hat{Y} - \hat{\widetilde{X}}_x \right\|_2^2 + \left\| \left(1 - \mathrm{e}^{-2\pi i \omega_2/N}\right) \hat{Y} - \hat{\widetilde{X}}_y \right\|_2^2 \\ & + \beta \left(\left\| \left(1 - \mathrm{e}^{-2\pi i \omega_1/N}\right) \hat{Y} \right\|_2^2 \right) + \beta \left(\left\| \left(1 - \mathrm{e}^{-2\pi i \omega_2/N}\right) \hat{Y} \right\|_2^2 \right) \\ & + \lambda \left\| \left(\hat{Y} - B\right) \mathbf{1}_\Omega \right\|_2^2 . \end{aligned} \tag{19}$$

Here $\mathbf{1}_\Omega$ denotes an indicator function which is 1 on Ω and 0 otherwise. Similarly, B is an $N \times N$ matrix that is equal to b on Ω and 0 otherwise. Based on this convenient alternative formulation of the problem, the following result was derived in [19]:

Proposition 3. *The least squares problem (19) can be solved element-wise by (20). Furthermore, if one lets $\lambda \to \infty$, then this solution will take the piecewise form (21).*

$$\hat{\tilde{X}}_{\omega_1,\omega_2} = \frac{\left(1 - \mathrm{e}^{2\pi i \omega_1/N}\right) (\hat{\widetilde{X}}_x)_{\omega_1,\omega_2} + \left(1 - \mathrm{e}^{2\pi i \omega_2/N}\right) (\hat{\widetilde{X}}_y)_{\omega_1,\omega_2} + \lambda B_{\omega_1,\omega_2} \mathbf{1}_\Omega}{(1+\beta)\left(\left|1 - \mathrm{e}^{-2\pi i \omega_1/N}\right|^2 + \left|1 - \mathrm{e}^{-2\pi i \omega_2/N}\right|^2 \right) + \lambda \mathbf{1}_\Omega}. \tag{20}$$

$$\hat{\tilde{X}}_{\omega_1,\omega_2} = \begin{cases} B_{\omega_1,\omega_2} & \text{if}(\omega_1,\omega_2) \in \Omega \\ \frac{\left(1 - \mathrm{e}^{2\pi i \omega_1/N}\right) (\hat{\widetilde{X}}_x)_{\omega_1,\omega_2} + \left(1 - \mathrm{e}^{2\pi i \omega_2/N}\right) (\hat{\widetilde{X}}_y)_{\omega_1,\omega_2}}{(1+\beta)\left(\left|1 - \mathrm{e}^{-2\pi i \omega_1/N}\right|^2 + \left|1 - \mathrm{e}^{-2\pi i \omega_2/N}\right|^2 \right)} & \text{otherwise} \end{cases} . \tag{21}$$

One can obtain $\tilde{X}$ by simply inverting the Fourier transform. Now observe that if $\lambda \to \infty$, $\beta = 0$, and the edge approximations are exact, i.e., $\widetilde{X_x} = X_x$ and $\widetilde{X_y} = X_y$, then it follows that $\tilde{X} = X$. In general, selecting $\beta > 0$ will only attenuate the magnitude of any Fourier coefficients outside the set Ω. If one lets $\beta \to \infty$ (with $\lambda = \infty$, then the solution becomes equivalent to that obtained by naive Fourier back-projection, i.e., selecting $\tilde{X} = \mathscr{F}_\Omega^* \mathscr{F}_\Omega X$. This produces poor results. As a result, it is prudent to simply leave $\beta = 0$.

With the choice of $\lambda = \infty$ and $\beta = 0$, it was shown that the solution to (21) satisfies the following reconstruction performance guarantee [19].

Proposition 4. *Given approximations $\widetilde{X_x}$ and $\widetilde{X_y}$ of X_x and X_y, then the solution $\tilde{X}$ of Eq. (21) will satisfy*

$$\|\tilde{X} - X\|_2 \leq O\left(\frac{N}{\sqrt{k_1^2 + k_2^2}}\right) \max\left(\left\|\widetilde{X_x} - X_x\right\|_2, \left\|\widetilde{X_y} - X_y\right\|_2\right),$$

where

$$(k_1, k_2) = \operatorname*{argmin}_{(\omega_1,\omega_2)\notin\Omega} \omega_1^2 + \omega_2^2.$$

Proof. *Observe that for each $(\omega_1, \omega_2) \in \Omega$,*

$$|\hat{\tilde{X}}_{\omega_1,\omega_2} - \hat{X}_{\omega_1,\omega_2}|^2 = 0$$

by definition. Outside of Ω,

$$\begin{aligned}
&|\hat{\tilde{X}}_{\omega_1,\omega_2} - \hat{X}_{\omega_1,\omega_2}|^2 \\
&= \left| \frac{(1 - e^{2\pi i\omega_1/N})\,(\widehat{\widetilde{X}_x})_{\omega_1,\omega_2} + (1 - e^{2\pi i\omega_2/N})\,(\widehat{\widetilde{X}_y})_{\omega_1,\omega_2}}{|1 - e^{-2\pi i\omega_1/N}|^2 + |1 - e^{-2\pi i\omega_2/N}|^2} \right. \\
&\quad \left. - \frac{(1 - e^{2\pi i\omega_1/N})\,(\widehat{X}_x)_{\omega_1,\omega_2} + (1 - e^{2\pi i\omega_2/N})\,(\widehat{X}_y)_{\omega_1,\omega_2}}{|1 - e^{-2\pi i\omega_1/N}|^2 + |1 - e^{-2\pi i\omega_2/N}|^2} \right|^2 \\
&= \left| \frac{(1 - e^{2\pi i\omega_1/N})\left((\widehat{\widetilde{X}_x})_{\omega_1,\omega_2} - (\widehat{X}_x)_{\omega_1,\omega_2}\right)}{|1 - e^{-2\pi i\omega_1/N}|^2 + |1 - e^{-2\pi i\omega_2/N}|^2} \right. \\
&\quad \left. + \frac{(1 - e^{2\pi i\omega_2/N})\left((\widehat{\widetilde{X}_y})_{\omega_1,\omega_2} - (\widehat{X}_y)_{\omega_1,\omega_2}\right)}{|1 - e^{-2\pi i\omega_1/N}|^2 + |1 - e^{-2\pi i\omega_2/N}|^2} \right|^2.
\end{aligned}$$

Now utilize the fact that for any $a, b \in \mathbb{C}$, $|a+b|^2 = |a|^2 + |b|^2 + 2\mathrm{Re}(ab) \leq 4$ $\max(|a|^2, |b|^2)$, and assume without loss of generality (so it isn't necessary to keep on writing out maximums) that

$$\frac{\left|1-\mathrm{e}^{2\pi i k_1/N}\right|}{\left|1-\mathrm{e}^{-2\pi i k_1/N}\right|^2+\left|1-\mathrm{e}^{-2\pi i k_2/N}\right|^2}\left|(\widehat{\widetilde{X}_x})_{\omega_1,\omega_2}-(\widehat{X_x})_{\omega_1,\omega_2}\right|$$
$$\geq \frac{\left|1-\mathrm{e}^{2\pi i k_2/N}\right|}{\left|1-\mathrm{e}^{-2\pi i k_1/N}\right|^2+\left|1-\mathrm{e}^{-2\pi i k_2/N}\right|^2}\left|(\widehat{\widetilde{X}_y})_{\omega_1,\omega_2}-(\widehat{X_y})_{\omega_1,\omega_2}\right|$$

to obtain that:

$$|\hat{\tilde{X}}_{\omega_1,\omega_2}-\hat{X}_{\omega_1,\omega_2}|^2$$
$$\leq \frac{4\left|1-\mathrm{e}^{-2\pi i k_1/N}\right|^2\left|(\widehat{\widetilde{X}_x})_{\omega_1,\omega_2}-(\widehat{X_x})_{\omega_1,\omega_2}\right|^2}{\left(\left|1-\mathrm{e}^{-2\pi i k_1/N}\right|^2+\left|1-\mathrm{e}^{-2\pi i k_2/N}\right|^2\right)^2}$$
$$\leq \frac{4\left|(\widehat{\widetilde{X}_x})_{\omega_1,\omega_2}-(\widehat{X_x})_{\omega_1,\omega_2}\right|^2}{\left|1-\mathrm{e}^{-2\pi i k_1/N}\right|^2+\left|1-\mathrm{e}^{-2\pi i k_2/N}\right|^2}.$$

Now simply use the Taylor expansion of the denominator and Parseval's theorem to get the desired result. □

As can be seen, the performance of this method depends on the selection of Ω. If Ω contains all the low frequencies within some radius $r = \sqrt{k_1^2 + k_2^2}$, then the final reconstruction error will be $O(N/r)$ times worse than the maximum edge reconstruction error. In general, if Ω contains mostly low frequencies, then this method will generate better results than if Ω contained mostly high frequencies. As a result, this "integration" is very appropriate in applications such as CT where Ω will consist of radial lines that congregate near the DC frequency. For the same reason, it may also be useful in MRI applications where the Fourier space is sampled according to a spiral trajectory (see [16, 25]). This method is referred to as GradientRec-LS in [19].

5.1 Numerical Examples

Figure 4 shows the reconstruction of a 512×512 Shepp-Logan phantom image using naive Fourier back-projection (i.e., selecting $\tilde{X} = \mathscr{F}_\Omega^* \mathscr{F}_\Omega X$), the L1TV method [16], and GradientRec-LS [19]. Only 5 % of its Fourier coefficients were used. The Fourier coefficients were restricted to a radial sampling pattern as

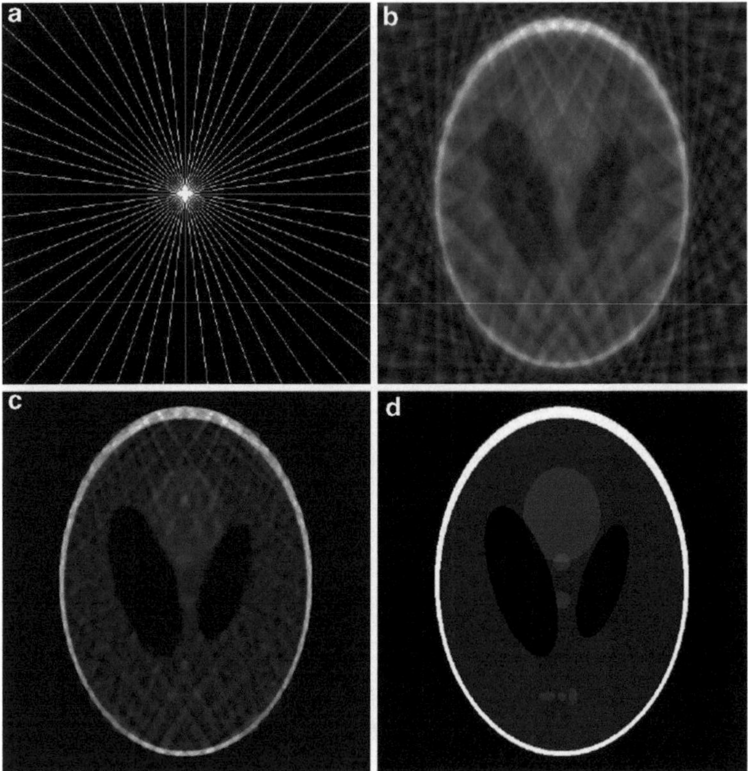

Fig. 4 512×512 Shepp-Logan phantom example. (**a**) Fourier domain sampling pattern. (**b**) Back-projection. (**c**) L1TV reconstruction. (**d**) GradientRec-LS reconstruction

shown in Fig. 4a. Figure 4b shows the result of classical Fourier back-projection which gives a relative error equal to 0.4953. The reconstruction from a TV-minimization method using the L1TV method [16] is shown in Fig. 4c, which gives a relative error equal to 0.2346. Figure 4d shows the GradientRec-LS reconstruction. The recovery is near perfect with relative errors obeying 1.59×10^{-5}.

The reason why TV minimization fails to recover the image perfectly is the following: As can be seen from Fig. 3, $\|X_x\|_0 = 2{,}972$, $\|X_y\|_0 = 2{,}132$, and $\|TV(X)\|_0 = 4{,}386$. It has been observed by many researchers that in practice, for a good reconstruction, the number of Fourier samples should be about three to five times the number of sparse coefficients [3, 16]. This means that GradientRec-LS can recover gradients perfectly from $13{,}107 = 0.05 \times 512 \times 512$ compressive measurements, which is approximately $6.15 \times \|X_y\|_0$ and $4.4 \times \|X_x\|_0$. Whereas for a good reconstruction, TV minimization requires about $4 \times 4{,}386 = 17{,}544$ measurements. Hence, 13,107 measurements are not enough for the TV minimization to recover the underlying sparse-gradient image.

Acknowledgments This work was partially supported by an ARO MURI grant W911NF0910383. Furthermore, the authors would like to thank Ray Maleh and Anna Gilbert for the discussion on image recovery from partial Fourier measurements.

References

1. Agrawal, A.K., Chellappa, R., Raskar, R.: An algebraic approach to surface reconstruction from gradient fields. In: ICCV, pp. 174–181 (2005)
2. Agrawal, A.K., Raskar, R., Chellappa, R.: What is the range of surface reconstructions from a gradient field? In: ECCV (1), pp. 578–591 (2006)
3. Candès, E., Romberg, J.: Signal recovery from random projections. In: Proceedings of SPIE computational imaging III **5674**, 76–86 (2005)
4. Candès, E., Romberg, J., Tao, T.: Robust uncertainty principles: exact signal reconstruction from highly incomplete frequency information. IEEE Trans. Inf. Theor. **52**(2), 489–509 (2006)
5. Fattal, R., Lischinski, D., Werman, M.: Gradient domain high dynamic range compression. ACM Trans. Graph. **21**, 249–256 (2002)
6. Frankot, R.T., Chellappa, R.: A method for enforcing integrability in shape from shading algorithms. IEEE Trans. Pattern Anal. Mach. Intell. **10**(4), 439–451 (1988)
7. Georghiades, A., Belhumeur, P., Kriegman, D.: From few to many: illumination cone models for face recognition under variable lighting and pose. IEEE Trans. Pattern Anal. Mach. Intell. **23**(6), 643–660 (2001)
8. Ghiglia, D.C., Pritt, M.D.: Two-Dimensional Phase Unwrapping: Theory, Algorithms, and Software. Wiley, New York (1998)
9. Horn, B.: Robot Vision. McGraw-Hill, New York (1986)
10. Horn, B.K.P.: Height and gradient from shading. Int. J. Comput. Vision **5**, 37–75 (1990)
11. Hsieh, J.W., Liao, H.Y.M., Ko, M.T., Fan, K.C.: Wavelet-based shape from shading. Graph. Models Image Process. **57**(4), 343–362 (1995)
12. Ikeuchi, K., Horn, B.K.P.: Shape from shading. chap. Numerical shape from shading and occluding boundaries, pp. 245–299 (1989)
13. Karaçali, B., Snyder, W.: Noise reduction in surface reconstruction from a given gradient field. Int. J. Comput. Vision **60**, 25–44 (2004)
14. Klette, R., Schluns, K., Koschan, A.: Computer Vision: Three-Dimensional Data from Images. Springer, Germany (1998)
15. Kovesi, P.: Shapelets correlated with surface normals produce surfaces. In: ICCV, vol. 2, pp. 994–1001 (2005)
16. Lustig, M., Donoho, D., Pauly, J.M.: Sparse mri: The application of compressed sensing for rapid mr imaging. Magn. Reson. Med. **58**(6), 1182–1195 (2007)
17. Ng, H.S., Wu, T.P., Tang, C.K.: Surface-from-gradients without discrete integrability enforcement: A Gaussian kernel approach. IEEE Trans. Pattern Anal. Mach. Intell. **32**(11), 2085–2099 (2010)
18. Patel, V.M., Easley, G.R., D. M. Healy, J., Chellappa, R.: Compressed synthetic aperture radar. IEEE J. Selected Topics Signal Process. **4**(2), 244–254 (2010)
19. Patel, V., Maleh, R., Gilbert, A., Chellappa, R.: Gradient-based image recovery methods from incomplete Fourier measurements. IEEE Trans. Image Process. **21**(1), 94–105 (2011)
20. Pérez, P., Gangnet, M., Blake, A.: Poisson image editing. In: ACM SIGGRAPH 2003 Papers, SIGGRAPH '03, pp. 313–318 (2003)
21. Reddy, D., Agrawal, A.K., Chellappa, R.: Enforcing integrability by error correction using l1-minimization. In: CVPR, pp. 2350–2357 (2009)
22. Rudelson, M., Vershynin, R.: On sparse reconstruction from Fourier and Gaussian measurements. Comm. Pure Appl. Math. **61**, 1025–1045 (2008)

23. Simchony, T., Chellappa, R., Shao, M.: Direct analytical methods for solving Poisson equations in computer vision problems. IEEE Trans. Pattern Anal. Mach. Intell. **12**(5), 435–446 (1990)
24. Sun, J., Jia, J., Tang, C.K., Shum, H.Y.: Poisson matting. In: ACM SIGGRAPH 2004 Papers, SIGGRAPH '04, pp. 315–321 (2004)
25. Trzasko, J., Manduca, A.: Highly undersampled magnetic resonance image reconstruction via homotopic ell_0 -minimization. IEEE Trans. Medical Imaging **28**(1), 106–121 (2009)
26. Tumblin, J., Agrawal, A.K., Raskar, R.: Why i want a gradient camera. In: CVPR, pp. 103–110 (2005)
27. Wei, T., Klette, R.: A wavelet-based algorithm for height from gradients. Lecture Notes in Computer Science **1998**, 84–90 (2001)
28. Woodham, R.J.: Shape from shading. chap. Photometric method for determining surface orientation from multiple images, pp. 513–531. MIT Press, Cambridge (1989)
29. Zhang, R., Tsai, P.S., Cryer, J., Shah, M.: Shape-from-shading: a survey. IEEE Trans. Pattern Anal. Mach. Intell. **21**(8), 690–706 (1999)

FM Perturbations due to Near-Identity Linear Systems

Anthony Teolis

Abstract Considered here is the disturbance induced in the frequency modulation (FM) of a signal passed through a linear system. For the case that the linear system is close to an identity system, closed-form approximations to the induced FM perturbations are derived. Taking the form of a linear operator acting on a system impulse response, the operator represents a well-behaved approximation to the true ill-behaved underlying non-linear operator. In its approximate linear form, the operator is amenable to standard linear theory. In addition, practical constraints including passband smoothness, zero group delay, and time localization are facilitated in the analysis. In general, the theory is suitable to determine practical and useful performance bounds in terms of system parameters and induced FM deviation. As an example, the FM perturbation analysis is applied in this chapter to determine systems that would produce maximal FM distortions in particular signals.

Keywords Frequency modulation • Perturbation • Near-identity linear system • Singular value decomposition • Passband smoothness • Zero group delay • Induced FM deviation • Maximal FM distortion • Non-linear operator • RF systems

1 Introduction

Radio-frequency (RF) system designers strive to build devices that exhibit ideally flat transfer function responses over all operating conditions of interest. For a given operating condition, e.g., center frequency and power input or output, the system

A. Teolis (✉)
Naval Research Laboratory, Washington, DC, USA

Teolis Consulting, Glenn Dale, MD, USA
e-mail: teolis@nrl.navy.mil; tony@teolis.org

T.D. Andrews et al. (eds.), *Excursions in Harmonic Analysis, Volume 1*, Applied and Numerical Harmonic Analysis, DOI 10.1007/978-0-8176-8376-4_20,

Fig. 1 A linear system H with input p and output r

may be modeled[1] as a linear system H. Typically the front-end analog components are presented to a digitizer so that subsequent processing is conducted in the digital domain. The system designer would like the analog signal to arrive at the digitizer in a form uncorrupted by the analog components. This translates into the requirement that H be ideally flat over the bandwidths of operation. Since the goal of an ideally flat response is a practically unachievable one, H, will always exhibit some systematic deviation from the desired ideal system.

1.1 Arbitrary Linear Systems

One important aspect of a signal to be processed in an RF system is its frequency modulation (FM). Many RF systems rely either explicitly or implicitly on the FM characteristics of the signals with which they operate. Figure 1 shows the very basic linear system block diagram of the composite system H.

Passage of a signal p through the composite system H will result in the output signal r with a necessarily different FM from that of the input p. The relevant question to be addressed is

> What is the frequency modulation of a signal after it is passed through an arbitrary linear system?

A closed-form solution to this problem for arbitrary linear systems is complicated by the non-linear and unbounded nature of the operators involved. The distortion of frequency-modulated and phase-modulated waveforms due to linear systems has been previously addressed in early work including [1–4]. These analyses provide solutions in terms of asymptotic expansions. For special cases of linear system disturbances the issue has been examined in [5–8]. In general, the results take a highly nonlinear form and do not directly lend themselves to tractable analyses for RF signal and system design. Fortunately, constraining the perturbing system to be *well-behaved* in the sense that it is not too far from an identity (non-perturbing) system provides a tractable way forward.

[1] RF systems commonly used in radar and communications process signals through various analog components to perform their function. Depending on operating conditions, the overall concatenation of these components may be modeled as a composite linear system, H.

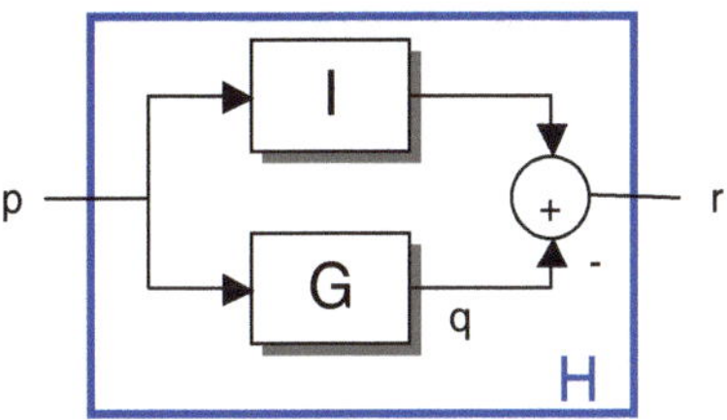

Fig. 2 A near-identity system model

1.2 *Near-Identity Linear Systems*

If the perturbing system in question is close to an identity system (as is the case when the intent of the designer is to have no effect on the FM), a tractable solution is possible. A near-identity system model is depicted in Fig. 2. In this case the question may be recast as

> How is the frequency modulation of a signal perturbed after passage through a linear system that is close to an identity system?

This is the main question that is addressed in this chapter.

Accordingly, the analysis conducted here limits consideration to linear systems that are near-identity systems, i.e., those that are small deviations from ideal (flat) systems. It is demonstrated in this chapter (see Sect. 6.3) that even systems that deviate only slightly (e.g., less than 1 dB amplitude variation) from ideal can induce relatively large perturbations in the frequency modulation of a pulsed signal that is passed through them.

Concisely, the analysis presented here is based on the linear approximation

$$(\partial \ln(I-G) \quad - \quad \partial \ln)p \quad \approx \quad D_p g,$$

where ∂ is a derivative operator [9], p is a signal, G is a linear system with impulse response g such that $\|G\| \ll 1$, and D_p is a linear operator. It will be seen that the induced FM deviation due to a systematic disturbance G is directly related to the imaginary part of the above approximation.

2 Preliminaries

This section provides the notation and mathematical constructs that are used in the rest of the chapter.

2.1 Basic Symbols and Spaces

The symbols $\mathbb{Z}$, $\mathbb{R}$, and $\mathbb{C}$ denote the integers, real numbers, and complex numbers, respectively. $L^p(\mathbf{R})$ is the space of complex-valued (finite energy for $p=2$) signals defined on the real line $\mathbf{R}$.

2.2 Fourier Transform

The (continuous-time) *Fourier transform* is a mapping $L^1(\mathbf{R}) \mapsto L^2(\mathbf{R})$ and is well defined as

$$\hat{f}(\gamma) = \int f(t)\mathrm{e}^{-j2\pi\gamma t}\,\mathrm{d}t$$

since the integral converges absolutely and $\gamma \in \mathbf{R}$. The "$\hat{}$" notation is used to indicate the forward Fourier transform and " $^\vee$ " to indicate the inverse Fourier transform.

2.3 Analytic Signals

An analytic[2] signal $p(t)$ has a representation in terms of its magnitude and phase

$$p(t) \;=\; A(t)\cdot \mathrm{e}^{j\phi(t)}.$$

The frequency modulation (FM) associated with $p(t)$ is defined as the time derivative of the phase. Thus, p also admits the representation

$$p(t) = A(t)\cdot \mathrm{e}^{j2\pi\left(f_0+\int_{t_0}^{t} f(s)\,\mathrm{d}s\right)}, \qquad t \geq t_0, \tag{1}$$

where f is the signal FM and f_0 is its starting frequency and t_0 is an arbitrary start time. Clearly, the phase and frequency are related as

$$\phi(t) \;=\; 2\pi\left(f_0 + \int_{t_0}^{t} f(s)\,\mathrm{d}s\right) \quad \text{and} \quad f(t) = \frac{1}{2\pi}\cdot\dot{\phi}(t).$$

Thus, an analytic signal p may be specified as a triple (A, f, f_0). Due to the fact that pure phase shifts are often inconsequential (in single-pulse analyses) the value of the constant f_0 may be set to zero in order to simplify the discussion. In the sequel it is assumed that $f_0 = 0$. With this understanding, an analytic signal p is specified

---

[2] Here an analytic signal is one whose Fourier transform vanishes for all negative frequencies.

by the AM/FM pair (A, f), i.e.,

$$p \equiv (A, f). \tag{2}$$

2.4 Frequency Modulation

Given an analytic signal $p \equiv (A, f)$ a useful expression for the FM is

$$f(t) = \frac{1}{2\pi} \cdot \operatorname{Im}\left\{\frac{\mathrm{d}}{\mathrm{d}t} \ln p(t)\right\}. \tag{3}$$

This expression may be validated by direct computation using (1). In terms of an operator theoretic view of the modulation process, this expression exposes the FM operator as

$$\mathscr{F} \triangleq \frac{1}{2\pi} \cdot \operatorname{Im} \partial \ln .$$

With this definition the FM of the signal p may be written as $f = \mathscr{F} p$. In general, $\mathscr{F}$ is not a well-behaved operator. It is unbounded and non-linear (it is not even homogeneous on the complex numbers).

2.5 Soft Inverse

The soft inverse is defined with respect to a threshold value $\delta > 0$

$$p^{\dagger}(t) \triangleq \begin{cases} p^{-1}(t), & |p(t)| > \delta, \\ (3\delta^{-2}|p(t)| - 2\delta^{-3}|p(t)|^2)\mathrm{e}^{-j\phi_p(t)}, & |p(t)| \leq \delta, \end{cases}$$

where ϕ_p is the phase of the signal p. The intent of the soft inverse is to provide a bounded and smooth approximation to the inverse of signals with finite support. The approximation is exact when the magnitude of the argument is greater than the threshold δ. For magnitudes that are smaller than the δ the soft inverse follows a parabola $f(x) = c_2 x^2 + c_1 x$ where c_1 and c_2 are chosen such that $f(\delta) = \delta^{-1}$ and $f'(\delta) = -\delta^{-2}$ (to match the derivative of $1/x$ at δ).

It has the important properties that

$$p(t) \cdot p^{\dagger}(t) = 1, \quad \forall t \ni |p(t)| \geq \delta,$$

and

$$|p(t) \cdot p^{\dagger}(t)| < 1, \quad \forall t \ni |p(t)| < \delta.$$

2.6 Window Functions

A window function, w_T, for the set T is one that has the following properties:

1. $0 \leq w_T(t) \leq 1, \ t \in T$
2. $w_T(t) = 0, \ t \in T^c$
3. $w'_T(t) < B, \ \forall t,$

where w' denotes the derivative of w, $B < \infty$ is a scalar constant, and T^c denotes the complement of the set T.

3 FM Perturbation Theory

3.1 Problem Setting

Consider the near-identity linear system H depicted in Fig. 2 with analytic input p and output r. The input and output both admit the AM/FM representations [c.f., (2)]

$$p \equiv (A_p, f_p) \quad \text{and} \quad r \equiv (A_r, f_r).$$

We are interested in determining the change in frequency modulation

$$\Delta f = \mathscr{F} r - \mathscr{F} p = f_r - f_p$$

from input to output due to the action of the linear system H. The quantity Δf is the *FM perturbation* due to H.

In general, H may be arbitrary as shown in Fig. 1, though it is constrained here to come from the class of "near-identity" systems. That is to say that H is supposed to differ from an identity system I by only a small systematic deviation G so that

$$r = Hp = (I - G)p.$$

Here r is the perturbed version of p and $q = Gp$ is the deviation. Figure 2 shows the deviation-based block diagram of the system H. Formally, a system H is defined to be *near-identity* if

$$H = I - G, \quad \|G\| \ll 1,$$

where $\|G\|$ is the operator norm of G

$$\|G\| \triangleq \sup_p \frac{\|Gp\|}{\|p\|}.$$

Clearly, the smaller the $\|G\|$ the closer H is to the identity system. It can be shown [10] that for a linear system G with impulse response g, $\|G\| = \|g\|$, i.e., the operator norm of the system G is the same as the L^2 norm of the impulse response g.

Because H and G are linear operators (stemming from linear systems) they may be specified by a convolution in the time domain between the input signal and the impulse responses, h and g of the filters, respectively. Explicitly,

$$(Hp)(t) = (p*h)(t) = \int p(t-s)\cdot h(s)\,\mathrm{d}s,$$
$$(Gp)(t) = (p*g)(t) = \int p(t-s)\cdot g(s)\,\mathrm{d}s,$$

where "*" denotes the convolution operator.

3.2 Main Result

It shall be seen that under the near-identity system assumption, the FM perturbation Δf is given by an input-signal-dependent integral linear operator $\mathscr{D}_p$ acting on the impulse response g of the deviation from ideal system. In symbols,

$$\Delta f \approx \frac{1}{2\pi}\cdot \mathrm{Im}\left\{\mathscr{D}_p g\right\} \quad \text{on supp } p, \tag{4}$$

where $\mathscr{D}_p$ is the linear integral operator

$$(\mathscr{D}_p g)(t) \triangleq \int D_p(s,t)\cdot g(s)\,\mathrm{d}s$$

having the kernel

$$D_p(s,t) = \frac{\partial}{\partial t}\frac{p(t-s)}{p(t)}.$$

It is worth stressing that this approximation is an instantaneous one as it provides an estimate of the FM perturbation at each time value t. Clearly, (4) is not well defined when $|p(t)|$ is very small. For this reason the formula is restricted to values of t for which the pulse p does not vanish, i.e., $t \in \text{supp } p$.

The approximation of (4) is remarkable because it provides a kind of "factorization" of the FM perturbation (dependent on an ill-behaved operator $\mathscr{F}$) in terms of a linear (well-behaved) operator. One factor (the linear operator) is dependent only on the signal input to the system, and the other factor (the system deviation impulse response) is dependent only on the system itself. In the sequel, $\mathscr{D}_p$ is called the perturbation operator, and its composition with the imaginary projection, Im $\mathscr{D}_p$, is called the FM perturbation operator.

3.3 Proof of Result

This section presents a proof of Eq. (4). Under the assumption of a small systematic perturbation $\|G\| \ll 1$, it shall be shown that

$$\Delta f(t) = \frac{1}{2\pi} \cdot \mathrm{Im}\left\{ \int D_p(s,t) \cdot g(s) \, \mathrm{d}s \right\} \; + \; O\left(\left| \frac{Gp(t)}{p(t)} \right|^2 \right), \tag{5}$$

where $(\mathscr{D}_p g)(t) \triangleq \int D_p(s,t) \cdot g(s) \, \mathrm{d}s$ is the integral operator with kernel $D_p(s,t)$. Two signal-dependent kernels, $R_p(s,t)$ and $D_p(s,t)$, are of interest here. The first is the instantaneous power-normalized autocorrelation function[3]

$$R_p(s,t) \triangleq \frac{p(t-s)}{p(t)},$$

and the second is its derivative with respect to time, i.e.,

$$D_p(s,t) \triangleq \frac{\partial}{\partial t} R_p(s,t).$$

It should be noted that the quantity $|Gp(t)/p(t)|$ determines the instantaneous quality of the approximation (smaller is better). The condition that $\|G\| \ll 1$ does not explicitly bound the instantaneous approximation error, but it does imply that it cannot be too large over significant periods of time in the support of p.

Proof. Using (3) the FM perturbation may be written as

$$\begin{aligned} 2\pi \cdot \Delta f(t) &= \mathrm{Im}\left\{ \frac{\mathrm{d}}{\mathrm{d}t} \ln r(t) \right\} - \mathrm{Im}\left\{ \frac{\mathrm{d}}{\mathrm{d}t} \ln p(t) \right\} \\ &= \mathrm{Im}\left\{ \frac{\mathrm{d}}{\mathrm{d}t} \ln r(t) - \frac{\mathrm{d}}{\mathrm{d}t} \ln p(t) \right\} \\ &= \mathrm{Im}\left\{ \frac{\mathrm{d}}{\mathrm{d}t} (\ln r(t) - \ln p(t)) \right\} \\ &= \mathrm{Im}\left\{ \frac{\mathrm{d}}{\mathrm{d}t} \ln \left(\frac{r(t)}{p(t)} \right) \right\}. \end{aligned}$$

A Taylor expansion of the logarithm yields

$$\ln\left(\frac{r}{p}\right) = \ln\left(1 - \frac{Gp}{p}\right) = \sum_{k=1}^{\infty} \frac{1}{k} \cdot \left(\frac{Gp}{p}\right)^k = \frac{Gp}{p} + O\left(\left|\frac{Gp}{p}\right|^2\right).$$

[3] It is called the *power-normalized autocorrelation* because $R_p(s,t) = p(t-s) \cdot \overline{p}(t)/|p(t)|^2$.

Since

$$\frac{(Gp)(t)}{p(t)} = \frac{1}{p(t)} \int g(s)p(t-s)\,\mathrm{d}s = \int R_p(t,s) \cdot g(s)\,\mathrm{d}s,$$

this gives, in turn, that

$$\begin{aligned} \frac{\mathrm{d}}{\mathrm{d}t} \ln\left(\frac{r(t)}{p(t)}\right) &\approx \frac{\mathrm{d}}{\mathrm{d}t} \int R_p(t,s) \cdot g(s)\,\mathrm{d}s \\ &= \int D_p(t,s) \cdot g(s)\,\mathrm{d}s. \end{aligned}$$

□

3.4 Utility of the FM Perturbation Operator

Equation (4) provides the means to ask and answer a number of important questions regarding the design and operation of systems that use frequency modulation in a fundamental way. It shows that the FM perturbation through a near-identity linear system H is linear with respect to the systematic deviation. For a given signal p, the analysis is supported by the linear theory. For example, some pertinent questions in RF system design and operation are the following:

1. *(Worst systematic deviation):* What near-identity linear system H will produce the greatest FM perturbation for a given signal p?
2. *(Worst signal):* For a given near-identity system, H, what signal p will incur the greatest FM perturbation when passed through H?
3. *(Bounded perturbation):* What constraints should be placed on the systematic deviation G so that the resulting FM perturbation incurred by any signal is bounded by some given value?

As posed, these questions are qualitative in nature. In this chapter, the first question is explicitly considered. A more precise version of question 1 is stated formally as Problem 1.

Problem 1 (Worst systematic deviation). Under the constraint that $\|G\| < a < 1$, find a linear system $H = I - G$ such that the FM perturbation norm $\|\Delta f\|$ is maximized.

For small enough values of a, the solution to this problem is directly facilitated via the singular value decomposition (SVD) of the perturbation operator $\mathscr{D}_p$. The next section discusses the SVD and the associated solution of Problem 1.

4 Singular Value Decomposition

Recall that the SVD [11, 12] factors a bounded linear operator T into three terms as $T = USV^*$ where U and V are unitary[4] operators and S is a diagonal matrix with real entries. The real values $\{s_k\}$ from the diagonal of S are the *singular* values. As unitary operators U and V define corresponding orthonormal bases $\{u_k\}$ and $\{v_k\}$ for $L^2(\mathbb{R})$ such that

$$Tv_k = s_k u_k, \quad k \in \mathbb{Z}.$$

In words, T operating on v_k yields a scaled version of u_k. In terms of energy, $\|Tv_k\|^2 = \|s_k u_k\|^2 = s_k^2$, i.e., the energy in the input is scaled by s_k^2. If s_{k^*} is the largest singular value,[5] then passing v_{k^*} through T yields the response with the largest energy of all possible inputs with unit energy.

As applied to the perturbation operator $\mathscr{D}_p$ the SVD would result in an orthonormal set of systematic deviations $\{v_k\}$ and an orthonormal set of corresponding perturbations $\{u_k\}$. The singular value s_k has the interpretation as a multiplier to the perturbation. That is to say that if the systematic deviation is taken to be av_k then the perturbation induced on p by passing it through the system $H = 1 + a\hat{v}_k$ is $as_k u_k$. If k^* denotes the index of the largest singular value then $g = v_{k^*}$ is the unit norm systematic deviation that results in the largest perturbation.

The remainder of this section details the use of the SVD to address Problem 1. Equation (7) represents the key result. It describes the FM perturbation in terms of the singular values associated with the real and imaginary projected versions of the perturbation operator.

First, Sect. 4.1 considers the SVD of the perturbation operator $\mathscr{D}_p$ as a full complex operator. This establishes the technique with a minimum of extraneous manipulations but yields a result that is not directly applicable to the problem outlined above. Second, a singular value analysis is carried out on the imaginary projection of the perturbation operator in Sect. 4.2. This approach yields results that directly solve the problem.

Section 5 goes on to discuss additional constraints on the solution to the problem so as to eliminate cases where the solution has little relevance to practical systems. Unconstrained solutions may result in perturbation systems that are not characteristic of those encountered in real systems. Maximally perturbing systems as determined by the unconstrained solution will likely have unrealistic group delays. Additionally, the energy in the resulting perturbation may be concentrated in time intervals that are not of practical importance to RF system performance, e.g., in the leading and trailing edges of a pulse. Thus, the incorporation of constraint mechanisms is very important in order to yield results that are of practical significance.

---

4 Recall a unitary operator, U, has the properties that it is self-adjoint and its own inverse, i.e., $U^*U = I$.

5 It should be mentioned that the largest singular value s^* of an operator T is the same as its operator norm, i.e., $s^* = \sup_x \|Tx\|/\|x\|$.

4.1 Full Complex Perturbation

For convenience, we drop the explicit dependence on the signal p of the perturbation operator and let $D = \mathscr{D}_p$ be its matrix representation.[6] Let the SVD of D be

$$D = USV^*.$$

The SVD of the perturbation operator D is illuminating because the V matrix contains an orthonormal set of systems $V = \{v_k\}$ and (the imaginary portion of) the U matrix contains the associated set of FM perturbations $U = \{u_k\}$. This is so since $DV = US$ and, in particular,

$$Dv_k = s_k u_k, \quad \forall k \in \mathbb{Z}.$$

Relating this to the FM perturbation, the systems v_k are the impulse responses associated with the systematic deviation G from the ideal system. Since $\|v_k\| = 1$ and the near-identity assumption requires $\|G\| \ll 1$, linear systems H appropriate for consideration have transfer functions of the form

$$\hat{h} = 1 - a \cdot \hat{v}_k,$$

where $a \in \mathbf{R}$ is a positive scalar free parameter and $\hat{h}$ and $\hat{v}_k$ are the Fourier transform of the impulse response h and v_k, respectively. Hence, we identify $\hat{g} = a\hat{v}_k$ where $a > 0$ is chosen such that $\|G\| = \|\hat{g}\| = a \cdot \|\hat{v}_k\| = a \ll 1$. A typical value is $a = 0.2$.

Thus, if the signal p is passed through the linear system H determined by v_k then the FM of the output will deviate from that of the input (the FM perturbation) as

$$\Delta f \approx \frac{1}{2\pi} \cdot a \cdot s_k \cdot \operatorname{Im}\{u_k\}.$$

From this it is seen that candidate systems for creating large FM perturbations are those with the largest singular values. Unfortunately, the analysis guarantees only u_k has unit energy and provides no constraint on how much of that energy resides in the imaginary portion. No firm conclusion can be made about the magnitude of the resulting FM perturbation due to this situation. For this reason, the imaginary portion of the operator is explicitly considered in the next section.

---

[6] Bounded linear operators admit matrix representations. Technically, the perturbation operator $\mathscr{D}_p$ need not be bounded due to the fact that p is pulsed (decays to zero off a finite interval). Practically, we constrain $\mathscr{D}_p$ to be bounded by defining it as $D_p(s,t) = \frac{\partial}{\partial t} p^\dagger(t) \cdot p(t-s)$ where $p^\dagger(t)$ is the soft inverse of $p(t)$, viz., 2.5.

4.2 Imaginary Projection

Explicitly including the imaginary projection into the singular value analysis results in dual sets of singular values stemming from the real and imaginary portions of the perturbation operator. Since the FM perturbation is determined from the imaginary portion of D only, it is convenient to write out all operators in their explicit complex form, e.g., $D = D_r + jD_i$ where the elements of $D_r \triangleq \mathrm{Re}\{D\}$ and $D_i \triangleq \mathrm{Im}\{D\}$ are all real values. Similarly, $g = g_r + jg_i$. Using this notation, we have

$$\mathrm{Im}\{Dg\} \;=\; D_i g_r + D_r g_i \;=\; U_i S_i V_i^* g_r \;+\; U_r S_r V_r^* g_i,$$

where $D_i = U_i S_i V_i^*$ and $D_r = U_r S_r V_r^*$ are the SVDs of D_i and D_r with singular values $\{(s_i)_m\}$ and $\{(s_r)_n\}$, respectively. Let a_i and a_r be such that $a_i^2 + a_r^2 = a^2 < 1$. If one picks the systematic deviation to be

$$g \;=\; a_i \cdot (v_i)_m \;+\; j \cdot a_r \cdot (v_r)_n \tag{6}$$

for some m and $n \in \mathbb{Z}$ then $\|g\| = a$, then the induced FM perturbation is

$$\Delta f \approx \frac{1}{2\pi} \cdot \left(a_i \cdot (s_i)_m (u_i)_m \;+\; a_r \cdot (s_r)_n (u_r)_n\right). \tag{7}$$

Equation (6) will prove useful for controlling the symmetry properties of the transfer function associated with the systematic deviation. This and other constraints are the topic of the Sect. 5.

4.3 Unconstrained Solution

As touched on in the previous sections, the fact that it is the imaginary part of the response to the perturbation operator that yields the FM deviation is a slight complication in the application of the SVD analysis. This complication may be addressed by considering a composite operator $\mathbb{D}$ made up of the real and imaginary portions of the full complex perturbation operator, i.e.,

$$\mathbb{D} \triangleq [D_i \;\, D_r].$$

More precisely, the solution to Problem 1 is given by the SVD of the composite operator $\mathbb{D}$ since

$$\mathrm{Im}\{Dg\} \;=\; D_i g_r + D_r g_i \;=\; [D_i \;\, D_r] \begin{bmatrix} g_r \\ g_i \end{bmatrix} \;=\; \mathbb{D}g,$$

where g is the corresponding composite vector $g \triangleq [g_r \;\; g_i]'$. If the SVD of $\mathbb{D}$ is

$$\mathbb{D} = USV^*,$$

then the maximal FM perturbing systemic deviation is determined by v_1. Thus, for a given value $a < 1$, the near-identity linear system $\hat{h} = 1 - a \cdot \hat{v}_1$ is the one that most disturbs the FM of the signal p and that maximal FM perturbation is

$$\Delta f \approx \frac{1}{2\pi} \cdot s_1 \cdot Lu_1, \tag{8}$$

where L is the *half shift and add* operator that folds a length $2N$ vector $x \in \mathbb{R}^{2N}$ into a length N vector $y \in \mathbb{R}^N$ by $y_n = x_n + x_{(n+N)}$ for $n = 1, 2, \ldots, N$. It has the matrix representation $[I \;\; I]$.

5 Constraints on the Systematic Deviation

Because real RF systems have characteristics that are somewhat well behaved, it is likely that the SVDs as presented above will yield maximal perturbing systems that have unrealistic properties. Since such systems are not of a practical concern the extremal results obtained using them are likely of little practical concern as well.

A more practical analysis may be gained by constraining the candidate systems to have properties more aligned to those found in real systems. Properties considered here are amplitude deviation, group delay, symmetry, amplitude smoothness, and time localization. Such constraints are also useful in determining general properties of systematic deviations that may or may not contribute to FM perturbations.

5.1 Amplitude Deviation

Given a systematic deviation g, the amplitude peak-to-peak deviation of the associated near-identity system may be controlled by simple scaling. This action is facilitated by the selection of the parameter a in the near-identity system model

$$\hat{h} = 1 - a \cdot \hat{g}.$$

Note that small values of $a > 0$ are needed to satisfy the near-identity assumption.

5.2 Symmetry

The explicit complex decomposition of systematic deviation g given in (6) provides a mechanism to control the symmetry of its transfer function $\hat{g}$ (and, hence, the symmetry of $\hat{h}$). For example, if g is chosen to be purely real ($a_i = 0$), then $\hat{g}$ is necessarily an even function (symmetric around 0). Similarly, choosing g to be purely imaginary ($a_r = 0$) results in odd symmetry for $\hat{g}$.

5.3 Zero Group Delay

Another important constraint is one placed on the allowable group delay of the perturbing system. It is not surprising that systems that possess large and greatly varying group delays would adversely Effect the FM of signals passed through them. Since a near-identity system as defined here has no explicit constraint on the behavior of its group delay, it is likely that unconstrained SVDs will yield maximal perturbing systems that have unrealistic group delays. The extremal constraint in this regard is to allow no variation in the group delay.

To constrain a system to have zero group delay it is sufficient to require that the impulse response function g be involutive,[7] i.e., that $g = \tilde{g}$ where $\tilde{g}$ is the *involution* of g given as

$$\tilde{g}(t) \stackrel{\triangle}{=} \overline{g}(-t).$$

Such a function g will be denoted g_0 and may be defined in terms of a second function h as

$$g_0 \stackrel{\triangle}{=} h + \tilde{h}.$$

The action of the perturbation on the involutive function g_0 is

$$\begin{aligned}
\mathscr{D}_p g_0 &= \mathscr{D}_p(h + \tilde{h}) \\
&= \mathscr{D}_p h + \mathscr{D}_p \tilde{h} \\
&= \mathscr{D}_p h + \int \tilde{h}(s) D_p(s,t)\, \mathrm{d}s \\
&= \mathscr{D}_p h + \int \overline{h}(-s) D_p(s,t)\, \mathrm{d}s \\
&= \mathscr{D}_p h + \int \overline{h}(u) D_p(-u,t)\, \mathrm{d}u \\
&= \mathscr{D}_p h + \overline{(\tilde{\mathscr{D}}_p h)},
\end{aligned}$$

where the new operator $\tilde{\mathscr{D}}_p$ has the kernel

$$\tilde{D}_p(s,t) \stackrel{\triangle}{=} \overline{D}_p(-s,t).$$

In other words, the kernel of $\tilde{\mathscr{D}}_p$ is the involuted (in s) version of the kernel $D_p(s,t)$.

[7] An involutive function has a Fourier transform that is purely real.

Applying the projection onto the imaginary portion yields

$$\begin{aligned}\operatorname{Im}\{\mathscr{D}_p g_0\} &= \operatorname{Im}\{\mathscr{D}_p h\} - \operatorname{Im}\{\tilde{\mathscr{D}}_p h\} \\ &= \operatorname{Im}\{(\mathscr{D}_p - \tilde{\mathscr{D}}_p)\, h\}.\end{aligned}$$

Thus, the analysis of Sect. 4.2 follows identically with the following modifications:

1. The SVDs are conducted on the operator $\mathscr{D}_p - \tilde{\mathscr{D}}_p$ to yield (U_r, S_r, H_r) and (U_i, S_i, H_i).
2. The systemic perturbations $\{(v_r)_k\}$ and $\{(v_i)_k\}$ are taken to be

$$(v_r)_k = \frac{1}{2}\left(h_r + \tilde{h}_r\right), \quad \text{and} \quad (v_i)_k = \frac{1}{2}\left(h_i + \tilde{h}_i\right).$$

Accordingly, Eqs. (7) and (8) also hold under these modifications.

5.4 *Time Localization*

It has been seen that the SVD analysis of Sect. 4.2 will result in two sets of orthonormal FM perturbation bases. Though any one of these basis elements will have unit energy, how that energy is distributed in time is constrained only by the support of the signal p under analysis. It is desirable in many instances to focus the analysis on particular regions in time in a given pulse.

Time localization is facilitated directly by modifying the perturbation kernel $D(s,t)$ to have zero value outside time values of interest. Letting T denote the set of times of interest, the kernel modification would be

$$D_T(s,t) = D(s,t) \cdot w_T(t)$$

where w_T is a window function as defined in Sect. 2.6. Windowing the kernel in time forces the SVD to consider only those u_k that place all their energy in the set T, i.e., supp $u_k = T$.

5.5 *Amplitude Smoothness*

Amplitude smoothness of the system H refers to the variation of the transfer function $\hat{h}$. For $\hat{h}$ to smooth it is sufficient that g be will be localized in the frequency domain. In this sense amplitude smoothness may be viewed as the dual to time localization.

Amplitude smoothness may be facilitated directly by modifying the perturbation kernel $D(s,t)$ to have zero value for shifts that are greater in absolute value than a given value s_0. Letting $S = \{s : |s| < s_0\}$ denote the set of shifts of interest, the kernel modification would be

$$D_S(s,t) = D(s,t) \cdot w_S(s),$$

where w_S is a window function as defined in Sect. 2.6.

6 Numerical Examples

Two synthetic signal types are defined and used to illustrate the FM perturbation theory developed in the previous sections. Specific instances of the types are specified for use in a numerical simulation. The signals are created with identical magnitudes and hence only differ in their instantaneous frequencies.

The next two subsections describe each of the synthetic signals, respectively. In addition to a specific realization of a signal of that form, key computational graphics that illustrate the FM perturbation theory are also presented. Various constraints are placed on the kernels as described in Table 1. The two types of signals are labeled in terms of their underlying frequency modulation as linear (a chirp) and chirped (chirping FM). For these signals the following graphics are presented:

1. Full complex kernel D_p
   (a) First two modes arranged in a 2×3 array of plots where each row $i = 1,2$ corresponds to the mode and each column represents the induced deviation u_i, the magnitude of the perturbing system, $|H|$, and its group delay, respectively, e.g., Fig. 4.
   (b) Approximation performance of the predicted systematic deviation when passed through the near-identity system, $H = 1 - a\hat{v}_i$ where $a = 0.1$.
2. Zero group delay constrained kernel ZD_p
   (a) First two modes arranged in a 2×2 array of plots where each row $i = 1,2$ corresponds to the mode and each column represents the induced deviation u_i and the magnitude of the perturbing system, $|H|$, e.g., Fig. 8.
   (b) Approximation performance of the predicted systematic deviation when passed through the near-identity system, $H = 1 - a\hat{v}_i$ where $a = 0.1$.

Two additional kernel constraints are illustrated for the case of the signal ChirpFM. These are as follows:

Table 1 Kernel modifiers

Kernel modifier	Description
$T_{a,b}$	Time-limit in the interval (a,b)
S_d	Delay-limit of $d > 0$
Z	Zero group delay constraint

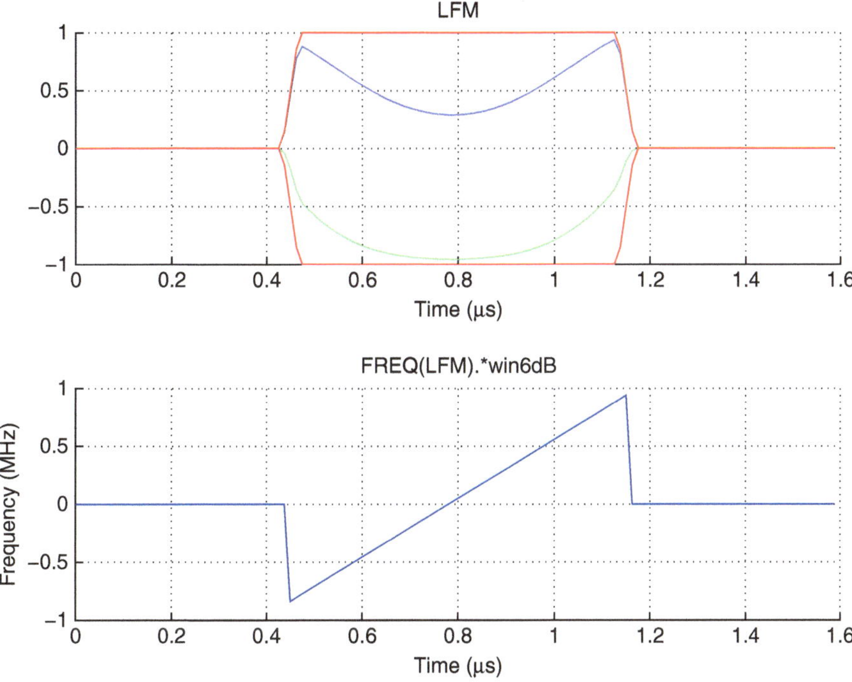

Fig. 3 The signal LFM: (*top*) real, imaginary, and magnitude; (*bottom*) instantaneous frequency

1. Delay limited and zero group delay constrained kernel $S_d Z D_p$ where $d = 0.2\,\mu\text{s}$.
2. Time limited, delay limited, and zero group delay constrained kernel $T_{a,b} S_d Z D_p$ where $d = 0.2\,\mu\text{s}$, and $(a,b) = (0.6, 1.0)\,\mu\text{s}$.

Each of these kernels has additional graphics identical to those described in items 2 (a) and (b) above.

6.1 Chirp (LFM)

A chirp is an analytic signal $p \equiv (A, f)$ with an FM that has the form

$$f(t) = \alpha t + \beta,$$

where $\alpha, \beta \in \mathbf{R}$. The parameter-$\alpha$ is called the chirp rate and has units of MHz/μs. Figure 3 shows a $1\,\mu\text{s}$ long chirp signal with parameters $\alpha = 2.5$ MHz/μs and $\beta = -1\,\text{MHz}$. The top plot shows the real, imaginary, and magnitude portions overlayed, and the bottom plot shows the FM.

6.1.1 Full Perturbation SVD (LFM) (Figs. 4 and 5)

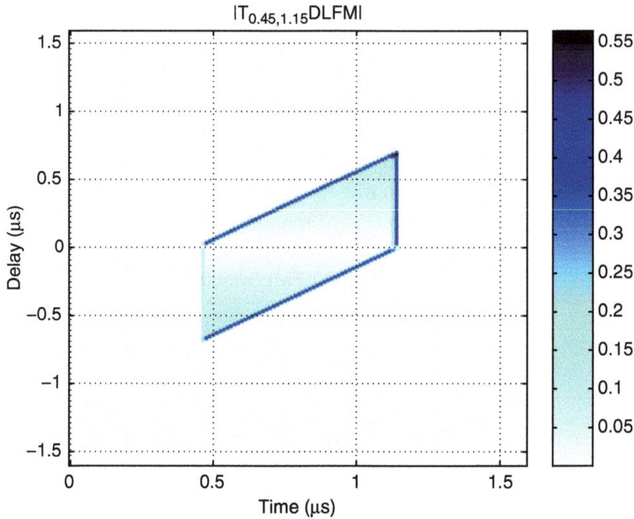

Fig. 4 Full perturbation kernel magnitude for the signal LFM

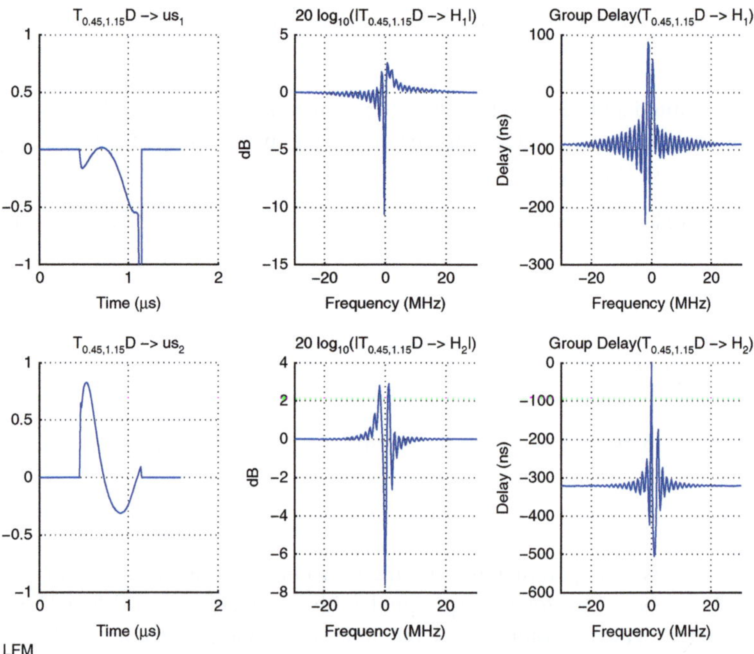

Fig. 5 First two SVD modes ($i = 1, 2$) for the LFM signal: (*left*) FM perturbations u_i; (*middle*) magnitude of $H = 1 - \hat{v}_i$; (*right*) group delay of H

6.1.2 Full Perturbation Approximation (LFM) (Figs. 6 and 7)

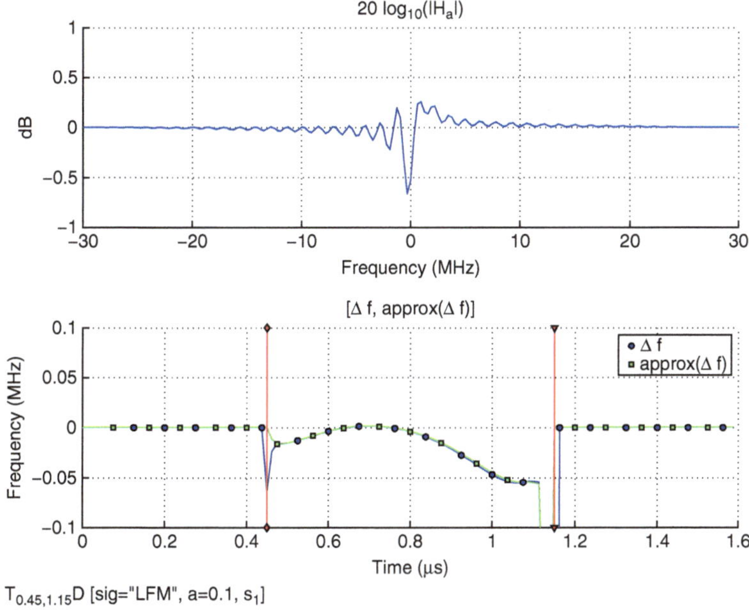

Fig. 6 FM deviation associated with largest singular value for signal LFM: (*top*) magnitude of $H = 1 - a\hat{v}_1$; (*bottom*) actual and approximated perturbations due to H

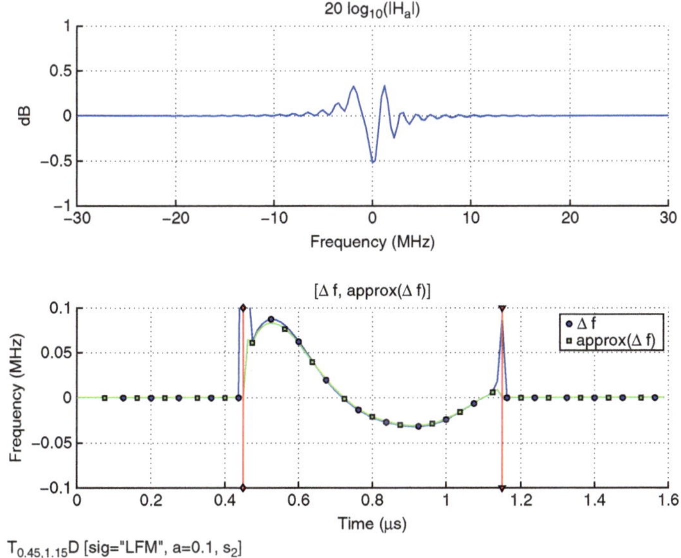

Fig. 7 FM deviation associated with second largest singular value for signal LFM: (*top*) magnitude of $H = 1 - a\hat{v}_2$; (*bottom*) actual and approximated perturbations due to H

6.1.3 Zero Group Delay Constrained Perturbation SVD (LFM) (Figs. 8 and 9)

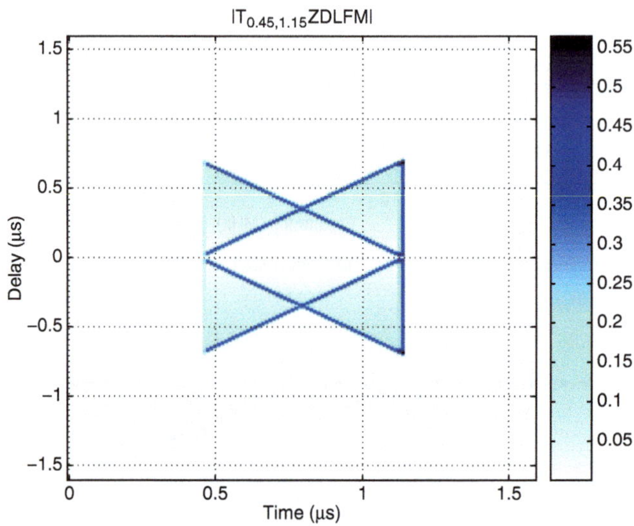

Fig. 8 Perturbation kernel magnitude constrained to zero group delay for the signal LFM

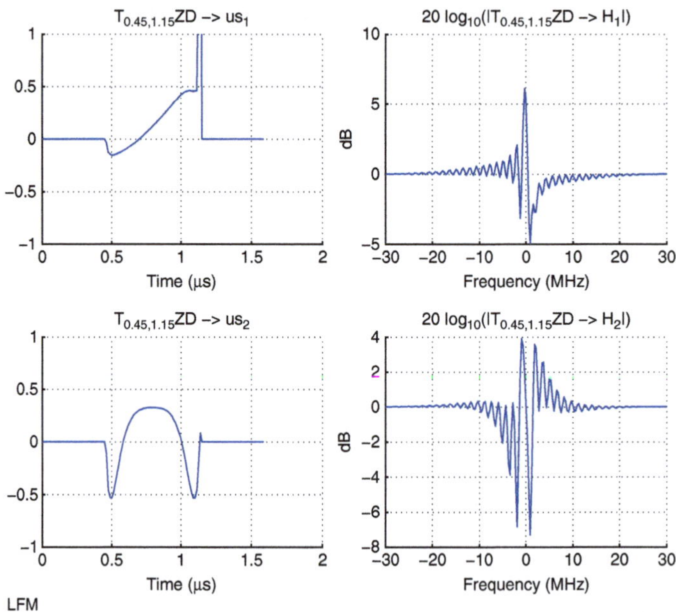

Fig. 9 First two, $i = 1,2$, SVD (constrained to zero group delay) elements for the LFM signal: (*left*) FM perturbations u_i; (*right*) magnitude of $H = 1 - \hat{v}_i$

6.1.4 Zero Group Delay Constrained Perturbation Approximation (LFM) (Figs. 10 and 11)

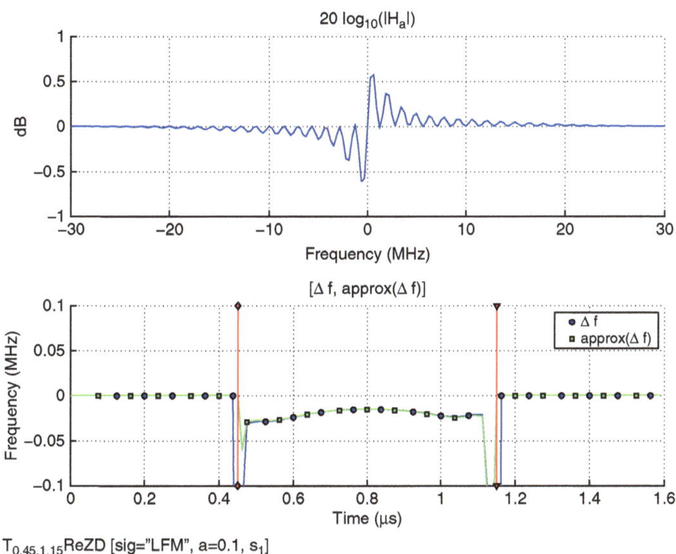

Fig. 10 FM deviation with zero group delay associated with largest singular value for signal LFM: (*top*) magnitude of $H = 1 - a\hat{v}_1$; (*bottom*) actual and approximated perturbations due to H

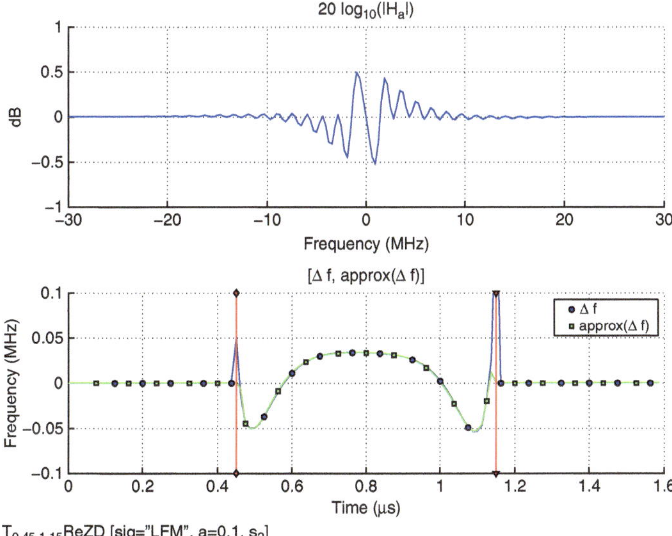

Fig. 11 FM deviation with zero group delay associated with second largest singular value for signal LFM: (*top*) magnitude of $H = 1 - a\hat{v}_2$; (*bottom*) actual and approximated perturbations due to H

6.2 Chirping Modulation (ChirpFM)

Standard chirp signals have FMs that are linear. An *FM chirp* has an FM that is itself chirping. An FM chirp is an analytic signal $p \equiv (A, f)$ with an FM that has the form

$$f(t) = M \sin\left(\pi(\alpha t^2 + \beta t + \gamma)\right),$$

where $M > 0$, and $\alpha, \beta, \gamma \in \mathbb{R}$. Figure 12 shows a chirped FM signal with parameters $M = 1$, $\alpha = 32$, and $\beta = \gamma = 0$.

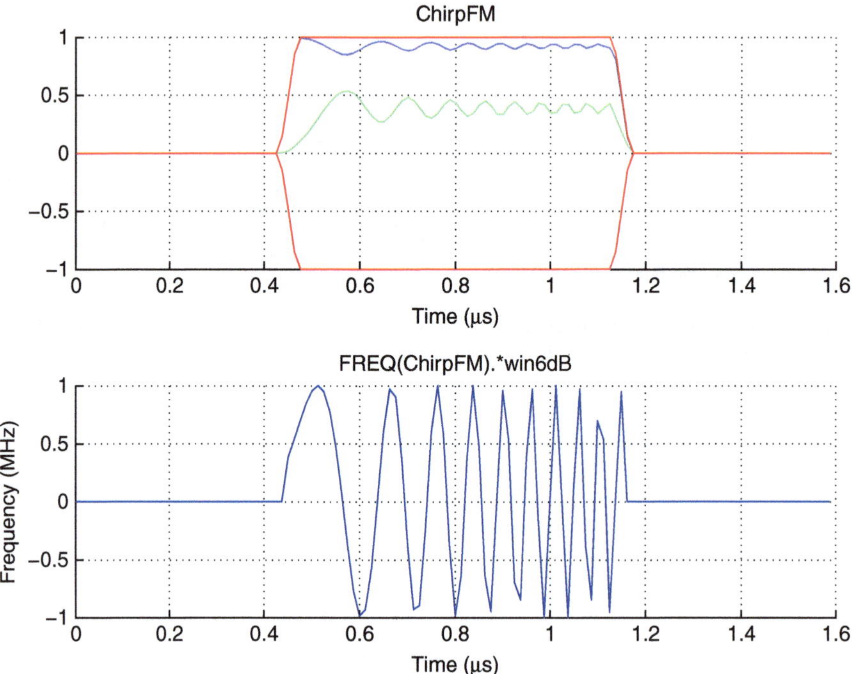

Fig. 12 The signal ChirpFM: (*top*) real, imaginary, and magnitude; (*bottom*) instantaneous frequency

6.2.1 Full Perturbation SVD (ChirpFM) (Figs. 13 and 14)

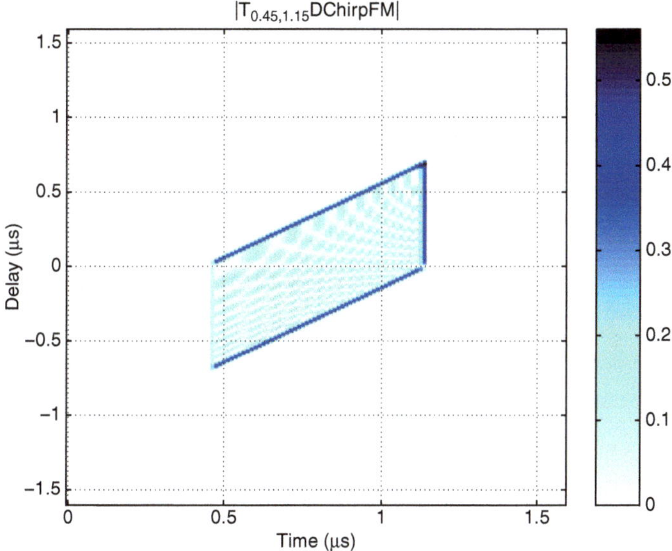

Fig. 13 Full perturbation kernel magnitude for the signal ChirpFM

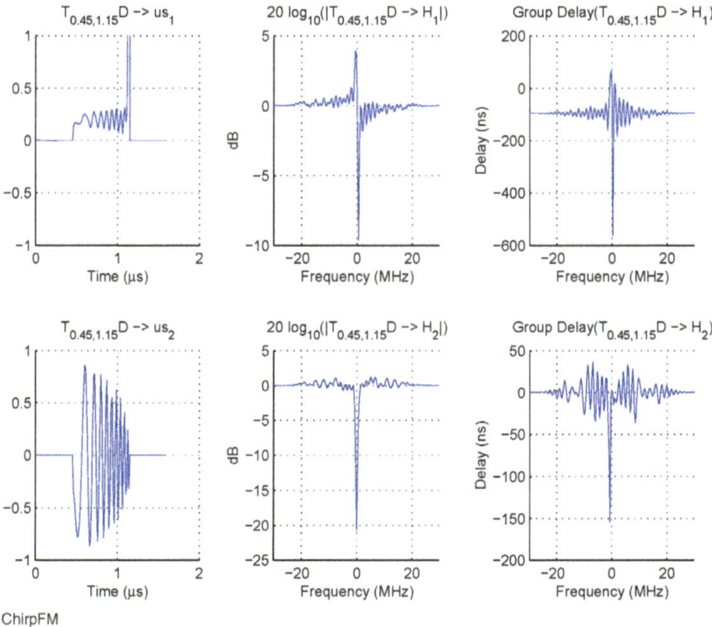

Fig. 14 First two SVD modes ($i = 1,2$) for the ChirpFM signal: (*left*) FM perturbations u_i; (*middle*) magnitude of $H = 1 - \hat{v}_i$; (*right*) group delay of H

6.2.2 Full Perturbation Approximation (ChirpFM) (Figs. 15 and 16)

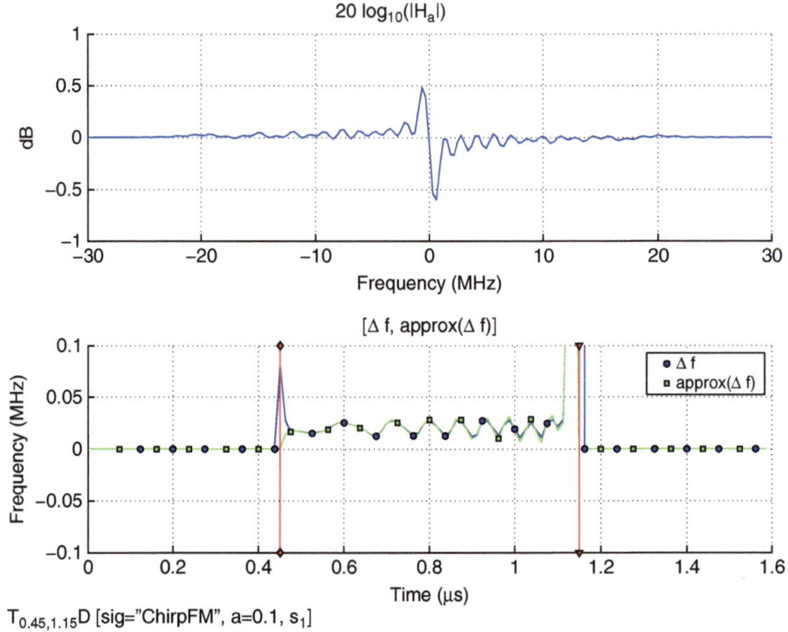

Fig. 15 FM deviation associated with largest singular value for signal ChirpFM: (*top*) magnitude of $H = 1 - a\hat{v}_1$; (*bottom*) actual and approximated perturbations due to H

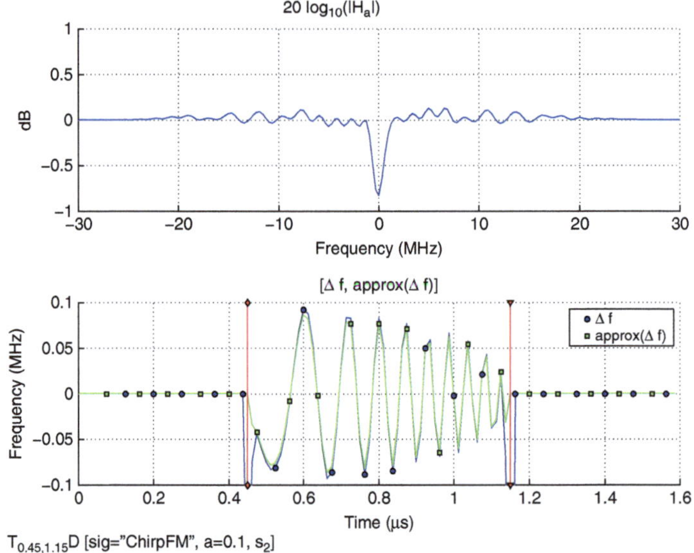

Fig. 16 FM deviation associated with second largest singular value for signal ChirpFM: (*top*) magnitude of $H = 1 - a\hat{v}_2$; (*bottom*) actual and approximated perturbations due to H

6.2.3 Zero Group Delay Constrained Perturbation SVD (ChirpFM) (Figs. 17 and 18)

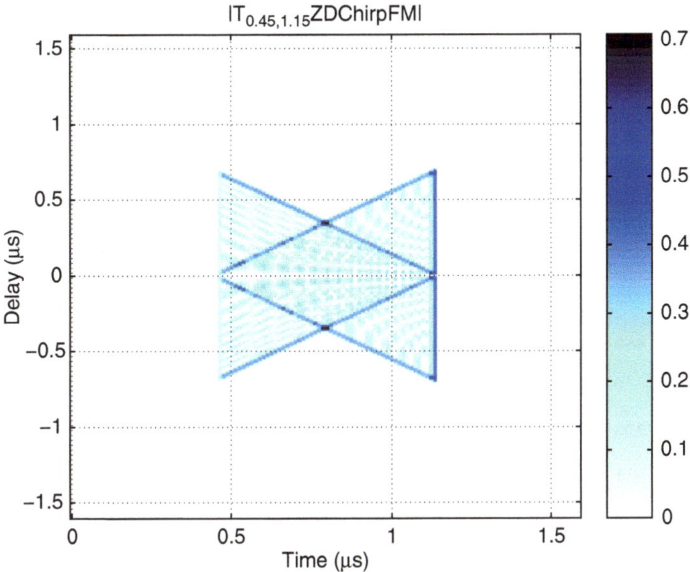

Fig. 17 Perturbation kernel magnitude constrained to zero group delay for the signal ChirpFM

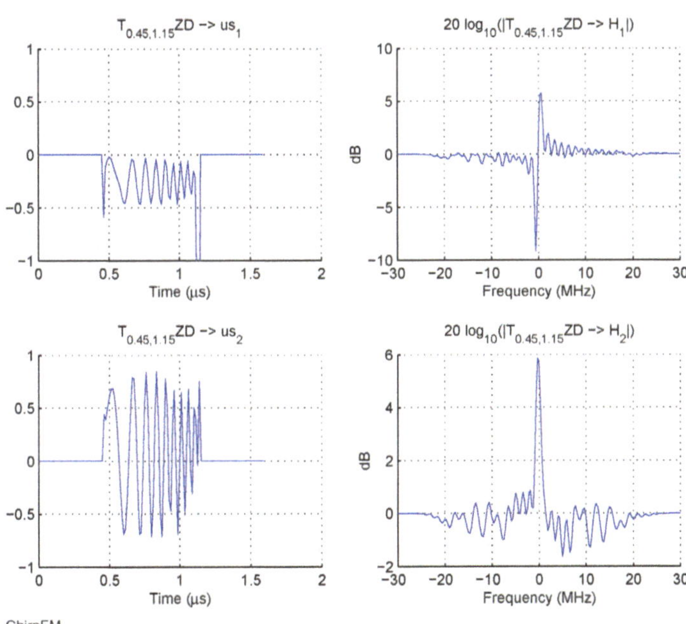

Fig. 18 First two, $i = 1,2$, SVD (constrained to zero group delay) elements for the ChirpFM signal: (*left*) FM perturbations u_i; (*right*) magnitude of $H = 1 - \hat{v}_i$;

6.2.4 Zero Group Delay Constrained Perturbation Approximation (ChirpFM) (Figs. 19 and 20)

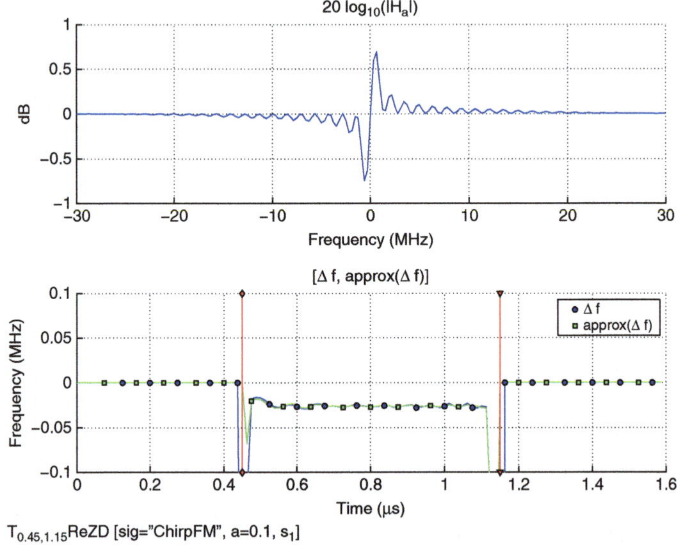

Fig. 19 FM deviation with zero group delay associated with largest singular value for signal ChirpFM: (*top*) magnitude of $H = 1 - a\hat{v}_1$; (*bottom*) actual and approximated perturbations due to H

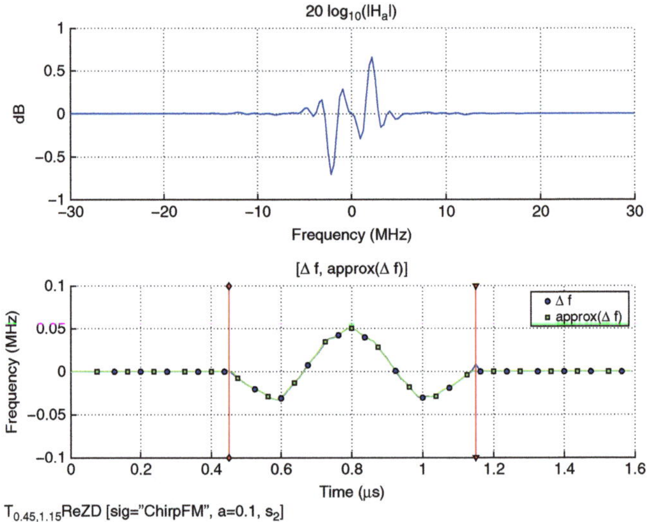

Fig. 20 FM deviation with zero group delay associated with second largest singular value for signal ChirpFM: (*top*) magnitude of $H = 1 - a\hat{v}_2$; (*bottom*) actual and approximated perturbations due to H

6.2.5 Delay Limit Constrained Perturbation SVD (ChirpFM) (Figs. 21 and 22)

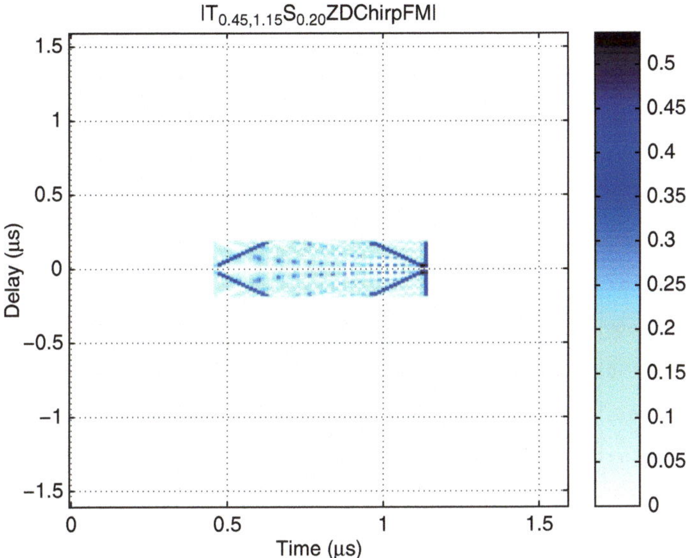

Fig. 21 Delay limit constrained perturbation kernel magnitude for the signal ChirpFM

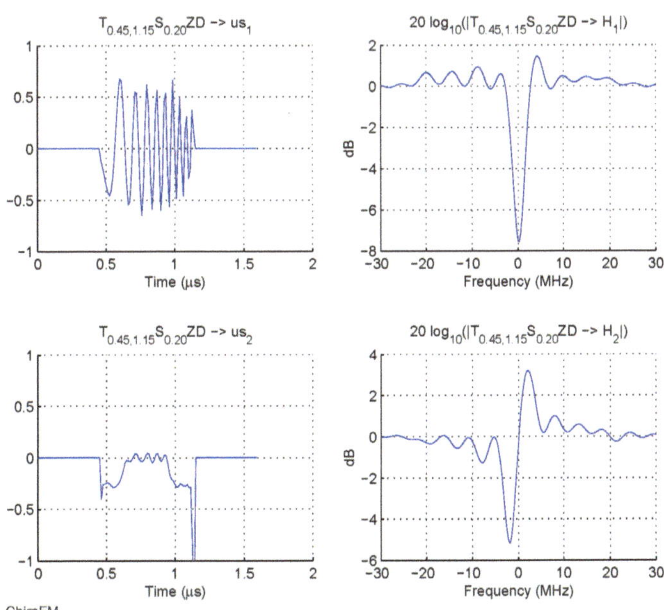

Fig. 22 First two SVD elements under the delay limit constraint for the ChirpFM signal: (*left*) FM perturbations u_i; (*right*) magnitude of $H = 1 - \hat{v}_i$

6.2.6 Delay Limit Constrained Perturbation Approximation (ChirpFM) (Figs. 23 and 24)

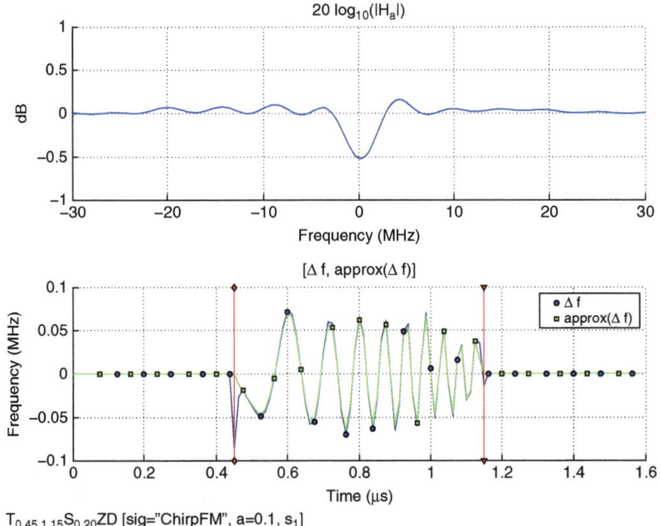

Fig. 23 FM deviation associated with largest singular value of delay limit constrained kernel for signal ChirpFM: (*top*) magnitude of $H = 1 - a\hat{v}_1$; (*bottom*) actual and approximated perturbations due to H

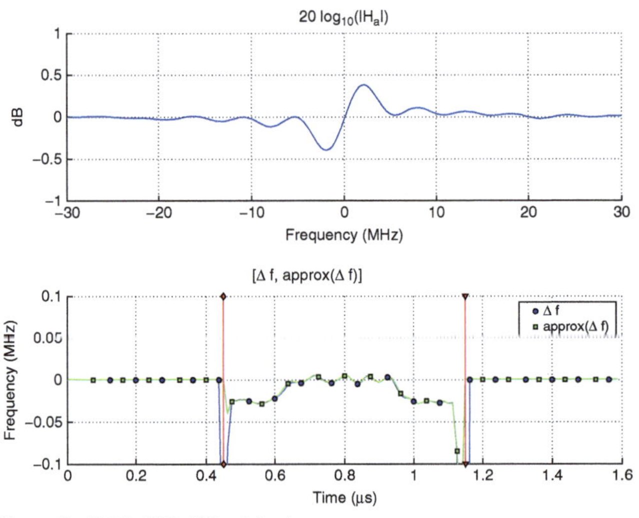

Fig. 24 FM deviation associated with second largest singular value of delay limit constrained kernel for signal ChirpFM: (*top*) magnitude of $H = 1 - a\hat{v}_2$; (*bottom*) actual and approximated perturbations due to H

6.2.7 Time-Delay Limit Constrained Perturbation SVD (ChirpFM) (Figs. 25 and 26)

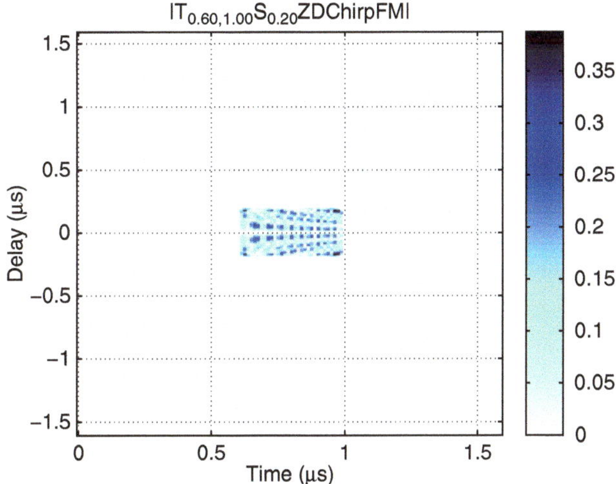

Fig. 25 Time-delay limit constrained perturbation kernel magnitude for the signal ChirpFM

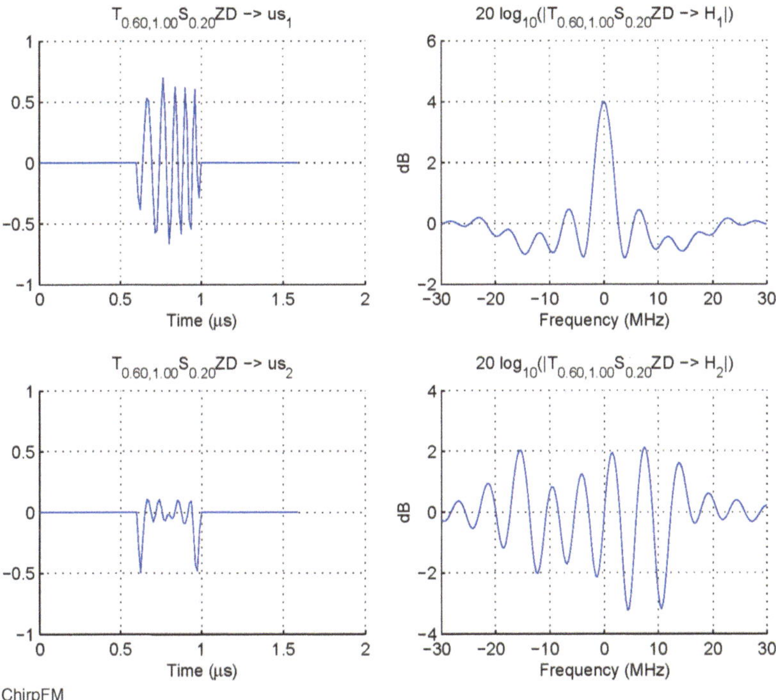

Fig. 26 First two SVD elements under the time-delay limit constraint for the ChirpFM signal: (*left*) FM perturbations u_i; (*right*) magnitude of $H = 1 - \hat{v}_i$

6.2.8 Time-Delay Limit Constrained Perturbation Approximation (ChirpFM) (Figs. 27 and 28)

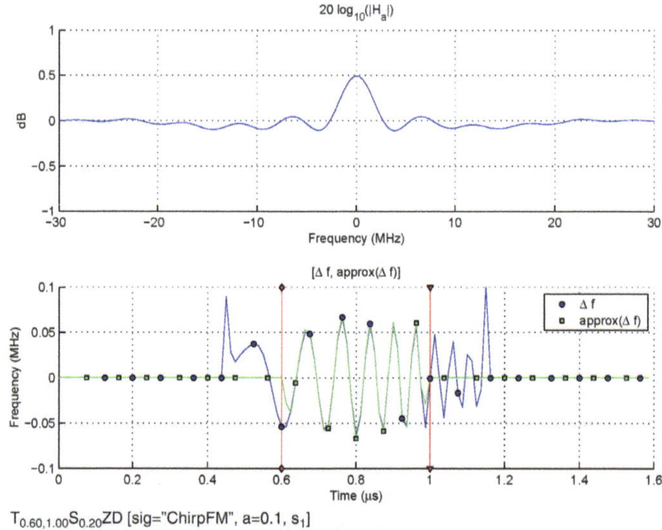

Fig. 27 FM deviation associated with largest singular value of time-delay limit constrained kernel for signal ChirpFM: (*top*) magnitude of $H = 1 - a\hat{v}_1$; (*bottom*) actual and approximated perturbations due to H

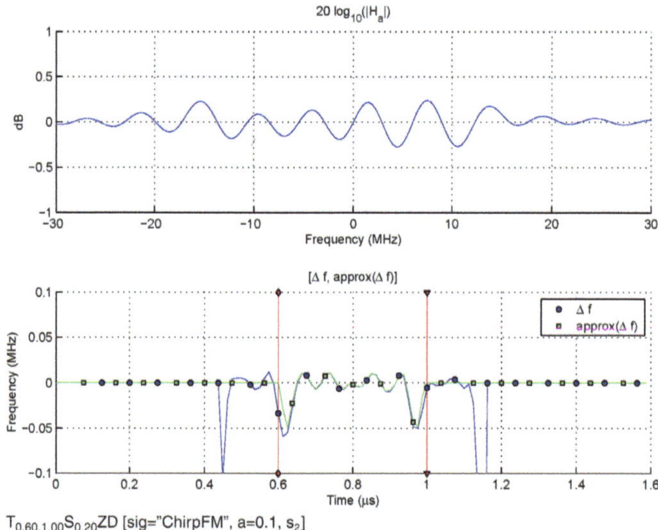

Fig. 28 FM deviation associated with second largest singular value of time-delay limit constrained kernel for signal ChirpFM: (*top*) magnitude of $H = 1 - a\hat{v}_2$; (*bottom*) actual and approximated perturbations due to H

6.3 Discussion

A review of the numerical examples presented in Sect. 6 yields the following observations:

1. *Approximation validation:* In the cases tested, the expected FM deviation agrees with the approximation (4) within an error[8] of about 10^{-2}.
2. *Full kernel perturbations:* The worst-case systematic deviations resulting from a full kernel SVD analysis are associated with systems that may have unrealistic group delays. Constraints that reign in the possible space of all perturbations are needed to yield solutions that are realistic, the need for constraints that reign in the possible space of all perturbations. Various constraints on the maximally perturbing system have been formulated as specific preprocessing operations on the Kernel function. System group delay and the energy concentration and rate of variation of the resulting FM perturbation are two quantities that may be constrained by specific operations on the kernel function. A group delay constraint is discussed in the following item.
3. *Zero group delay limited perturbations:* Group delay in a system is one contributing source of FM perturbation. To completely remove the possibility of the SVD analysis to yield maximal perturbing systems with unrealistic group delays one may enforce a zero group delay constraint. This is an extremal constraint in that it allows no group delay to occur in the perturbing system. Even so, the numerical examples presented here support the notion that a zero group delay constraint does not substantially impact the magnitude of the FM perturbations associated with the maximally perturbing system as computed using the SVD analyses. Typical reductions are anecdotally seen to be given by about a factor of 2 when comparing the full kernel SVD analysis to the zero group delay constrained kernel SVD analysis. This situation suggests that constraining the group delay to be zero is a reasonable method to achieve near worst-case performance. An area of future investigation is the allowance of nonzero group delay with a specified bound.
4. *FM sensitivity:* It is desirable to create a measure of FM sensitivity that quantifies how much a system will perturb the FM of a signal in terms of peak-to-peak deviations. As a first attempt, one might define the FM sensitivity as the ratio of the peak-to-peak variations in the magnitude (measured in dB) of the perturbing system $|H|$ and that of the percentage of induced FM deviation with units of %/dB.

   As an example, consider the chirpFM signal and the system and perturbation corresponding to the largest singular value. Figure 23 on page 426 displays the pertinent curves. From the figure it can be seen that the peak-to-peak variation in the magnitude H of the system is about 0.6 dB. Similarly the peak-to-peak variation in the resulting FM deviation is $0.15\,\text{MHz}$, or $0.15/2 \times 100\,\% = 7.5\,\%$

[8] To be precise this is only true over an interval strictly contained in the support of p.

(recall the unperturbed peak-to-peak FM variation is 2 MHz for all the synthetic signals). Thus, for the case of the chirpFM signal, the FM sensitivity is

$$\frac{7.5}{0.6} = 12.5 \quad \left(\frac{\%}{\mathrm{dB}}\right).$$

Due to the linearity of the FM perturbation for small perturbations, this relation will scale linearly as well. It is reasonable to conclude, for instance, that scaling the systematic deviation so that the peak-to-peak variation in H is 4 dB will lead to an FM deviation of 50 %.

5. *Performance significant modes:* It is important to note that the modes corresponding to the largest singular values may not be those that are most important to specific performance criteria.

   Because of the pulsed nature of the signals, the FM is most sensitive at the leading and trailing edges. This can be seen in the examination of a typical kernel plot, e.g., Fig. 4 on page 416. The largest contributions to the kernel are indicated by the trapezoidal outline displayed in the kernel magnitude. For this reason, the SVD analysis may favor perturbations that spike in these areas. This leads to the situation where the largest singular value may correspond to a case where all FM perturbation energy is concentrated in the leading and/or trailing edges. In this case, the FM perturbation is very small throughout the duration of the pulse. For many RF systems leading and trailing edge perturbations may not be a cause for concern. Thus, the full unconstrained kernel solutions corresponding to the largest singular values may not be the ones that are most detrimental to an RF system's performance. RF system performance and kernel constraints are discussed in the next item.
6. *Largest SVD versus peak-peak variation:* One potentially practically useful measure of RF system parameter may be defined in terms of peak-to-peak quantities. In particular, the peak-to-peak variation induced in the FM of signal passed through a linear system is a measure of the systems FM integrity. Because the SVD theory yields the perturbation with the largest energy (an L^2 norm), it does not translate directly to peak-to-peak variation (an L^∞ norm). However, it has been seen that placing kernel constraints on time duration and delay provides a mechanism to focus perturbation energy to regions that are significant with respect to RF system performance. In this way, the inherent smoothing associated with the kernel constraints provides maximal perturbations determined by the SVD theory that are more closely coupled to their peak-to-peak variations.

7 Conclusion

It has been seen that under the assumption of a near-identity system model the FM deviation induced by the system is given by a signal-dependent *linear* operator.

This operator has been called the FM perturbation operator. The linearization of the problem allows standard linear analyses to be applied.

As such, a SVD of the FM perturbation operator has been used to yield the most FM perturbing (worst-case) system for a given signal. The validity of the approximation has been numerically verified and applied to the problem of determining the worst case system in terms of induced FM deviation.

Because the unconstrained solution may yield maximally perturbing systems that have properties (group delay and amplitude deviation) that do not reflect practical systems, further realistic constraints have been embodied in formulation. Under such constraints, the FM perturbation theory has been used to construct realistic near-identity systems that induce relatively large FM perturbations. In particular, it has been seen that systems having FM sensitivities as great as 12 %/dB may be constructed.

Acknowledgments The author gratefully acknowledges the support of the Naval Research Laboratory (Code 5720).

References

1. Carson, J.R., Fry, T.C.: Variable-frequency electric circuit theory with application to the theory of frequency modulation. Bell Sys. Tech. J. **16**, 513–540 (1937)
2. Salinger, H.: Transients in frequency modulation. Proc. IRE **30**(8), 378–383 (1942)
3. Gladwin, A.S.: The distortion of frequency-modulated waves by transmission networks. Proc. IRE **35**(12), 1436–1445 (1947)
4. Bedrosian, E., Rice, S.O.: Distortion and crosstalk of linearly filtered, angle-modulated signals. Proc. IEEE **56**(1), 2–13 (1968)
5. Corrington, M.S.: Frequency-modulation distortion caused by multipath transmission. Proc. IRE **33**(12), 878–891 (1945)
6. Abuelma'atti, M.T.: Nonlinear distortion of an FM signal by a passive network. Proc. IEE **131**(1), 61–64 (1991)
7. Mazzaro, G.J., Steer, M.B., Gard, K.G., Walker, A.L.: Response of RF networks to transient waveforms: interference in frequency-hopped communications. IEEE Trans. Microw. Theor. Tech. **56**(12), 2808–2814 (2008)
8. Bassoo, V., Linton, L., Faulkner, M.: Analysis of distortion in pulse modulation converters for switching radio frequency power amplifiers. IET Microw. Antennas Propag. **4**(12), 2088–2096 (2010)
9. Gohberg, I., Goldberg S.: Basic Operator Theory. Birkhauser, Boston (1980)
10. Meinsma, G., Mirkin, L.: Sampling from a system-theoretic viewpoint: part I-concepts and tools. IEEE Trans. Signal Process. **58**(7), 3578–3590 (2010)
11. Golub, G.H., Van Loan, C.F.: Matrix Computations, 3rd edn. Johns Hopkins, Baltimore (1996)
12. Strang, G.: Introduction to Linear Algebra, 3rd edn. Wellesley-Cambridge Press, Wellesley (1988)

Eddy Current Sensor Signal Processing for Stall Detection

Carole Teolis, David Gent, Christine Kim, Anthony Teolis, James Paduano, and Michelle Bright

Abstract This chapter presents algorithms that use data from eddy-current sensors mounted in the engine casing for the purpose of gas turbine engine stability monitoring. To date, most signal-processing techniques using blade tip sensors have been limited to simple parametric measurements associated with the sensor waveform, for example measurement of zero-crossing locations for time of arrival information or maxima for tip clearance information. Using this type of parametric information, many computations require more than one sensor per stage. The use of a minimal number of sensors is an extremely important practical consideration since each pound that is added to an aircraft engine adds considerable costs over the life cycle of the engine. Because of this we have focused on developing algorithms that allow the reduction in the number of sensors needed for fault prognosis. Using our algorithms we have been able to demonstrate the detection of stall cell precursors using a single ECS. These algorithms have been demonstrated in real time in tests at

C. Teolis (✉)
Organization TRX Systems, Inc Greenbelt, MD, USA
e-mail: carole@trxsystems.com

D. Gent • C. Kim
Organization Techno-Sciences, Inc Beltsville, MD, USA
e-mail: gent@technosci.com; Kim.l.Christine@gmail.com

A. Teolis
Organization Teolis Consulting, Glenn Dale, MD, USA
e-mail: tony@teolis.org

J. Paduano
Organization Aurora Flight Sciences, Cambridge, MA, USA
e-mail: jpaduano@aurora.aero

M. Bright
Organization NASA Glenn Research Center, Cleveland, OH, USA
e-mail: Michelle.M.Bright@mail.nasa.gov

T.D. Andrews et al. (eds.), *Excursions in Harmonic Analysis, Volume 1*,
Applied and Numerical Harmonic Analysis, DOI 10.1007/978-0-8176-8376-4_21,

the NASA Glenn W8 single-stage axial-flow compressor facility. The rotor tested, designated NASA Rotor 67, is a fan with 22 blades.

Keywords Eddy current sensor • Gas turbine engine • Engine health monitoring • Stall detection • Stall precursors (prestall behavior) • Harmonic analysis • Real-time implementation • Direct FM • Fast Fourier transform (FFT) • Instantaneous frequency

1 Introduction

Development of a system to detect and compensate for potentially catastrophic engine failures and instability is the primary objective of this work. Such a system would enable increased performance, reliability, and safety of gas turbine (jet) engines through active and automatic control. The system would be suitable for integration into both military and commercial jet airplanes and provide a level of air safety heretofore unattainable.

Current gas turbine engine design practice is to base fan and compressor stall margin requirements on a worst-case scenario of destabilizing factors with an added margin for engine-to-engine variability. These factors include external destabilizing factors such as inlet distortion, as well as internal factors such as large tip clearances. This approach results in larger than necessary design stall margin requirement with a corresponding reduction in performance and/or increase in weight [1]. The availability of a sensor system that could detect the onset of stall could allow these margins to be safely decreased.

NASA Glenn Research Center, in their high stability engine control (HISTEC) program, has pursued two approaches to avoiding engine instability (stall and surge) while maximizing performance. The first, which has already been demonstrated in flight tests, is distortion tolerant control [1, 2]. The idea is to increase the stall margin on line as the engine face pressure distortion is encountered. The HISTEC implementation adjusts the engine operating point based on distortion inputs from the disturbance estimation system (DES) to maintain sufficient stall margin through stability management and control [2]. The second approach, active stall control, has been demonstrated in research compressors, for example see [3–6], but not yet in flight tests.

Our goal, in the NASA tests described in this chapter as follows: first, to demonstrate that precursors to onset of stall could be detected reliably with eddy-current sensors (ECSs) and, second, to prevent the onset of stall using NASA designed discrete tip injections [7]. We were unable to complete the second goal but are still pursuing this objective.

Using two ECS, one at the leading and one at the trailing edge, it has already been demonstrated by GDATS that the ECS can detect rotating stall cells (Fig. 1). Using the same test data, we have extended this result to detect stall cells using only one ECS, thus demonstrating the feasibility of using the ECS information to enhance and/or replace the pressure sensors for stall disturbance estimation and with the additional benefit that vibration and tip clearance can also be measured [8].

This chapter is organized as follows: Sects. 2 and 3 give background material on stall inception, stall detection, and control in gas turbine engines, Sect. 4 gives background on the GDAIS ECS. Sections 5 through 7 present the main results of the chapter. Section 5 proposes methods for stall detection using a single GDAIS ECS. Section 6 discusses the direct "FM" implementation of the stall algorithms and presents some test results from the real-time algorithms. Section 7 presents post-processing results from a single-sensor parametric stall detection algorithm.

2 Stall in Gas Turbine Engines

Day [9] showed through experimental studies on two different compressors that the modal perturbations predicted by the Moore–Greitzer model [10] are not always present prior to stall. He showed that there are two routes to rotating stall in axial compressors: two-dimensional long-length scale "modal waves" that extend axially through the compressor and three-dimensional short-length scale, "spike" or "pip," disturbances. Subsequent stall experiments in test rigs [11–14] and in engines [15, 16] have served to confirm Day's findings and expand understanding of prestall behavior.

The modal oscillations[1] were predicted by Moore and Greitzer [10] before being observed by McDougall et al. [17]. Modal oscillations are small amplitude, essentially two-dimensional long-wavelength disturbances that extend axially through the compressor[2] and appear close to the peak of the total-to-static pressure rise characteristic. They typically rotate at 0.2–0.5 times the rotor speed [6, 11], though they have been observed at higher proportional frequencies in high-speed machines [18]. Two-dimensional linearized stability analysis [10, 19] has been shown to give an approximation of the modal wave shape, phase speed, and growth rate for many compressors.

The three dimensional spike disturbances are localized to the tip region of a specific blade row in a multistage compressor and have a length scale on the order of the blade pitch. When the spike first emerges, it is small in circumferential extent and thus propagates quickly around the annulus. Spikes typically initially rotate at 0.7–0.8 times the rotor speed. As the spike begins to propagate it rapidly increases in size and its speed of rotation reduces to 0.2–0.5 the rotor speed before developing into a full stall cell [13]. The inception process is fundamentally non-linear, in contrast

[1]A first-order mode would have a wavelength equal to the circumference of the compressor; a second-order mode would have a wavelength equal to half of the circumference of the compressor, and so on.

[2]Stage mismatching can limit the disturbance to a particular axial region.

to the essentially linear behavior seen in modal stall inception. The time between detecting a spike and the development of the full stall cell is typically much shorter than for modal stall development.

The three-dimensional nature of spike disturbances makes them more difficult to detect since the best axial placement for sensors is unclear at this point. In low-speed compressors, the first rotor row seems to be the most susceptible to spike initiation. It is thought that this is because the deviation angle of the flow leaving the inlet guide vanes is approximately constant as stall is approached, whereas the deviation angles from the downstream stator rows increase. Thus near the point of stall the first rotor operates at higher incidence than the downstream rotors, and therefore it is this row that first succumbs to flow separation [13].

However, spike-type stalling is not confined to the first stage. The formation of spikes in the rear stages of high-speed compressors has also been observed [15]. In a high-speed compressor, stage matching changes automatically as the speed of rotation changes. Due to compressibility effects, the position of highest loading shifts from the front to the rear of the compressor as the speed of rotation increases. Thus, at low speed the front stages are heavily loaded and most likely to stall, at medium speed all stages are evenly matched near the stall point, and at high speed the rear stages are the most likely to stall.

As discussed by Moore and Greitzer [10], modal oscillations appear close to the peak of the total-to-static pressure rise characteristic. At this point some or all of the blade rows are operating close to their stalling limit; hence any modal-induced velocity deficit may be sufficient to initiate flow separation. Transition to stall from a modal oscillation can happen in a couple of different ways. When a modal oscillation develops smoothly into full stall, detailed measurements have shown that the low velocity trough in the modal pattern initiates flow breakdown over a wide sector of the annulus. This results in a broad stall cell, which, because of its size, rotates comparatively slowly, at a speed close to that of the fully developed cell. The formation of a broad, slow-moving cell is thought to be associated with flow separation near the hub [13]. In other situations, the low-velocity trough in the modal wave will trigger flow separation in a localized region near the tip of one particular blade row [13]. In such cases the transition from modal oscillation to rotating stall occurs via a spike disturbance.

Modal waves do not necessarily precede spike formation. Hoss et al. reported stall inception measurements taken from the compressor section of a two-spool turbofan engine at various power settings [16]. In their tests they observed that at low engine speed, stall originates from spike-type precursors, while modal waves were observed prior to stall at mid-speed for undistorted inlet flow. At high engine speed, the rotor shaft unbalancing dominates the stall inception process as an external forcing function. In the case of distorted inlet flow, they found spike-type stall inception behavior dominates throughout the speed range.

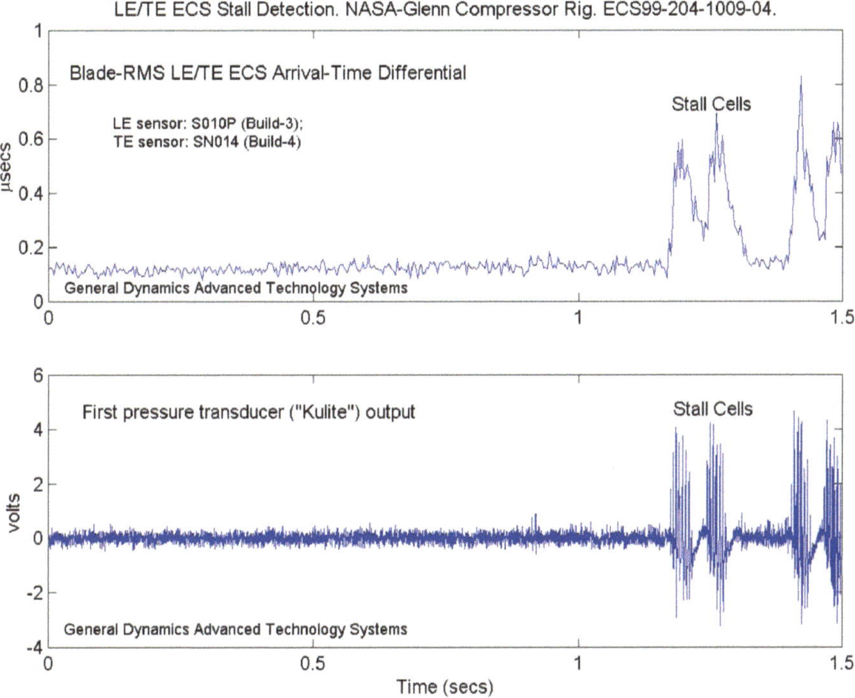

Fig. 1 Eddy-current sensor versus kulite pressure transducer

3 Current Detection Methods

There are several methods in the literature for prestall detection, a few of which are summarized in this section. Predominantly the detection methods rely on data from high-bandwidth pressure sensor data with the exception of the general dynamics results where two ECS are positioned, one on the leading and one on the trailing edge of the blade row. After processing the two ECS are shown to get similar results of a single pressure sensor [20]; see Fig. 1 which has been reproduced from that paper.

3.1 Traveling Wave Energy

Tryfondis et al. [11] introduce the idea of using traveling wave energy (TWE) as a real-time measure of compressor stability. The idea is motivated by a linearization of the hydrodynamic theory of compressor stability developed by Moore and Greitzer[3] [10]. The Moore–Greitzer model predicts the development of circumferential waves of perturbation pressure, termed modal perturbations, which become under-damped near the stall inception point. The TWE analysis provides a means of detecting the growth of these small-amplitude perturbations.

In the development of the TWE analysis, the circumferential pressure perturbations are written as a Fourier series, i.e., broken into a sum of sinusoidal components. According to the linearized hydrodynamic theory of compressor stability, the spatial Fourier coefficients of the pressure perturbations evolve independently and thus constitute the fundamental states of the system.[4] Several sensors are needed to resolve the circumferential sinusoidal harmonics. Given N circumferential measurements of the engine pressure perturbations $\delta P_x(\theta,t)$, i.e., measurements taken from the same axial location x, an approximation of the Fourier coefficients $a_k(t)$ is given by

$$a_k(t) = \frac{1}{N}\sum_{n=1}^{N} \delta P(\theta_n,t)\mathrm{e}^{ik\theta_n},$$

where $N \geq 2k+1$ as dictated by the Nyquist criterion. The sensor locations, θ_n, should be evenly (or nearly evenly) spaced to insure that harmonics below the Nyquist frequency do not alias into the estimate of $a_k(t)$.

In the work of Hoss et al. [15,16], five wall-static pressure transducers were used in front of each of the two engine stages of their test rig. The five sensors allowed the circumferentially distributed pressure fluctuations to be resolved up to the second-order sinusoidal harmonic. It was thought that 2 harmonics were sufficient since in most cases where the modal waves have been observed the first two spatial harmonics dominated the inception process. Other researchers have used up to 8 transducers in order to resolve the first 3 harmonics.

Further information about flow phenomena in the circumferential direction can be extracted from the pressure signals by computing the power spectral density (PSD) of their spatial Fourier coefficients. The PSD gives the development of the power of separate frequencies as a function of time without the phase information. In choosing the length of the spectral window a trade-off must be made: the longer the window, the less the variance due to noise. However, choosing a window that is too long will smooth out the transients of interest. Tryfondis et al. used a 50 rotor revolution window.

The TWE is obtained by overlaying the negative half of the PSD spectrum[5] with the positive and integrating the resulting differences over a fixed frequency range for every step in time. Tryfondis et. al. integrated TWE for frequencies between

[3]Moore and Greitzer developed a model for low-speed compressors which was later extended to for analysis of high-speed machines by Feullner et al. [30] and others.

[4]In contrast, the pressure perturbation at a given circumferential position is correlated to that at any other position.

[5]Because the Fourier coefficients are complex functions of time, the PSD's are not symmetric with respect to zero frequency.

12 % and 125 % of the rotor frequency. Since standing wave phenomenon yields symmetric PSD's with respect to zero frequency, they do not contribute to the TWE. Thus in cases where the compressor exhibits stall via modal wave, a stall warning indication could be given when the TWE crossed a certain threshold. The TWE could be computed essentially as is in real time.

Since the TWE analysis is based on a linearized model, the analysis only applies to the "small-signal region" preceding stall. As the waves grow to large amplitude, the assumption that the dynamics of each spatial harmonic and each mode evolve independently is no longer valid.

3.2 Chaotic Time Series Analysis

Bright et al. [12, 14] introduce two different chaotic time series analysis methods, correlation integral and structure function, for prestall detection using a single sensor upstream of the stage. These methods, which were developed in the nonlinear dynamics and chaos communities [21, 22], attempt to distinguish between low-dimensional dynamics and randomness in measured time series.

Bright et al. [12] deduce that a single pressure transducer carries the underlying dynamics of the compressor by computing the Kolmogorov entropy for the pressure sensor data and showing that it is different from the Kolmogorov entropy of a surrogate data set. A surrogate data set is a data set with the same Fourier magnitude but randomized phase. The entropy in the sensor data is found to be an order of magnitude lower than the entropy in the surrogate data. Differences in the statistic are indications of nonlinearities in the sensor data.

The correlation integral has been shown to be an effective measure of changes in the prestall behavior in high-speed compressors and has been able to detect both pips and modal disturbances [14]. However, Bright et al. suggest that due to the computational complexity of the correlation integral, this method is best suited for post processing of data and that the structure function algorithm [22] is more suited for real-time computations.

The structure function can be used to detect modal stall precursors as well as pip structures. The structure function algorithm is closely related to the reconstruction signal strength statistic from chaos theory. It is also related to the correlation function. The structure function (SF) is often more convenient to use in cases when one is not concerned with absolute quantities but only with pressure differences at distinct instants of time. This is very effective for noting local disturbances rather than monitoring magnitude pressure rise [23, 24].

3.3 Frequency of Prestall Dynamics

In order to determine the frequency of the dynamics of interest Bright et al. [14] examined pressure sensor data sampled at 20 kHz and anti-alias filtered at 10 kHz while the engine was transitioned into stall through slow closure of the throttle. This data was then filtered in three different frequency regions and the CI results compared: (1) The data were low-pass filtered at 500 Hz. In this case, the CI value begins to rapidly decrease approximately 292 rotor revolutions before stall. (2) The data were low-pass filtered at 100 Hz. In this case, the CI value does not begin to rapidly decrease until approximately 140 rotor revolutions before stall, coincident with the observation of modes in the pressure traces. (3) The data were band-pass filtered between 4,000 and 7,000 Hz commensurate with blade-passage frequencies. In this case the CI value does not begin to rapidly decrease until approximately 100 rotor revolutions before stall.

The analysis of data sets in 2 and 3 above points to the occurrence of blade-passage frequency events in addition to low-frequency events that occur simultaneously before rotating stall. While these results imply that it is the events in the 100–500 Hz range that give the first indication of stall, there is some indication that blade motion, which is what we need for ECS detection, may be significant in advance of stall.

3.4 *Wavelet Analysis*

The great advantage of the wavelet transform over the windowed Fourier transform is that while in the Fourier transform the same time window is used (and therefore the maximum resolution is limited by the window length), the wavelet transform uses a window in which the time resolution varies with frequency. When dealing with lower frequencies the time window length is larger, and with higher frequencies the length is smaller. Among the first investigations of wavelets to detect stall precursors were performed by Liao and Chen [25]. Cheng et al. [26] continued to pursue the wavelet detection method. Cheng et al. use two-dimensional wavelet image-processing techniques to analyze data from a single pressure sensor located near the blade tip sampled at 237,500 Hz. A pressure image created by forming a 2-D array with the data by taking a fixed number of samples (nominally one rotation of sensor data at a fixed speed) to form each subsequent row. The data is then analyzed at three different length scales, where the three different scales correspond to the half blade-passage scale (small), the blade-passage scale (medium), and the several fold blade-passage scale (large). For each scale, the intensity of the wavelet transform of the pressure image is summed over frequency for each time to yield a stall indicator curve. Cheng et al. found that for the small scale, the indicator curve began to increase in magnitude at 230 revolutions before stall, while at medial scale the indicator curve began to decrease 230 revolutions before stall. At the large scale or stall cell scale, an increase was seen abruptly as the engine stalled.

Hoss et al. [16] pursue one-dimensional wavelet processing as a method of detecting stall precursors. Because of its similarity to the footprint of a spike in the pressure signal, the Daubechies wavelet is used as the analyzing wavelet. Considering the magnitude of the wavelet transform they generate a stall indicator function by first enforcing a low-magnitude threshold to reduce the contribution of noise,[6] and then breaking each frequency band[7] into time windows (they use 0.1 s) and essentially averaging the magnitude values within the time window. More recently Lin et al. [27, 28] have extended these results using continuous wavelet transform and addressed practical issues of choice of wavelet basis functions.

3.5 *Stall Control*

In recent years there has been a flurry of research on compressor modeling and stability control. Most of the theoretical efforts focus on modal stall inception mechanism. The controls are typically based on the model developed by Moore and Greitzer [10, 29], a simple three-state nonlinear model that describes the basic dynamics of modal stall and surge and their interaction in low-speed compressors. There are many modifications/extensions to the Moore–Greitzer model, for example, to extend it to higher speed compressors [30], to allow non-cubic compressor characteristics [31, 32], or to allow distorted flow [33, 34]. Most controls are developed based on a bifurcation theoretic approach that changes the characteristic of the pitchfork bifurcation at the stall inception point from hard subcritical to soft supercritical, thus avoiding an abrupt transition into rotating stall [35, 36]. The control proposed by Liaw and Abed modifies the throttle characteristic and can be realized experimentally through the use of a bleed valve. To reduce the rate requirement on the bleed valve actuator, it may be coupled with air injection [37,38].

Both rotating stall and surge impose limits on the low flow operating range of compressors. Surge is characterized by violent oscillations in the annulus-averaged flow throughout the compression system. Rotating stall is a two-dimensional (modal oscillations) or three-dimensional (pip) disturbance localized to the compressor and characterized by regions of reduced or reversed flow that rotate around the annulus of the compressor. Many of the control papers focus on stall independently of surge; however they are not independent in high-speed compressors operating regimes. Eveker et al. address the integrated control of rotating stall and surge [39].

In the applications cited above which use sensors (including the HISTEC flight test [2]) the sensors used are high-response static pressure sensors (6 +). In Yeung and Murray's work amplitude and phase of the first and second mode of the pressure perturbation are needed for active stall control [37, 38]. In the HISTEC controller, the pressure sensors are needed to give a disturbance estimate [1, 2].

---

[6]Since noise is generally a broadband, low-power disturbance, the wavelet transform of the noise would have a low-amplitude contribution for each frequency and time. Thus setting to zero any value below a low-amplitude threshold is usually an effective means of reducing noise.

[7]The wavelet transform can be implemented as a bank of filters, where the frequency bands are the outputs associated with a particular filter.

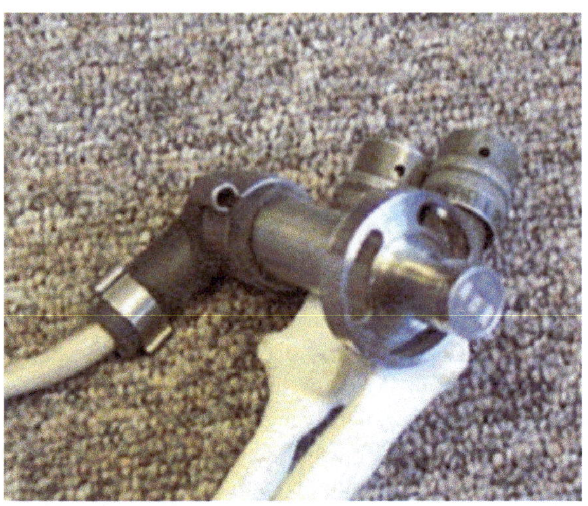

Fig. 2 GDAIS eddy-current sensor

4 The GDAIS ECS

ECSs for engine monitoring were originally developed as a tip clearance sensor. Since their development it has been demonstrated that they are capable of measuring foreign object damage (FOD), blade vibration, and stall/surge in addition to tip clearance. Thus an ECS is a very valuable multifunctional sensor to have in an engine.

Previous methods for stall detection with ECS require two sensors. By positioning sensors at the leading and trailing edge of the blade passage, the sensors are able to measure the blade twist that occurs as the blades pass through the low-pressure stall cell. Thus it has been demonstrated that the blade motion during stall is sufficient to be an indicator of stall.

Two important questions that we will address here are the following: (1) can the number of sensors used be reduced? (2) is the blade motion sufficient to detect prestall events? Here we present two methods of detecting stall with only one ECS and we present preliminary evidence of the capability to detect stall precursors based on the ECS signatures.

4.1 The Eddy-Current Sensor

In our tests we use the general dynamics advanced information systems (GDAIS) ECS; see Fig. 2. The GDAIS ECS is well suited for permanent installation in operational units and can measure a large number of blade parameters while remaining light in weight, small, and relatively inexpensive [40].

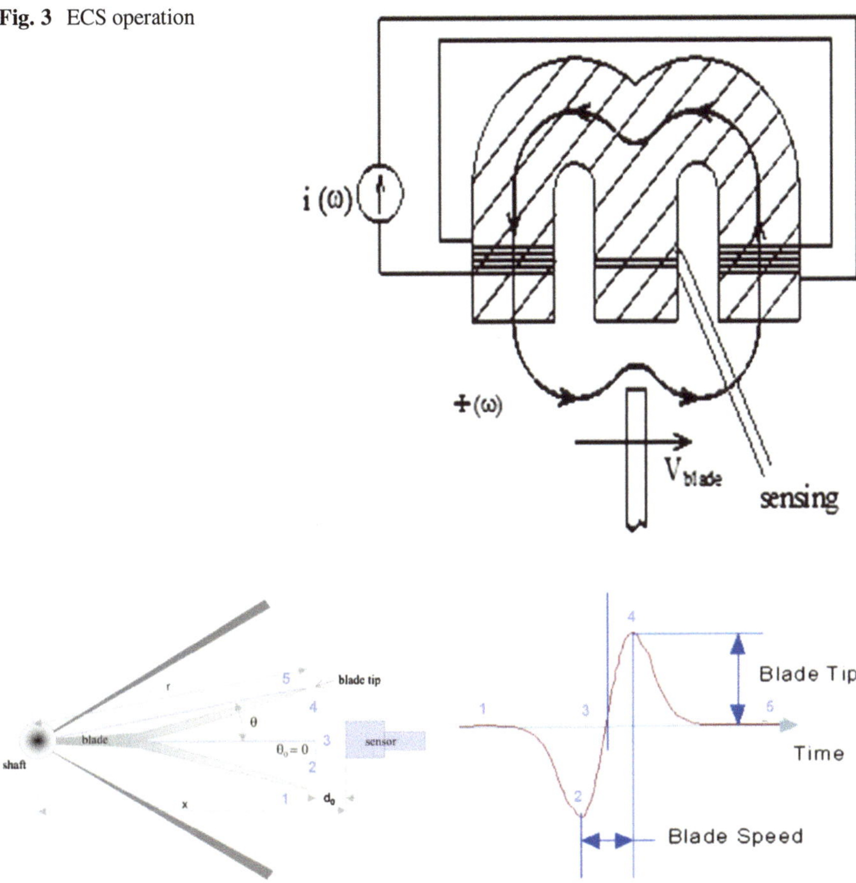

Fig. 3 ECS operation

Fig. 4 ECS signature

The sensor operates by creating a time-varying magnetic field, then measuring the field change caused by induced eddy currents which are created in a conducting object, for example, a turbine blade, when it enters the field. The sensor is rpm independent and relatively insensitive to temperature. It can monitor individual blades from the fan through the turbine with only materials changes. Initial tests on the fan blades of a next-generation gas turbine engine have proved the sensor capable of accurately detecting blade vibration, tip clearance, bent blades, missing blade tips, flutter, and pitch angle. Sensor power consumption is less than $\frac{1}{4}$ watt, and the S/N ratio achieved was better than 60 dB [40]. Figure 3 shows the principle of operation of the blade tip ECS, and Fig. 4, a sample of sensor signal output.

5 ECS for Prestall Detection

Degradation of transient stall margin due to mechanical wear of the engine has been shown to be potentially significant, and procedures for early detection of this degradation have been demonstrated. As described above, deducing precursors to rotating stall and flutter using a single sensor has been attempted by various researchers [12, 14, 26], but these researchers always used information that was "continuously available" from the sensors. Thus for instance static pressure at the casing wall or over the rotor, measured by a single sensor, has been used in full-scale engines to deduce the transient stall margin of the engine.

Performing these functions using ECSs adds a significant challenge to overall algorithm development. First, ECSs indirectly measure blade displacement, rather than wall-static pressure, which is a quantity more sensitive to flow perturbations. The second challenge associated with ECSs is that they measure discrete passage events of multiple blades as they spin past the sensor. Establishing a link between pre-stall perturbations and blade motion has been one of the goals of this work.

5.1 ECS Engine Test Data

GDAIS has given us access to various sets of ECS test data. The data used in the first analysis is data from an experiment performed by General Dynamics in collaboration with Pratt and Whitney to assess the effectiveness of the ECS in detecting (and classifying) blade features and engine performance [40]. We obtained samples of the data from these experiments and conducted a preliminary analysis of the signals using wavelet transform methods. The results, while very preliminary, indicated the potential of the methods for diagnostics of the blades and for detection of prestall conditions.

The traces we consider are from an ECS mounted in the outer casing of a Pratt and Whitney experimental gas turbine engine viewing the first set of rotor fan blades. The engine was operated over a significant portion of its performance envelope, and measurements were taken using the ECS. In this section we shall only discuss a small sample of the total experimental data. One 5 s data file from each of steady, accelerating, and prestall operating conditions was available. The signals in Fig. 5 correspond to the engine in steady-state operation: (1) ECS012, steady operation at 5,700 rpm; (2) ECS802, accelerating from 5,770 to 9,100 rpm and (3) ECS552, engine in prestall at 9,800 rpm. The traces correspond to a single revolution of the turbine; the elements of each curve correspond to the signatures of individual blades passing the sensor.

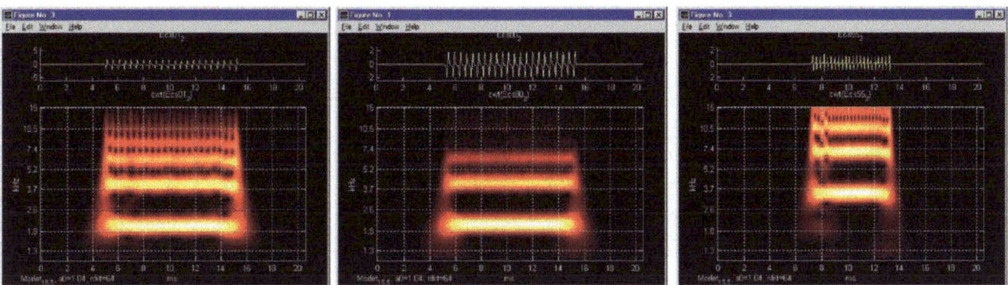

Fig. 5 OCWT of ECS test data: (**a**) constant speed, (**b**) acceleration, (**c**) prestall

5.2 *Wavelet Analysis of ECS Data*

Wavelet analyses of a small sample of the ECS data from the Pratt and Whitney engine were performed using the Morlet analyzing wavelet.[8] The preliminary analyses show the potential of the wavelet transform to differentiate between normal operation modes and prestall conditions.

The results of the analysis are displayed in Fig. 5. The top plot displays the time signal, and the bottom plot the over complete wavelet transform (OCWT). In the OCWT, time is plotted on the horizontal axis, and wavelet filter-bank center frequency is plotted on the vertical axis. A color map is used to indicate the magnitude of the (2D) transform signal. Low magnitudes are dark, and high magnitudes are light.

Each of the data sets consists of one revolution of data. The data was taken from the beginning of each of the files. Harmonics of the blade-passage speed are seen as horizontal bands. Some information from the individual blade signatures can be seen at the higher frequencies. The steady speed and acceleration data are taken at approximately the same rotor speed. It is interesting that in the acceleration data most of the energy is concentrated in the primary harmonic, and there is almost no energy in the higher frequency filter bands where the information from the individual blade signatures is observable. The prestall data, the right-hand plot in Fig. 5, is taken a higher rotor speed. In the prestall data a time-localized disturbance is visible in all of the harmonics of the blade speed near blade 4.[9]

[8] For more details on wavelet analysis refer to [41].

[9] The blade numbering begins with 1 after the rising edge of the synchronization pulse. This numbering may be inconsistent with the physical blade numbers from the experiment, which are unavailable to us at this time.

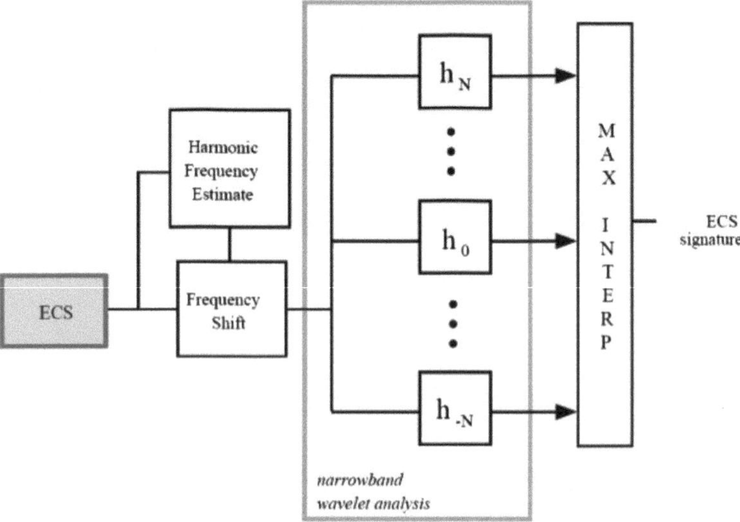

Fig. 6 Harmonic signature generation

5.3 Harmonic Processing

From the wavelet transform in Fig. 5, the embedded harmonic structure can plainly be seen as stripes. Note the deviations in each stripe. What are brought out by a time-frequency transform are these fine-frequency fluctuations over time associated with each harmonic.

The underlying harmonic structure of the ECS data leads us to consider a harmonic-based approach to extracting an engine-dependent signature. The fine fluctuations associated with each harmonic are tied to the physical makeup of the fan blades and their motions. For this reason it was reasonable to expect that changes in the physical makeup or their motions would lead to changes in the fine harmonic structure of the sensed ECS data.

One of the techniques we implemented to extract the harmonic signature is described pictorially in Fig. 6. It was predicated on the idea of a time-frequency analysis and subsequent frequency extraction at a specific harmonic (or harmonics) of the data. The primary computational element of the scheme is a filter bank with fine-frequency resolution. Such a bank may be configured to track very small, i.e., fine, frequency deviations in particular harmonics. If we assume a constant engine speed, the first step in the process is to estimate the fundamental harmonic and then shift the input signal in frequency such that the chosen harmonic, say the second, is in the center of the filter-bank frequency coverage range. Passing the shifted signal through the wavelet filter bank yields the wavelet transform restricted to a small band of frequencies, say 100 Hz. This followed by an arg-maximum operation yields a fine-frequency fluctuation curve around the given harmonic.

Underlying this approach is the (analytic) signal model

$$s(t) = \sum_{k=1}^{N} A_k(t) \mathrm{e}^{j2\pi \int_0^t (kf_0 + f_k(s))\,\mathrm{d}s},$$

where f_0 is the fundamental harmonic and the $f_k(t)$ describe, the characteristic fine-frequency variation in the kth harmonic. The hope is that one or some combination of these functions f_k will be sufficient to detect and characterize fault conditions in a given fan blade assembly.

Signature extraction may be accomplished using either wavelet or WFFT time-frequency representations. In addition, the signature extraction may be implemented in a "direct FM" method described below which sacrifices robustness to noise for increased computational speed.

During stall tests on the NASA Glenn compressor rig taken in 1999, GDAIS collected data from both an ECS and a pressure sensor simultaneously. Here we use the data set 99-204-1009-04, the same one used to demonstrate the two sensor detection in Fig. 1. The tests allow the comparison of the capability of the ECS to that of a pressure sensor, the current standard for stall prediction research. Figure 7 shows a direct comparison of the ECS data to that of the pressure sensor data. The stall cell is clearly evident in the pressure data (bottom), whereas it is not obvious from the raw ECS data (top).

Results of the signature analysis using the direct FM method in the figure below show potential for the prediction of stall from a single ECS. In the top plot is shown the fourth-harmonic signature extracted from a single ECS as the engine approaches stall. In the bottom plot is the data from a Kulite pressure sensor over the same interval. As can be seen from the figure, the stall cells are evident in the ECS signals by their effect on the frequency fluctuations of the blades.

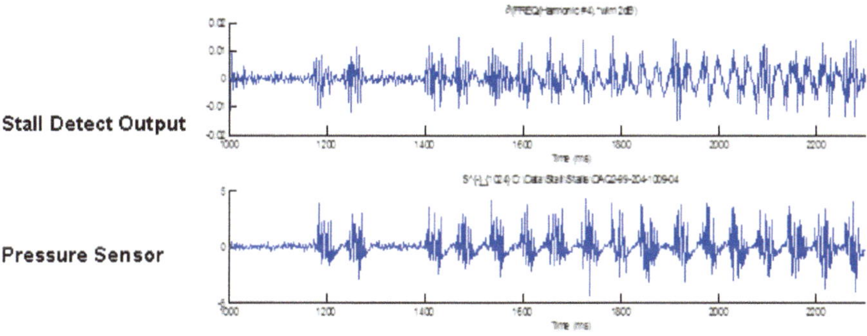

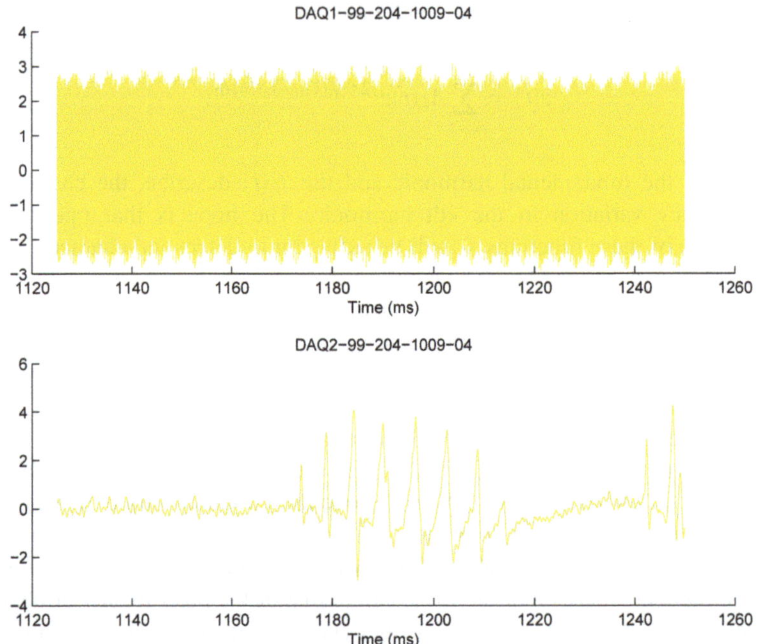

Fig. 7 ECS data versus pressure sensor data (zoom)

6 FM Algorithm Implementation

The direct FM algorithm is described first in its post processing form. In this case we take in all the data including the stall and process it in batch mode. In order to implement the algorithm for real-time detection of stall precursors the algorithm is modified to work on smaller segments of data. In the real-time implementation sampling rate becomes a critical issue because the time it takes to compute the FFT grows significantly with the number of sample points. The real-time implementation is discussed next along with determinations of the minimum acceptable sampling rate.

6.1 Basic FM Algorithm

The basic flow diagram for the direct FM implementation of the algorithm is shown in the figure stall detection processing I below.

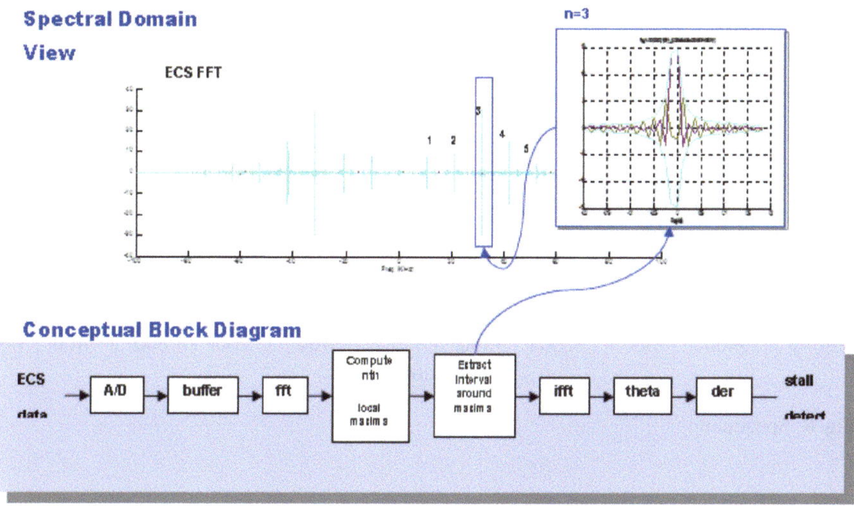

The ECS data is sampled by an A/D converter. The FFT (fast Fourier transform) of the data is taken. The frequency of the mth local maximum of the FFT is computed. This is the frequency of the mth harmonic of the ECS data. An interval of data around the maxima is extracted, and the inverse FFT of this data is taken. While the original sampled data was real, because we have created an odd frequency spectrum by extracting an interval around the maxima only on the positive frequency axis, the signal $\{x_k\}$ resulting from the inverse FFT of the interval is complex. From the complex data we can compute the angle at each sample as $\theta_k = \tan^{-1}(\Im(x_k)/\Re(x_k))$. The derivative of this angle is the instantaneous frequency. The derivative of the instantaneous frequency is our stall detector output.

High-bandwidth pressure sensors are the industry standard sensors for stall detection. The output of our detector versus a high-bandwidth kulite pressure sensor is shown on the previous page.

6.2 Real-Time Algorithm

In order to obtain quick detection information we segmented the data into blocks for processing. The block size is directly related to the delay in obtaining the detector output. There is a trade-off because a larger block tends to yield better results but smaller blocks lead to less delay in detection. Besides the actual time to collect the block of data, a main constraint was that the blocks be small enough that the FFT could be computed in a reasonable amount of time. We used blocks of size 2^k, where $k \ll 18$. We were able, increase processing speed also by reducing the sampling rate. Figure 8 shows the Fourier spectrum of ECS data plotted on a dB scale. The ECS data was taken in a NASA W8 Compressor test facility, described briefly in Sect. 6,

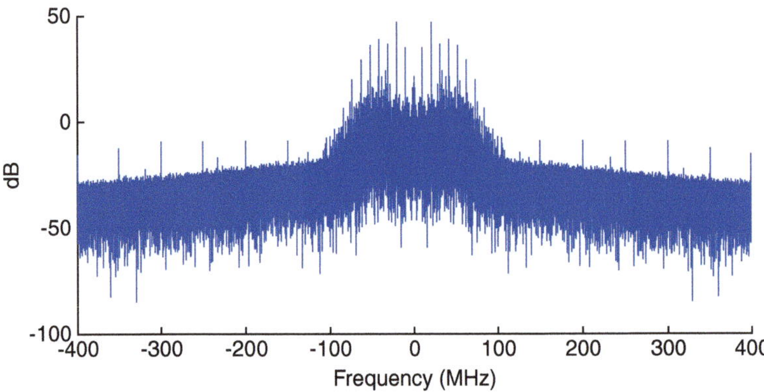

Fig. 8 Spectrum

and sampled at 800 kHz. From the figure it is clear that the bandwidth of the signal is 100 KHz. Thus reducing the sampling rate to 200 KHz appears to be sufficient to capture all harmonics.

A flow diagram of the "segmented stall detection algorithm" that was used for the real-time implementation of the algorithm is shown below.

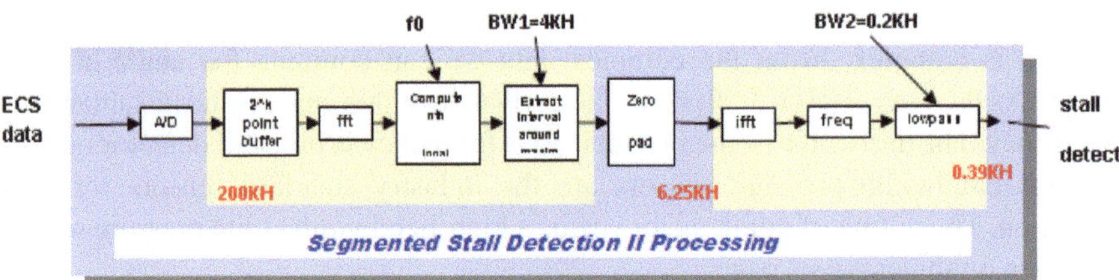

The algorithm proceeds as follows. First, we collect a 2^k point buffer of data sampled at 200 KHz. As above, the FFT of the data is taken. The frequency of the mth local maximum of the FFT is computed. A 4 Khz interval of data around the maxima is extracted. In the real-time implementation, we then zero pad the extracted interval data to the next highest power of 2. The effect of the zero padding is to (1) improve the performance of the FFT because the FFT is optimized to work on data sets that are powers of 2 in size and (2) interpolate the inverse FFT results using Sync interpolation. At this point the data rate is 6.25 KHz. Next the inverse FFT of this data is taken. From the complex results of the inverse FFT, we compute the angle at each sample time and then the derivative of this angle to obtain the instantaneous frequency. Following this we low-pass filter the data which reduces the final data rate to 0.39 KHz from the original 200 KHz.

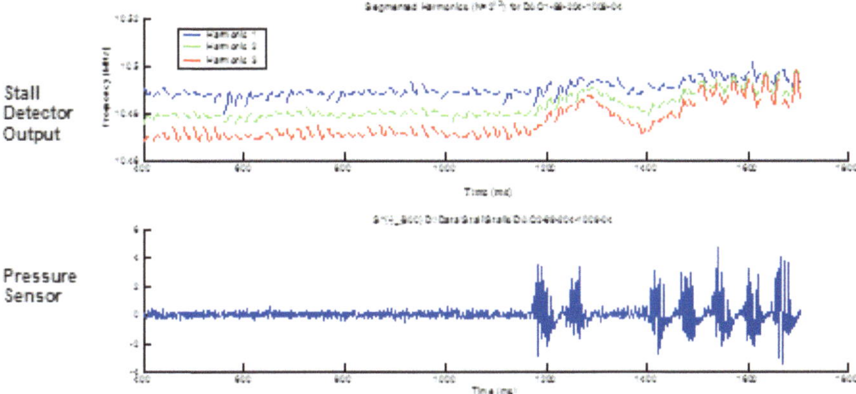

Fig. 9 Segmented stall detection

The results of the segmented (real-time) processing algorithms are shown in Fig. 9. The bottom plot is again the high bandwidth pressure sensor data. The top plot shows the stall detector output for each of the first three harmonics. We have found that in general harmonic 3 seems to be the best indicator, but the others will also work. In the stall detector output, each segment covers 10 ms and consists of 8 samples. Leading and trailing samples contain transient response of lowpass filter but the overall shape of harmonic segments follows the envelope of non-segmented stall detector output.

We implemented the FM algorithm in C on a RT Linux platform, using one of the many FFT libraries available. We used a high-speed DAQ board capable of sampling 4 channels at 200 KHz.

6.3 *Testing*

The FM algorithm was tested in real time at NASA Glenn W8 compressor test facility; see Fig. 10. The test article used in the testing was a research fan designated Rotor 67. The fan has a relatively low aspect ratio of 1.56, a tip speed of 429 m/s, and an inlet relative tip Mach number of 1.38.

The W8 facility is well suited for stall testing. A circumferential array of high-response pressure sensors placed upstream of the compressor was available for verification purposes. In addition, a throttle that can be moved very smoothly and slowly has been developed, allowing slow, accurate approaches to stall.

We completed 3 separate sets of tests in December 2003, February 2004, and June 2004 respectively. The first two tests were completed in a rotor only configuration, i.e., there was not a complete engine stage; there was a rotor but no stator. In this configuration data was collected with both clean and radially distorted inlet flows.

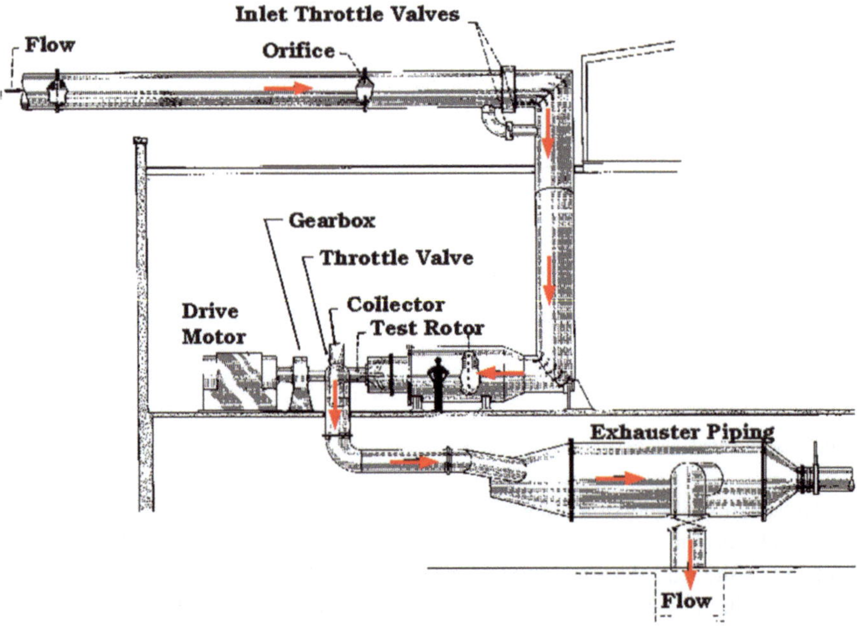

Fig. 10 NASA W8 compressor test facility schematic

The data taken was a more complete set including:

(1) Stalls at 70, 80, 80, and 100 % of design speed (induced by closing the throttle valve to reduce the mass flow) with clean and distorted inlet flow.
(2) Stalls induced by acceleration and deceleration with clean and distorted inlet flow.

The third tests were completed with a full engine stage and casing-mounted injectors. The casing-mounted injectors (Fig. 11) were located upstream of the rotor. The injectors were designed to generate a jet along the casing wall. The injected flow was aligned with the inlet annulus flow in the downstream axial direction to minimize mixing between the annulus flow and the injected flow. The injectors were connected to recirculation vent that takes air from behind the stage and reinjects it in front of the stage. Stall measurements were taken with the recirculation vent open and with it closed off. The effect of the recirculation was increased rotor stability.[10] The detection algorithms were not adversely affected by the recirculation (Fig. 11). In all of the tests at constant speed, modal waves were observed in the pressure data preceding stall. Surge was not observed. In the tests where stall was entered by increasing or decreasing rotor speed our algorithms had problems with detection.

[10]The recirculation tests were performed as part of another testing effort and the results of the stability performance are reported elsewhere.

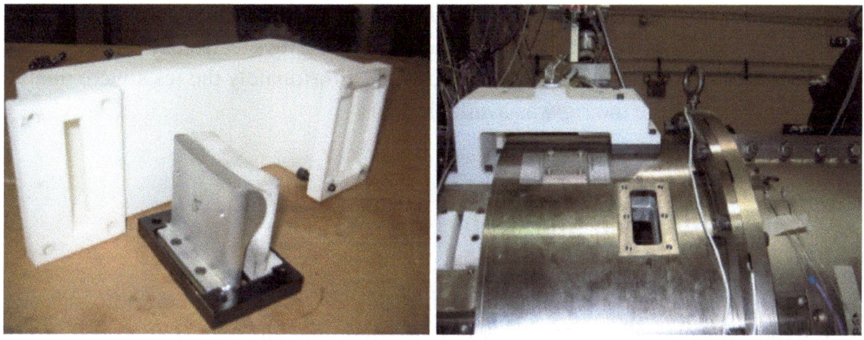

Fig. 11 *Left*—injector and recirculation apparatus. *Right*—mounted on rig casing

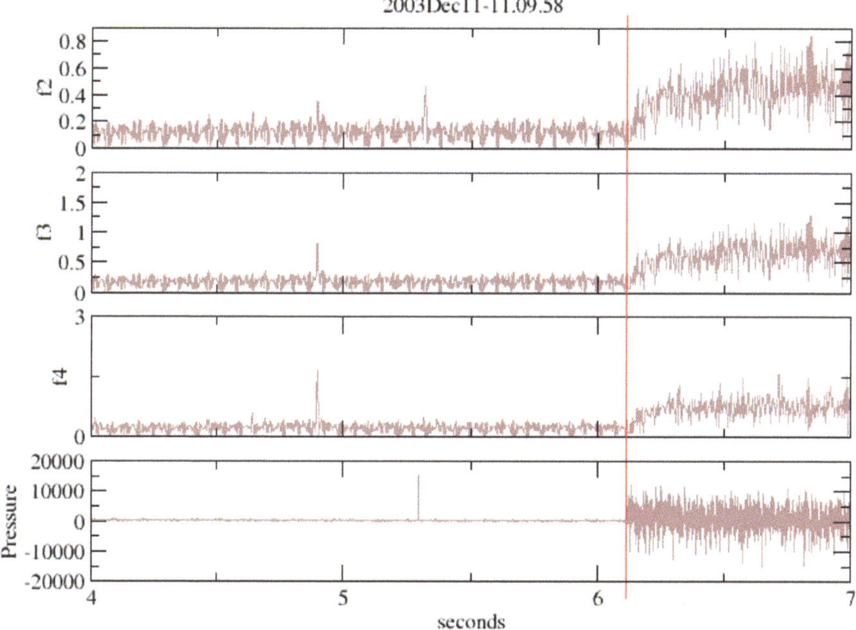

Fig. 12 Real-time processor output: *top* 3—harmonics of ECS output and *bottom*—pressure sensor data

Figure 12 shows output of the real-time processor for clean inlet flow. The top 3 plots are the first 3 harmonics of the ECS data. The bottom plot is raw pressure sensor data.

Our goal was to attempt to actively control the injectors, i.e., to detect stall precursors (or at worst stall onset) and only then turn on the injectors. In the active control setup, the injectors were connected to a torus that was supplied with air at ambient temperature from an external source. During active stall control, the injectors would be able to be turned on and off using an electromechanical valve, and the flow rate would be able to be adjusted. Unfortunately the tests were unable to be completed while the rig was available to us.

7 Threshold-Based Stall Detection

While we were able to demonstrate with the direct FM algorithm the capability to reliably detect stall using one ECS, we felt that the algorithms used were complex and questioned whether the detection could be accomplished with one ECS using a computationally simpler method, for example, a parametric method. To understand which signature parameters would be most effective for stall detection a good understanding of the blade motion during prestall was needed as well as an understanding of how the sensor signature changes because of the motion.

It was our contention that some measurement of blade twist due to the modal variations in pressure would be the key information to extract. Blade twist affects the signature shape by changing the width of the positive and negative lobes of the signature (as well as the time of arrival depending on sensor location relative to the twist axis).

Instead of measuring lobe width directly, i.e., signature start to zero crossing and then zero crossing to signature end, we picked an arbitrary positive threshold greater than the noise floor and considered the widths between threshold crossings. Threshold crossings have the advantage that they can be easily measured in hardware, whereas extrema and zero crossings (due to noise) are more difficult to accurately detect with simple hardware.

A threshold point is defined as a point where the absolute value of the sensor signature crosses the threshold, where the threshold value is a positive real number. As shown in Fig. 13, the GDAIS ECS blade signature has four threshold points, two points corresponding to the threshold value and two points corresponding to the negative of the threshold value.

The time elapsed from when the blade signature hits its first threshold point to when it hits its last is what we have termed the *approximate sensor aperture*. The time elapsed between the first and second threshold points is termed *threshold width (1)* or δt_L, and the time elapse between the third and fourth threshold points is termed *threshold width (2)* or δt_R.

For the GDAIS ECS, the threshold widths as well as other functions of the threshold width, for example their difference and their ratio, are good indicators of stall. We expect that functions of thresholds will be useful for other types of blade tip sensors such as other ECS designs and capacitive sensors both of which typically have a single lobe for each blade passage.

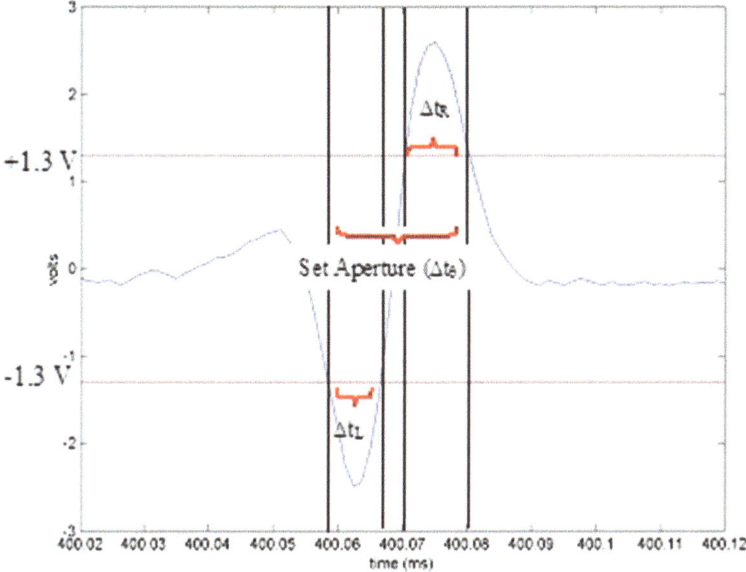

Fig. 13 ECS blade signature with threshold at +/− 1.3 V

7.1 Effect of Stall on Threshold Widths

Interestingly our tests showed that preceding a stall event, threshold width (1) responds in the opposite direction to threshold width (2). Threshold width (2) decreases abruptly with each stall cell, while threshold width (1) increases. We hypothesize that this is related to blade twist.

Another property of the two threshold widths in each blade signature is that they are usually unequal. In other words, the blade signature is asymmetric. This has to do with the imperfection in alignment of the ECS with the blade. In addition, dynamic loading causes the alignment to change with speed because the loading causes the blades to twist from their rest state. The results from the NASA data indicate that the threshold ratio can be used to predict stall in situations with both clean and distorted inlet flow.

While extrema remain essentially constant until 9.514 s, Fig. 14 shows that moving averages of the threshold widths show trends of slowly increasing or decreasing, beginning approximately at 6 s into the data set.

Building on this result, a detector is developed that looks at the variance of the threshold ratio to find a stall indicator. The variance of the threshold ratio gives a good stall warning indicator at about 2 s, or 540.54 revolutions before the stall event see Fig. 15. The data is taken at 100 % of design speed as the compressor is throttled into stall. The top plot of Fig. 15 shows the threshold ratio, the middle plot shows the variance of the threshold ratio, and the bottom plot the stall detector output. Results

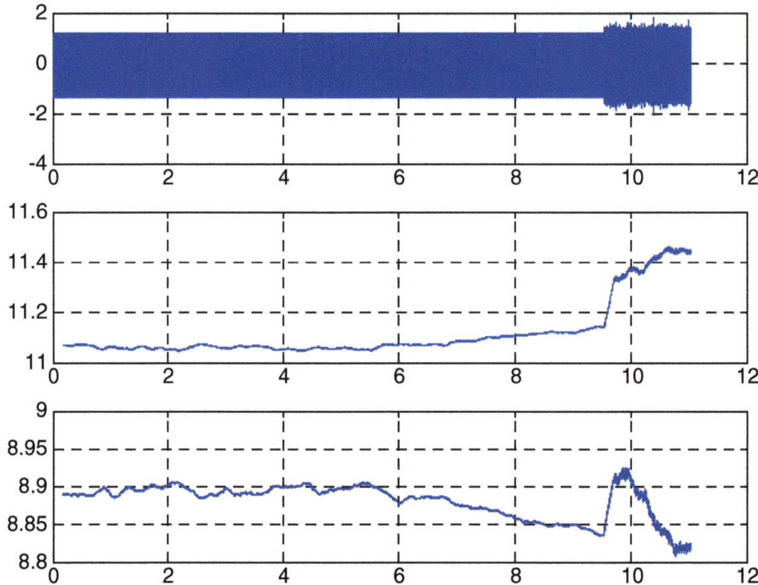

Fig. 14 *Top*: raw ECS data, *Middle*: moving average of threshold width (1). *Bottom*: moving average of threshold width (2)

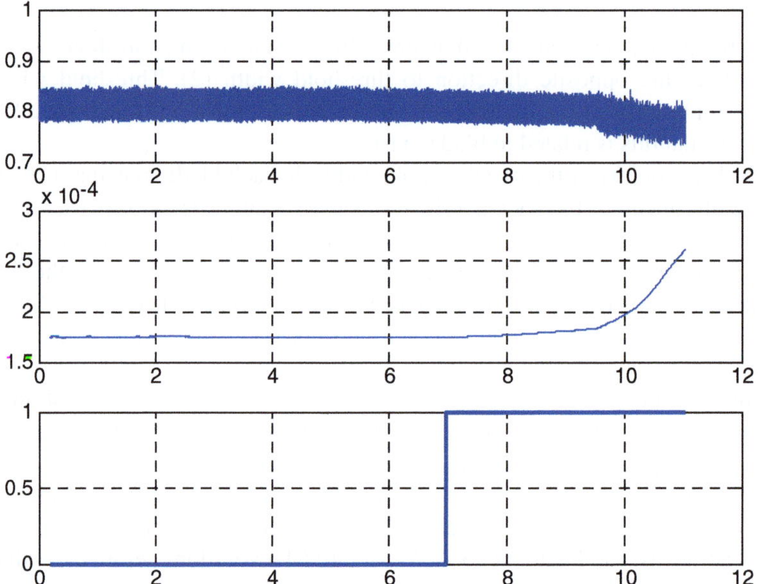

Fig. 15 Threshold ratio, variance of threshold ratio, and stall detector performance

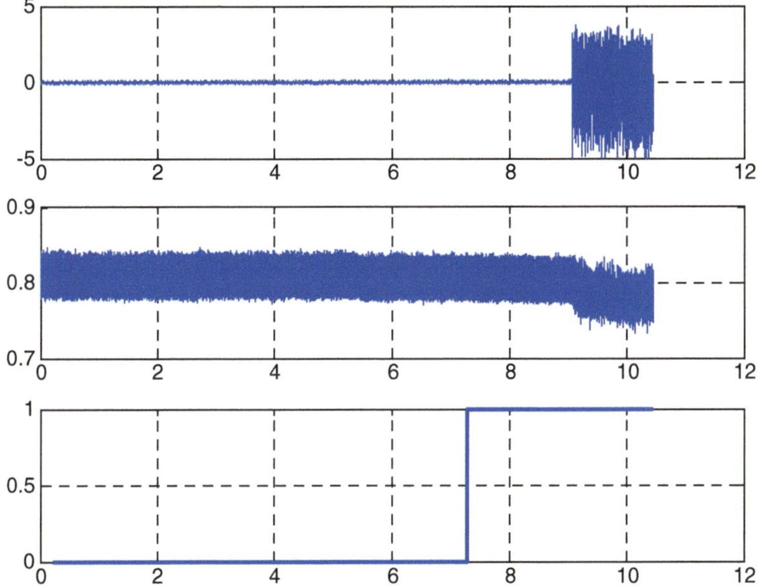

Fig. 16 Pressure sensor, threshold ratio, and stall detector output at 90 % of design speed

as the compressor is throttled into stall at 90 % and 70 % of design speed are shown in Figs. 16 and 17, respectively. In the figures, the top plots are the pressure sensor data, the middle plots are the threshold ration data, and the bottom plots are the stall detector output.

8 Conclusions

We have demonstrated in testing that a single ECS can be used to detect stall precursors. We have also shown the ECS data can be processed to output a data stream that closely follows pressure sensor data. This makes the ECS a possible candidate for replacing a pressure sensor in a stall control scenario, with the added advantage that other engine health data can be collected from the ECS as well.

Because the blade motion is a secondary affect of the pressure and flow variations that precede stall, algorithms to detect stall with ECS will most likely not be as sensitive to stall precursors, though a thorough study has not yet been performed to compare the results of single pressure-sensor detection algorithms to those of the ECS.

Because the ECS data can be processed to yield a data stream that closely follows pressure sensor data some of the techniques for stall detection using pressure sensors may apply.

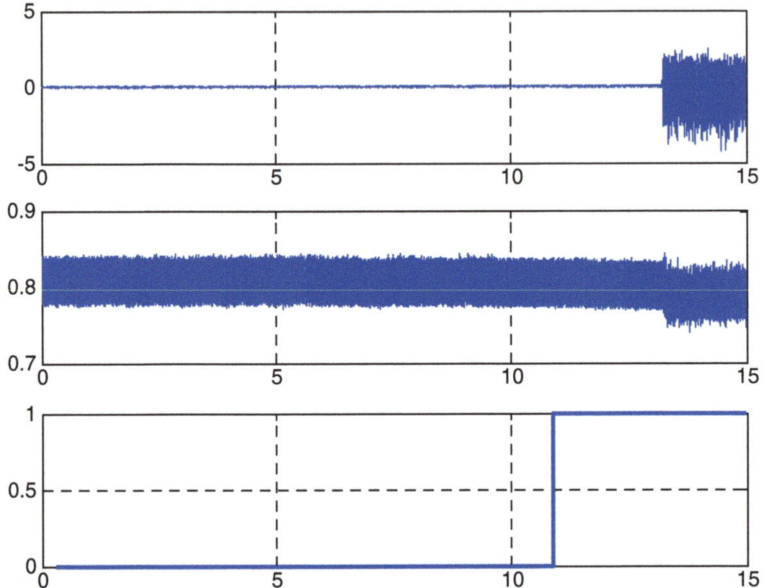

Fig. 17 Pressure sensor, threshold ratio, and stall detector output at 70 % of design speed

Acknowledgments There are very few engine test facilities and fewer opportunities for a small company to participate in testing. Making this testing opportunity possible involved the help of many people. The authors would like to thank Carole Ginty, Vithal Dalsania, Michael Hathaway, Joe Veres, Sanjay Garg, and Osvlado Rivera for allowing us to piggyback on their tests under the UEET Program and also Tony Strazisar for coordination of the testing and testing advice and Dennis Culley, Jose Gonzales, Helmi Abulaban, Scott Thorp, Sue Prahst, Rick Brokopp, and Bruce Wright for their essential roles in preparing for and conducting the test. We would also like to thank Erdal Unver at GDAIS and Ravi Ravindranath at NAVAIR for lending sensors and electronics for the testing. Also Mike Dowell at GDAIS who provided data from previous engine tests to facilitate algorithm development. This work has been funded by NASA and NSF SBIR programs, contract NAS#-99002, and NSF-0110316, NSF-0216021, respectively.

References

1. DeLaat, J.C., Southwick, R.D., Gallops, G.W.: High stability engine control (HISTEC). In: 32nd Joint Propulsion Conference (1996)
2. DeLaat, J.C., et al.: High stability engine control (HISTEC) program: flight demonstration phase. In: 34nd Joint Propulsion Conference (1998)
3. Day, I.J.: Active suppression of rotating stall and surge in axial compressors. J. Turbomachinery **115**, 40–47 (1993)
4. Paduano, J.D., et al.: Active control of rotating stall in a low speed axial compressor. ASME J. Turbomachinery **115**, 48–56 (1993)
5. Behnken, R.L., D'Angrea, R., Murray, R.M.: Control of rotating stall in a low speed axial flow compressor using pulsed air injection: modeling, simulations, and experimental validation. In: 34th Conference on Decision and Control (1995)

6. Haynes, J.M., Hendricks, G.J., Epstein, A.H.: Active stabilization of rotating stall in a three-stage axial compressor. ASME J. Turbomachinery **116**, 226–239 (1994)
7. Suder, K.L., et al.: Compressor stability enhancement using discrete tip injection. In: ASME Turbo Expo (2000)
8. Dowell, M., Sylvester, G.: Turbomachinery prognostics and health management via eddy current sensing: Current developments, Proceedings IEEE Aerospace Conference, **3**, 1–9 (1999)
9. Day, I.J.: Stall inception in axial flow compressors. J. Turbomachinery **115**, 1–9 (1993)
10. Moore, F.K., Greitzer, E.M.: A theory of post stall transients in axial compression systems: part 1-development of equations. J. Eng. Gas, Turbines, Power **108**, 68–76 (1986)
11. Tryfonidis, M., et al.: Prestall behavior of several high-speed compressors. J. Turbomachinery **117**, 62–80 (1995)
12. Bright, M.M., et al.: Stall precursor identification in high-speed compressor stages using chaotic time series analysis methods. J. Turbomachinery **119**, 491–499 (1997)
13. Camp, T.R., Day, I.J.: A study of spike and modal stall phenomena in a low speed axial compressor. J. Turbomachinery **120**, 393–401 (1998)
14. Bright, M.M., Qammar, H.K., Wang, L.: Investigation of pre-stall mode and pip inception in high-speed compressors through the use of correlation integral. J. Turbomachinery **121**, 743–750 (1999)
15. Day, I.J., et al.: Stall inception and the prospects for active control in four high-speed compressors. J. Turbomachinery **121**, 18–27 (1999)
16. Hoss, B., Leinhos, D., Fottner, L.: Stall inceptions in the compressor system of a turbofan engine. J. Turbomachinery **122**, 32–44 (2000)
17. McDougall, N.M., Cumpsty, N.A., Hynes, T.P.: Stall inception in axial compressors. ASME J. Turbomachinery **112**, 116–125 (1990)
18. Hendricks, G.J., et al.: Analysis of rotating stall onset in high speed axial flow compressors. In: AIAA Paper No (1993)
19. Weigl, H.J., et al.: Active stabilization of rotating stall and surge in a transonic single-stage axial compressor. J. Turbomachinery **120**, 625–636 (1998)
20. Dowell, M., et al.: Progress in turbomachinery prognostics and health management via eddy current sensing: current. In: IEEE Aerospace Conference (2001)
21. Grassberger, P., Procaccia, I.: Measuring the strangeness of strange attractors. Physica D **9**, 189–208 (1983)
22. Provenzale, A., et al.: Distinguishing between low-dimensional dynamics and randomness in measured time series. Physica D **58**, 31–49 (1992)
23. Bright, M.M., et al.: Rotating pip detection and stall warning in high speed compressors using structure function. In: AGARD RTO AVT Conference (1998)
24. Bright, M.M.: Chaotic time series analysis tools for identification and stabilization of rotating stall precursor events in high speed compressors, PhD thesis, University of Akron (2000)
25. Liao, S.F., Chen, J.Y.: Time frequency analysis of rotating stall by means of wavelet transform. In: ASME 96-GT-57 (1996)
26. Cheng, X., Chen, J., Nie, C.: Investigation on the precursor behavior of compressor rotating stall through two-dimensional wavelet transform. Institute of Engineering Thremophysics, Chinese Academy of Sciences, Beijing (1999)
27. Lin, F., Chen, J., Li, M.: Practical issues of wavelet analysis of unsteady rotor tip flows in compressors. In: AIAA (2002)
28. Lin, F., Chen, J., Li, M.: Experimental investigation of unsteady rotor tip flows in a high speed compressor throttled to stall. In: ASME Turbo Expo GT-2002-30360 (2002)
29. Greitzer, E.M., Moore, F.K.: A theory of post stall transients in axial compression systems: part II-applications. J. Eng. Gas, Turbines, Power **108**, 231–239 (1986)
30. Feullner, M.R., Hendricks, G.J., Paduano, J.D.: Modeling for control of rotating stall in high speed multistage axial compressors. ASME J. Turbomachinery **118**(1), 1–10 (1996)
31. Sepulcre, R., Kokotovic, P.V.: Shape signifiers for control of surge and stall in jet engines. IEEE Trans. Automat. Contr. **43**, 1643–1648 (1995)

32. Wang, H.H.: A control-oriented model for aeroengine compressors with deep hysteresis. (1996, Preprint)
33. van Schalkwyk, M.C., et al.: Active stabilization of axial compressors with circumferential inlet distortion. J. Turbomachinery, **120**, 431–439 (1998)
34. Hynes, T.P., Greitzer, E.M.: A method for assessing effects of circumferential flow distortion on compressor stability. ASME J. Turbomachinery **109**, 371–379 (1987)
35. Liaw, D.C., Abed, E.H.: Active control of compressor stall inception: a bifurcation theoretic approach. Automatica **32**, 109–115 (1996)
36. McCaughan, F.E.: Bifurcation analysis of axial flow compressor stability. SIAM J. Appl. Math **50**, 1232–1253 (1990)
37. Yeung, S., Murray, R.M.: Nonlinear control of rotating stall using axisymmetric bleed with continuous air injection on a low-speed, single stage axial compressor. in AIAA joint propulsion conference (1997)
38. Yeung, S., Murray, R.M.: Reduction of bleed valve rate requirements for control of rotating stall using continuous air injection. In: Conference on Control Applications (1998)
39. Eveker, K.M., et al.: Integrated control of rotating stall and surge in high-speed multistage compression systems. J. Turbomachinery **120**, 440–445 (1998)
40. Terpay, G., Zipfel, G.: Measuring blade condition in a gas turbine engine using eddy currents. In: International Conference3 on Adaptive Structures and Technologies (1998)
41. Teolis, A.: Computational Signal Processing with Wavelets. Applied and Numerical Harmonic Analysis. Birkhauser, Boston (1998)

State-Dependent Channels: Strong Converse and Bounds on Reliability Function

Himanshu Tyagi and Prakash Narayan

Abstract We consider an information theoretic model of a communication channel with a time-varying probability law. Specifically, our model consists of a state-dependent *discrete memoryless channel*, in which the underlying state process is independent and identically distributed with known probability distribution, and for which the channel output at any time instant depends on the inputs and states only through their current values. For this channel, we provide a strong converse result for its capacity, explaining the structure of optimal transmission codes. Exploiting this structure, we obtain upper bounds for the reliability function when the transmitter is provided channel state information causally and noncausally. Instrumental to our proofs is a new technical result which provides an upper bound on the rate of codes with code words that are "conditionally typical over large *message-dependent* subsets of a typical set of state sequences." This technical result is a nonstraightforward extension of an analogous result for a discrete memoryless channel without states; the latter provides a bound on the rate of a good code with code words of a fixed composition.

Keywords State dependent channel • Channel state information • Strong converse • Reliability function • Capacity • Probability of error • Gelfand–Pinsker channel • Type • Typical set • DMC

H. Tyagi • P. Narayan (✉)
Department of Electrical and Computer Engineering and the Institute for Systems Research, University of Maryland, College Park, MD 20742, USA
e-mail: tyagi@umd.edu; prakash@umd.edu

T.D. Andrews et al. (eds.), *Excursions in Harmonic Analysis, Volume 1*, Applied and Numerical Harmonic Analysis, DOI 10.1007/978-0-8176-8376-4_22,

1 Introduction

The information theoretic model of a communication channel for message transmission is described by the conditional probability law of the channel output given the input. For instance, the binary symmetric channel is a model for describing the communication of binary data in which noise may cause random bit-flips with a fixed probability. A reliable encoded transmission of a message generally entails multiple uses of the channel. In several applications, such as mobile wireless communication, digital fingerprinting, and storage memories, the probability law characterizing the channel can change with time. This time-varying behavior of the channel probability is described typically in terms of the evolution of the underlying channel condition, termed "state." The availability of *channel state information* (CSI) at the transmitter or receiver can enhance overall communication performance (cf. [1, 6, 7]).

We consider a state-dependent *discrete memoryless channel* (DMC), in which the underlying state process is independent and identically distributed (i.i.d.) with known probability mass function (PMF), and for which the channel output at any time instant depends on the inputs and states only through their current values. We address the cases of *causal and noncausal* CSI at the transmitter. In the former case, the transmitter has knowledge of all the past channel states as well as the current state; this model was introduced by Shannon [8]. In the latter case, the transmitter is provided access at the outset to the entire state sequence prevailing during the transmission of a message; see Gelfand–Pinsker [5]. We restrict ourselves to the situation where the receiver has no CSI, for receiver CSI can be accommodated by considering the states, too, as channel outputs.

Two information theoretic performance measures are of interest: *Channel capacity* and *reliability function*. The channel capacity characterizes the largest rate of encoded transmission for reliable communication. The reliability function describes the best exponential rate of decay of decoding error probability with transmission duration for coding rates below capacity. The capacities of the models above with causal and noncausal CSI were characterized in classic papers by Shannon [8] and Gelfand–Pinsker [5]. The reliability function is not fully characterized even for a DMC without states; however, good upper and lower bounds are known, which coincide at rates close to capacity [3, 9, 10].

Our contributions are twofold. First, we provide a *strong converse* for the capacity of state-dependent channels, which explains the structure of optimal codes. Second, exploiting this structure, we obtain upper bounds for the reliability functions of the causal and noncausal CSI models. Instrumental to our proofs is a new technical result which provides an upper bound on the rate of codes with code words that are "conditionally typical over large *message-dependent* subsets of a typical set of state sequences." This technical result is a nonstraightforward analog of [3, Lemma 2.1.4] for a DMC without states; the latter provides a bound on the rate of a good code with codewords of a fixed composition. A preliminary conference version of this work is in [11].

In the next section, we compile pertinent technical concepts and tools that will be used to prove our results. These standard staples can be found, for instance, in [2, 3]. The channel models are described in Sect. 3. Sections 4–6 contain our main results.

2 Preliminaries: Types, Typical Sets and Image Sets

Let $\mathscr{X}$ be a finite set. For a sequence $\mathbf{x} \in \mathscr{X}^n$, the type of $\mathbf{x}$, denoted by $Q_{\mathbf{x}}$, is a pmf on $\mathscr{X}$, where $Q_{\mathbf{x}}(x)$ is the relative frequency of x in $\mathbf{x}$. Similarly, *joint types* are pmfs on product spaces. For example, the joint type of two given sequences $\mathbf{x} \in \mathscr{X}^n$ $\mathbf{s} \in \mathscr{S}^n$ is a pmf Q on $\mathscr{X} \times \mathscr{S}$, where $Q_{\mathbf{x},\mathbf{s}}(x,s)$ is the relative frequency of the tuple (x,s) among the tuples (x_t,s_t), $t = 1,\ldots,n$. Joint types of several n-length sequences are defined similarly.

The number of types of sequences in $\mathscr{X}^n$ is bounded above by $(n+1)^{|\mathscr{X}|}$. Denoting by $\mathscr{T}_Q^{(n)}$ the set of all sequences in $\mathscr{X}^n$ of type Q, we note that

$$(n+1)^{-\|\mathscr{X}\|}\exp[nH(Q)] \le \left\|\mathscr{T}_Q^{(n)}\right\| \le \exp[nH(Q)]. \tag{1}$$

For any pmf P on $\mathscr{X}$, and type Q on $\mathscr{X}^n$,

$$\begin{aligned}\mathrm{P}^n(\mathbf{x}) &= \prod_{t=1}^{n}\mathrm{P}(x_t) = \prod_{x\in\mathscr{X}}\mathrm{P}(x)^{nQ(x)}\\ &= \exp[-n(D(\mathrm{P}\|Q)+H(Q))], \quad \mathbf{x}\in\mathscr{T}_Q^{(n)},\end{aligned}$$

from which, along with (1), it follows that

$$(n+1)^{-\|\mathscr{X}\|}\exp[-n(D(\mathrm{P}\|Q)] \le \mathrm{P}^n\left(\mathscr{T}_Q^{(n)}\right) \le \exp[-n(D(\mathrm{P}\|Q)].$$

Next, for a pmf P on $\mathscr{X}$ and $\delta > 0$, a sequence $\mathbf{x} \in \mathscr{X}^n$ is P typical with constant δ if

$$\max_{x\in\mathscr{X}} |Q_{\mathbf{x}}(x) - \mathrm{P}(x)| \le \delta,$$

and $\mathrm{P}(x) = 0$ implies $Q_{\mathbf{x}}(x) = 0$. The set of all P-typical sequences with constant δ, is called the P-typical set, denoted $\mathscr{T}_{[\mathrm{P}]}^{(n)}$ (where the dependence on δ is not displayed explicitly). Thus,

$$\mathscr{T}_{[\mathrm{P}]}^{(n)} = \bigcup_{\substack{\text{types } Q:\\ \max_{x\in\mathscr{X}}|Q_{\mathbf{x}}(x)-\mathrm{P}(x)|\le\delta}} \mathscr{T}_Q^{(n)}.$$

In general, $\delta = \delta_n$ and is assumed to satisfy the "δ-convention" [3], namely

$$\delta_n \to 0, \quad \sqrt{n}\delta_n \to \infty \text{ as } n \to \infty. \tag{2}$$

The typical set has large probability. Precisely, for $\delta = \delta_n$ as in (2),

$$\mathrm{P}^n\left(\mathscr{T}_Q^{(n)}\right) \geq 1 - \frac{\|\mathscr{X}\|}{4n\delta^2}. \tag{3}$$

Consider sequences $\mathbf{x} \in \mathscr{X}^n$, $\mathbf{y} \in \mathscr{Y}^n$ of joint type $Q_{\mathbf{x},\mathbf{y}}$. The sequence $\mathbf{y} \in \mathscr{Y}^n$ has conditional type V if $Q_{\mathbf{x},\mathbf{y}} = Q_{\mathbf{x}}V$, for some stochastic matrix $V : \mathscr{X} \to \mathscr{Y}$. Given a stochastic matrix $W : \mathscr{X} \to \mathscr{Y}$, and $\mathbf{x} \in \mathscr{X}^n$, a sequence $\mathbf{y} \in \mathscr{Y}^n$ of conditional type V is W-conditionally typical if for all $x \in \mathscr{X}$:

$$\max_{y \in \mathscr{Y}} |V(y \mid x) - W(y \mid x)| \leq \delta,$$

and $W(y \mid x) = 0$ implies $V(y \mid x) = 0$. The set of all W-conditionally typical sequences conditioned on $\mathbf{x} \in \mathscr{X}^n$ is denoted by $\mathscr{T}_{[W]}^{(n)}(\mathbf{x})$. In a manner similar to (3), it holds that

$$W^n\left(\mathscr{T}_{[W]}^{(n)}(\mathbf{x}) \mid \mathbf{x}\right) \geq 1 - \frac{\|\mathscr{X}\|\|\mathscr{Y}\|}{4n\delta^2}.$$

For a subset A of $\mathscr{X}$, we shall require also estimates of the minimum cardinality of sets in $\mathscr{Y}$ with significant W-conditional probability given $x \in A$. Precisely, a set $B \subseteq \mathscr{Y}$ is an ε-image $(0 < \varepsilon \leq 1)$ of $A \subseteq \mathscr{X}$ under $W : \mathscr{X} \to \mathscr{Y}$ if $W(B \mid x) \geq \varepsilon$ for all $x \in A$. The minimum cardinality of ε-images of A is termed the image size of A (under W), and is denoted by $g_W(A, \varepsilon)$. Coding theorems in information theory use estimates of the rates of the image size of $A \subseteq \mathscr{X}^n$ under W^n, i.e., $(1/n)\log g_{W^n}(A, \varepsilon)$. In particular, for multiterminal systems, we compare the rates of image sizes of $A \subseteq \mathscr{X}^n$ under two different channels W^n and V^n. Precisely, given stochastic matrices $W : \mathscr{X} \to \mathscr{Y}$ and $V : \mathscr{X} \to \mathscr{S}$, for every $0 < \varepsilon < 1$, $\delta > 0$ and for every $A \subseteq \mathscr{T}_{[\mathrm{P}_X]}^{(n)}$, there exists an auxiliary rv U and associated pmfs $\mathrm{P}_{UXY} = \mathrm{P}_{U|X}\mathrm{P}_X W$ and $\mathrm{P}_{UXZ} = \mathrm{P}_{U|X}\mathrm{P}_X V$ such that

$$\left| \frac{1}{n} \log g_{W^n}(B(m_0), \varepsilon) - H(Y|U) - t \right| < \delta, \tag{4}$$

$$\left| \frac{1}{n} \log g_{V^n}(B(m_0), \varepsilon) - H(S|U) - t \right| < \delta,$$

where $0 \leq t \leq \min\{I(U \wedge Y), I(U \wedge S)\}$.

3 Channels with States

Consider a state-dependent DMC $W : \mathscr{X} \times \mathscr{S} \to \mathscr{Y}$ with finite input, state, and output alphabets $\mathscr{X}$, $\mathscr{S}$, and $\mathscr{Y}$, respectively. The $\mathscr{S}$-valued state process $\{S_t\}_{t=1}^{\infty}$ is i.i.d. with known pmf P_S. The probability law of the DMC is specified by

$$W^n(\mathbf{y} \mid \mathbf{x}, \mathbf{s}) = \prod_{t=1}^{n} W(y_t \mid x_t, s_t), \quad \mathbf{x} \in \mathscr{X}^n, \mathbf{s} \in \mathscr{S}^n, \mathbf{y} \in \mathscr{Y}^n.$$

An (M,n)-code with encoder CSI consists of the mappings (f, ϕ) where the encoder mapping $f = (f_1, \ldots, f_n)$ is either *causal*, i.e.,

$$f_t : \mathscr{M} \times \mathscr{S}^t \to \mathscr{X}, \quad t = 1, \ldots, n,$$

or *noncausal*, i.e.,

$$f_t : \mathscr{M} \times \mathscr{S}^n \to \mathscr{X}, \quad t = 1, \ldots, n.$$

with $\mathscr{M} = \{1, \ldots, M\}$ being the set of messages. The decoder ϕ is a mapping

$$\phi : \mathscr{Y}^n \to \mathscr{M}.$$

We restrict ourselves to the situation where the receiver has no CSI. When the receiver, too, has CSI, our results apply in a standard manner by considering an associated DMC with augmented output alphabet $\mathscr{Y} \times \mathscr{S}$.

The rate of the code is $(1/n) \log M$. The corresponding (maximum) probability of error is

$$e(f, \phi) = \max_{m \in \mathscr{M}} \sum_{\mathbf{s} \in \mathscr{S}^n} \mathrm{P}_S(\mathbf{s}) W^n((\phi^{-1}(m))^c \mid f(m, \mathbf{s}), \mathbf{s}), \tag{5}$$

where $\phi^{-1}(m) = \{\mathbf{y} \in \mathscr{Y}^n : \phi(\mathbf{y}) = m\}$ and $(\cdot)^c$ denotes complement.

Definition 1. Given $0 < \varepsilon < 1$, a number $R > 0$ is ε-achievable if for every $\delta > 0$ and for all n sufficiently large, there exist (M,n)-codes (f, ϕ) with $(1/n) \log M > R - \delta$ and $e(f, \phi) < \varepsilon$. The supremum of all ε-achievable rates is denoted by $C(\varepsilon)$. The capacity of the DMC is

$$C = \lim_{\varepsilon \to 0} C(\varepsilon).$$

If $C(\varepsilon) = C$ for $0 < \varepsilon < 1$, the DMC is said to satisfy a strong converse [12]. This terminology reflects the fact that for rates $R > C$, $e(f, \phi) > \varepsilon$ for $n \geq N(\varepsilon)$, $0 < \varepsilon < 1$. (In contrast, a standard converse shows that for $R > C$, $e(f, \phi)$ cannot be driven to 0 as $n \to \infty$.)

For a given pmf $P_{\tilde{X}\tilde{S}}$ on $\mathscr{X} \times \mathscr{S}$ and an rv U with values in a finite set $\mathscr{U}$, let $\mathscr{P}(P_{\tilde{X}\tilde{S}}, W)$ denote the set of all pmfs P_{UXSY} on $\mathscr{U} \times \mathscr{X} \times \mathscr{S} \times \mathscr{Y}$ with

$$X = h(U, S) \tag{6}$$

for some mapping h,

$$U -\!\circ\!- X, S -\!\circ\!- Y, \qquad P_{X,S,Y} = P_{\tilde{X}\tilde{S}}W. \tag{7}$$

For $\gamma \geq 0$, let $\mathscr{P}_\gamma(P_{\tilde{X}\tilde{S}}, W)$ be the subset of $\mathscr{P}(P_{\tilde{X}\tilde{S}}, W)$ with $I(U \wedge S) \leq \gamma$; note that $\mathscr{P}_0(P_{\tilde{X}\tilde{S}}, W)$ corresponds to the subset of $\mathscr{P}(P_{\tilde{X}\tilde{S}}, W)$ with U independent of S.

The classical results on the capacity of a state-dependent channel are due to Shannon [8] when the encoder CSI is causal and Gelfand and Pinsker [5] when the encoder CSI is noncausal.

Theorem 1. *For the case with causal CSI, the capacity is*

$$C_{Sh} = \max_{P_{X|S}} \max_{\mathscr{P}_0(P_{X|S}P_S, W)} I(U \wedge Y),$$

and holds with the strong converse.

Remark. The capacity formula was derived by Shannon [8], and the strong converse was proved later by Wolfowitz [12].

Theorem 2 (Gelfand–Pinsker [5]). *For the case with noncausal CSI, the capacity is*

$$C_{GP} = \max_{P_{X|S}} \max_{\mathscr{P}(P_{X|S}P_S, W)} I(U \wedge Y) - I(U \wedge S).$$

One main result below is to show that the previous result, too, holds with a strong converse.

Definition 2. The *reliability function* $E(R)$, $R \geq 0$, of the DMC W is the largest number $E \geq 0$ such that for every $\delta > 0$ and for all sufficiently large n, there exist n-length block codes (f, ϕ) with causal or noncausal CSI as above of rate greater than $R - \delta$ and $e(f, \phi) \leq \exp[-n(E - \delta)]$ (see, for instance, [3]).

4 A Technical Lemma

For a DMC without states, the result in [3, Corollary 6.4] provides, in effect, an image size characterization of a good codeword set; this does not involve any auxiliary rv. In the same spirit, our key technical lemma below provides an image size characterization for good codeword sets for the causal and noncausal DMC models, which now involves an auxiliary rv.

Lemma 1. *Let $\varepsilon, \tau > 0$ be such that $\varepsilon + \tau < 1$. Given a pmf $P_{\tilde{S}}$ on $\mathscr{S}$ and conditional pmf $\tilde{P}_{X|S}$, let (f, ϕ) be a (M,n)-code as above. For each $m \in \mathscr{M}$, let $A(m)$ be a subset of $\mathscr{S}^n$ which satisfies the following conditions:*

$$A(m) \subseteq \mathscr{T}^n_{[P_{\tilde{S}}]}, \tag{8}$$

$$\|A(m)\| \geq \exp\left[n\left(H(P_{\tilde{S}}) - \frac{\tau}{6}\right)\right], \tag{9}$$

$$f(m,\mathbf{s}) \in \mathscr{T}^n_{[P_{\tilde{X}|\tilde{S}}]}(\mathbf{s}), \quad \mathbf{s} \in A(m). \tag{10}$$

Furthermore, let (f, ϕ) satisfy one of the following two conditions:

$$W^n(\phi^{-1}(m) \mid f(m,\mathbf{s}),\mathbf{s}) \geq 1 - \varepsilon, \quad \mathbf{s} \in A(m), \tag{11a}$$

$$\frac{1}{\|A(m)\|} \sum_{\mathbf{s} \in A(m)} W^n(\phi^{-1}(m) \mid f(m,\mathbf{s}),\mathbf{s}) \geq 1 - \varepsilon. \tag{11b}$$

(a) *In the causal CSI case, for $n \geq N(\|\mathscr{X}\|, \|\mathscr{S}\|, \|\mathscr{Y}\|, \tau, \varepsilon)$,*[1] *it holds that*

$$\frac{1}{n}\log M \leq I(U \wedge Y) + \tau,$$

for some $P_{UXSY} \in \mathscr{P}_\tau(P_{\tilde{X}|\tilde{S}} P_{\tilde{S}}, W)$.

(b) *In the noncausal CSI case, for $n \geq N(\|\mathscr{X}\|, \|\mathscr{S}\|, \|\mathscr{Y}\|, \tau, \varepsilon)$, it holds that*

$$\frac{1}{n}\log M \leq I(U \wedge Y) - I(U \wedge S) + \tau,$$

for some $P_{UXSY} \in \mathscr{P}(P_{\tilde{X}|\tilde{S}} P_{\tilde{S}}, W)$.

Furthermore, in both cases it suffices to restrict the rv U to take values in a finite set $\mathscr{U}$ with $\|\mathscr{U}\| \leq \|\mathscr{X}\|\|\mathscr{S}\| + 1$.

Proof. Our proof below is for the case when (11a) holds. The case when (11b) holds can be proved similarly with minor modifications; specifically, in the latter case, we can find subsets $A'(m)$ of $A(m)$, $m \in \mathscr{M}$, that satisfy (8)–(10) and (11a) for some $\varepsilon', \tau' > 0$ with $\varepsilon' + \tau' < 1$ for all n sufficiently large.

With (11a) holding, set

$$B(m) = \{(f(m,\mathbf{s}),\mathbf{s}) \in \mathscr{X}^n \times \mathscr{S}^n : \mathbf{s} \in A(m)\}, \quad m \in \mathscr{M}.$$

[1] In our assertions, we indicate the validity of a statement "for all $n \geq N(.)$" by showing the explicit dependency of N; however, the standard picking of the "largest such N" from (finitely many) such Ns is not indicated.

Let $\mathrm{P}_{\tilde{Y}} = \mathrm{P}_{\tilde{X}\tilde{S}}W$ be a pmf on $\mathscr{Y}$ defined by

$$\mathrm{P}_{\tilde{Y}}(y) = \sum_{s,x} \mathrm{P}_{\tilde{S}\tilde{X}}(s,x) W(y \mid x,s), \quad y \in \mathscr{Y}.$$

Consequently,

$$W^n(\mathscr{T}^n_{[\mathrm{P}_{\tilde{Y}}]} \mid f(m,\mathbf{s}),\mathbf{s}) > \varepsilon + \tau, \quad \mathbf{s} \in A(m), \tag{12}$$

for all $n \geq N(\|\mathscr{X}\|, |\mathscr{S}\|, |\mathscr{Y}\|, \tau, \varepsilon)$ (not depending on m and $\mathbf{s}$ in $A(m)$). Denoting

$$C(m) = \phi^{-1}(m) \cap \mathscr{T}^n_{[\mathrm{P}_{\tilde{Y}}]},$$

we see from (11a) and (12) that

$$W^n(C(m) \mid f(m,\mathbf{s}),\mathbf{s}) > \tau > 0, \quad ,(f(m,\mathbf{s}),\mathbf{s}) \in B(m),$$

so that

$$\|C(m)\| \geq g_{W^n}(B(m),\tau),$$

where $g_{W^n}(B(m),\tau)$ denotes the smallest cardinality of a subset D of $\mathscr{Y}^n$ with

$$W^n(D \mid (f(m,\mathbf{s}),\mathbf{s})) > \tau, \quad (f(m,\mathbf{s}),\mathbf{s}) \in B(m). \tag{13}$$

With $m_0 = \mathtt{arg}\min_{1 \leq m \leq M} \|C(m)\|$, we have

$$M\|C(m_0)\| \leq \sum_{m=1}^{M} \|C(m)\| = \|\mathscr{T}^n_{[\mathrm{P}_{\tilde{Y}}]}\| \leq \exp n\left(H(\mathrm{P}_{\tilde{Y}}) + \frac{\tau}{6}\right).$$

Consequently,

$$\frac{1}{n}\log M \leq H(\mathrm{P}_{\tilde{Y}}) + \frac{\tau}{6} - \frac{1}{n}\log g_{W^n}(B(m_0),\tau). \tag{14}$$

The remainder of the proof entails relating the "image size" of $B(m_0)$, i.e., $g_{W^n}(B(m_0),\tau)$, to $\|A(m_0)\|$, and is completed below separately for the cases of causal and noncausal CSI.

First consider the causal CSI case. For a rv $\hat{S}^n$ distributed uniformly over $A(m_0)$, we have from (9) that

$$\frac{1}{n}H(\hat{S}^n) = \frac{1}{n}\log\|A(m_0)\| \geq H(\mathrm{P}_{\tilde{S}}) - \frac{\tau}{6}. \tag{15}$$

Since

$$\frac{1}{n}H(\hat{S}^n) = \frac{1}{n}\sum_{i=1}^{n} H(\hat{S}_i \mid \hat{S}^{i-1}) = H(\hat{S}_I \mid \hat{S}^{I-1}, I),$$

where the rv I is distributed uniformly over the set $\{1,\dots,n\}$ and is independent of all other rvs, the previous identity, together with (15), yields

$$H(\mathrm{P}_{\tilde{S}}) - H(\hat{S}_I \mid \hat{S}^{I-1}, I) \leq \frac{\tau}{3}. \tag{16}$$

Next, denote by $\hat{X}^n$ the rv $f(m_0, \hat{S}^n)$ and by $\hat{Y}^n$ the rv which conditioned on $\hat{X}^n$, $\hat{S}^n$, has (conditional) distribution W^n, i.e., $\hat{Y}^n$ is the random output of the DMC W when the input is set to $(\hat{X}^n, \hat{S}^n)$. Then, using [3, Lemma 15.2], we get

$$\frac{1}{n}\log g_{W^n}(B(m_0), \tau) \geq \frac{1}{n}H(\hat{Y}^n) - \frac{\tau}{6}, \tag{17}$$

for all n sufficiently large. Furthermore,

$$\begin{aligned}\frac{1}{n}H(\hat{Y}^n) &= \frac{1}{n}\sum_{i=1}^{n} H(\hat{Y}_i \mid \hat{Y}^{i-1}) \\ &\geq H(\hat{Y}_I \mid \hat{X}^{I-1}, \hat{S}^{I-1}, \hat{Y}^{I-1}, I) \\ &= H(\hat{Y}_I \mid \hat{X}^{I-1}, \hat{S}^{I-1}, I) \\ &= H(\hat{Y}_I \mid \hat{S}^{I-1}, I),\end{aligned}$$

where the last-but-one equality follows from the DMC assumption, and the last equality holds since $\hat{X}^{I-1} = f(m_0, \hat{S}^{I-1})$. The inequality above, along with (17) and (14) gives

$$\frac{1}{n}\log M \leq H(\mathrm{P}_{\tilde{Y}}) - H(\hat{Y}_I \mid \hat{S}^{I-1}, I) + \frac{\tau}{3}. \tag{18}$$

Denote by $\hat{U}$ the rv $(\hat{S}^{I-1}, I)$ and note that the following Markov property holds:

$$\hat{Y}_I -\!\circ\!- \hat{X}_I, \hat{S}_I -\!\circ\!- \hat{U}.$$

Also, from the definition of $B(m_0)$,

$$\begin{aligned}\mathrm{P}_{\hat{X}_I,\hat{S}_I}(x,s) &= \frac{1}{n}\sum_{i=1}^{n} \mathrm{P}_{\hat{X}_i,\hat{S}_i}(x,s) \\ &= \frac{1}{n}\sum_{i=1}^{n}\sum_{\mathbf{x},\mathbf{s}\in B(m_0)} \frac{\mathbf{1}(x_i = x, s_i = s)}{\|B(m_0)\|} \\ &= \frac{1}{\|B(m_0)\|}\sum_{\mathbf{x},\mathbf{s}\in B(m_0)} Q_{\mathbf{x},\mathbf{s}}(x,s),\end{aligned}$$

where $Q_{\mathbf{x},\mathbf{s}}(x,s)$ is the joint type of $\mathbf{x},\mathbf{s}$, and the last equation follows upon interchanging the order of summation. It follows from (8) and (10) that $\|\mathrm{P}_{\hat{X}_I,\hat{S}_I} - \mathrm{P}_{\tilde{X}\tilde{S}}\| \le \delta_n$ for some $\delta_n \to 0$ satisfying the delta convention. Furthermore,

$$\begin{aligned}\|\mathrm{P}_{\hat{Y}_I} - \mathrm{P}_{\tilde{Y}}\| &= \sum_y \left| \sum_{x,s} W(y|x,s)\mathrm{P}_{\tilde{X}\tilde{S}}(x,s) - \sum_{x,s} W(y|x,s)\mathrm{P}_{\hat{X}_I\hat{S}_I}(x,s) \right| \\ &\le \sum_{x,s}\sum_y W(y|x,s)\left|\mathrm{P}_{\tilde{X}\tilde{S}}(x,s) - \mathrm{P}_{\hat{X}_I\hat{S}_I}(x,s)\right| \\ &= \|\mathrm{P}_{\tilde{X}\tilde{S}} - \mathrm{P}_{\hat{X}_I\hat{S}_I}\| \le \delta_n.\end{aligned}$$

Let the rvs $\tilde{X},\tilde{S},\tilde{Y}$ have a joint distribution $\mathrm{P}_{\tilde{X}\tilde{S}\tilde{Y}}$. Define a rv U which takes values in the same set as $\hat{U}$, has $\mathrm{P}_{\hat{U}|\hat{X}_I\hat{S}_I}$ as its conditional distribution given X,S, and satisfies the Markov relation

$$Y -\circ- X,S -\circ- U.$$

Then using the continuity of the entropy function and the arguments above, (18) yields

$$\frac{1}{n}\log M \le I(U \wedge Y) + \tau,$$

and (16) yields

$$I(U \wedge S) \le \tau,$$

for all n sufficiently large, where $\mathrm{P}_{UXSY} \in \mathscr{P}_\tau(\mathrm{P}_{\tilde{X}\tilde{S}}, W)$.

Turning to the case with noncausal CSI, define a stochastic matrix $V : \mathscr{X} \times \mathscr{S} \to \mathscr{S}$ with

$$V(s' \mid x,s) = \mathbf{1}(s' = s),$$

and let g_{V^n} be defined in a manner analogous to g_{W^n} above with $\mathscr{S}^n$ in the role of $\mathscr{Y}^n$ in (13). For any $m \in \mathscr{M}$ and subset E of $\mathscr{S}^n$, observe that

$$V^n(E \mid f(m,\mathbf{s}),\mathbf{s}) = \mathbf{1}(s \in E), \quad \mathbf{s} \in \mathscr{S}^n.$$

In particular, if E satisfies

$$V^n(E \mid f(m,\mathbf{s}),\mathbf{s}) > \tau, \quad \mathbf{s} \in A(m), \tag{19}$$

it must be that $A(m) \subseteq E$, and since $E = A(m)$ satisfies (19), we get that

$$\|A(m)\| = g_{V^n}(B(m),\tau) \tag{20}$$

using the definition of $B(m)$. Using the image size characterization in (4) [3, Theorem 15.11], there exists an auxiliary rv U and associated pmf $\mathrm{P}_{UXSY} = \mathrm{P}_{U|XS}\mathrm{P}_{\tilde{X}\tilde{S}}W$ such that

$$\left|\frac{1}{n}\log g_{W^n}(B(m_0),\tau) - H(Y|U) - t\right| < \frac{\tau}{6},$$

$$\left|\frac{1}{n}\log g_{V^n}(B(m_0),\tau) - H(S|U) - t\right| < \frac{\tau}{6}, \tag{21}$$

where $0 \le t \le \min\{I(U \wedge Y), I(U \wedge S)\}$. Then, using (14), (20), and (21) we get

$$\frac{1}{n}\log M \le I(U \wedge Y) + H(S \mid U) - \frac{1}{n}\log \|A(m_0)\| + \frac{\tau}{2},$$

which, by (9), yields

$$\frac{1}{n}\log M \le I(U \wedge Y) - I(U \wedge S) + \tau.$$

In (21), $\mathrm{P}_{UXSY} \in \mathscr{P}(\mathrm{P}_{\tilde{X}|\tilde{S}}\mathrm{P}_{\tilde{S}}, W)$ but need not satisfy (6). Finally, the asserted restriction to $\mathrm{P}_{UXSY} \in \mathscr{P}(\mathrm{P}_{\tilde{X}|\tilde{S}}\mathrm{P}_{\tilde{S}}, W)$ follows from the convexity of $I(U \wedge Y) - I(U \wedge S)$ in $\mathrm{P}_{X|US}$ for a fixed P_{US} (as observed in [5]).

Lastly, it follows from the support lemma [3, Lemma 15.4] that it suffices to consider those rvs U for which $\|\mathscr{U}\| \le \|\mathscr{X}\|\|\mathscr{S}\| + 1$. □

5 The Strong Converse

Theorem 3 (Strong converse). *Given $0 < \varepsilon < 1$ and a sequence of (M_n, n) codes (f_n, ϕ_n) with $e(f_n, \phi_n) < \varepsilon$, it holds that*

$$\limsup_n \frac{1}{n}\log M_n \le C,$$

where $C = C_{Sh}$ and C_{GP} for the cases of causal and noncausal CSI, respectively.

Proof. Given $0 < \varepsilon < 1$ and a (M,n)-code (f,ϕ) with $e(f,\phi) \le \varepsilon$, the proof involves the identification of sets $A(m)$, $m \in \mathscr{M}$, satisfying (8)–(10) and (11a). The assertion then follows from Lemma 1. Note that $e(f,\phi) \le \varepsilon$ implies

$$\sum_{\mathbf{s} \in \mathscr{S}^n} \mathrm{P}_S(\mathbf{s})\, W^n(\phi^{-1}(m) \mid f(m,\mathbf{s}), \mathbf{s}) \ge 1 - \varepsilon$$

for all $m \in \mathscr{M}$. Since $\mathrm{P}_S\left(\mathscr{T}^n_{[\mathrm{P}_S]}\right) \to 1$ as $n \to \infty$, we get that for every $m \in \mathscr{M}$,

$$\mathrm{P}_S\left(\left\{\mathbf{s} \in \mathscr{T}^n_{[\mathrm{P}_S]} : W^n(\phi^{-1}(m) \mid f(m,\mathbf{s}), \mathbf{s}) > \frac{1-\varepsilon}{2}\right\}\right) \ge \frac{1-\varepsilon}{3} \tag{22}$$

for all $n \geq N(\|\mathscr{S}\|, \varepsilon)$. Denoting the set $\left\{\cdot\right\}$ in (22) by $\hat{A}(m)$, clearly for every $m \in \mathscr{M}$,

$$W^n(\phi^{-1}(m) \mid f(m,\mathbf{s}),\mathbf{s}) \geq \frac{1-\varepsilon}{2}, \quad \mathbf{s} \in \hat{A}(m),$$

and

$$\mathrm{P}_S\left(\hat{A}(m)\right) \geq \frac{1-\varepsilon}{3}$$

for all $n \geq N(\|\mathscr{S}\|, \varepsilon)$, whereby for an arbitrary $\delta > 0$, we get

$$\|\hat{A}(m)\| \geq \exp\left[n(H(\mathrm{P}_S) - \delta)\right]$$

for all $n \geq N(\|\mathscr{S}\|, \delta)$. Partitioning $\hat{A}(m)$, $m \in \mathscr{M}$, into sets according to the (polynomially many) conditional types of $f(m,\mathbf{s})$ given $\mathbf{s}$ in $\hat{A}(m)$, we obtain a subset $A(m)$ of $\hat{A}(m)$ for which

$$f(m,\mathbf{s}) \in \mathscr{T}_m^n(\mathbf{s}), \quad \mathbf{s} \in A(m),$$
$$\|A(m)\| \geq \exp\left[n(H(\mathrm{P}_S) - 2\delta)\right],$$

for all $n \geq N(\|\mathscr{S}\|, \|\mathscr{X}\|, \delta)$, where $\mathscr{T}_m^n(\mathbf{s})$ represents a set of those sequences in $\mathscr{X}^n$ that have the same conditional type (depending only on m).

Once again, the polynomial size of such conditional types yields a subset $\mathscr{M}'$ of $\mathscr{M}$ such that $f(m,\mathbf{s})$ has a fixed conditional type (not depending on m) given $\mathbf{s}$ in $A(m)$, and with

$$\frac{1}{n}\log\|\mathscr{M}'\| \geq \frac{1}{n}\log M - \delta$$

for all $n \geq N(\|\mathscr{S}\|, \|\mathscr{X}\|, \delta)$. Finally, the strong converse follows by applying Lemma 1 to the subcode corresponding to $\mathscr{M}'$ and noting that $\delta > 0$ is arbitrary. □

6 Outer Bound on Reliability Function

An upper bound for the reliability function $E(R)$, $0 < R < C$, of a DMC without states is derived in [3] using a strong converse for codes with codewords of a fixed type. The key technical Lemma 1 gives an upper bound on the rate of codes with codewords that are conditionally typical over large *message-dependent* subsets of the typical set of state sequences and serves, in effect, as an analog of [3, Corollary 6.4] for state-dependent channels to derive an upper bound on the reliability function.

Theorem 4 (Sphere packing bound). *Given $\delta > 0$, for $0 < R < C$, it holds that*

$$E(R) \leq E_{SP}(1+\delta)+\delta,$$

where

$$E_{SP} = \min_{P_{\tilde{S}}} \max_{P_{\tilde{X}|\tilde{S}}} \min_{V \in \mathscr{V}(R, P_{\tilde{X}\tilde{S}})} \left[D(P_{\tilde{S}} \| P_S) + D(V \| W \mid P_{\tilde{X}\tilde{S}})\right] \tag{23}$$

with

$$\mathscr{V}(R, P_{\tilde{X}\tilde{S}}) = \mathscr{V}_{Sh}(R, P_{\tilde{X}\tilde{S}}) = \left\{V : \mathscr{X} \times \mathscr{S} \to \mathscr{Y} : \max_{P_{UXSY} \in \mathscr{P}_0(P_{\tilde{X}\tilde{S}}, V)} I(U \wedge Y) < R\right\}, \tag{24}$$

and

$$\begin{aligned} \mathscr{V}(R, P_{\tilde{X}\tilde{S}}) &= \mathscr{V}_{GP}(R, P_{\tilde{X}\tilde{S}}) \\ &= \left\{V : \mathscr{X} \times \mathscr{S} \to \mathscr{Y} : \max_{P_{UXSY} \in \mathscr{P}(P_{\tilde{X}\tilde{S}}, V)} I(U \wedge Y) - I(U \wedge S) < R\right\}, \end{aligned} \tag{25}$$

for the causal and noncausal CSI cases, respectively.

Remark. In (23), the terms $D(\mathrm{P}_{\tilde{S}} \| \mathrm{P}_S)$ and $D(V \| W \mid \mathrm{P}_{\tilde{S}} \mathrm{P}_{\tilde{X}|\tilde{S}})$ account, respectively, for the shortcomings of a given code for corresponding "bad" state pmf and "bad" channel.

Proof. Consider sequences of type $\mathrm{P}_{\tilde{S}}$ in $\mathscr{S}^n$. Picking $\hat{A}(m) = \mathscr{T}^n_{\mathrm{P}_{\tilde{S}}}$, $m \in \mathscr{M}$, in the proof of Theorem 3, and following the arguments therein to extract the subset $A(m)$ of $\hat{A}(m)$, we have for a given $\delta > 0$ that for all $n \geq N(\|\mathscr{S}\|, \|\mathscr{X}\|, \delta)$, there exists a subset $\mathscr{M}'$ of $\mathscr{M}$ and a fixed conditional type, say $\mathrm{P}_{\tilde{X}|\tilde{S}}$ (not depending on m), such that for every $m \in \mathscr{M}'$,

$$\begin{aligned} A(m) &\subseteq \hat{A}(m) = \mathscr{T}^n_{\mathrm{P}_{\tilde{S}}}, \\ \|A(m)\| &\geq \exp\left[n(H(\mathrm{P}_{\tilde{S}}) - \delta)\right], \\ f(m, \mathbf{s}) &\in \mathscr{T}^n_{\mathrm{P}_{\tilde{X}|\tilde{S}}}(\mathbf{s}), \qquad \mathbf{s} \in A(m), \\ \frac{1}{n} \log \|\mathscr{M}'\| &\geq R - \delta. \end{aligned}$$

Then for every $V \in \mathscr{V}(R, \mathrm{P}_{\tilde{X}\tilde{S}})$, we obtain using Lemma 1 (in its version with condition (11b)), that for every $\delta' > 0$, there exists $m \in \mathscr{M}'$ (possibly depending on δ' and V) with

$$\frac{1}{\|A(m)\|} \sum_{\mathbf{s} \in A(m)} V^n((\phi^{-1}(m))^c \mid f(m, \mathbf{s}), \mathbf{s}) \geq 1 - \delta'$$

for all $n \geq N(\|\mathscr{S}\|, \|\mathscr{X}\|, \|\mathscr{Y}\|, \delta')$. Since the average V^n-(conditional) probability of $\left(\phi^{-1}(m)\right)^c$ is large, its W^n-(conditional) probability cannot be too small. To that end, for this m, apply [3, Theorem 10.3, (10.21)] with the choices

$$Z = \mathscr{Y}^n \times A(m),$$
$$S = (\phi^{-1}(m))^c \times A(m),$$
$$Q_1(\mathbf{y}, \mathbf{s}) = \frac{V^n(\mathbf{y} \mid f(m, \mathbf{s}), \mathbf{s})}{\|A(m)\|},$$
$$Q_2(\mathbf{y}, \mathbf{s}) = \frac{W^n(\mathbf{y} \mid f(m, \mathbf{s}), \mathbf{s})}{\|A(m)\|},$$

for $(\mathbf{y}, \mathbf{s}) \in Z$, to obtain

$$\frac{1}{\|A(m)\|} \sum_{\mathbf{s} \in A(m)} W^n((\phi^{-1}(m))^c \mid f(m, \mathbf{s}), \mathbf{s}) \geq \exp\left(-\frac{nD(V\|W \mid \mathrm{P}_{\tilde{X}|\tilde{S}} \mathrm{P}_{\tilde{S}}) + 1}{1 - \delta'}\right).$$

Finally,

$$\begin{aligned} e(f, \phi) &\geq \sum_{\mathbf{s} \in A(m)} \mathrm{P}_S(\mathbf{s}) W^n\left((\phi^{-1}(m))^c \mid f(m, \mathbf{s}), \mathbf{s}\right) \\ &\geq \exp[-n(D(\mathrm{P}_{\tilde{S}}\|\mathrm{P}_S) + D(V\|W \mid \mathrm{P}_{\tilde{X}|\tilde{S}} \mathrm{P}_{\tilde{S}})(1 + \delta) + \delta)] \end{aligned}$$

for $n \geq N(\|\mathscr{S}\|, \|\mathscr{X}\|, \|\mathscr{Y}\|, \delta, \delta')$, whereby it follows for the noncausal CSI case that

$$\begin{aligned} \limsup_n -\frac{1}{n} \log e(f, \phi) \leq \min_{\mathrm{P}_{\tilde{S}}} \max_{\mathrm{P}_{\tilde{X}|\tilde{S}}} \min_{V \in \mathscr{V}(R, \mathrm{P}_{\tilde{X}\tilde{S}})} &[D(\mathrm{P}_{\tilde{S}}\|\mathrm{P}_S) \\ &+ D(V\|W \mid \mathrm{P}_{\tilde{X}|\tilde{S}} \mathrm{P}_{\tilde{S}})(1 + \delta) + \delta] \end{aligned}$$

for every $\delta > 0$. Similarly, for the case of causal CSI, for $\tau > 0$, letting

$$\mathscr{V}_\tau(R, \mathrm{P}_{\tilde{X}\tilde{S}}) = \left\{V : \mathscr{X} \times \mathscr{S} \to \mathscr{Y} : \max_{P_{UXSY} \in \mathscr{P}_\tau(\mathrm{P}_{\tilde{X}\tilde{S}}, V)} I(U \wedge Y) < R\right\}, \tag{26}$$

we get

$$\limsup_n -\frac{1}{n} \log e(f, \phi) \leq \min_{\mathrm{P}_{\tilde{S}}} \max_{\mathrm{P}_{\tilde{X}|\tilde{S}}} \min_{V \in \mathscr{V}_\tau(R, \mathrm{P}_{\tilde{X}\tilde{S}})} [D(\mathrm{P}_{\tilde{S}}\|\mathrm{P}_S) + D(V\|W \mid \mathrm{P}_{\tilde{X}|\tilde{S}} \mathrm{P}_{\tilde{S}})].$$

The continuity of the right side of (26), as shown in the Appendix, yields the claimed expression for E_{SP} in (23) and (24). □

Acknowledgements The work of H. Tyagi and P. Narayan was supported by the U.S. National Science Foundation under Grants CCF0830697 and CCF1117546.

Appendix: Continuity of the Right Side of (26)

Let

$$f(R, \mathrm{P}_{U\tilde{X}\tilde{S}}) = \min_{\substack{V: I(U \wedge Y) < R \\ \mathrm{P}_{Y|\tilde{X}\tilde{S}} = V}} D(\mathrm{P}_{\tilde{S}} \| \mathrm{P}_S) + D(V \| W \mid \mathrm{P}_{\tilde{X}|\tilde{S}} \mathrm{P}_{\tilde{S}}). \tag{27}$$

Further, let

$$g(\mathrm{P}_{\tilde{S}}, \tau) = \max_{\substack{\mathrm{P}_{U\tilde{X}|\tilde{S}}: I(U \wedge \tilde{S}) \leq \tau \\ U -\!\circ\!- \tilde{X}, \tilde{S} -\!\circ\!- Y}} f(R, \mathrm{P}_{U\tilde{X}\tilde{S}}), \tag{28}$$

and

$$g(\tau) = \min_{\mathrm{P}_{\tilde{S}}} g(\mathrm{P}_{\tilde{S}}, \tau). \tag{29}$$

To show the continuity of $g(\tau)$ at $\tau = 0$, first note that $g(\tau) \geq g(0)$ for all $\tau \geq 0$. Next, let $\mathrm{P}^0_{\tilde{S}}$ attain the minimum in (29) for $\tau = 0$. Clearly,

$$g(\mathrm{P}^0_{\tilde{S}}, \tau) \geq g(\tau). \tag{30}$$

Also, let $\mathrm{P}^\tau_{U\tilde{X}|\tilde{S}}$ attain the maximum of $g(\mathrm{P}^0_{\tilde{S}}, \tau)$ in (28). For the associated joint pmf $\mathrm{P}^0_{\tilde{S}} \mathrm{P}^\tau_{U\tilde{X}|\tilde{S}}$, let P^τ_U denote the resulting U-marginal pmf, and consider the joint pmf $\mathrm{P}^\tau_U \mathrm{P}^0_{\tilde{S}} \mathrm{P}^\tau_{\tilde{X}|U\tilde{S}}$. Then, using (28) and (29) and the observations above,

$$f(R, \mathrm{P}^\tau_U \mathrm{P}^0_{\tilde{S}} \mathrm{P}^\tau_{\tilde{X}|U\tilde{S}}) \leq g(0) \leq g(\tau) \leq g(\mathrm{P}^0_{\tilde{S}}, \tau) = f(R, \mathrm{P}^0_{\tilde{S}} \mathrm{P}^\tau_{U\tilde{X}|\tilde{S}}).$$

The continuity of $g(\tau)$ at $\tau = 0$ will follow upon showing that

$$f(R, \mathrm{P}^0_{\tilde{S}} \mathrm{P}^\tau_{U\tilde{X}|\tilde{S}}) - f(R, \mathrm{P}^\tau_U \mathrm{P}^0_{\tilde{S}} \mathrm{P}^\tau_{\tilde{X}|U\tilde{S}}) \to 0 \text{ as } \tau \to 0.$$

The constraint on the mutual information (28) gives by Pinsker's inequality [3, 4] that,

$$\tau \geq D\left(\mathrm{P}^\tau_{U|\tilde{S}} \mathrm{P}^0_{\tilde{S}} \| \mathrm{P}^\tau_U \mathrm{P}^0_{\tilde{S}}\right) \geq 2 \left\| \mathrm{P}^\tau_{U|\tilde{S}} \mathrm{P}^0_{\tilde{S}} - \mathrm{P}^\tau_U \mathrm{P}^0_{\tilde{S}} \right\|^2,$$

i.e.,

$$\left\| \mathrm{P}^{\tau}_{U|\tilde{S}} \mathrm{P}^{0}_{\tilde{S}} - \mathrm{P}^{\tau}_{U} \mathrm{P}^{0}_{\tilde{S}} \right\| \leq \sqrt{\frac{\tau}{2}}. \tag{31}$$

For $\mathrm{P}_{U\tilde{X}\tilde{S}} = \mathrm{P}^{\tau}_{U}\mathrm{P}^{0}_{\tilde{S}}\mathrm{P}^{\tau}_{\tilde{X}|U\tilde{S}}$, let V^0 attain the minimum in (26), i.e.,

$$\mathrm{P}_{Y|\tilde{X}\tilde{S}} = V^0, \quad I(U \wedge Y) < R, \text{ and}$$
$$f(R, \mathrm{P}^{\tau}_{U}\mathrm{P}^{0}_{\tilde{S}}\mathrm{P}^{\tau}_{\tilde{X}|U\tilde{S}}) = D(\mathrm{P}_{\tilde{S}}\|\mathrm{P}_S) + D(V^0\|W \mid \mathrm{P}_{\tilde{X}|\tilde{S}}\mathrm{P}_{\tilde{S}}).$$

By (31), for $\mathrm{P}_{U\tilde{X}\tilde{S}} = \mathrm{P}^{0}_{\tilde{S}}\mathrm{P}^{\tau}_{U\tilde{X}|U\tilde{S}}$ and $\mathrm{P}_{Y|\tilde{X}\tilde{S}} = V^0$, by standard continuity arguments, we have

$$I(U \wedge Y) < R + \nu,$$

and

$$D(\mathrm{P}_{\tilde{S}}\|\mathrm{P}_S) + D(V^0\|W \mid \mathrm{P}_{\tilde{X}|\tilde{S}}\mathrm{P}_{\tilde{S}}) \leq f(R, \mathrm{P}^{\tau}_{U}\mathrm{P}^{0}_{\tilde{S}}\mathrm{P}^{\tau}_{\tilde{X}|U\tilde{S}}) + \nu,$$

where $\nu = \nu(\tau) \to 0$ as $\tau \to 0$. Consequently,

$$f(R, \mathrm{P}^{0}_{\tilde{S}}\mathrm{P}^{\tau}_{U\tilde{X}|U\tilde{S}}) \leq D(\mathrm{P}_{\tilde{S}}\|\mathrm{P}_S) + D(V^0\|W \mid \mathrm{P}_{\tilde{X}|\tilde{S}}\mathrm{P}_{\tilde{S}}) \leq f(R, \mathrm{P}^{\tau}_{U}\mathrm{P}^{0}_{\tilde{S}}\mathrm{P}^{\tau}_{\tilde{X}|U\tilde{S}}) + \nu.$$

Finally, noting the continuity of $f(R, \mathrm{P}_{U\tilde{X}\tilde{S}})$ in R [3, Lemma 10.4], the proof is completed. □

References

1. Biglieri, E., Proakis, J., Shamai, S.(Shitz).: Fading channels: information-theoretic and communications aspects. IEEE Trans. Inf. Theor. **44**(6), 2619–2692 (1998)
2. Csiszár, I.: The method of types. IEEE Trans. Inf. Theor. **44**(6), 2505–2523 (1998)
3. Csiszár, I., Körner, J.: Information Theory: Coding Theorems for Discrete Memoryless Channels, 2nd edn. Cambridge University Press, Cambridge (2011)
4. Fedotov, A.A., Harremoës, P., Topsøe, F.: Refinements of Pinsker's inequality. IEEE Trans. Inf. Theor. **49**(6), 1491–1498 (2003)
5. Gelfand, S.I., Pinsker, M.S.: Coding for channels with random parameters. Prob. Contr. Inf. Theor. **9**(1), 19–31 (1980)
6. Keshet, G., Steinberg, Y., Merhav, N.: Channel coding in presence of side information. Foundations and Trends in Commun. Inf. Theor. **4**(6), 445–586 (2008)
7. Lapidoth, A., Narayan, P.: Reliable communication under channel uncertainty. IEEE Trans. Inf. Theor. **44**(6), 2148–2177 (1998)
8. Shannon, C.E.: Channels with side information at the transmitter. IBM J. Res. Dev. **2**, 289–293 (1958)
9. Shannon, C.E., Gallager, R.G., Berlekamp, E.R.: Lower bounds to error probability for coding on discrete memoryless channels-i. Inf. Contr. 65–103 (1966)

10. Shannon, C.E., Gallager, R.G., Berlekamp, E.R.: Lower bounds to error probability for coding on discrete memoryless channels-ii. Inf. Contr. 522–552 (1967)
11. Tyagi, H., Narayan, P.: The Gelfand-Pinsker channel: strong converse and upper bound for the reliability function. In: Proceedings of the IEEE International Symposium on Information Theory, Seoul, Korea (2009)
12. Wolfowitz, J.: Coding Theorems of Information Theory. Springer, New York (1978)

Index

T.D. Andrews et al. (eds.), *Excursions in Harmonic Analysis, Volume 1*, Applied and Numerical Harmonic Analysis, DOI 10.1007/978-0-8176-8376-4,

MIX
Papier aus verantwortungsvollen Quellen
Paper from responsible sources
FSC® C105338

If you have any concerns about our products, you can contact us on
ProductSafety@springernature.com

In case Publisher is established outside the EU, the EU authorized representative is:
Springer Nature Customer Service Center GmbH
Europaplatz 3, 69115 Heidelberg, Germany

Printed by Libri Plureos GmbH
in Hamburg, Germany